Ecological Aspects of Soil and Land Preservation

Olga Anatolievna Pasko
National Open Institute, St. Petersburg, Russia

Nadezhda Anatolievna Lebedeva
International Personnel Academy, Germany

A volume in the Advances in Environmental Engineering and Green Technologies (AEEGT) Book Series

IGI Global
PUBLISHER of TIMELY KNOWLEDGE

Published in the United States of America by
IGI Global
Engineering Science Reference (an imprint of IGI Global)
701 E. Chocolate Avenue
Hershey PA, USA 17033
Tel: 717-533-8845
Fax: 717-533-8661
E-mail: cust@igi-global.com
Web site: http://www.igi-global.com

Library of Congress Cataloging-in-Publication Data

CIP Pending

Ecological Aspects of Soil and Land Preservation
Olga Pasko, Nadezhda Lebedeva
2024 Engineering Science Reference

ISBN: 979-8-3693-3374-7
eISBN: 979-8-3693-3375-4

British Cataloguing in Publication Data
A Cataloguing in Publication record for this book is available from the British Library.

For electronic access to this publication, please contact: eresources@igi-global.com.

Advances in Environmental Engineering and Green Technologies (AEEGT) Book Series

Sang-Bing (Jason) Tsai
Zhongshan Institute, University of Electronic Science and Technology of China, China & Wuyi University, China
Ming-Lang Tseng
Lunghwa University of Science and Technology, Taiwan
Yuchi Wang
University of Electronic Science and Technology of China Zhongshan Institute, China

ISSN:2326-9162
EISSN:2326-9170

Mission

Growing awareness and an increased focus on environmental issues such as climate change, energy use, and loss of non-renewable resources have brought about a greater need for research that provides potential solutions to these problems. Research in environmental science and engineering continues to play a vital role in uncovering new opportunities for a "green" future.

The **Advances in Environmental Engineering and Green Technologies (AEEGT)** book series is a mouthpiece for research in all aspects of environmental science, earth science, and green initiatives. This series supports the ongoing research in this field through publishing books that discuss topics within environmental engineering or that deal with the interdisciplinary field of green technologies.

Coverage

- Alternative Power Sources
- Biofilters and Biofiltration
- Cleantech
- Electric Vehicles
- Green Transportation
- Policies Involving Green Technologies and Environmental Engineering

IGI Global is currently accepting manuscripts for publication within this series. To submit a proposal for a volume in this series, please contact our Acquisition Editors at Acquisitions@igi-global.com or visit: http://www.igi-global.com/publish/.

The Advances in Environmental Engineering and Green Technologies (AEEGT) Book Series (ISSN 2326-9162) is published by IGI Global, 701 E. Chocolate Avenue, Hershey, PA 17033-1240, USA, www.igi-global.com. This series is composed of titles available for purchase individually; each title is edited to be contextually exclusive from any other title within the series. For pricing and ordering information please visit http://www.igi-global.com/book-series/advances-environmental-engineering-green-technologies/73679. Postmaster: Send all address changes to above address.

Titles in this Series

For a list of additional titles in this series, please visit: www.igi-global.com/book-series

Fostering an Ecological Shift Through Effective Environmental Education
Tatjana Kochetkova (O.P. Jindal Global University, India)
Engineering Science Reference • copyright 2024 • 404pp • H/C (ISBN: 9798369325773) • US $255.00 (our price)

Biodiversity Loss Assessment for Ecosystem Protection
Ashok Kumar Rathoure (Saraca Research Inc., India)
Engineering Science Reference • copyright 2024 • 328pp • H/C (ISBN: 9798369333303) • US $325.00 (our price)

Revolutionizing Automated Waste Treatment Systems IoT and Bioelectronics
Alex Khang (Global Research Institute of Technology and Engineering, USA) Hajimahmud Vugar Abdullayev (Azerbaijan State Oil and Industry University, Baku, Azerbaijan) Eugenia Litvinova (Kharkov National University of Radio Electronics, Ukraine) Gadirova Elmina Musrat (Baku State University, Azerbaijan) and Zoran Ž. Avramovic (University of Belgrade, Serbia)
Engineering Science Reference • copyright 2024 • 460pp • H/C (ISBN: 9798369360163) • US $325.00 (our price)

Sensors for Environmental Monitoring, Identification, and Assessment
Khursheed Ahmad Wani (Government Degree College Thindim Kreeri, India)
Engineering Science Reference • copyright 2024 • 482pp • H/C (ISBN: 9798369319307) • US $295.00 (our price)

Reshaping Environmental Science Through Machine Learning and IoT
Rajeev Kumar Gupta (Pandit Deendayal Energy University, India) Arti Jain (Jaypee Institute of Information Technology, India) John Wang (Montclair State University, USA) and Rajesh Kumar Pateriya (Maulana Azad National Institute of Technology, India)
Engineering Science Reference • copyright 2024 • 438pp • H/C (ISBN: 9798369323519) • US $295.00 (our price)

Advancements in Climate and Smart Environment Technology
Jamal Mabrouki (Faculty of Science, Mohammed V University in Rabat, Morocco) and Mourade Azrour (Faculty of Sciences and Techniques, Moulay Ismail University of Meknes, Errachidia, Morocco)
Engineering Science Reference • copyright 2024 • 292pp • H/C (ISBN: 9798369338070) • US $295.00 (our price)

IGI Global
PUBLISHER of TIMELY KNOWLEDGE

701 East Chocolate Avenue, Hershey, PA 17033, USA
Tel: 717-533-8845 x100 • Fax: 717-533-8661
E-Mail: cust@igi-global.com • www.igi-global.com

Table of Contents

Detailed Table of Contents

spatial analysis, feature extraction, classification, network analysis, temporal analysis, and integration with other datasets to quantify green infrastructure and assess urban green space quality. OSM data offers substantial opportunities with global coverage, open access, rich attribute information, timeliness, and customizability, enabling nuanced analyses and comparative studies. However, limitations such as data quality variability, inconsistent tagging, and biases in coverage exist, necessitating careful consideration and validation methods. Despite challenges, judicious utilization of OSM data can advance understanding of urban green spaces, aiding evidence-based decision-making in urban planning and environmental management.

The development of the production of unmanned aerial vehicles (UAVs), their reduction in cost and mass production, and the ability to carry various measuring devices have led to an expansion of the capabilities of remote sensing of agricultural fields. The development of methods for monitoring agricultural fields has led to the emergence of several areas of application of optical devices, which are mostly aimed at studying vegetation. The characteristics of the soil are studied much less frequently, since most of the year it is covered with snow or vegetation, and only in the spring or after plowing does the soil become available for study using unmanned aerial vehicles. In this regard, the authors do not consider ground-based methods for studying soil, since they are well developed and widely known. Therefore, it is necessary to systematically and comprehensively study the application of UAV remote sensing of soil systems that are feasible for analysis. In this article, we review the available literature and present the results of our own research in the field of using UAVs to study the color characteristics of soil. The presented results will be useful for improving the understanding of UAV applications for monitoring soil characteristics.

In recent decades, there has been a steady trend in Russia towards a decrease in the fertility of arable soils, which is recorded by changes in their agrochemical, agrophysical, and biological indicators. In addition, most regions, in particular the Leningrad region, are characterized by the presence of limiting climatic factors that limit the possibility of obtaining high yields, despite favorable natural conditions for the development of agricultural production. Knowledge of the properties of arable land makes it possible to optimize the nutrition of agricultural crops as much as possible, taking into account the necessary and sufficient doses of fertilizers, regardless of suitable conditions.

Drought and soil erosion stand as predominant drivers of land degradation within arid regions. The infrequent occurrence of rainfall events and subsequent flash floods significantly diminish agri-food productivity and exacerbate soil erosion rates. Thus, employing a modeling approach becomes imperative for forecasting flash floods and soil losses. The aim of this book chapter is to provide an introduction to soil erosion modeling and to delve into the performance evaluation of the kinematic runoff and erosion model (KINEROS2) through a case study conducted in Egypt. The model utilizes satellite imagery of digital elevation maps, soil maps, and annual weather datasets. Runoff discharge data observed at the outlet of the watershed is integrated into the model. In conclusion, it is strongly advised to integrate satellite imagery with the KINEROS2 model, a step that promises significant benefits in detecting environmental changes and understanding their impact on the watershed.

The purpose of the research is to identify color patterns of different types of soils to assess the conditions for identifying types using RGB images. To achieve this goal, the following tasks were solved: 1) Identify problems in the agricultural plan where soil plays an important aspect; 2) identify the characteristics of the soil composition and the relationship between the composition and color characteristics; 3) determine the features of processing RGB images during preliminary and thematic processing; 4) evaluate the color characteristics of several types of soils; and 5) determine criteria for distinguishing the color characteristics of different types of soils. The object of research is the soil type and its RGB images. Research methods included the collection of statistical information about brightness variations in RGB spectral channels for different types of soils.

Green plants are capable of producing their own sustenance. They accomplish this through a process known as photosynthesis, which makes use of a chlorophyll-based green pigment. A pigment is a chemical that has a specific color and, depending on that color, may absorb light at various wavelengths. The chlorophyll of the leaf may be predicted from leaf color, which has the potential to be an excellent indicator of plant health. Chlorophyll a (Chl-a) and Chlorophyll b are the two main types of chlorophyll found in the photosystems of green plants (Chl-b). Each kind has a unique function as well as advantages and a unique chemical composition. So, it is crucial to anticipate the chlorophyll of the takes off as it provides valuable information regarding plant health and edit management. In the proposed work, a non-destructive image processing technique is employed to estimate the chlorophyll content of tomato plants using images of the leaves. The chlorophyll content of the acquired image is predicted using a deep learning (DL) algorithm. Using the camera of an Android smartphone, samples of tomato plant leaves were collected from the field, and the color vegetation indices for all three RGB channels were then abstracted. The Chlorophyll detector, which is aimed at both small- and large-scale farms, has finally been built. For easy deployment, the mobile app combines an image from a smartphone camera with cloud-hosted software.

An analysis of the possibilities of using intelligent systems to study the condition of soils was carried out. The use of intelligent systems scheme for studying the condition of soils using the example of agrogenic and anthropogenically disturbed soils is proposed in this chapter. It includes a number of successive stages, containing field research, analysis of results and selection of deciphering features and conditions. This work focuses on the first stages, since they are specific to soil research. Other stages of statistical data processing are unified for a wide variety of objects in different fields of knowledge. They include creating a training sample, calculating statistical indicators and analyzing the decipherability of object properties from training samples, adjustment, segmentation and the use of artificial intelligence to solve average problems in the chosen direction.

The chapter presents the results of the study related to the solid waste landfill resource management models used in various regions of the world. Studied parameters include usage of land, collection and utilization of landfill gas and control of landfill's negative impact on the environment. The relevance of the study is due to the fact that all over the world, solid waste landfills are among the largest environmental polluters; they have a negative impact not only on the condition of the ground water and surface water and soils of adjacent territories, but also on the efficiency of use of land resources in the region. The authors consider technological and organizational innovations that contribute to this optimization of solid waste landfill management. The drivers of these innovations are both changes in legislation and technologies that are proven, effective from an economic, environmental and social point of view, and available to the landfill staff. Another important factor is the availability of up-to-date and reliable information about all types of resources involved in the operation process. The suggestions developed by the authors make it possible to take into account the dynamics of investments, as well as the inequality of costs and results carried out and received in different periods of time. The results are of interest both to the solid waste landfill staff and to regional and local authorities, as well as to investors and specialists involved in the development of regional waste management infrastructure.

The chapter discusses the problems of forming specially protected natural areas in the Arctic zone of the Russian Federation. Research shows the need to preserve and restore the natural environment, awareness of the importance of environmental protection of nature, especially in the context of economic globalization. As part of the study, data on specially protected natural areas of the Arctic zone of the Russian Federation were collected from open sources of information. Data processing and analysis was carried out, which made it possible to determine the current state of the system of specially protected natural areas. The authors identified a number of problems in the system of specially protected natural areas of the Arctic zone of the Russian Federation and proposed practical ways to solve them.

The consequences of the virgin lands development are still felt in the Republic of Kazakhstan. In addition to the environmental burden caused by the expansionary cultivation in the north of the Kazakh Soviet Socialist Republic (KazSSR), modern Kazakhstan has perceived the social and demographic effects of the all-Union Virgin Land program since the early 1990s. This chapter examines the motives of the Soviet leadership, initiators and opponents of the Virgin Land program (1953-1958), and critically analyzes the plans and implementation based on the experience gained. The impact on economic and social conditions is subjected to historical analysis. Further environmental and demographic results of the program, factors influencing the current soils and agricultural conditions in the Republic of Kazakhstan are reviewed and demonstrated. This allows giving an objective, scientifically based assessment of the Virgin Land program, taking into account multi-vector economic and functional influences.

In modern language education, when teaching foreign languages to master's students in environmental specialization and land management, as well as for students of other non-linguistic universities, the term ESP (English for specific purposes) is used. A competency-based approach should be implemented, in line with which the acquisition of knowledge is combined with the experience of using the acquired

Preface

The urgent task of our time is to maintain a balance between natural processes and human activity. Without a doubt, the environmental aspects of protecting soil and land resources are of great importance for the conservation of biodiversity and ensuring the sustainable development of human society, in maintaining environmental stability in the modern world. Therefore, the subject of the collective monograph has attracted great attention from scientific researchers from more than ten countries in different regions of the world: Vietnam, Germany, Egypt, India, Indonesia, Kazakhstan, Pakistan, Russia, the USA, Tajikistan and Turkey. Their efforts created a picture of the state of the problem being studied in general and showed its regional features in particular.

Conventionally, the chapters of the monograph can be combined into three sections: "Soil and land resources as an integral part of ecosystems, their interaction with other components of the natural environment"; "Monitoring the condition of soils, lands and vegetation using modern technological and hardware-software systems" and "Problems associated with ecosystem disturbances, deficiencies in scientific and linguistic communication and ways to solve them."

The section "Soil and land resources as an integral part of ecosystems, their interaction with other components of the natural environment" shows that soils play a vital role in maintaining life on Earth. They are the main source of nutrition for plants, providing them with the nutrients and water they need to grow. In addition, soils serve as an accumulator of organic carbon, important for climate regulation, as well as a filter that purifies water from contaminants. The role of soils in terms of biodiversity conservation, especially about climate change, is less studied. The mechanisms of relationships between soil microorganisms, plants and animals have been poorly studied. Methods to improve these interactions to maintain healthy ecosystems have received little attention. Of interest in this regard is chapter "Exploring the Synergy of Soil-Animal Interactions: A Blueprint for Biodiversity Preservation" by V. M. Makwana, K. Thakkar, P.P. Dodia. This scientific work describes the complex interactions between soil, plants and animals, highlights the role of animal biodiversity in ensuring ecosystem health and resilience, and proposes strategies for conserving biodiversity through scientific management of soil ecosystems.

Land resources serve as the basis, means and subject of labour in agriculture and forestry. Soil and land resources interact with other components of the natural environment, such as the atmosphere, hydrosphere and biosphere. Thus, deforestation can lead to soil erosion and pollution of water bodies, air pollution can lead to acid rain, which destroys the soil and destroys biota.

Farmland around the world is deteriorating, becoming acidic and losing essential humus content. This affects the quality of food around the world. Causes include extreme natural events and harmful human activities such as deforestation. The consequences are severe-shrinking farmland, decreasing fertility, decreasing crop yields and increasing costs. Developing countries are facing food shortages, declining populations and shorter life expectancies.

Kovyazin V., Nguyen A., and Zubov A. devoted their chapter "Zoning of Vietnam's Forest Fund Based on Climatic Conditions and Organic Mass Growth for Environmental Improvement" to the results of climatic conditions studies in three forest growth zones of Vietnam, identified to assess the increase in organic mass of wood depending on changes in climatic conditions. The authors established a connection between the increase in timber reserves and the climatic conditions of forest growth on various forest lands and proposed zoning of the forest fund of Vietnam taking into account soil quality.

The interaction of water and land resources is analyzed in the chapter "The Utilization of Water Resources in the Applications of the Future Projects" by S. Mohan. It is noted that water resources are used in agricultural, industrial, domestic, recreational and environmental activities, and today most countries experience a huge water demand. By 2030, it is predicted that the world could face a 40% shortfall between projected water demand and available water supply. Chronic water shortages, hydrological unpredictability and extreme weather events (floods and droughts) are seen as some of the greatest threats to global prosperity, and stability and causes of increased instability and conflict.

In arid regions, the main causes of land degradation are drought and soil erosion in "Utilizing Soil Erosion Modeling and GIS Technologies to Develop Future Soil and Water Conservation Strategies" by H. M.Salem, M. A. I Abdalla. Infrequent rainfall gives way to flash floods, which worsen the rate of soil erosion and significantly reduce agricultural productivity. To predict flash floods and soil loss, the authors propose an evaluation of the performance of a kinematic runoff and erosion model, which is illustrated using a study from Egypt. They use satellite imagery of digital elevation maps, soil maps and annual weather datasets and integrate runoff data at the watershed outlet into the model.

Mathematical models are also presented in a study by Russian scientists "Spatio-Temporal Variability of Soil and Methodological Approaches to its Assessment by the Example of the Leningrad Region". They analyze the spatial and temporal variability of agrochemical indicators that determine the fertility of agricultural lands in the Leningrad region and establish that the content of available phosphorus is maximum in farms with alkaline soil conditions. Medium and strong positive correlations were found between agrochemical indicators - phosphorus and humus content ($r = 0.57$), as well as pH values ($r = 0.54$); the number of metabolic bases and humus content increases ($r = 0.4$). A pattern was noted of a decrease in the number of exchangeable bases with an increase in the hydrolytic acidity of soils ($R2 = -0.82$).

These patterns make it possible to zone agricultural land and minimize impacts on soils. Such generalizations reduce the economic cost of soil and agrochemical surveys and the application of mineral fertilizers.

The work by scientists in Tajikistan also draws attention to the need for the most efficient use of land and soil resources in extreme conditions of agricultural production (mountainous terrain, desert zone, arid climate and poor grey soils). The authors rely on the introduction of promising, highly productive crops, maximally adapted to local conditions.

It is also proposed to take into account the ecological characteristics of soils and plants in the process of agricultural production in the chapter by E. C. Aytar, A. Durmaz, E. İ. İ. Torunoğlu. Turkish scientists emphasize that agriculture is a complex process involving various methods of producing food, feed, fibre and other products by raising plants and animals. Scientists are focusing on the transition to organic agriculture, which aims to more effectively and sustainably manage soil, water and biological resources while preserving ecosystem processes and biodiversity.

Its main approaches are: reducing the use of chemical fertilizers and pesticides, natural pest control, diversification of crop rotations and sustainable water management, contributing to the sustainability of agricultural areas and the conservation of global biodiversity. Ecological agriculture contributes to the sustainable development of territories and the conservation of global biodiversity.

The presented review indicates the need for more careful monitoring of the condition of soils and land to prevent negative anthropogenic impacts on nature. The task is complicated by climate change and increasing resource consumption by society. An important link in assessing the condition and protection of soil and land resources is their monitoring. It is a system of observations, assessments and forecasts aimed at obtaining reliable information about the state of lands, their quantitative and qualitative characteristics, their use and the state of soil fertility.

The data obtained allows us to identify problem areas and timely take effective operational measures to preserve and improve the environmental and economic value of soil and land resources.

The section "Monitoring the state of soils, land and vegetation using modern technological and hardware-software systems" presents modern problems in this scientific field and the results of their solution.

To carry out monitoring, modern technological and hardware-software systems are successfully used around the world. They provide objective information about the condition of soils and lands, including determining the level of their pollution, degree of erosion and fertility. These data can be used to develop soil and land conservation measures.

Thus, Siberian researchers M. Y. Kataev, E. Y. Kartashov, I. B. Sorokin note that the development of the production of unmanned aerial vehicles (UAVs), their cheapness and mass availability, and the possibility of the emergence of various measuring instruments have led to an expansion of the capabilities of radio sounding of agricultural fields. In the chapter, the authors provide an analytical review of the literature and share the results of their research in the field of using UAVs to study the colour characteristics of soil. In the chapter "Determination of Soil Color Characteristics Using RGB Images" they describe the colour characteristics of some types of soils in the regions of Western Siberia, obtained from RGB images. The use of UAVs allows for rapid mapping of soil types by colour indicators and, based on the resulting images, assessing changes in soil characteristics.

The topic of using non-destructive methods to assess the condition of soil and vegetation cover continues Y. Dharshan, D. Devasena, B. Sharmila in "Automated Detection of Plant Diseases and Chlorophyl with Image Processing using Mobile Application". The authors point out that remote methods are less accurate, but do not require much time and money. They estimate foliar chlorophyll and nitrogen content in tomato plants using digital colour image analysis based on the RGB colour model and show its suitability, in particular, for detecting nutrient deficiencies in plants.

Another application of Earth remote sensing is OpenStreetMap, which is designed to analyze urban green spaces around the world. The chapter titled "Unveiling Urban Oasis Insights from OSM Data on Green Spaces" provides an extensive literature review demonstrating the diverse applications of OSM data to understand the distribution, availability, and performance of green infrastructure. Researchers use a variety of methodologies, including spatial analysis, feature extraction, classification, network analysis, temporal analysis, and integration with other datasets to quantify green infrastructure and the quality of urban green spaces.

OSM data offers significant opportunities through its global coverage, open access, rich attribute information, timeliness, and customizability, allowing for detailed analysis and benchmarking. However, there are limitations such as variability in data quality, inconsistent labelling, and coverage bias that require careful consideration and validation methods. Despite the challenges, wise use of OSM data can improve understanding of urban green spaces, helping to make informed decisions in urban planning and environmental management.

The increase in the volume of information and the complication of models describing the processes of transformation of soil and plant resources necessitate the involvement of intelligent systems for analysis. In "Algorithm for Using Artificial Intelligence to Monitor Agrogenic and Anthropogenically Disturbed Soils" it is proposed to use the scheme of intelligent systems to study the state of soils using the example of agrogenically and anthropogenically disturbed soils. It includes several successive stages containing field research, analysis of results and selection of deciphering features and conditions. The main attention is paid to the first stages since they are specific to soil research.

The remaining stages of statistical data processing are unified for a wide range of objects in different fields of knowledge. They include the creation of a training sample, calculation of statistical indicators and analysis of the decipherability of object properties from the training sample, adjustment, segmentation and the use of artificial intelligence to solve average problems in the chosen direction.

The following chapters of the collective monograph are devoted to problems associated with ecosystem disturbances, shortcomings of scientific and linguistic communication and ways to solve them.

The protection of soil and land resources requires solving several applied problems, which include optimizing the model for managing soil and land resources, analyzing historical experience and training specialists in the field of ecology, etc. For example, the negative consequences of virgin land development are still felt in the Republic of Kazakhstan, which is studied in "Virgin Lands Development in Kazakhstan. Causes, Implementation, Consequences". Extensive farming, violation of agricultural technology, etc. led to an aggravation of the environmental situation and the deterioration of social and demographic processes on a huge scale. The historical analysis carried out in the chapter allows us to give an objective, scientifically based assessment of the Virgin Land program, taking into account the multi-vector economic and functional impacts.

Scientific work "Optimizing the Solid Waste Landfill Resource Management Model: Land Usage Efficiency Increase and Negative Environmental Impact Mitigation" by V.S. Merkusheva, M.A. Lyubarskaya, N.S. Klunko, T.M. Regent describes negative processes at a more local level. They relate to municipal solid waste (MSW) landfills and the development of models for managing related resources used in various regions of the world. Parameters being studied include landfill gas utilization and control of the negative impact of the landfill on the environment. The authors propose an economic and mathematical model for optimizing the use of solid waste landfill land for organizing a biogas collection system for solid waste management and illustrate it using the example of the Krasnoyarsk Territory of Russia. The introduction of such models at existing and future landfills can be considered as a possible method for increasing the efficiency of land use and mitigating the negative consequences of anthropogenic pollution and littering of soils and lands near cities and other populated areas.

Success and efficiency in solving environmental problems. arising from the use of soil and land resources largely depends on the speed of information exchange and good communication between scientists from different countries. An excellent example in this regard is this monograph. Communications are ensured, among other things, by the level of proficiency in scientific Russian and English languages as a means of international communication.

From this point of view, the material of the chapter "ESP in the Educational Process of Master's Degree Students and Majoring in Ecology and Land Management" is valuable, interesting and useful. It is dedicated to teaching foreign languages to undergraduates and graduate students of environmental specialities, agricultural chemistry, soil science and land management, as well as students of other non-linguistic universities. The author uses the term ESP (English for Specific Purposes) and implements a competency-based approach. In this approach, the acquisition of knowledge is combined with the experience of using the acquired knowledge in situations close to real life.

In this approach, the acquisition of knowledge is combined with the experience of using the acquired knowledge in situations close to real life. Lexical, grammatical, sociocultural and other competencies developed in the process of studying a foreign language will help the graduate to effectively collaborate with native speakers and feel free in the country of the language being studied.

The global nature of environmental problems related to the state of land and soil resources and the urgency of their solution require international cooperation and new scientific and technological achievements. The collective monograph "Ecological Aspects of Soil Conservation" explores this critical issue, emphasizing the need for collective action and sustainable solutions, which is characteristic of IGI's 35-year unifying scientific and information policy.

The book serves as a resource for scholars, students, and professionals committed to addressing the crisis. The book is intended for a diverse audience passionate about environmental sustainability. From undergraduate and graduate students to experienced academics, lecturers, and industry professionals, the book is a comprehensive resource. This is a compass for those who navigate the intricacies of rational land use, study modern approaches to the dynamics of soil fertility, use GIS technologies and delve into educational approaches in the field of ecology and agriculture.

Optimizing soil and land management requires an integrated approach and the involvement of various parties in the decision-making process. It is important to strive for sustainable use of resources and the development of environmentally sustainable technologies to maintain a balance between ecosystems and human needs. Only in this way will we be able to provide future generations with favourable conditions for life on the planet.

Thus, the results of labour research described in this collective monograph have practical and theoretical significance, as well as prospects for further study of the problems raised by the authors of the book. This collective work by an international team of authors provides not only valuable material but also a topic for further research in the field of conservation of invaluable land and soil resources around the world.

Officials and policymakers seeking to develop evidence-based policy will find this work an invaluable guide in promoting collaboration between academia and industry to pave the way for sustainable land management. In the pages of *Ecological Aspects of Soil and Land Preservation*, readers will find not only a critical exploration of the issues but also a roadmap for collective action that provides a glimpse into a future where the delicate balance of our lands is preserved and nurtured for generations to come.

Chapter 1
Theoretical Foundations of Soil Ecology and Land Resources

Erdi Can Aytar
http://orcid.org/0000-0001-6045-0183
Faculty of Agriculture Department of Horticulture, Usak University, Turkey

Alper Durmaz
http://orcid.org/0000-0001-6927-3283
Ali Nihat Gokyigit Botanical Garden Application and Research Center, Turkey

Emine İncilay İncilay Torunoğlu
http://orcid.org/0000-0003-4641-0067
Department of Medical Biochemistry, Necmettin Erbakan University, Turkey

Betül Aydın
http://orcid.org/0000-0002-9092-1350
Department of Biology, Gazi University, Turkey

Ayşen Melda Çolak
Faculty of Agriculture Department of Horticulture, Usak University, Turkey

ABSTRACT

This chapter examines the interaction of plant ecology with the environment and the environmental impacts of agriculture. Plant ecology is a scientific field focused on understanding how plants adapt to their growing environments, their roles within ecosystems, relationships among plants, and responses to environmental factors. Agriculture is a comprehensive process involving various methods for producing food, feed, fiber, and other products through plant cultivation and animal husbandry. Ecological agriculture aims to manage soil, water, and biological resources more effectively and sustainably while preserving ecosystem processes and biodiversity. Principles such as reducing the use of chemical fertilizers and pesticides, natural pest control, diversified crop rotations, and sustainable water management contribute to the sustainability of agricultural areas and the conservation of global biodiversity. This study analyzes the environmental impacts of agriculture on a large scale and emphasizes the contributions of ecological agriculture to the sustainability of agricultural areas and the preservation of global biodiversity.

DOI: 10.4018/979-8-3693-3374-7.ch001

INTRODUCTION

Plant ecology is a scientific field that studies how plants interact with their surroundings. Researchers in this discipline seek to understand how plants adapt to their growing environments, their roles within ecosystems, relationships among plants, and responses to environmental factors. Plant ecology covers diverse topics including biodiversity, ecosystem dynamics, soil plants, and plant-community interactions. Plant ecologists investigate how plants grow, reproduce, and respond to conditions in different habitats. They explore the contributions of plant species to water and nutrient cycles, as well as their interactions with other organisms, ensuring the balanced functioning of ecosystems. The insights from plant ecology are essential in both theoretical and practical applications. The principles derived from plant ecology play a vital role in developing sustainable strategies for natural resource use and promoting ecosystem health in areas such as agriculture, forestry, natural area management, and ecosystem restoration. Agriculture is a comprehensive process that involves cultivating plants and raising animals to produce food, feed, fiber, and other products. It serves the purposes of meeting basic nutritional needs, providing animal care, engaging in trade, and producing various plant and animal products for industrial use. Agriculture utilizes diverse methods and practices, adapting to varied climatic, soil, and geographical conditions globally.

Plant Cultivation: This category involves the production of various plant-based products like cereals, vegetables, fruits, legumes, and industrial crops. The process of field farming includes activities such as planting seeds, cultivating, irrigating, fertilizing, and harvesting. Livestock Farming: This category focuses on breeding and caring for animals, aiming to produce items like meat, milk, eggs, wool, and leather. Livestock farming covers essentials like feeding, sheltering, healthcare, and reproductive control. Agriculture encompasses diverse approaches such as organic farming, integrated farming, sustainable farming, and biological farming. Over time, agricultural practices have evolved with advancements in technology, genetic engineering, and modern equipment. Today, agriculture not only contributes to global food security but also serves as a crucial source of raw materials for various industries. Studies in this field aim to understand the development of vegetation in response to both living and non-living environmental factors. Investigations into the growth and distribution of plants in natural habitats emphasize key physical factors, with changes in soil composition being a notable consideration. Studies conducted in areas with prevalent vegetation reveal clear patterns in the distribution of flowering plants. The gradients in physicochemical factors, such as nutrients and water levels, shaped by the combined influence of various environmental parameters, play a crucial role in determining the settlement of species and the characteristics of vegetation.

Ecological Agriculture is a contemporary and efficient farming approach that blends principles from ecology and economics. It provides specific pathways for implementing Sustainable Development Goals. The distribution of ecological farming patterns reveals features such as variations in regional agricultural development, combinations, and diversity. A thorough examination of these distributions supports the creation of reliable development templates for similar regions at the micro-level, aiding in the formulation of macro-level development plans and strategy adjustments. Uncovering these ecological farming patterns necessitates the integration of detailed spatial, temporal, and typological information. Currently, the examination of ecological farming patterns can be categorized into three main methods: field research, literature analysis, and administrative statistics. Field research identifies ecological farming patterns through interviews with local farmers. The summary of these interviews can be employed to create local farming patterns and historical trends. Literature analysis pinpoints ecological farming patterns using typical case studies and categorizes them accordingly. The administrative statistics method qualifies ecological farming patterns starting from lower-level administrative units, utilizing effective implementation orders. These methods generally operate at different spatial scales. Field research is often suitable for local-scale investigations, such as villages and towns, due to the limited data collection efficiency of interviews. On the other hand, the literature analysis method typically favors a spatial scale at the city or state level, as it relies on typical case studies at this level. The administrative statistics method theoretically allows for investigating patterns at any scale.

There is strong experimental evidence indicating the negative impact of agricultural land use on various organisms. While positive effects are exceptions, most large-scale studies have found detrimental effects on biota in areas adjacent to rivers with increased agricultural land use. Agriculture has been recognized as a significant contributor to the poor ecological condition of freshwater. The proportion of agricultural land use in the watershed area is already an indicator of the applied agricultural stress to river biota. However, it does not account for variations in agricultural intensities. Depending on agricultural density, the effects on aquatic communities can significantly differ, as shown in numerous small-scale studies. For instance, the maximum nitrogen and phosphorus application is up to two times less in average corn fields compared to wheat fields, three times less in vegetables, and at least ten times less in perennial crops. On the flip side, perennial crops receive, on average, three times higher pesticide applications than vegetables. These variations in agricultural density reflect the stress imposed on their surroundings. Perennial crops, in particular, are linked to higher pesticide concentrations. Corn cultivation in surrounding areas can lead to a substantial entry of phosphorus, especially in regions with intensive livestock farming, which is prone to nutrient influx after severe rainfall events in the watershed. It is well-established that these stress factors trigger various adverse effects on the environment. When exploring the environmental impacts of agricultural density from different crop types, a comprehensive study of organism groups is crucial.

Moreover, the potential differences in the relationships between agricultural intensity and the environment across ecoregions are noteworthy. Studies relying on the agricultural share alone to gauge agricultural stress have consistently observed more pronounced effects in mountainous regions compared to lowlands. This might be attributed to the smaller impact sizes of disrupted gradients in lowlands, although differences in land-use legacies could also contribute. Notably, low valleys in Europe have been significantly disturbed since the 19th century, primarily due to pesticides, agricultural fertilizers, and morphological changes, which likely still reflect the extent of hydromorphological alterations. It is also acknowledged that the environmental impact of agricultural stress is contingent on the diversity of components used in agriculture and the level of awareness in their utilization. The variations in agricultural intensity and biological effects arising from different crop types could be responsible for the heterogeneity observed in large-scale studies. In such investigations, it becomes imperative to assess crop type data, often unavailable at larger spatial scales but essential for quantifying agricultural stress. High-resolution global maps of crop types can serve as valuable tools for extensive analyses of agricultural impacts, facilitating crop-specific evaluations of environmental effects on a broader scale. These maps provide a complementary and scalable approach to existing small-scale case studies on the impacts of nutrients and agricultural chemicals.

In this context, this large-scale study facilitates the analysis of the impact of agricultural land use on environmental ecology. Considering the heightened sensitivity of ecological balances in the environment to chemicals used in agriculture, the effects of fertilizers on the living groups around agricultural areas are also among the topics that need to be ecologically investigated. The impact of agricultural intensity varies significantly across ecoregions, with the most pronounced effects being felt in mountainous and alpine environments.

Ecological farming is an approach that contributes to the sustainability of agriculture and the preservation of global biodiversity. It aims to manage soil, water, and biological resources more efficiently and in a balanced manner by preserving natural ecosystem processes and biodiversity. Here are some key points on how ecological farming contributes to the sustainability of agricultural areas and the preservation of biodiversity: Reduction in the Use of Chemical Fertilizers and Pesticides: Ecological farming minimizes or completely eliminates the use of chemical fertilizers and pesticides. This helps reduce soil and water pollution, preserving the health of natural habitats and aquatic ecosystems. Additionally, this approach contributes to the conservation of natural microorganisms in the soil and beneficial insect populations. Natural Pest Control: Reliance on natural enemies, insect-eating birds, and beneficial insect species is employed to control pest insects. This contributes to the preservation of ecosystems rich in biodiversity and reduces dependence on chemical pest control. Diversified Crop Rotations: It enhances soil fertility through crop rotations. This mitigates the negative impacts of monoculture farming based on a single crop and ensures sustainable soil health. Reduction of Soil Erosion: Techniques such as cover crops, green manure, and natural landscape elements are used to minimize erosion. This increases soil fertility and prevents natural habitat losses due to erosion. Sustainable Water Management: Ecological farming encourages the efficient use of water and adopts sustainable irrigation methods. This contributes to the conservation of water resources and the sustainability of aquatic ecosystems. Preservation of Genetic Diversity: Traditional seed diversity is preserved, supporting genetic diversity. This prevents the negative impacts of agriculture based on a uniform genetic structure and enhances the long-term sustainability of farming. Increase in Organic Matter: It promotes the use of organic fertilizers and materials. This improves soil structure, increases water retention capacity, and supports soil fertility. Ecological farming methods make a significant contribution to the sustainability of agricultural areas and the preservation of global biodiversity by enhancing natural resource utilization and supporting the long-term health of agriculture. To achieve this, particularly in agricultural areas or areas planned for agricultural practices, both the public and policymakers need to be educated and informed in a way that maximally understands the ecological dynamics of the environment and the needs of agricultural areas. The use of an ecological perspective is essential in every step taken to minimize environmental damage and prevent irreversible damage. This can only be achieved through ecological education and awareness reaching a reflective level.

Plant Ecology: Interactions with Surroundings, Biodiversity in Plant Ecology

Plant ecology is the scientific discipline that investigates the mutual interactions between plants and environmental factors. Ecologists, on the other hand, study the growth conditions, distribution, diversity, adaptations, competitions, and interactions of plants. Plant ecology examines and interprets various aspects of plant life, such as their life cycle, nutrient uptake, absorption of water and mineral substances, and mechanisms of seed dispersal, in association with environmental factors. Furthermore, it encompasses the role of plants in different ecosystems, their adaptations, and their interactions with other organisms, all of which are fundamental topics within ecology. Ecological studies aid in understanding the strategies plants develop for adaptation to their environments and to other organisms, and help in comprehending the dynamics of specific ecosystems, thus contributing to a better understanding of agricultural areas and the characteristics of cultivated crops.

The entire process of succession, from transforming a barren and desolate piece of land into a green and thriving area with a large population, can be examined through the lens of ecology. During this process, pioneer species initially colonize the area, arriving through seed dispersal and the feces of other animals. These pioneer plants, with the assistance of their roots and other organisms, gradually alter the soil structure of the land. As conditions become more favourable, they make way for taller plants. This transition occurs because species with a highly competitive ability outcompete the pioneer species and become dominant in the area. This process continues until a stable community is established, known as climax vegetation. If the area undergoes disturbance due to natural disasters or human activities, secondary succession begins, and the process continues until climax vegetation is restored. Understanding this entire process without considering the interactions of plants with their environment and other organisms, i.e., without an ecological perspective, is not feasible (Figure 1).

Figure 1. Succession: transforming a barren land into a green area

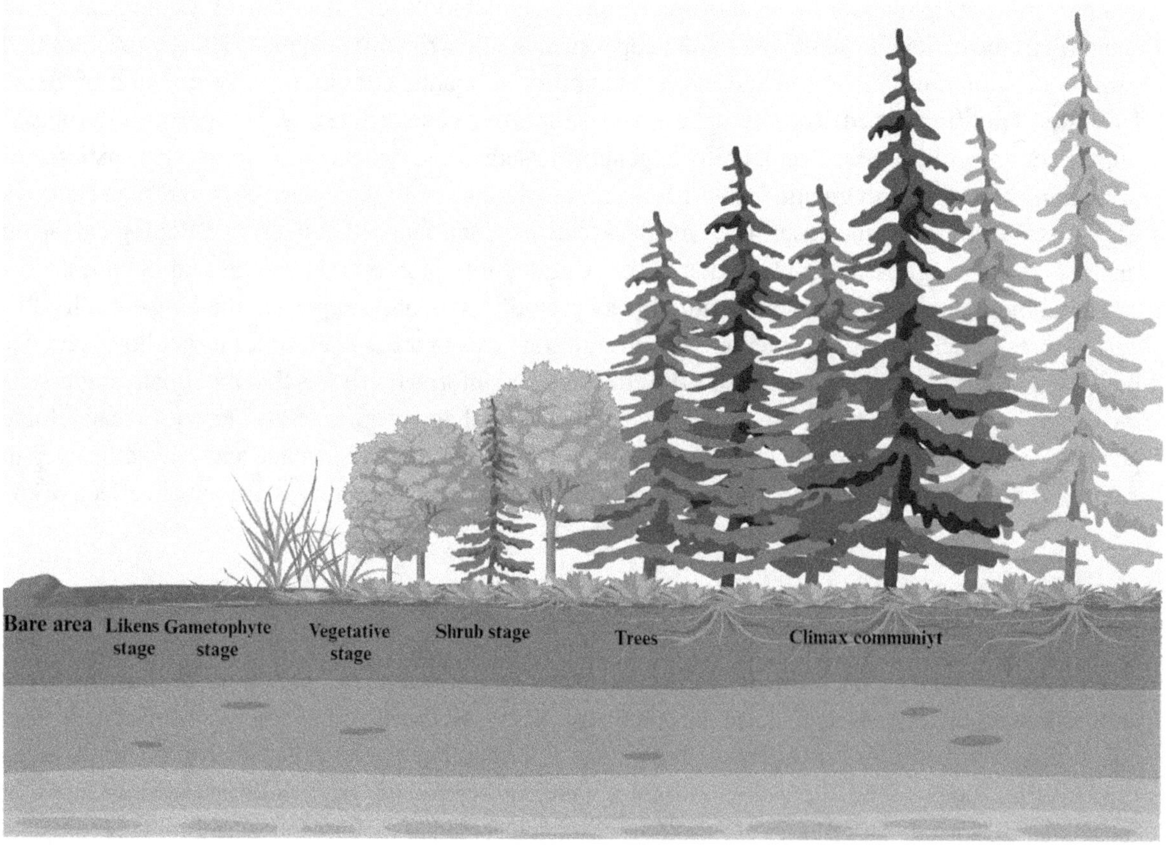

Plant ecology employs various methods to investigate how plants respond to edaphic factors. Among the commonly used techniques in ecological studies are fieldwork, laboratory experiments where abiotic and biotic factors are controlled, statistical analysis of ecological data and biological variables, and environmental geographical modelling. For instance, an ecologist can determine the distribution of a particular plant species and the environmental conditions in which it thrives through fieldwork. By conducting ecological simulation experiments in the laboratory, they can obtain data to understand how plants respond to environmental stress factors. They can process all collected data using various analysis programs and methods, and through environmental modelling, they can comprehend the general principles of plant ecology using large-scale data sets.

Plant ecology forms the foundation of basic scientific approaches to understand and conserve ecosystems. This is because plants are integral components of ecosystems, and their interactions with other organisms are considered a fundamental factor in maintaining the delicate balance of ecosystems. Additionally, knowledge of plant ecology is essential for evaluating the impacts of human activities on ecosystems, such as agriculture, forestry, and the conservation of natural habitats.

Plant ecology examines not only the physiological characteristics of plants but also their responses to environmental factors. For instance, a plant's growth rate, photosynthetic efficiency, or seed production may vary depending on environmental factors such as climate conditions and soil composition. Additionally, interactions between plants and other organisms are crucial. For example, plants may depend on insects or wind-borne pollen for pollination, and these interactions can influence the diversity and distribution of plant populations.

When anthropogenic influences combine with environmental factors, highly complex systems emerge, making it challenging to understand and determine variables. Through plant ecology, the effects of edaphic factors, climatic data, and human impact on plants and their habitats are deciphered. Consequently, optimal living conditions for any plant species or plant community can be determined, or the optimal dynamics of a damaged ecosystem can be predicted for restoration planning.

Plant life profoundly influences habitats for animals and serves as the cornerstone of food webs. This significance was paramount even during the ancestral periods of human hunter-gatherer societies. However, the advent of plant domestication and the onset of agriculture initiated a transformative process wherein human societies altered vegetation and facilitated the migration of plant taxa to novel geographical regions. These human-plant interactions catalysed demographic expansions, technological advancements in farming practices, and the global dispersal of diverse plant species. Research efforts within archaeology and plant genomics have enriched our comprehension of how domestication reshaped the intricate dynamics between humans and plants. Notably, such investigations underscore the protracted nature of domestication, emphasizing that plant adaptations were often unintended consequences of human economic activities. Moreover, domestication unfolded across diverse cultural landscapes, yielding insights into convergent evolution among different cropping types globally. In contemporary contexts, it is imperative to glean lessons from the past. Understanding the paramount importance of biodiversity emerges as a central tenet. While the diversity of plant and animal species may undergo declines over time, avenues for recuperation remain viable. Furthermore, contemporary observations reveal diminishing agricultural biodiversity and the inadvertent relegation of certain plant species to oblivion. Nonetheless, factors such as trade dynamics and human mobility offer promising prospects for the rediscovery and revitalization of erstwhile forgotten plant species. (Fuller et al., 2023).

Plant biodiversity refers to the variety of plant species within a specific geographical area, indicating the richness of different plant types. It encompasses the coexistence of diverse plant species from various taxonomic groups. Plant biodiversity is critical for maintaining the health and functionality of ecosystems because each plant species contributes uniquely to ecosystem dynamics. For example, plants play pivotal roles in sustaining food chains and ensuring ecological balance, thereby facilitating the stable distribution and completion of life cycles for other organisms. Additionally, different plant species significantly influence human activities in sectors such as agriculture, pharmaceuticals, and food security. Agricultural practices rely on various plant species for crop cultivation and the development of resilient varieties. The pharmaceutical industry utilizes natural plants for discovering novel medicinal compounds. Furthermore, diverse plant species contribute to nutritional diversity and balance, thereby enhancing food security. However, human-induced activities and environmental alterations pose significant threats to plant biodiversity. Deforestation, uncontrolled urbanization, and habitat degradation jeopardize the habitats of plant species and increase the risk of extinction (Figure 2). Conserving and sustainably managing plant biodiversity are paramount for maintaining ecosystem health and human well-being. Therefore, initiatives such as habitat conservation, ecosystem restoration, and environmental mitigation measures are essential for safeguarding plant biodiversity.

Figure 2. Ecological cycle of life and human impact

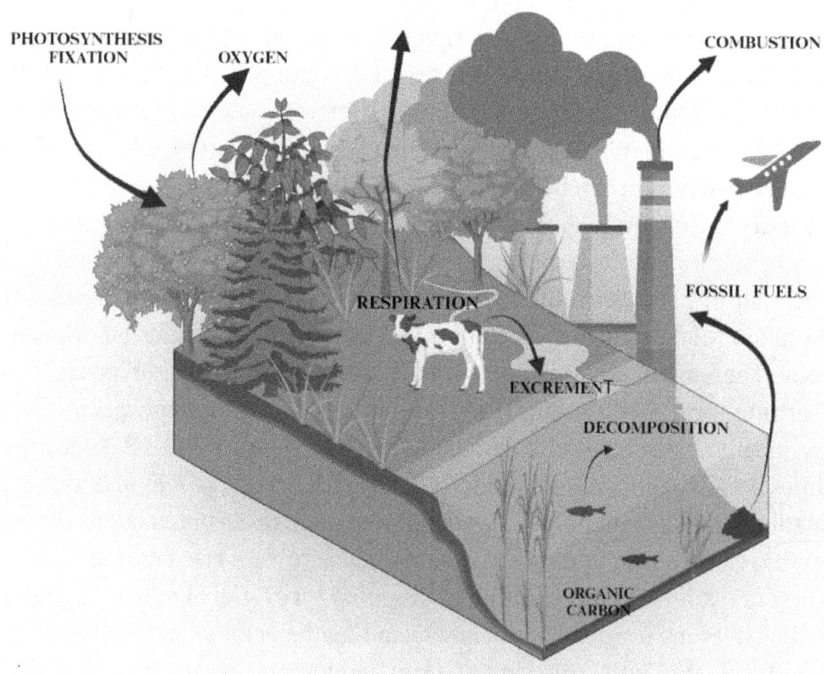

Ecology and biodiversity studies employ a multidisciplinary approach, incorporating diverse methodologies. Here's a breakdown of the commonly utilized methods in such studies: Fieldwork entails conducting on-site investigations to evaluate biodiversity within a specific region. This involves direct observation and examination to identify plant species. For instance, to assess species diversity within a plant community in a defined area, it's crucial to accurately delineate the boundaries of the plant community with distribution. Homogeneous areas representing the plant community are identified within the study area, along with observations based on topography.

Sampling is a prevalent method during fieldwork. After determining the sampling area, systematic or random sampling methods are employed to collect samples. Sampling areas are determined based on the characteristics of the dominant plant group, and the abundance of plant species is recorded using an appropriate method.

Data collection in ecology and biodiversity studies employs various methods tailored to the research objectives. This may involve taking field notes, capturing photographs and videos, collecting specimens, and obtaining numerical data using satellite imagery.

Genetic analyses play a significant role in these studies, used to understand interspecies relationships, population structures, and genetic diversity. Techniques such as PCR, microsatellite analysis, and DNA sequencing are employed for genetic analyses.

Modelling and data analysis involve evaluating collected data using appropriate techniques (e.g., CCA, DCA, GIS). This includes species distribution models, quantitative analyses, diversity indices, and similar techniques, aiding in interpreting the data and drawing conclusions (Mora & Iverson, 2002).

These methods and tools encompass various aspects of ecology and biodiversity studies and are often combined and adapted based on research questions and study focus. Biodiversity research is crucial for comprehending the distribution, diversity, interactions, and conservation of organisms in natural environments.

Plants within ecological communities engage in various interactions both with their surroundings and with each other, playing a fundamental role in shaping the dynamics of these communities. The relationships among plant species within the same geographical area can manifest in diverse forms, profoundly influencing the composition, structure, and functioning of plant communities. Several prevalent types of interactions are observed: Competition: Plants vie for essential resources such as nutrients and space. For instance, coexisting plants in a shared soil environment engage in competitive interactions for root space and access to vital nutrients. These competitive dynamics significantly impact the distribution and diversity of plant species

Symbiosis: Certain plant species form symbiotic associations with other organisms, resulting in mutually beneficial relationships. Mycorrhizal fungi, for example, form symbiotic associations with plant roots, facilitating mineral uptake and nutrient exchange. Similarly, specific plant species engage in symbiosis with nitrogen-fixing bacteria, enhancing nutrient availability and promoting plant growth.

Predation and Herbivory: Plant communities harbour herbivorous organisms and predators that exert selective pressures on plant populations. Grazing animals and insect herbivores, for instance, regulate plant populations and influence species abundance, thereby shaping community composition and structure.

Mutualistic Interactions with Pollinators: Many plant species establish mutually beneficial relationships with pollinating organisms, such as bees and butterflies. These organisms aid in pollen transfer between flowers while benefiting from the nectar and resources provided by plants, fostering pollination and contributing to plant reproduction.

Allelopathy: Some plants release chemical compounds into their environment, inhibiting the growth of neighbouring plant species a phenomenon known as allelopathy. This mechanism influences species coexistence and community dynamics by altering resource availability and interspecific interactions.

Understanding the intricacies of these interactions is vital for elucidating ecosystem dynamics and conserving biodiversity within plant communities. Such insights are essential for informing conservation strategies and ecosystem management practices aimed at preserving the integrity and functioning of natural ecosystems.

PRINCIPLES OF AGRICULTURE: CULTIVATING PLANTS

Roughly 12,000 years ago, since the inception of agriculture, farmers across diverse regions of the globe embarked on endeavors to augment crop yields. This involved meticulous selection of the most prolific plants, the innovation of novel tools, and experimentation with various agricultural methodologies. Successful innovations garnered wider acceptance, and certain technologies and practices diffused to disparate corners of the world. However, this trajectory of innovation remained gradual until the 19th century. The majority of advancements in agricultural production primarily originated from the expansion of arable land rather than enhancements in crop yield. Instances of famine and malnutrition were prevalent. Even in Western Europe, a considerable portion of the populace grappled with food insecurity and insufficient access to calories and nutrients until the 19th century. Towards the close of the 18th century, Thomas Robert Malthus, an English clergyman and economist, foresaw an impending famine due to the escalating pace of population growth and food demand surpassing the potential for expanding land for food production (Qaim, 2020).

The race between population growth and food production ushered in a new era in the latter half of the 19th century. Agricultural research underwent a paradigm shift towards a more scientific approach. Fresh insights into plant genetics, plant nutrition, and breakthroughs in the chemical industry significantly hastened the agricultural innovation process. The evolution and dissemination of improved crop varieties, coupled with the adoption of chemical fertilizers and other contemporary inputs, yielded substantial increments in agricultural productivity in the early 20th century across the United States and Europe. Consequently, food insecurity and malnutrition witnessed rapid declines in these regions. However, the assimilation and broader utilization of modern technologies in developing nations took a considerable amount of time. In the 1950s and 1960s, particularly in Asia, apprehensions about an impending famine surfaced as population growth began to outpace food production (Qaim, 2016).

Traditional agricultural practices often encompass methods closely tied to the soil, potentially resulting in various adverse ecological impacts. The term 'traditional' historically refers to the cultivation of crops in soil, outdoors, typically involving irrigation and the application of active nutrients. The detrimental effects of traditional agriculture extend beyond the growth conditions of crops, encompassing high and inefficient water demands, extensive land requirements, fertilizer usage, soil degradation, and the loss of biodiversity within natural ecosystems. With the rising global issue of hunger, the escalation of food production needs to adapt to these challenges. Particularly, the judicious use of water and soil resources justifies the development of technologies fostering optimized crop yield. Information regarding plant water and nutrient uptake in greenhouse systems plays a pivotal role in the formulation of sustainable strategies that holistically support plant growth. In comparison to alternative strategies, traditional agriculture consumes significant quantities of freshwater and fertilizers, yielding relatively low efficiency. Hydroponic, aeroponic, and aquaponic systems represent agricultural approaches that integrate nutrient solutions with inert substrates, bypassing the need for traditional soil-based plant nutrition. The necessity to cultivate efficient soils for food production could be mitigated through these approaches, thereby opening up new realms for cultivation. The potential to further enhance yield per unit area through vertical farming approaches holds equal ecological promise. Additionally, continuous, controlled environment-based production can be sustained throughout the year (Killebrew & Wolff, 2010; Serio et al., 2022).

When delving into soilless agricultural production systems, they are commonly referred to as 'soilless farming' techniques. The term hydroponics is derived from the Greek words 'hydro,' meaning water, and 'ponos,' meaning labor. It entails a method of cultivating plants using soilless mineral nutrient solutions. Terrestrial plants can thrive either in a mineral nutrient solution or within an inert medium such as perlite, gravel, or mineral wool. Hydroponics, therefore, represents a technique for nurturing plants in soilless conditions, immersing their roots in a nutrient solution without relying on any traditional aggregates (Fuentes-Peñailillo et al., 2024; Sahmat et al., 2024).

A seed serves as the fundamental structure responsible for the continuation of a plant's lineage, playing a crucial role in preserving genetic diversity, inter-species richness, and enhancing production capacity. Defined as such, germination involves a series of processes starting with the passive absorption of water by a dry seed, culminating in the development of the embryonic axis. Consequently, seed germination stands as a critical phase in the life of a plant, susceptible to both internal and external influences. Key factors influencing germination include water, temperature, oxygen, and light (Chenyin et al., 2023). Under favorable conditions, seed germination and seed formation occur rapidly. However, under extreme conditions, there is an internal barrier called 'dormancy,' hindering germination; while in unsuitable ecological conditions, there exists a mechanism slowing down germination. Therefore, breaking the 'dormancy' is essential for the initiation of seed germination. The simplest method used to eliminate dormancy is to control factors such as salinity, humidity, and temperature in the environment where the seeds are kept (Rifna et al., 2019) .

In recent years, various studies have been conducted on different seed germination methods. These investigations have yielded promising results not only in crop plants but also among medicinal plants. In the study, Song et al. (2024) employed the response surface methodology to investigate the impact of pulsed electric fields on the processing and cultivation of *Scutellaria baicalensis* seeds. The results obtained indicated that the application of pulsed electric fields significantly enhanced seed germination and boosted the metabolic activity and stress resistance of seedlings. Specifically, the optimal parameters for the pulsed electric field application (0.5 kV·cm -1, 120 μs, and 99 pulses) led to a notable increase of 29.25% in seed germination potential and 20.65% in the germination index. Moreover, there was a significant rise in the activities of SOD, POD, and α-amylase, along with an increase in the content of Pro, soluble sugars, and soluble proteins in seedlings from the 5th to the 15th day, compared to the control group.By establishing a theoretical basis for enhancing the germination and seedling growth of medicinal plants like *S.baicalensis*, this study provides insights into practical applications for their cultivation (Song et al., 2024). In a separate investigation, Yadav et al. (2024) isolated bacterial endophytes from the shoots and roots of wheat, exhibiting diverse salt tolerance ranging from 0.5 M to 4.0 M. The endophytic isolates showcased varied salt tolerance potentials, with a mere few demonstrating the ability to withstand elevated salt concentrations (≥3.0 M NaCl). These endophytic isolates comprised 65 distinct Operational Taxonomic Units (OTUs) recovered from both the roots (53.85%) and shoot tissues (46.15%) of the wheat plant. Analysis of the 16S rRNA gene sequences unveiled the preeminence of Firmicutes (81.54%), Proteobacteria (13.85%), and Actinobacteria (4.62%) at the phylum level. Within these phyla, Bacillus (63.08%), Priestia (13.85%), Pseudomonas (10.77%), along with Exiguobacterium and Curtobacterium (each 3.08%), emerged as the dominant genera. Furthermore, this study marks the inaugural report on specific bacterial species, such as *Bacillus australimaris*, *B. vallismortis*, *B. haynesii*,

Rossellomorea marisflavi, *Exiguobacterium undae*, *Curtobacterium pusillum*, serving as endophytes in wheat cultivated in salt-affected regions. The selected endophytes significantly alleviated the deleterious impacts of salt stress and heightened various growth parameters, including root length, shoot length, fresh weight, and dry weight, in the salt-sensitive wheat variety 'HD 2733,' in comparison to the control group. This study holds promise for enhancing soil fertility and mitigating crop losses under saline stress conditions (Yadav et al., 2024). In recent times, orchid species have found themselves in jeopardy due to excessive harvesting and habitat degradation. Similar to their flowering counterparts, orchids rely on seed production as the primary strategy for reproduction. Orchids generate minute seeds composed of a seed coat and an underdeveloped embryo. Notably, these seeds lack the endosperm, a typical energy source crucial for germination. The exclusive avenue for orchid seed germination involves acquiring nutrients externally, a feat achieved in nature through mycorrhizal symbiosis. Aytar and Özdener Kömpe (2023) explored the efficacy of various in vitro germination and seedling development environments for Anacamptis laxiflora (Lam.) seeds, employing *Ceratobasidium* sp. inoculation alongside diverse organic substrates. The most noteworthy germination rate (35.78%) manifested in a medium enriched with young hazelnut leaves. Following this, a soilless *ex vitro* symbiotic germination was executed utilizing young hazelnut leaves as the most potent organic substrate. In this setting, seed germination registered at 19.01%, with 14.87% of seedlings displaying well-developed leaves and roots. This groundbreaking achievement marks the inaugural instance of successfully cultivating *A. laxiflora* from seed in *ex vitro* conditions, without soil, thereby acclimatizing it to soilless circumstances and integrating it into the natural habitat (Aytar & Kömpe, 2023).

ECOLOGICAL AGRICULTURE: BLENDING ECOLOGY AND ECONOMICS FOR SUSTAINABILITY

The search for new agricultural methods aims to find ways to feed the world's population while also preserving ecosystem functions. Two main approaches have emerged: sustainable agricultural practices and ecologically-based agricultural methods (Figure 3). Sustainable approaches aim to balance producing higher agricultural value with minimizing resource consumption globally. Meanwhile, ecologically-based methods prioritize designing agricultural practices and cultivation areas to work harmoniously with ecosystems, advocating for efficient use of natural resources to enable their self-renewal (Tittonell, 2014).

Figure 3. Sustainable agricultural practices and ecologically-based agricultural methods

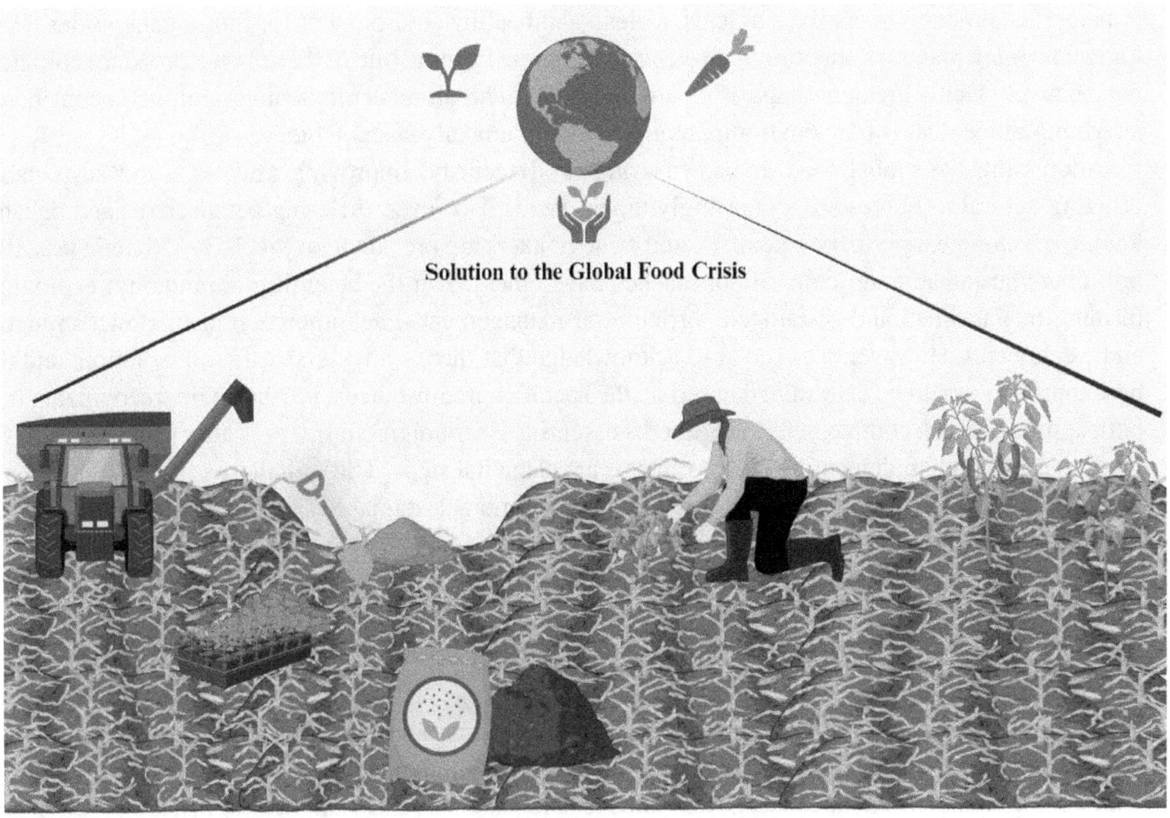

Ecological agriculture is an approach to farming that focuses on sustainability, economic reforms, and meeting global food needs. It aims to prevent irreversible changes in ecological parameters and ecosystem variables from reaching unsustainable levels, unlike traditional farming methods. This approach involves practices that prioritize maintaining soil dynamics, water balance, aquatic ecosystems, and biological diversity in a sustainable manner, with a focus on natural ecosystem functioning.

Furthermore, ecological agriculture seeks to provide farmers with long-term income security while ensuring the food needs of the human population are met. In regions where ecological farming is practiced, farmers must follow certain fundamental principles. These include measures to enhance soil health, preserve natural habitats, promote biological diversity, and control harmful organisms using environmentally friendly methods.

Efficient resource management is encouraged through waste recycling, composting, and cyclical farming practices. Local seed varieties and species are preferred, along with marketing strategies that support the local economy. Education and knowledge sharing on ecological farming principles and practices are provided to farmers and communities.

Diverse cultivation of plant species is advocated to maintain natural balance, increase productivity, and prevent the spread of harmful organisms. Water conservation and optimization of irrigation methods are also emphasized. Additionally, principles are in place to protect individuals involved in agricultural production while preserving ecosystem dynamics and natural habitats.

The goal of ecological agriculture is to boost productivity and profitability while safeguarding environmental sustainability. This approach aims to carry out agricultural practices without causing harm to natural ecosystems and strives to leave a clean and healthy environment for future generations. Ecological farming practices are often seen as part of organic farming but are based on a broader ecological perspective, placing greater emphasis on sustainability. Therefore, ecological agriculture is considered a farming model that is both environmentally and economically sustainable.

Addressing the global food crisis, ensuring food security, improving nutrition, and sustainably utilizing natural resources are increasingly urgent needs. However, differing perspectives and debates among development partners, scientists, and policymakers are prevalent on this issue. Nonetheless, two prominent sustainable agricultural approaches have emerged in the scientific community: ecological farming management and sustainable agricultural management. Each approach has its own strengths and weaknesses. However, it's crucial to acknowledge that there's no one-size-fits-all solution, and the best approach should be determined based on the specific circumstances. Furthermore, recognizing that both approaches can complement each other is essential. Encouraging farmers to adopt suitable agricultural practices and implementing policy reforms are also vital steps. Through these efforts, we can work towards developing a collective approach for achieving more sustainable agriculture and ensuring food security (Mockshell & Kamanda, 2018).

Within the realm of ecological farming, there exist several distinct models that vary depending on factors such as geographical location, agricultural product type, or societal preferences (Sumberg & Giller, 2022). However, certain fundamental models are commonly observed. Agroecology, for instance, is an agricultural approach centered around principles like diversity, efficiency, recycling, and regulation of agricultural areas. It encourages farming practices that harmonize with nature and seeks to minimize farmers' impacts on soil, water, and biological diversity. Another notable model is Organic farming, which limits synthetic chemical usage and aims to safeguard soil and plant health. Instead of relying on chemical fertilizers, pesticides, and hormones, organic farming employs organic fertilizers, natural pest control methods, and organic cultivation techniques. Diversified farming systems, commonly found in culturally diverse regions, involve cultivating various plant and animal species together rather than relying on monoculture farming. These systems enhance soil fertility and deter pest proliferation. In regions prioritizing ecological awareness and traditional societal elements, models inspired by Nature Mimicry are prevalent. These models advocate for agricultural systems designed to mirror natural ecosystems, supporting practices that regulate the water cycle, mitigate soil erosion, and maintain biological equilibrium (Reganold & Wachter, 2016).

In areas boasting biological reserves or rich biodiversity, Conservation Agriculture takes precedence, emphasizing soil and environmental conservation. By reducing soil tillage frequency, preserving natural habitats, and fostering biological diversity, this approach plays a vital role in erosion prevention and ecological balance. Finally, in regions dominated by mountainous or forested ecosystems, Agroforestry integrates trees and other woody plants into agricultural land. This practice aids in soil erosion reduction, water cycle regulation, shade provision, and biodiversity enhancement. Overall, these ecological farming models strive to conserve natural resources, promote biodiversity, and ensure the long-term sustainability of agriculture. Consequently, while meeting global food demand, they also contribute to the preservation of natural habitats and biodiversity, ultimately benefiting the human population, whether directly or indirectly (Figure 4).

Figure 4. Importance of sustainability agriculture and education

TEACHING APPROACHES AND EXAMPLES FOR ECOLOGICAL AGRICULTURE PRACTICES AND METHODS

Introduction to Ecological Agriculture Principles

Agroecology stands out as a dynamic concept in scientific, agricultural, and political discussions in recent years. The United Nations (UN) Special Rapporteur on the Right to Food underscores agroecology as a practical approach for sustainable agriculture. Despite its increased visibility over the past two decades, agroecology boasts a rich history in agricultural ecology. The term has evolved significantly since its inception in the early 20th century, with diverse meanings, definitions, interpretations, and approaches emerging. The proliferation of definitions for agroecology in recent times stems from various institutions and countries defining it based on their distinct concerns and priorities. These definitions collectively recognize the interdisciplinary nature of the agroecological approach, spanning science, diverse applications, and a social movement applicable to all aspects of agricultural food systems (Wezel et al., 2020).

Commonly used definitions for agroecology as a scientific discipline include: (i) an integrative study of the ecology of the entire food system, encompassing ecological, economic, and social dimensions ; or simply, the ecology of the food system (ii) the application of ecological concepts and principles to the design and management of sustainable food systems ; and more recently, (iii) research, education, action, and integration of change bringing sustainability to all parts of the food system (ecological, economic, and social) (Gliessman, 2018).

As a suite of agricultural practices, agroecology aims to enhance agricultural systems by leveraging natural processes, fostering beneficial biological interactions and synergies among components of agricultural ecosystems, minimizing synthetic and toxic external inputs, and augmenting ecological processes and ecosystem services. While agroecology has predominantly concentrated on defining sustainable agricultural practices at the field scale, the discipline is increasingly acknowledging multiple embedded ecological and managerial scales, from fields to landscapes (Duru et al., 2015; Jeanneret et al., 2021). This shift is prompted by the realization that many ecosystem functions providing services required by agroecological approaches, such as crop pollination and pest control, depend on processes occurring at larger landscape scales. Notably, attention has shifted towards the redesign of agricultural landscapes to serve agriculture in providing services, a subject that has not previously held a prominent place in political and research agendas. Achieving this goal necessitates collaborative efforts from ecologists, agricultural scientists, and social scientists (Bürgi et al., 2017; Jeanneret et al., 2021; Sayer et al., 2013).

Hands-On Field Practices in Ecological Agriculture

Agroecological practices aim to produce food while prioritising ecological processes and ecosystem services as fundamental elements in their development. This is achieved by integrating these processes and services, rather than relying solely on ordinary techniques like chemical fertilizers and synthetic pesticides, or technological solutions such as genetically modified organisms (Wezel et al., 2014).

Ecological agriculture field practices encompass a range of techniques aimed at sustainable and regenerative farming systems. These practices emphasise holistic management approaches that integrate biodiversity, soil health and ecosystem services to increase agricultural productivity while minimising environmental impacts (Hawes et al., 2021).

Crop Rotation and Polyculture: Crop rotation involves systematically alternating crops within a specific area over successive seasons to disrupt pest and disease cycles, improve soil fertility, and reduce reliance on external inputs (Shah et al., 2021). Polyculture, on the other hand, entails growing multiple crops simultaneously in the same field, mimicking natural ecosystems to enhance resilience and ecosystem stability. Crop rotation and polyculture are two agricultural practices that promote biodiversity, enhance nutrient cycling, and mitigate weed and pest pressures (Alves et al., 2023).

Cover Cropping and Green Manure: Cover cropping is the planting of non-crop crops during fallow periods to protect and improve the quality of the soil, prevent erosion, suppress weeds and improve the retention of nutrients (Puech et al., 2021). Green manure is the cultivation of certain crops, often legumes, which are then incorporated into the soil to enrich it with nitrogen and organic matter (Tomar & Pali, 2023). Both practices support sustainable crop production by contributing to soil fertility, structure and moisture retention.

Agroforestry Systems: Agroforestry involves the integration of trees and shrubs into agricultural landscapes to enhance ecological resilience, diversify production, and provide multiple ecosystem services (Buratti-Donham et al., 2023). Common agroforestry practices include alley cropping (Garrett et al., 2021), silvopasture (Poudel et al., 2022), and windbreaks (Király et al., 2023), which improve soil fertility, regulate microclimate, sequester carbon, and conserve biodiversity. These systems offer opportunities for sustainable intensification while fostering ecosystem functionality.

Conservation Tillage and No-Till Farming: Conservation tillage aims to minimize soil disturbance and erosion by reducing or eliminating plowing and cultivating practices (Jug et al., 2021). No-till farming takes this a step further by avoiding soil disturbance altogether, which helps to preserve soil structure, organic matter, and microbial communities (Calegari et al., 2020). These practices have been shown to enhance soil health, water infiltration, and carbon sequestration, while also reducing energy inputs and greenhouse gas emissions associated with conventional tillage.

Integrated Pest Management (IPM): To manage pests while minimizing reliance on synthetic pesticides, IPM integrates a range of pest management strategies, including biological, cultural, and mechanical methods (Deguine et al., 2021). This approach focuses on prevention by diversifying crops, manipulating the environment, and using biological control to keep pest populations below economic levels. IPM promotes ecological balance and sustainable crop protection by encouraging natural pest regulation and reducing chemical inputs (Petit et al., 2020).

Water Management and Conservation: Efficient water management practices, such as rainwater harvesting, drip irrigation, and soil moisture conservation, are essential for ecological agriculture (Abdallah et al., 2021). These practices optimize water use efficiency, mitigate drought stress, and minimize soil erosion and nutrient runoff (Anantha et al., 2021). By enhancing water availability and resilience to climate variability, efficient water management contributes to sustainable crop production and ecosystem health.

Sustainable Water Management

Despite oceans containing 97% of the world's water, it remains saline. The remaining 3% is freshwater, of which 2-3% is stored as glaciers. Approximately half of this freshwater is surface water, while the rest is stored underground. Human survival relies on this freshwater, with half of it distributed across lakes, rivers, and other reservoirs. Agriculture accounts for nearly 75% of freshwater usage. (Motoshita et al., 2018) The demand for crops has surged, reaching up to 75%, especially in countries like Brazil. However, various constraints and the absence of suitable monitoring tools have led to the temporary application of irrigation water. For example, the United States utilizes about 80% of its freshwater for irrigation purposes. The United Nations Convention to Combat Desertification (UNCCD) in 2013 identified irrigation water shortages in around 168 countries globally. Studies project a significant worldwide water scarcity for irrigation by the year 2030 (Ayaz et al., 2019). The availability of freshwater, due to the escalating demand for irrigation, will be restricted to countries implementing optimal water resource management practices. Hence, awareness programs promoting the adoption of effective irrigation practices need to be encouraged to alleviate water scarcity. Several new irrigation techniques, such as sprinkler and drip irrigation, are being deployed to minimize losses and prevent water shortages. Traditional irrigation methods, like furrow and flood irrigation, contribute to water wastage. Moreover, conventional irrigation approaches (e.g., flooding) result in nutrient reduction in the soil through water losses, adversely affecting crop productivity. Robust technologies and tools, particularly smart irrigation techniques, are imperative. Smart irrigation techniques aid in predicting crop water needs based on soil types, previous soil moisture, and climatic conditions. In this context, numerous cutting-edge tools, including the Internet of Things (IoT), play a pivotal role in measuring air and soil moisture, ultimately contributing to the enhancement of crop development. The augmentation of crop productivity and quality can be realized through effective soil and water management. The application of the Internet of Things (IoT) for irrigation management based on the Crop Water Stress Index (CWSI) is anticipated to markedly boost crop productivity (Devaux et al., 2014; Khan et al., 2021). Therefore, calculating CWSI necessitates obtaining product canopy data at various intervals and temperatures. CWSI functions as a system to predict irrigation water requirements, incorporating diverse product parameter measurement systems within a wireless sensor-based monitoring framework (Ayaz et al., 2018). In their 2023 research, Patel and colleagues investigate the enhancement of water utilization in wheat cultivation within a semi-arid region of India characterized by clayey-loam soils and a growing water supply-demand gap. The primary focus lies in the optimization of planning for drip irrigation (DI) in wheat crops, utilizing the Crop Water Stress Index (CWSI) based on the Canopy-Air Temperature Differential (CATD). Field experiments were carried out over two winter seasons at Junagadh Agricultural University (JAU) in the Gujarat region of India, employing high-discharge drip irrigation (DI) on wheat crops (variety - GW 366). The experiment comprised nine treatments randomly replicated three times as blocks. DI applications labeled T1 to T6 were irrigated at CATD values of -4, -3, -2, -1, 0, and 1 °C, respectively. T7 represented border irrigation. Well-irrigated treatments (T8) and excessively dry treatment (T9) in DI were considered to establish the upper and lower baseline of CWSI. The drip irrigation system (DIS) had an application rate of 7.78 mm/hour with a uniformity coefficient of 90.40%. In comparison to surface irrigation, DI resulted in water savings ranging from 3.52% (T1) to 43.24% (T6). Real-time CATD data were utilized to compute the theoretical Crop Water Stress Index (CWSI Th) and the vapor pressure deficit (VPD)-based semi-empirical CWSI (CWSI VPD), measured as 0.3351 and 0.3668, respectively. CATD values exhibited significant differences in CWSI among treatments, demonstrating sensitivity to water stress during critical growth stages. The study surpasses border irrigation outcomes with a 36.48% higher crop yield (4744 kg/ha), 30.23% water savings (388.42 mm), and a noteworthy 95.67% increase in water use efficiency (12.21 kg/ha-mm), compared to border irrigation results (3476 kg/ha, 556.74 mm, and 6.24 kg/ha-mm). These findings hold important implications for addressing concerns related to food and water security for farmers and policymakers. This research underscores the efficacy of CATD-based irrigation planning combined with drip irrigation to optimize wheat crop performance. In the context of critical water management, this study provides fundamental insights for sustainable agricultural practices (Patel et al., 2023).

Environmental catastrophes, the relentless surge in population, and urbanization collectively pose supplementary threats to the sustainable management of water resources and the assurance of a secure water supply. Principal contributors to water pollution encompass agricultural practices, waste disposal sites, mining operations, and the discharge of industrial and urban wastewater. These effluents discharge substantial amounts of hazardous substances such as heavy metal ions, organics, and micro-pollutants directly into water reservoirs. To tackle this urgent predicament, a variety of water treatment technologies have been suggested, predominantly relying on distillation, membrane filtration, and adsorption. Despite their efficacy in pollutant removal, these methodologies are seldom deemed sustainable due to their exorbitant investment and operational expenditures. Establishing sustainable technologies, grounded in the all-encompassing integration of environmental, techno-economic, and social facets, becomes imperative to meet current requirements without compromising accessibility for future generations (Peydayesh & Mezzenga, 2021).

In this framework, the adsorption technique, specifically employing protein-based materials for pollutant removal, emerges as a compelling technology. Proteins, serving as adsorbents in water purification, demonstrate remarkable proficiency in binding a diverse array of pollutants, positioning them as exemplary eco-friendly materials for water treatment. Additionally, proteins can be easily customized and enhanced by converting them into specific formats, such as thin films or nanofibers with elevated aspect ratios, a pivotal factor in adsorption processes (Ibrahim et al., 2018). Crucially, proteins can be promptly procured in significant quantities and at minimal costs by recovering or transforming them from waste generated across diverse sectors, including agriculture, livestock, dairy, and food industries. Despite this potential, the global value chain of protein-based products lacks effective sustainable management, given that around one-third of global food production and 20% of total dairy products for human consumption are discarded (Bolisetty et al., 2019).

Furthermore, numerous industrial wastes or by-products, like whey or rapeseed waste (comprising 40% protein), contain substantial protein content. Repurposing protein wastes into higher-value products not only curtails overall waste streams but also augments the value of by-products from the food industry, thereby mitigating the industry's overarching environmental impact. Consequently, the efficient utilization and reassessment of protein-based by-products and wastes for water treatment not only offer a viable and eco-friendly solution for ensuring clean water but also yield substantial benefits in terms of waste management and environmental conservation (Peydayesh & Mezzenga, 2021).

Incorporating Organic Matter: Composting and Fertilization

The environmental degradation caused by the excessive use of fertilizers has led to the development of plant-nutrient management and an integrated approach to agriculture. Organic farming, as a technology that promotes recycling, soil fertility sustainability, and fertilizer conservation, is increasingly gaining popularity. In comparison to the application of fertilizers alone, the feasibility of applying organics alone results in lower yield production. However, the combined application of fertilizers and organics may enhance agricultural production and the applicability of organic matter in the soil. Fertilizers and organics can offer different advantages to soil management. The combined application of fertilizers and organics may result in a positive interaction effect in agricultural production (Wei et al., 2016).

Compost is defined as a stabilized and sterilized composting product (Senesi, 1989). Compost represents a significant type of high-organic soil improvement, providing nutritive fertility to the soil (Weindorf et al., 2011). Fresh compost is defined as a partially decomposed, unstabilized intermediate product that can undergo further decomposition and, when necessary, achieve sanitization. Mature compost is the final product of the stabilization stage and is considered as organic fertilizer. Cured compost is defined as a highly stabilized product resulting from an extended stabilization and mineralization stage (Senesi, 1989).

Composting is the process of breaking down organic matter by microorganisms. Microorganisms break down organic materials, creating nutrients for plants. It transforms organic materials into a fertilizer by decomposing them to improve soil. Composting reduces environmental pollution and is an eco-friendly solid waste processing technology. Composting provides sufficient aerobic fermentation for nitrogen stability, decreases nitrication and denitrification, reduces greenhouse gas emissions, and improves contaminated soils by addressing organic pollutants such as phthalates (Xie et al., 2023).

Composting organic waste and the utilization of compost in agriculture will help recover plant nutrients and organic matter. The most significant benefit of compost is to increase soil organic matter. It enhances soil fauna and biomass, stimulates enzyme activity, leads to increased soil mineralization, and develops resistance against diseases and pests, both of which are fundamental to organic farming. A substantial increase in soil organic carbon content can contribute to reducing the greenhouse effect. The increase in soil organic matter has a significant impact on soil structure, improving physical characteristics such as aggregate stability, bulk density, porosity, water-holding capacity, and infiltration (Erhart & Hartl, 2010).

Apart from causing environmental pollution, ammonia emissions also impact the quality of compost products. In a study conducted by Zhou et al. (2023), a composting system (Condensation Return Composting System, CRCS) was designed with the aim of reducing ammonia emissions. It was reported that CRCS reduced ammonia emissions by 59.3%. When compared with the control group, this composting system was found to increase the total nitrogen content by 19.4%. Consequently, it has been suggested that CRCS could be a promising strategy for reducing ammonia emissions and producing nitrogen-rich organic fertilizers (Zhou et al., 2023).

Preservation of Genetic Diversity in Ecological Agriculture

Genetic diversity is crucial for ecological agriculture as it provides resilience to environmental stresses, pests, and diseases, ensuring sustainable food production. Genetic diversity refers to genetic variation within species. Plants need to adapt and evolve to changing environmental conditions. In ecological agriculture, which emphasizes the use of natural resources and ecological processes, genetic diversity plays a key role in maintaining the stability of ecosystems, increasing the resilience of crops, and reducing the dependence on external inputs such as pesticides and fertilizers (Pathirana & Carimi, 2022). Monocultures (Bourke et al., 2021), industrial agriculture (Tian et al., 2021), climate change (Henry, 2020) and seed patents and intellectual property rights (de Mévius, 2022) are just some of the factors that continue to threaten genetic diversity in agricultural systems. Several strategies can be implemented to overcome these challenges and maintain genetic diversity in ecological agriculture. These include some international agreements and some approaches in national or international use (Salgotra and Chauhan, 2023). Some of these are as follows: the International Treaty on Plant Genetic Resources for Food and Agriculture (Esquinas-Alcázar et al., 2012), Nagoya Protocol (Ambler et al., 2021), Svalbard Global Seed Vault (Asdal, 2021), the Cartagena Protocol on Biosafety (Zhao, 2020), in-situ conservation approach (Wambugu and Henry, 2022), ex-situ conservation approach (Engels and Ebert, 2021), in vitro propagation (Kulak et al., 2022), cryopreservation (Rani et al., 2021), DNA banks (González-Toral and Cires, 2022), digital sequence information (Cowell et al., 2022).

Educating on Reduced Chemical Usage: Fertilizers and Pesticides

Worldwide, the use of chemical fertilization and pesticides has become an integral part of modern agricultural systems to meet global food demands. Fertilization provides essential nutrients to crops, while pesticides protect them from diseases and pests (Dhankhar & Kumar, 2023). Due to the limited availability of cultivation areas and the increasing global population, there has been a noticeable increase in the use of fertilization and pesticides. Fertilization plays a crucial role in enhancing crop productivity (Xu et al., 2023). Over time, the growing reliance on these chemicals may have negative effects on human health. These chemical substances often possess toxicity and can disrupt ecological balance by causing the undesired death of organisms (Dhankhar & Kumar, 2023). Excessive use of agricultural chemicals is deteriorating the quality of soil and groundwater. Contamination of groundwater poses a threat to human health. Sustainable farming practices not only aim to preserve crop yield and the environmental ecosystem but also work towards preventing groundwater pollution (Srivastav, 2020).

To ensure the continuity of life and the sustainability of ecosystems, soil health preservation and management are crucial for maintaining biological diversity and the sustainability of agricultural production. Although modern agriculture has a positive impact on crop productivity through fertilizer use, the continuous use of chemical fertilizers can have adverse effects on soil health. Excessive use of chemical fertilizers can lead to soil compaction, reduced productivity, and water, air, and soil pollution. Additionally, it decreases organic matter content, weakens microbial activity, and negatively impacts biodiversity. Therefore, when aiming for sustainable agriculture, the effects of chemical fertilizers should be considered, and a balanced approach should be adopted for the long-term health of the soil (Tahat et al., 2020).

Pesticides, defined as a chemical mixture used to prevent, eliminate, and repel pests, are integral to traditional agriculture for pest prevention. Organic farming serves as an alternative that rejects synthetic pesticides. Ecological pesticides, as an alternative to synthetic pesticides, are substances derived from organic materials, considered environmentally friendly, and safeguard human and animal health while causing no harm to habitats and ecosystems. For instance, linalool, a monoterpene found in coconut and basil, exhibits insecticidal properties. D-limonene, derived from orange peels, serves as an antibacterial substance and is the main component of orange oil. This substance is an effective insecticide for ants, cockroaches, and flies (Zowada et al., 2020).

Pudjiastuti et al. (2023) conducted studies to increase vegetable production and prevent soil degradation through the use of fertilization and pesticides. The research reported that the chemicals used in the cultivation of Brassica rapa and Apium graveolens did not affect production. However, in the cultivation of Capsicum annuum, excessive use of chemicals was found to decrease production. The study highlighted that chemical had an impact on the soil structure (Pudjiastuti et al., 2023). El-Bassiony et al. (2020) study reported that Conservation Agriculture and Integrated Pest Management practices could serve as sustainable and environmentally friendly alternatives to increase vegetable production efficiency. The research suggested that these practices, aimed at reducing pesticide use, could contribute to the preservation of human health and the environmental ecosystem (El-Bassiony et al., 2020).

Diversified Crop Rotations: Enhancing Soil Fertility

Diversified crop rotation worldwide has the potential to increase agricultural productivity and improve soil structure. As a result of the increasing global population, the demand for food is rising, and crop productivity is declining. Therefore, resolving the productivity crisis is of great importance (Wang et al., 2020).

Sustainable field management aims to increase soil organic matter content and prevent erosion through the implementation of crop rotation practices. Crop rotation affects the reproduction of insects and other pathogens by disrupting their life cycles. When certain plant species are included in the rotation, there is a reduced need for chemical fertilizers to support the plant's growth and nutrient requirements. Crop rotation is a crucial technique in the implementation of sustainable agriculture (Shah et al., 2021).

Diversified crop rotation is crucial not only for crop production but also for maintaining soil fertility and preventing the spread of soil-borne diseases. It is highly important for preventing soil-borne diseases. In the absence of rotation, infections will rapidly spread in the soil, making plant diseases inevitable. Rotation has recently become significant for reducing greenhouse gas emissions, gaining resistance against biotic and abiotic stresses, optimizing the microbial structure and function of the soil, and increasing carbon turnover (Liu et al., 2022).

Elderly soil left fallow can help increase the efficiency of crop rotation. The purpose of leaving the soil fallow is to enhance soil fertility and protect it from soil-borne diseases and pests (Wang et al., 2020).

Marcos-Perez et al. (2023) conducted a study where they planted broccoli and peas in the same cultivation area. The report indicated a reduction in fertilizer application when broccoli was planted alone. Furthermore, when broccoli and peas were planted together, fertilizer application decreased even further, and it was reported that soil nutrients such as N, P, K, and some micronutrients increased compared to broccoli planted alone. The interaction between the two species increased soil fertility, thus promoting high crop yields (Marcos-Pérez et al., 2023).

In conclusion, implementing teaching approaches such as incorporating organic matter through composting and fertilization, educating on reduced chemical usage for fertilizers and pesticides, and promoting diversified crop rotations are essential practices and methods for fostering ecological agriculture.

Promoting Ecological Awareness in Agricultural Communities

Ecological awareness involves understanding the complex relationships between organisms and their environment and recognizing the importance of ecological processes in agricultural systems (Despotović et al., 2021). In agricultural communities, fostering ecological awareness is crucial for transitioning towards sustainable farming practices that prioritize ecosystem health, biodiversity conservation, and resilience to environmental challenges. Ecological awareness requires an understanding of the dynamic interactions between biotic and abiotic factors within agricultural ecosystems. This includes an appreciation of the role of soil microorganisms, pollinators, natural predators, and other components in maintaining ecosystem balance and productivity. Ecological awareness involves adopting a holistic approach to agricultural management, considering the interdependence of environmental, social, and economic factors (Kpienbaareh et al., 2020). Farmers can improve the sustainability of their operations by prioritizing long-term environmental protection and ecosystem services (Meng et al., 2022). Education and training, demonstration farms, farmer-to-farmer networks, and policy support are important methods for raising ecological awareness (Albizua et al., 2021; Brown et al., 2021)

CONCLUSION

This chapter has addressed the interaction of plant ecology with the environment and the environmental impacts of agriculture. Plant ecology is a scientific field focusing on understanding the adaptation of plants to their growing environments, their roles within ecosystems, interplant relationships, and responses to environmental factors. Agriculture, on the other hand, encompasses a broad process involving the production of food, feed, fiber, and other products. Ecological farming aims to manage ecosystem processes and biological diversity, effectively and sustainably, while preserving soil, water, and biological resources. This study has analyzed the environmental impacts of agriculture on a large scale and emphasized the contributions of ecological farming to the sustainability of agricultural areas and the conservation of global biodiversity. In conclusion, understanding the environmental impacts of agricultural practices and promoting sustainable farming methods are vital for maintaining ecosystem health and supporting biodiversity.

REFERENCES

Abdallah, A. M., Jat, H. S., Choudhary, M., Abdelaty, E. F., Sharma, P. C., & Jat, M. L. (2021). Conservation agriculture effects on soil water holding capacity and water-saving varied with management practices and agroecological conditions: A Review. *Agronomy (Basel)*, 11(9), 1681. 10.3390/agronomy11091681

Albizua, A., Bennett, E. M., Larocque, G., Krause, R. W., & Pascual, U. (2021). Social networks influence farming practices and agrarian sustainability. *PLoS One*, 16(1), e0244619. 10.1371/journal.pone.024461933411756

Alves, A. L., da Silva Santos, A. C., de Mattos, J. L. S., Costa, P. M. O., de Souza Motta, C. M., Melo, R. F. R., & Tiago, P. V. (2023). Diversity of filamentous fungi communities in the soils of agroecological crop polycultures and the Atlantic Rain Forest. *Archives of Agronomy and Soil Science*, 69(3), 374–386. 10.1080/03650340.2021.1995717

Ambler, J., Dearden, P. K., Wilcox, P., Hudson, M., & Tiffin, N. (2021). Including digital sequence data in the Nagoya Protocol can promote data sharing. *Trends in Biotechnology*, 39(2), 116–125. 10.1016/j.tibtech.2020.06.00932654776

Anantha, K. H., Garg, K. K., Moses, D. S., Patil, M. D., Sawargaonkar, G. L., Kamdi, P. J., Malve, S., Sudi, R., Raju, K. V., & Wani, S. P. (2021). Impact of natural resource management interventions on water resources and environmental services in different agroecological regions of India. *Groundwater for Sustainable Development*, 13, 100574. 10.1016/j.gsd.2021.100574

Asdal, Å. (2021). The role of the Svalbard Global Seed Vault in preserving crop genetic diversity. In *Plant genetic resources* (pp. 217–225). Burleigh Dodds Science Publishing.

Ayaz, M., Ammad-Uddin, M., Baig, I., & Aggoune, E. H. M. (2018). Wireless sensor's civil applications, prototypes, and future integration possibilities: A review. *IEEE Sensors Journal*, 18(1), 4–30. 10.1109/JSEN.2017.2766364

Ayaz, M., Ammad-Uddin, M., Sharif, Z., Mansour, A., & Aggoune, E. H. M. (2019). Internet-of-Things (IoT)-based smart agriculture: Toward making the fields talk. *IEEE Access : Practical Innovations, Open Solutions*, 7, 129551–129583. 10.1109/ACCESS.2019.2932609

Aytar, E. C., & Kömpe, Y. Ö. (2023). Effect of different substrates on in vitro symbiotic seed germination for soilless production of Anacamptis laxiflora orchid. *Acta Botanica Croatica*, 82(2), 101–108. 10.37427/botcro-2023-010

Bolisetty, S., Peydayesh, M., & Mezzenga, R. (2019). Sustainable technologies for water purification from heavy metals: Review and analysis. *Chemical Society Reviews*, 48(2), 463–487. 10.1039/C8CS00493E30603760

Bourke, P. M., Evers, J. B., Bijma, P., van Apeldoorn, D. F., Smulders, M. J., Kuyper, T. W., & Bonnema, G. (2021). Breeding beyond monoculture: Putting the "intercrop" into crops. *Frontiers in Plant Science*, 12, 734167. 10.3389/fpls.2021.73416734868116

Brown, C., Kovács, E., Herzon, I., Villamayor-Tomas, S., Albizua, A., Galanaki, A., & Zinngrebe, Y. (2021). Simplistic understandings of farmer motivations could undermine the environmental potential of the common agricultural policy. *Land Use Policy*, 101, 105136. 10.1016/j.landusepol.2020.105136

Buratti-Donham, J., Venn, R., Schmutz, U., & Migliorini, P. (2023). Transforming food systems towards agroecology–a critical analysis of agroforestry and mixed farming policy in 19 European countries. *Agroecology and Sustainable Food Systems*, 47(7), 1023–1051. 10.1080/21683565.2023.2215175

Bürgi, M., Verburg, P. H., Kuemmerle, T., & Plieninger, T. (2017). Analyzing dynamics and values of cultural landscapes. *Landscape Ecology*, 32(11), 2077–2081. 10.1007/s10980-017-0573-0

Calegari, A., de Araujo, A. G., Tiecher, T., Bartz, M. L. C., Lanillo, R. F., dos Santos, D. R., & Friedrich, T. (2020). No-till farming systems for sustainable agriculture in South America. *No-till Farming Systems for Sustainable Agriculture: Challenges and Opportunities,* 533-565.

Chenyin, P., Yu, W., Fenghou, S., & Yongbao, S. (2023). Review of the Current Research Progress of Seed Germination Inhibitors. *Horticulturae, 9*(4), 462. 10.3390/horticulturae9040462

Cowell, C., Paton, A., Borrell, J. S., Williams, C., Wilkin, P., Antonelli, A., Baker, W. J., Buggs, R., Fay, M. F., Gargiulo, R., Grace, O. M., Kuhnhäuser, B. G., Woudstra, Y., & Kersey, P. J. (2022). Uses and benefits of digital sequence information from plant genetic resources: Lessons learnt from botanical collections. *Plants, People, Planet*, 4(1), 33–43. 10.1002/ppp3.10216

de Mévius, J. (2022). Impact of the European Union's Seed Legislation and Intellectual Property Rights on Crop Diversity. *European Energy and Environmental Law Review, 31*(3).

Deguine, J. P., Aubertot, J. N., Flor, R. J., Lescourret, F., Wyckhuys, K. A., & Ratnadass, A. (2021). Integrated pest management: Good intentions, hard realities. A review. *Agronomy for Sustainable Development*, 41(3), 38. 10.1007/s13593-021-00689-w

Despotović, J., Rodić, V., & Caracciolo, F. (2021). Farmers' environmental awareness: Construct development, measurement, and use. *Journal of Cleaner Production*, 295, 126378. 10.1016/j.jclepro.2021.126378

Devaux, A., Kromann, P., & Ortiz, O. (2014). Potatoes for Sustainable Global Food Security. *Potato Research*, 57(3–4), 185–199. 10.1007/s11540-014-9265-1

Dhankhar, N., & Kumar, J. (2023). Impact of increasing pesticides and fertilizers on human health: A review. *Materials Today: Proceedings*. 10.1016/j.matpr.2023.03.766

Duru, M., Therond, O., Martin, G., Martin-Clouaire, R., Magne, M. A., Justes, E., Journet, E. P., Aubertot, J. N., Savary, S., Bergez, J. E., & Sarthou, J. P. (2015). How to implement biodiversity-based agriculture to enhance ecosystem services: a review. *Agronomy for Sustainable Development, 35*(4), 1259–1281. 10.1007/s13593-015-0306-1

El-Bassiony, M. F., El-Hawy, A. S., El-Lamei, A. M., & Abd-Elazem, R. A. E. (2020). Conservation Agriculture and Integrated Pest Management Practices Improve Yield and Income while Reducing Labor, Pests, Diseases and Chemical Pesticide Use in Smallholder Vegetable Farms in Nepal. *Sustainability, 12*(16), 6418. 10.3390/su12166418

Engels, J. M., & Ebert, A. W. (2021). A critical review of the current global ex situ conservation system for plant agrobiodiversity. I. History of the development of the global system in the context of the political/legal framework and its major conservation components. *Plants*, 10(8), 1557. 10.3390/plants1008155734451602

Erhart, E., & Hartl, W. (2010). *Compost Use in Organic Farming*. Springer. 10.1007/978-90-481-8741-6_11

Esquinas-Alcázar, J., Hilmi, A., & Noriega, I. L. (2012). A brief history of the negotiations on the International Treaty on Plant Genetic Resources for Food and Agriculture. In *Crop Genetic Resources as a Global Commons* (pp. 135–149). Routledge.

Fuentes-Peñailillo, F., Gutter, K., Vega, R., & Silva, G. C. (2024). New Generation Sustainable Technologies for Soilless Vegetable Production. *Horticulturae, 10*(1), 49. 10.3390/horticulturae10010049

Fuller, D. Q., Denham, T., & Allaby, R. (2023). Plant domestication and agricultural ecologies. *Current Biology*, 33(11), R636–R649. 10.1016/j.cub.2023.04.03837279694

Garrett, H. E. G., Wolz, K. J., Walter, W. D., Godsey, L. D., & McGraw, R. L. (2021). Alley cropping practices. *North American agroforestry*, 163-204.

Gliessman, S. (2018). Defining Agroecology. *Agroecology and Sustainable Food Systems*, 42(6), 599–600. 10.1080/21683565.2018.1432329

González-Toral, C., & Cires, E. (2022). Relevance of DNA preservation for future botany and ecology. *Molecular Ecology*, 31(20), 5125–5131. 10.1111/mec.1665236214196

Hawes, C., Iannetta, P. P., & Squire, G. R. (2021). Agroecological practices for whole-system sustainability. *CABI Reviews.*

Henry, R. J. (2020). Innovations in plant genetics adapting agriculture to climate change. *Current Opinion in Plant Biology*, 56, 168–173. 10.1016/j.pbi.2019.11.00431836470

Ibrahim, Y., Arafat, H. A., Mezher, T., & AlMarzooqi, F. (2018). An integrated framework for sustainability assessment of seawater desalination. *Desalination*, 447, 1–17. 10.1016/j.desal.2018.08.019

Jeanneret, P., Aviron, S., Alignier, A., Lavigne, C., Helfenstein, J., Herzog, F., Kay, S., & Petit, S. (2021). Agroecology landscapes. *Landscape Ecology, 36*(8), 2235–2257. 10.1007/s10980-021-01248-0

Jug, I., Brozović, B., Đurđević, B., Wilczewski, E., Vukadinović, V., Stipešević, B., & Jug, D. (2021). Response of crops to conservation tillage and nitrogen fertilization under different agroecological conditions. *Agronomy (Basel)*, 11(11), 2156. 10.3390/agronomy11112156

Khan, N., Ray, R. L., Sargani, G. R., Ihtisham, M., Khayyam, M., & Ismail, S. (2021). Current Progress and Future Prospects of Agriculture Technology: Gateway to Sustainable Agriculture. *Sustainability, 13*(9), 4883. 10.3390/su13094883

Killebrew, K., & Wolff, H. (2010). *Environmental impacts of agricultural technologies*. IDEAS. https://ideas.repec.org/p/udb/wpaper/uwec-2011-01.html

Király, É., Keserű, Z., Molnár, T., Szabó, O., & Borovics, A. (2023). Carbon Sequestration in the Aboveground Living Biomass of Windbreaks—Climate Change Mitigation by Means of Agroforestry in Hungary. *Forests*, 15(1), 63. 10.3390/f15010063

Kpienbaareh, D., Bezner Kerr, R., Luginaah, I., Wang, J., Lupafya, E., Dakishoni, L., & Shumba, L. (2020). Spatial and ecological farmer knowledge and decision-making about ecosystem services and biodiversity. *Land (Basel)*, 9(10), 356. 10.3390/land9100356

Kulak, V., Longboat, S., Brunet, N. D., Shukla, M., & Saxena, P. (2022). In vitro technology in plant conservation: Relevance to biocultural diversity. *Plants*, 11(4), 503. 10.3390/plants1104050335214833

Liu, C., Plaza-Bonilla, D., Coulter, J. A., Kutcher, H. R., Beckie, H. J., Wang, L., Floc'h, J. B., Hamel, C., Siddique, K. H. M., Li, L., & Gan, Y. (2022). Diversifying crop rotations enhances agroecosystem services and resilience. *Advances in Agronomy*, 173, 299–335. 10.1016/bs.agron.2022.02.007

Marcos-Pérez, M., Sánchez-Navarro, V., & Zornoza, R. (2023). Intercropping systems between broccoli and fava bean can enhance overall crop production and improve soil fertility. *Scientia Horticulturae*, 312, 111834. 10.1016/j.scienta.2023.111834

Meng, L., & Si, W. (2022). Pro-Environmental Behavior: Examining the Role of Ecological Value Cognition, Environmental Attitude, and Place Attachment among Rural Farmers in China. *International Journal of Environmental Research and Public Health*, 19(24), 17011. 10.3390/ijerph192417011365548 98

Mockshell, J., & Kamanda, J. (2018). Beyond the agroecological and sustainable agricultural intensification debate: Is blended sustainability the way forward? *International Journal of Agricultural Sustainability*, 16(2), 127–149. 10.1080/14735903.2018.1448047

Mora, F., & Iverson, L. (2002). A spatially constrained ecological classification: Rationale, methodology and implementation. *Plant Ecology*, 158(2), 153–169. 10.1023/A:1015534615415

Motoshita, M., Ono, Y., Pfister, S., Boulay, A. M., Berger, M., Nansai, K., Tahara, K., Itsubo, N., & Inaba, A. (2018). Consistent characterisation factors at midpoint and endpoint relevant to agricultural water scarcity arising from freshwater consumption. *The International Journal of Life Cycle Assessment*, 23(12), 2276–2287. 10.1007/s11367-014-0811-5

Patel, R. J., Rank, P. H., Vekariya, P. B., Rank, H. D., Vadar, H. R., Parmar, H. V., Tiwari, M. K., Marviya, P. B., & Baca, M. M. L. (2023). *Enhancing Wheat (Triticum aestivum L.) Crop Yield and Water Use Efficiency: A Study on Canopy Air Temperature Difference-Based Drip Irrigation Scheduling in a Semi-Arid Region of Western India*. 10.21203/rs.3.rs-3376468/v1

Pathirana, R., & Carimi, F. (2022). Management and utilization of plant genetic resources for a sustainable agriculture. *Plants*, 11(15), 2038. 10.3390/plants1115203835956515

Petit, S., Muneret, L., Carbonne, B., Hannachi, M., Ricci, B., Rusch, A., & Lavigne, C. (2020). Landscape-scale expansion of agroecology to enhance natural pest control: A systematic review. *Advances in Ecological Research*, 63, 1–48. 10.1016/bs.aecr.2020.09.001

Peydayesh, M., & Mezzenga, R. (2021). Protein nanofibrils for next generation sustainable water purification. *Nature Communications, 12*(1), 1–17. 10.1038/s41467-021-23388-2

Poudel, S., Bansal, S., Podder, S., Paneru, B., Karki, S., Fike, J., & Kumar, S. (2022). Conversion of open pasture to hardwood silvopasture enhanced soil health of an ultisol. *Agroforestry Systems*, 96(8), 1237–1247. 10.1007/s10457-022-00783-2

Pudjiastuti, A. Q., Kaluge, D., & Widowati, W. (2023). Reallocation of the use of chemical fertilizers and pesticides to increase the income of vegetable farmers and prevent land degradation. *Journal of Degraded and Mining Lands Management*, 11(1), 5095–5103. 10.15243/jdmlm.2023.111.5095

Puech, C., Brulaire, A., Paraiso, J., & Faloya, V. (2021). Collective design of innovative agroecological cropping systems for the industrial vegetable sector. *Agricultural Systems*, 191, 103153. 10.1016/j.agsy.2021.103153

Qaim, M. (2016). *Genetically modified crops and agricultural development*. https://books.google.com/books?hl=en&lr=&id=skYDAAAQBAJ&oi=fnd&pg=PP1&ots=DSYSZE-vjq&sig=mFnSsqS3Qmi3lc-Cz7d8nN11e6U

Qaim, M. (2020). Role of New Plant Breeding Technologies for Food Security and Sustainable Agricultural Development. *Applied Economic Perspectives and Policy*, 42(2), 129–150. 10.1002/aepp.13044

Rani, A., Kashyap, R., & Azmi, W. (2021). Conservation of Biodiversity by Biotechnology. In *Basic Concepts in Environmental Biotechnology* (pp. 149–175). CRC Press. 10.1201/9781003131427-12

Reganold, J. P., & Wachter, J. M. (2016). Organic agriculture in the twenty-first century. *Nature Plants, 2*(2), 1–8. 10.1038/nplants.2015.221

Rifna, E. J., Ratish Ramanan, K., & Mahendran, R. (2019). Emerging technology applications for improving seed germination. *Trends in Food Science & Technology*, 86, 95–108. 10.1016/j.tifs.2019.02.029

Sahmat, S. S., Rafii, M. Y., Oladosu, Y., Jusoh, M., Hakiman, M., & Mohidin, H. (2024). Unravelling the dynamics of genotype and environment interactions on chilli (Capsicum annuum L.) yield-related attributes in soilless planting systems. *Scientific Reports, 14*(1), 1–17. 10.1038/s41598-023-50381-0

Salgotra, R. K., & Chauhan, B. S. (2023). Genetic diversity, conservation, and utilization of plant genetic resources. *Genes*, 14(1), 174. 10.3390/genes1401017436672915

Sayer, J., Sunderland, T., Ghazoul, J., Pfund, J. L., Sheil, D., Meijaard, E., Venter, M., Boedhihartono, A. K., Day, M., Garcia, C., Van Oosten, C., & Buck, L. E. (2013). Ten principles for a landscape approach to reconciling agriculture, conservation, and other competing land uses. *Proceedings of the National Academy of Sciences of the United States of America*, 110(21), 8349–8356. 10.1073/pnas.121059511023686581

Senesi, N. (1989). Composted materials as organic fertilizers. *The Science of the Total Environment*, 81–82(C), 521–542. 10.1016/0048-9697(89)90161-7

Serio, F., Manuel, R., Machado, A., Fussy, A., & Papenbrock, J. (2022). An Overview of Soil and Soilless Cultivation Techniques—Chances, Challenges and the Neglected Question of Sustainability. *Plants, 11*(9), 1153. 10.3390/plants11091153

Shah, K. K., Modi, B., Pandey, H. P., Subedi, A., Aryal, G., Pandey, M., & Shrestha, J. (2021). Diversified crop rotation: An approach for sustainable agriculture production. *Advances in Agriculture*, 2021, 1–9. 10.1155/2021/8924087

Shah, K. K., Modi, B., Pandey, H. P., Subedi, A., Aryal, G., Pandey, M., & Shrestha, J. (2021). Diversified Crop Rotation: An Approach for Sustainable Agriculture Production. *Advances in Agriculture*, 2021, 1–9. Advance online publication. 10.1155/2021/8924087

Song, Y., Zhao, W., Su, Z., Guo, S., Du, Y., Song, X., Shi, X., Li, X., Liu, Y., & Liu, Z. (2024). Effect of Pulsed Electric Field Treatment on Seed Germination and Seedling Growth of Scutellaria baicalensis. *Agriculture 2024, Vol. 14, Page 158, 14*(1), 158. 10.3390/agriculture14010158

Srivastav, A. L. (2020). Chemical fertilizers and pesticides: role in groundwater contamination. *Agrochemicals Detection, Treatment and Remediation: Pesticides and Chemical Fertilizers*, 143–159. 10.1016/B978-0-08-103017-2.00006-4

Sumberg, J., & Giller, K. E. (2022). What is 'conventional' agriculture? *Global Food Security*, 32, 100617. 10.1016/j.gfs.2022.100617

Tahat, M. M., Alananbeh, K. M., Othman, Y. A., & Leskovar, D. I. (2020). Soil Health and Sustainable Agriculture. *Sustainability, 12*(12), 4859. 10.3390/su12124859

Tian, Z., Wang, J. W., Li, J., & Han, B. (2021). Designing future crops: Challenges and strategies for sustainable agriculture. *The Plant Journal*, 105(5), 1165–1178. 10.1111/tpj.1510733258137

Tittonell, P. (2014). Ecological intensification of agriculture — Sustainable by nature. *Current Opinion in Environmental Sustainability*, 8, 53–61. 10.1016/j.cosust.2014.08.006

Tomar, D., & Pali, N. (2023). Advantages to agroecosystem by green manures and cover crops. *Journal of Pharmacognosy and Phytochemistry*, 12(2), 146–149.

Wambugu, P. W., & Henry, R. (2022). Supporting in situ conservation of the genetic diversity of crop wild relatives using genomic technologies. *Molecular Ecology*, 31(8), 2207–2222. 10.1111/mec.1640235170117

Wang, L., Zhao, Y., Al-Kaisi, M., Yang, J., Chen, Y., & Sui, P. (2020). Effects of Seven Diversified Crop Rotations on Selected Soil Health Indicators and Wheat Productivity. *Agronomy, 10*(2), 235. 10.3390/agronomy10020235

Wei, W., Yan, Y., Cao, J., Christie, P., Zhang, F., & Fan, M. (2016). Effects of combined application of organic amendments and fertilizers on crop yield and soil organic matter: An integrated analysis of long-term experiments. *Agriculture, Ecosystems & Environment*, 225, 86–92. 10.1016/j.agee.2016.04.004

Weindorf, D. C., Muir, J. P., & Landeros-Sánchez, C. (2011). *Organic Compost and Manufactured Fertilizers: Economics and Ecology*. Springer. 10.1007/978-94-007-1309-3_2

Wezel, A., Casagrande, M., Celette, F., Vian, J. F., Ferrer, A., & Peigné, J. (2014). Agroecological practices for sustainable agriculture. A review. *Agronomy for Sustainable Development*, 34(1), 1–20. 10.1007/s13593-013-0180-7

Wezel, A., Herren, B. G., Kerr, R. B., Barrios, E., Gonçalves, A. L. R., & Sinclair, F. (2020). Agroecological principles and elements and their implications for transitioning to sustainable food systems. A review. *Agronomy for Sustainable Development*, 40(6), 1–13. 10.1007/s13593-020-00646-z

Xie, S., Tran, H. T., Pu, M., & Zhang, T. (2023). Transformation characteristics of organic matter and phosphorus in composting processes of agricultural organic waste: Research trends. *Materials Science for Energy Technologies*, 6, 331–342. 10.1016/j.mset.2023.02.006

Xu, H., Fan, Z., Ahmad, F., & Zhang, D. (2023). Exploring the ecological protection impacts of cultivated land transfer: Explanation based on fertilizers and pesticides. *Ecological Indicators*, 154, 110681. 10.1016/j.ecolind.2023.110681

Yadav, J., Srivastva, A. K., & Singh, R. (2024). Diversity of halotolerant endophytes from wheat (Triticum aestivum) and their response to mitigate salt stress in plants. *Biocatalysis and Agricultural Biotechnology*, 56, 103000. 10.1016/j.bcab.2023.103000

Zhao, J. (2020). The role of international organizations in preventing conflicts between the SPS Agreement and the Cartagena Protocol on Biosafety. *Review of European, Comparative & International Environmental Law*, 29(2), 271–281. 10.1111/reel.12326

Zhou, S., Jia, P., Xu, W., Shane Alam, S., & Zhang, Z. (2023). A novel composting system for mitigating ammonia emissions and producing nitrogen-rich organic fertilizer. *Bioresource Technology*, 386, 129455. 10.1016/j.biortech.2023.12945537419288

Zowada, C., Frerichs, N., Zuin, V. G., & Eilks, I. (2020). Developing a lesson plan on conventional and green pesticides in chemistry education – a project of participatory action research. *Chemistry Education Research and Practice*, 21(1), 141–153. 10.1039/C9RP00128J

KEY TERMS AND DEFINITIONS

Agroforestry: That involves the intentional integration of trees and shrubs into farming systems to maximize land productivity, conserve natural resources, and enhance environmental sustainability.

Biodiversity: Refers to the richness and variety of life on Earth, including the diversity of species, genes, and ecosystems.

Climate Change: Refers to the long-term alteration of Earth's climate patterns, driven primarily by human activities that release greenhouse gases into the atmosphere, resulting in global warming and various environmental impacts.

Conservation Agriculture: Agricultural practice aimed at achieving sustainable and efficient crop production through minimal soil disturbance, maintaining soil cover with organic materials, and implementing diverse crop rotations to enhance soil health and ecosystem resilience.

Ecological Agriculture: Sustainable farming practice that emphasizes harmony with natural ecosystems, utilizing methods that promote biodiversity, soil health, and minimal environmental impact.

Plant Cultivation: The practice of growing and caring for plants to produce food, ornamental plants, or other useful products.

Plant Ecology: The study of how plants interact with their environment, including factors like climate, soil, water availability, and relationships with other organisms.

Succession: The process of succession refers to the sequential and orderly progression of changes in species composition and community structure within an ecosystem over time, typically following a disturbance or the creation of new substrate.

Chapter 2
Exploring the Synergy of Soil-Animal Interactions:
A Blueprint for Biodiversity Preservation

Vishal M. Makwana
http://orcid.org/0000-0001-6343-9449
Zoology Department, Sir P. P. Institute of Science, M. K. Bhavnagar University, Gujarat, India

Karan Thakkar
http://orcid.org/0000-0002-9078-1431
Consumer Education and Research Centre, Ahmedabad, India

Pravinsang P. Dodia
Zoology Department, Sir P. P. Institute of Science, M. K. Bhavnagar University, Gujarat, India

ABSTRACT

The chapter explores the intricate interactions between soil, plants, and animals, highlighting the essential role of animal biodiversity in ecosystem health. It discusses the impacts of climate change on soil-dwelling animals, emphasizing underestimated effects through case studies. Synergistic mechanisms, such as mutualistic relationships, are examined for their contribution to ecosystem resilience. Preservation strategies, challenges, and future directions in biodiversity conservation are outlined, providing insights for sustainable soil ecosystem management. Through interdisciplinary inquiry, the chapter offers a comprehensive blueprint for biodiversity preservation, fostering resilience in soil ecosystems.

INTRODUCTION

At the heart of terrestrial ecosystems lies a dynamic interplay among soil, flora, and fauna, shaping the intricate network of life crucial for the survival of all species. Soil plays a crucial role as the essential groundwork for the advancement of plant groupings, delivering important nutrients, moisture, and structural support. Through photosynthesis, plants convert carbon dioxide into organic matter and oxygen, while also stabilizing the soil and preventing erosion. Nevertheless, the often-underestimated role of animals completes this ecological triad, enhancing soil ecosystems through diverse interactions and contributing to the resilience and sustainability of terrestrial environments. The kingdom Animalia

DOI: 10.4018/979-8-3693-3374-7.ch002

encompasses an incredibly vast array of species, with 1,552,319 formally described species according to (Zhang, 2011). However, there are many more species yet to be formally documented and described. Some of these species' group like earthworms, nematodes, collembola, mites, centipedes, millepedes, coleopterans, etc. are closely associated with the soil ecosystem. They play a vital role in controlling microbial pathogens (Gorres & Amador, 2021), changing microbial community composition (Kicaj, 2023), enhancing primary production, and soil formation (Ertiban, 2019).

The animal population in soil habitats encompasses a wide array of organisms, ranging from minuscule nematodes and earthworms to burrowing mammals and ground-dwelling insects. These underground creatures play crucial roles in processes like nutrient cycling, decomposition, soil aeration, and seed dispersal, greatly impacting the operation and stability of ecosystems. For instance, earthworms enhance soil quality and fertility through their burrowing activities, facilitating better water infiltration and nutrient dispersion. Similarly, decomposers like dung beetles accelerate the breakdown of organic matter, recycling nutrients and enhancing soil fertility. Moreover, soil-dwelling animals represent essential components of intricate food chains, serving as both prey for higher trophic levels and predators of soil-borne pests and diseases. Predatory soil mites, for instance, control the populations of plant-parasitic nematodes, thereby reducing crop damage and enhancing plant vitality. By residing in diverse ecological niches and engaging in intricate trophic interactions, soil-dwelling creatures are essential for maintaining the stability and resilience of soil ecosystems, assisting in mitigating environmental fluctuations and safeguarding biodiversity.

The health and resilience of ecosystems are inextricably linked to the maintenance of animal biodiversity within soil. As mentioned earlier, activities of a wide variety of life forms contribute to the dynamics of soil health. Starting from the smallest life forms, i.e. microorganisms; Soils can contain a large amount of microbial biomass including protists, fungi, viruses and bacteria (Fierer, 2017) which together forms microbiome whose more than 40 soil microbiome functions effect the health of the soil as well as the biotic component of the ecosystem directly or indirectly (Banerjee & Van der Heijden, 2023). The larger life forms like mega-mammals, shapes up the dynamics of entire ecosystem, for instance, Rhinos in the Indo-Nepal Terai are known to create open spaces in a dense vegetation by grazing. This creates a habitat for the variety of grasses, while their wallowing behaviour in which a rhino rolls in the soil or water to get rid of the ectoparasites and creates depression in the ground which eventually collects water. This creates a space for various aquatic plants and animals to thrive. It is understood that a mammal of weighing more than one tonne strolling across a landscape is going to influence a significant change on soil dynamics and ultimately the vegetation of the area. Thus, it is clear that soil-plant-animal interactions range from dynamics at smallest of molecular level, to the largest of ecosystem level. Therefore, this subject is going to have a wide range of possibilities. However, even if there are numerous possibilities, ecosystems tend to remain stable if certain environmental factors (abiotic or biotic) remain relatively stable.

One such factor is animal biodiversity whose preservation contributes not only to the well-being of the individual species, but also to the overall health and resilience of the ecosystem (Erlewin, 2022). Let's understand why is it so important. Healthy Biodiversity ensures that an ecosystem has good variety of life forms performing different functions or roles (Valladares, 2022). Functions being pollination, nutrient cycling, seed dispersal, natural pest control (Bio-control), etc. Try to imagine an ecosystem after removing a set of species responsible for these functions; for example: Removing pollinators like Lepidopterans, Hymenopterans, and some species of birds from a landscape will not only compromise the floral fidelity, but also the plant reproductive functions (Borosi & Briggs, 2013), ultimately the ecosystem would begin to be loose it productivity. The density of vegetation would decrease in a landscape, the soil would start to lose it binding power, and it will ultimately erode giving a rise to barren or semi-barren landscape. Similar consequences follow the removal of set of species responsible for other functions.

The present chapter embarks on an exploration of the synergistic relationships between soil and animals. Further in the chapter we will see the significance of Biodiversity in various ecosystems through the perspective of soil-animal interactions. It is important to identify the roles of various animals in the ecosystem. The presence of every species in an ecosystem is not going to affect the ecosystem in the positive way. Some species would, through its activities will affect some of the factor in a negative or harmful way. For instance, the presence of termites in an agroecosystem would be not be ideal for the agriculture that for the humans is of a high importance. Similarly, some studies have shown that presence of a specific species of earthworm increases the surface runoff rate of the water, which might not be ideal case when we talk of soil erosion issue. Thus, it becomes important to carefully assess the role of different species in an ecosystem, especially the ones where there has been a chops and changes due to anthropogenic activity.

1. SIGNIFICANCE OF ANIMAL BIODIVERSITY FOR ECOSYSTEM HEALTH

1.1. Importance of Biodiversity in Agriculture

It is of utmost importance of any country for their agriculture to be safe, as to ensure the food safety of its people. Agriculture ecosystems are dynamic and complex systems governed by multiple factors, including both biotic and abiotic. Abiotic factors are the key players in the success of agriculture these days, since the biotic factors with the support of modern advancements are mostly secured when it comes to Pest control, Pollination, Nutrient availability, etc. The progressing agricultural science has mostly found a cure for pest control through chemical pesticides or the more recent and more sustainable solution in Pest Bio-control. Even then, losses of 10-15% of the total yield value of arable crops is caused by pests globally (Sawika & Egbuna, 2020).

1.2. Importance of Biodiversity in Forests

Forests have been named as the Lungs of earth, Carbon sink, Water filters, Soil stabilizer, Climate regulators etc. A good forest will naturally be rich with biodiversity. Rich biodiversity ensures the resilience against disturbances viz., diseases, insect outbreaks, and climate change. However, with the increasing human interruption in the forests for building roads, unsustainable tourism, deforestation has pushed a lot a biodiversity to the brink of extinction or has already been extinct.

This has also influenced the forest land immensely. According to a recent study (Chinnasamy & Honap, 2023), the western ghats of India, which is one of the seven Biodiversity hotspots in the Indian subcontinent has faced a tremendous soil loss. An average of 94% increase in soil erosion rate is reported in Western Ghats between 1990 to 2020. Subsequently, the forest loss in western ghats was estimated as 33,579 km^2 from 1920s to 2013 (Reddy et al., 2013). This could very well be a reason behind such a high rate of soil erosion in the region. With the loss of forest cover, the biodiversity is going to suffer due to habitat loss, increased competition for resources, etc.

Figure 1. Role of different animals in the Soil Ecosystem

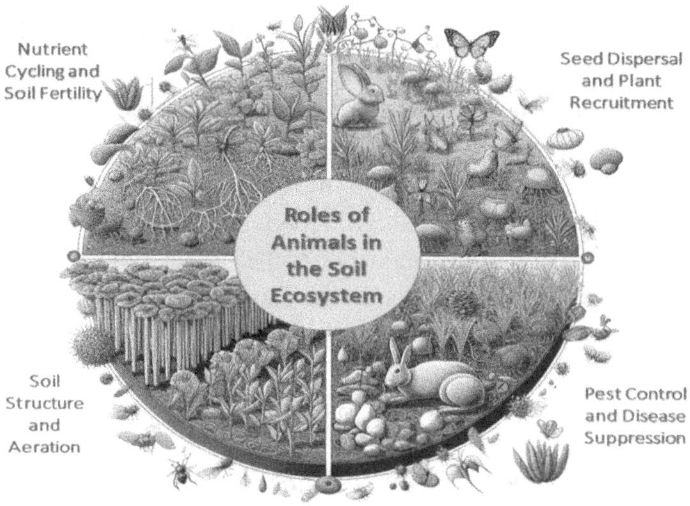

1.3. Importance of Preserving Animal Biodiversity in Soil Ecosystems

Preserving the diversity of animals in soil ecosystems is paramount for maintaining the functionality and sustainability of terrestrial habitats. Animal biodiversity encompasses not only species richness but also genetic and functional diversity, all of which contribute to ecological resilience and adaptability. Animals carry out a wide range of ecological functions that are vital for the health and productivity of soil ecosystems. A large diversity of soil fauna, approximately 85% in terms of species richness is represented by the members of phylum Arthropoda (Raza et. al., 2019) which also happen to be the most diverse phylum in the kingdom Animalia (Zhang, 2013). Table 1 emphasizes on roles of different animals in the soil ecosystem (Figure 1).

Table 1. Roles of different Arthropods in the Soil Ecosystem

Phylum	Animal	Interaction with	Influence	Effect	Reference
Arthropoda	Ants	Soil	Aeration	+ve	Raza et. al., 2019
			Water holding capacity	+ve	Raza et.al., 2019
			surface crusting and erosion	+ve	Raza et. al., 2019
			Water content	+ve	Chen et. al., 2019
			Microbial Biomass	+ve	Chen et. al., 2019
		Plants	Root penetration of plants	+ve	Raza et. al., 2019
			Nutrition supply to plants	+ve	Raza et. al., 2019
			Diversity	+ve	Majer & Agosti, 2016
	Termites	Soil	Microbial Diversity	-ve	Chen et. al., 2021
			Microbial Diversity	+ve	Baker et. al., 2020
			Clay fraction content	+ve	Fruet et. al., 2023
			Organic carbon	+ve	Fruet et. al., 2023
			Iron oxides	+ve	Fruet et. al., 2023
			Hydraulic conductivity	+ve	Cheik et. al., 2019
			Water infiltration	+ve	Cheik et. al., 2020
			Soil resistomes abundance and diversity	-ve	Yan et. al., 2021
		Plants	Damage to agricultural crops	-ve	Rouland-Lefèvre, 2010
			Plant density	+ve	Khan et. al., 2018
			Height	+ve	Khan et. al., 2018
			Species richness	+ve	Khan et. al., 2018
			Functional diversity	+ve	Khan et. al., 2018
			Reproductive output	+ve	Khan et. al., 2018
Annelida	Earthworms	Soil	Bacterial diversity	+ve	Sofo et. al., 2023
			Bacterial abundance	+ve	Sofo et. al., 2023
			Litter decomposition	+ve	Sofo et. al., 2023
			Fertility	+ve	Ahmed & Al-Mutairi, 2022
			Water infiltration	+ve	Chen et. al., 2022
			Surface Runoff rate	-ve	Chen et. al., 2022
			Soil erosion	+ve	Chen et. al., 2022
			Macroporosity	+ve	Reyes et. al., 2023
			Water holding capacity	+ve	Reyes et. al., 2023
		Plants	Plant growth	+ve	Sofo et. al., 2023
			Stomatal density	+ve	Reyes et. al., 2023
			Root growth	+ve	Sofo et. al., 2023
			Chemical defence compounds	-ve	Thakur et. al., 2021
			Pathogen infection	+ve	Thakur et. al., 2021
			Herbivory	+ve	Thakur et. al., 2021
			Plant Biomass	+ve	Mudrák et. al., 2012
			Plant Numbers	-ve	Mudrák et. al., 2012

1.3.1 Nutrient Cycling and Soil Fertility:

Soil-dwelling animals fulfil crucial roles in nutrient cycling by breaking down organic matter and returning nutrients to the soil. For instance, soil-dwelling invertebrates like springtails and millipedes break down leaf litter and organic debris, releasing crucial nutrients for plant growth and development. Animals enhance soil fertility and terrestrial ecosystem productivity by facilitating nutrient cycling.

1.3.2 Soil Structure and Aeration:

Burrowing animals, such as earthworms and ants, exert a significant influence on soil structure and aeration through their tunnelling activities. Earthworm burrows enhance soil porosity and water infiltration, reducing the risk of erosion. Likewise, ant colonies contribute to soil aeration and nutrient mixing through their extensive underground tunnels, promoting root growth and nutrient absorption by plants.

1.3.3 Seed Dispersal and Plant Recruitment:

Many soil-dwelling animals act as seed dispersers, aiding in the colonization of new areas and the regeneration of plant populations. Ants, for example, disperse seeds through a process known as myrmecochory, in which they collect and transport seeds to favourable microsites for germination. Across diverse ecosystems, soil-dwelling invertebrates foster genetic diversity and enhance plant recruitment by dispersing seeds far from parent plants.

1.3.4 Pest Control and Disease Suppression:

Predatory soil organisms, such as predatory mites and nematodes, are known to fulfil significant roles in the regulation of pest and disease populations. These predatory species contribute to the control of pest outbreaks through predation on pest organisms or parasitism of plant pathogens, thereby reducing the reliance on synthetic pesticides. For instance, entomopathogenic nematodes are harnessed as biocontrol agents for managing soil-dwelling insect pests, providing sustainable alternatives to conventional chemical pest control methods.

2. CLIMATE CHANGE IMPACT ON ANIMAL BIODIVERSITY

Climate change has emerged as a prominent topic of discussion in contemporary times. The frequency of natural climatic disasters is notably high worldwide. The alteration in the global climate, attributed to unsustainable human practices over the past few decades, has had a profound impact on animal biodiversity. The consequences of climate change include habitat loss, exemplified by the warming of the oceans due to the escalating global temperatures. Consequently, various species such as polar bears and numerous types of penguins are experiencing habitat depletion. The escalating ocean temperatures pose a significant threat to coral reefs as well. Given their high sensitivity, corals struggle to adapt to the rising water temperatures. Additionally, animals with calcareous shells face harm from the acidic ocean conditions resulting from the absorption of excess carbon dioxide by the oceans. Those with weaker shells become susceptible to both predators and physical harm.

While climate change poses a threat to numerous animal species, particularly those in polar regions and coral ecosystems, it disproportionately affects larger animals compared to smaller ones like rodents and insects that can thrive in warmer temperatures. Rodents, in particular, can cause soil degradation if their population remains unchecked. In the absence of natural predators such as snakes, birds of prey, canids, and felines, the rodent population can surge. The unchecked proliferation of species benefitting from climate change disrupts the intricate food webs, leading to the deterioration of various ecosystems.

2.1. Examining the Effects of Climate Change on Soil-Dwelling Animals

Climate change presents a notable risk to the intricate equilibrium of soil ecosystems and the diversity they uphold. Soil-dwelling organisms, ranging from microarthropods to earthworms, exhibit heightened sensitivity to climate change due to their dependency on environmental factors such as temperature, moisture, and habitat quality. This section delves into the diverse impacts of climate change on soil-dwelling species, utilizing scientific evidence and case studies to elucidate the extensive repercussions for biodiversity conservation.

2.1.1. Shifts in Distribution and Abundance:

The distribution and abundance of soil-dwelling animals are being modified by climate change, as they adjust to evolving climatic conditions and search for suitable habitats. Some species have expanded their ranges towards higher latitudes and altitudes in response to rising temperatures, whereas others have dwindled or faced extinction in regions impacted by severe weather events or habitat deterioration.

2.1.2. Altered Phenology and Life History Traits:

Changes in climatic patterns, such as modifications in seasonal temperature and precipitation schedules, possess the capacity to disturb the phenology and life history characteristics of subterranean organisms, thus exerting influence on reproductive cycles, developmental phases, and population trends. Research has demonstrated that milder winters and earlier springs can influence the timing of emergence and activity of subterranean insects and arthropods, consequently impacting predator-prey relationships and trophic dynamics (Gerlich et. al., 2023). Variations in precipitation patterns may also lead to alterations in soil moisture levels, ultimately influencing the survival and reproductive achievements of soil-dwelling organisms adapted to specific moisture conditions.

2.1.3. Impacts on Soil Microbial Communities:

Soil-dwelling fauna engage closely with soil microorganisms, establishing complex ecological networks that regulate nutrient circulation and ecosystem functions. The alteration of soil microbial communities by climate change can disrupt these interactions, leading to cascading effects on soil-dwelling fauna and ecosystem functionality. For instance, variations in temperature and precipitation patterns can influence microbial decomposition rates and nutrient availability, subsequently impacting the composition and distribution of detritivores and decomposers within the soil (Wang et. al., 2023).

2.1.4 Feedback Effects on Ecosystem Services:

Soil-dwelling organisms contribute various ecosystem services, including nutrient cycling, soil aeration, and pest control, essential for sustaining terrestrial ecosystem productivity and resilience. Disruptions in soil animal populations due to climate change-induced disturbances could have significant implications for ecosystem services, potentially affecting soil quality, plant development, and agricultural output. Droughts or severe weather conditions, for instance, may decrease earthworm populations, leading to reduced soil aeration and water infiltration rates, thereby compromising soil structure and fertility in agricultural settings (Lu et. al., 2023).

2.1.5 Adaptive Responses and Resilience:

Despite the obstacles posed by climate change, soil-dwelling species exhibit remarkable resilience and adaptability in response to environmental disruptions. Scientists have recognized evolutionary adjustments in soil animal populations, such as enhanced genetic diversity and phenotypic flexibility, enabling them to cope with changing environmental conditions and thrive in evolving landscapes. Additionally, soil animal communities may display behavioural adaptations, such as altering digging depths or feeding behaviours, to mitigate the impacts of climatic variations and sustain ecosystem functioning (Goncharov et. al., 2023).

2.2. Uncovering Underestimated Impacts and Notable Case Studies

Climate change is a substantial menace to global biodiversity, impacting ecosystems and species globally. Despite the considerable attention given to the impact of climate change on charismatic megafauna and iconic species, the effects on soil-dwelling creatures have at times been discounted. Following are the underestimated repercussions of climate change on soil-dwelling animal species, utilizing notable case studies to underscore the extensive implications for ecosystem resilience and biodiversity conservation.

2.2.1.Declines in Soil Arthropod Abundance and Diversity:

Recent studies have revealed notable declines in soil arthropod richness and diversity due to environmental stressors induced by climate change. For instance, research conducted in temperate forests indicated substantial reductions in the richness and diversity of soil arthropods, including springtails, mites, and beetles, following periods of elevated heat and drought. These declines stem from alterations in soil moisture levels, fluctuations in temperature, and shifts in plant composition, all of which disrupt habitat suitability and resource availability for soil-dwelling animals (Cours et. al. 2023).

An investigation carried out in grasslands found that prolonged drought and increasing temperatures led to a decrease in soil arthropod richness and diversity. Observations by the researchers revealed shifts in the composition of soil arthropod communities, characterized by fewer species adapted to cooler and moister conditions and an increase in generalist species tolerant to warmer and drier habitats. These changes in community structure have implications for ecosystem functioning and nutrient cycling, highlighting the susceptibility of soil-dwelling animals to the stresses induced by climate change (Barnett & Facey, 2016).

2.2.2. Disruption of Soil Microbial Communities:

The altering climate has the capacity to modify the intricate interactions between soil-dwelling animals and microbial communities, with profound implications for ecosystem processes and soil quality. Changes in temperature and precipitation influence microbial activity and composition, thereby affecting nutrient availability and resource dynamics for soil-dwelling animals. Variations in soil moisture levels and extreme weather events can exacerbate soil erosion and habitat degradation, posing a threat to soil ecosystem resilience.

A study conducted in Australian woodlands investigated the impact of prolonged drought on soil macrofauna, such as earthworms and millipedes. The scientists discovered that decreases in soil moisture and organic matter levels as a result of drought had negative impacts on the quantity and variety of soil macrofauna, causing a reduction in ecosystem functions such as decomposition and nutrient cycling. These discoveries highlight the susceptibility of soil-dwelling creatures to drought caused by climate change, along with the interconnection of soil biodiversity and ecosystem functionality (Maisnam et. al., 2023).

2.2.3. Range Shifts and Habitat Loss:

Climate change is resulting in shifts in ranges and habitat loss for numerous soil-dwelling animal species as they seek suitable environmental conditions and attempt to relocate. Nonetheless, habitat fragmentation and alterations in land use obstruct the ability of soil-dwelling species to disperse and establish populations in new areas, making them more vulnerable to local extinctions and population declines. Additionally, competition from invasive species and modifications in trophic relationships exacerbate the challenges faced by soil animal groups in a changing environment.

Antarctic springtails, minute arthropods residing in soil and adapted to frigid and ice-clad environments, are experiencing shifts in their geographical range and the loss of their habitats due to the warming induced by climate change on the Antarctic Peninsula. Research has demonstrated that the populations of Antarctic springtails are dwindling due to the escalation in temperatures and the thawing of ice, leading to the fragmentation and deterioration of their habitats. These modifications in distribution have implications for the processes within ecosystems and the cycling of nitrogen in Antarctic soils, underscoring the susceptibility of polar ecosystems to climate change (Chown et. al., 2022).

2.2.4. Additional Case Studies and Examples

In a recent study by Goncharov et al. (2023), the responses of six major groups of soil-dwelling fauna to controlled variations in environmental factors such as air temperature, precipitation, and CO_2 levels were assessed. The study revealed a robust correlation between air temperature, precipitation levels, and the population dynamics of soil invertebrates. Specifically, a 1°C rise in air temperature correlated with a 12.5% increase in Acari abundance, while the population of collembola simultaneously decreased by 2.6%. Moreover, a 10% increase in precipitation led to a rise in Nematoda abundance by 1.4% and Oligochaeta by 34.7%.

Wu et al. (2012) observed a concerning decline in earthworm cocoon production under simulated warming conditions, indicating potential population reductions in these crucial decomposers. Such declines could disrupt the decomposition of organic matter and nutrient cycling within grassland ecosystems. Additionally, Liu et al. (2018) noted a direct association between drought occurrences and heightened mortality rates among springtail populations. These findings underscore the susceptibility of soil fauna to alterations in precipitation patterns, which are expected to become more irregular with climate change. Apart from temperature and precipitation changes, modifications in plant communities due to climate change can trigger cascading effects. Ayres et al. (2009) studied the effects of exchanging endemic wildflowers with non-native grasses in a meadow habitat, indicating a worrying 30% drop in mite prevalence.

Climate change presents significant threats to soil-dwelling animal species, leading to extensive consequences for the stability of ecosystems and the conservation of biodiversity. The intricate connections between climate change and soil biodiversity can be understood through the identification of the underestimated impacts of climate change on soil animal populations and the examination of key case studies from various environments. In order to mitigate these impacts and enhance the preservation of biodiversity, it is imperative to collaborate on efforts aimed at addressing climate change, safeguarding habitats, and implementing adaptive management strategies that ensure the health and resilience of soil ecosystems. By engaging in cooperative research, monitoring, and conservation initiatives, a framework for preserving biodiversity can be developed to safeguard the enduring existence of soil-dwelling animal communities and the ecosystems they populate.

3. SYNERGISTIC MECHANISMS

3.1. Mutualistic Relationships among Soil, Plants, and Animals

The complex interconnections among soil, vegetation, and fauna within ecosystems orchestrate a harmonious network of mutually beneficial relationships crucial for supporting life and conserving biodiversity. These relationships, which are based on ecological interactions, serve as the foundation upon which ecosystems flourish and adjust to changes in the environment. Soil microbes, often underestimated yet significantly impactful, establish intricate partnerships with plants to improve their ability to obtain nutrients and withstand challenges. An excellent illustration of this is the symbiotic relationship between mycorrhizal fungi and plant roots. By colonizing plant roots, mycorrhizal fungi extend their hyphal networks into the soil, effectively expanding the plant's nutrient absorption capacity, specifically for phosphorus and nitrogen, in return for a portion of the plant's photosynthetic products. This mutually beneficial interaction not only enhances plant growth and vitality but also enhances their ability to cope with stressors like drought and nutrient deficiencies.

Moreover, nitrogen-fixing bacteria belonging to the genus Rhizobium establish nodules on the roots of leguminous plants, transforming atmospheric nitrogen into a usable form for plants. This phenomenon, termed biological nitrogen fixation, not only enriches the soil with vital nitrogenous compounds but also provides a sustainable nitrogen source for plant growth, reducing the dependency on artificial fertilizers and decreasing environmental contamination. In return, plants offer a refuge for soil microorganisms abundant in organic material and nutrients, facilitating their diversification and proliferation. Root secretions, comprising various organic compounds, serve as a fuel source for soil bacteria, enhancing their metabolic functions and optimizing nutrient circulation within the soil environment. Additionally, the breakdown of plant debris by soil organisms releases crucial nutrients back into the soil, perpetuating a cycle of nutrient renewal crucial for maintaining plant health and ecosystem efficiency.

3.2. The Critical Role of Animal Biodiversity in Enhancing Ecosystem Resilience

Animals, spanning from earthworms to elephants, contribute to the intricate network of interactions between soil and plants through a variety of behaviours. Earthworms, often recognized as pivotal players in ecosystems, enhance both the structure and fertility of soil by promoting the mixing of nutrients and improving aeration. Within terrestrial environments, the movement and feeding patterns of herbivores have a significant impact on plant communities, ultimately influencing the composition and diversity of species. For instance, browsing by ungulates can stimulate the growth of herbaceous plants by constraining the dominance of woody vegetation, leading to a varied mixture of plant communities. Furthermore, animals play a crucial role in dispersing seeds, enabling the colonization of new habitats and the regeneration of plant populations. Particularly, birds contribute significantly to the spatial distribution of plant species and the preservation of genetic diversity within plant populations by consuming fruits and subsequently depositing seeds in distant areas. The mutualistic connections play a pivotal role not only in individual species interactions but also in the overall functionality and resilience of ecosystems. Interactions among soil, plants, and animals contribute to ecosystem stability and sustainability amidst environmental changes by facilitating nutrient cycling, enhancing soil fertility, and promoting plant diversity.

Empirical research and ecological modelling have underscored the significant implications of these synergistic pathways for ecosystem resilience and adaptability. For instance, research has illustrated that the disappearance of crucial animal species like pollinators or seed dispersers can trigger a domino effect across entire ecosystems, leading to reductions in plant diversity and ecosystem productivity (Murphy et. al., 2022). The intricate web of mutualistic interactions involving soil, plants, and animals serves as the cornerstone of ecosystem well-being and resilience. The identification and preservation of these synergistic mechanisms are imperative for biodiversity conservation and the enduring health of ecosystems in a dynamic environment. Consequently, by elucidating the complexities of soil-plant-animal interactions, we lay the foundation for a conservation strategy that transcends disciplinary boundaries and promotes interdisciplinary cooperation in fostering harmonious cohabitation between humanity and the environment.

4. PRESERVATION STRATEGIES FOR ANIMAL BIODIVERSITY

A diverse array of solutions is necessary to safeguard animal biodiversity and enhance the intricate network of interactions between soil and animals. Illuminating the path towards sustainable coexistence and biodiversity conservation involves examining effective conservation strategies, targeted efforts to enhance soil-dwelling animal populations, and informative case studies.

4.1 Implementing Effective Conservation Practices

Efficient conservation methodologies form the basis of endeavours to minimize animal species loss and safeguard ecosystems. These endeavours encompass a broad spectrum of activities, ranging from preserving habitats to enacting laws aimed at mitigating human-induced threats.

4.1.1. Habitat Preservation:

Preservation of Habitats is crucial to safeguarding critical habitats and minimizing the impacts of habitat fragmentation and degradation. An exemplary instance is the Galápagos Marine Reserve, a UNESCO World Heritage Site that safeguards vast marine habitats supporting a diverse array of species, including iconic marine mammals like whales and dolphins. By establishing marine protected areas (MPAs) such as these, sanctuaries are created where species can flourish without the threats of overfishing and habitat destruction.

4.1.2. Sustainable Land Management:

Sustainable Land Management practices such as agroforestry and rotational grazing can enhance habitat connectivity and biodiversity in agricultural landscapes. In Costa Rica, initiatives like Payment for Ecosystem Services (PES) incentivize landowners to conserve biodiversity-rich ecosystems on their lands. By compensating farmers for ecological services like carbon sequestration and watershed protection, these programs not only preserve biodiversity but also foster rural socioeconomic development (Murguia et. al., 2022).

4.1.3. Legislative Measures:

Legislative Actions play a crucial role in strengthening wildlife protection laws and implementing measures to combat illicit wildlife trade, essential for tackling poaching and trafficking of endangered species. The African Elephant Coalition, comprising 32 African nations, advocates for a total ban on ivory trade to combat elephant poaching and ensure the survival of these iconic animals. The coalition aims to safeguard elephant populations and habitats across Africa through advocacy for legislative changes and mobilization of global support (Hoare et. al., 2022).

4.2. Strategies to Enhance and Protect Soil-Dwelling Animal Populations

Soil-dwelling animal populations play a significant role in ecosystem functioning, yet they are sometimes overlooked in conservation endeavours. Targeted interventions to enhance and protect these vital inhabitants are paramount for upholding biodiversity and enhancing ecosystem resilience.

4.2.1. Habitat Restoration:

Habitat Restoration initiatives aimed at enhancing soil biodiversity can rejuvenate degraded ecosystems and provide a refuge for soil-dwelling animals. The Grain for Green Programme on China's Loess Plateau incentivizes farmers to reforest marginal agricultural lands and restore ecosystems through grassland revitalization. These efforts not only combat soil erosion and desertification but also promote habitat restoration and biodiversity conservation in a highly vulnerable region (Li et. al., 2023).

4.2.2. Soil Management Practices:

Soil Management techniques such as limited tillage and organic farming can boost soil health and biodiversity by reducing soil disturbance and increasing organic matter levels. In the Midwest Corn Belt of the United States, conservation tillage practices have led to enhanced soil structure and fertility, providing habitat for soil-dwelling species like earthworms and beneficial microbes (Sangotayo et. al., 2023).

4.2.3. Biodiversity Monitoring:

Using state-of-the-art technology such as DNA metabarcoding and remote sensing, researchers are able to assess the abundance, diversity, and spatial distribution of subterranean animal communities in soil with remarkable precision. In the Brazilian Cerrado region, scientists utilize DNA metabarcoding techniques to study soil microbial populations in reaction to alterations in land use and deforestation. Through monitoring changes in soil biodiversity over time, researchers can pinpoint key areas for conservation and advise on land management strategies to mitigate the impacts of habitat destruction and deterioration (Bröring et. al., 2023).

4.3. Case Studies

4.3.1. Beaver Reintroduction in Scotland:

The initiative to reintroduce beavers in Scotland, known as the Scottish Beaver Trial, commenced in 2009 with the aim of reintegrating Eurasian beavers into their native habitats in Scotland after an absence of over four centuries. By leveraging the ecosystem engineering capabilities of beavers, such as constructing dams and felling trees, the project strives to restore degraded wetland environments and enhance biodiversity. Initial findings suggest that reintroducing keystone species has a positive effect on habitat formation, water retention, and species richness, underscoring their potential for reviving ecosystems and conserving biodiversity (Finger et. al., 2023). Efficient conservation of animal biodiversity requires a comprehensive approach encompassing habitat preservation, sustainable land management, and specialized conservation methods tailored to soil-dwelling animal communities. By drawing insights from prosperous case studies and implementing science-based conservation tactics, a blueprint for biodiversity preservation can be devised to ensure ecosystem durability and vitality for future generations.

5. CHALLENGES AND FUTURE DIRECTIONS

In order to maintain biodiversity and enhance interactions between soil and animals, it is imperative to acknowledge the prevailing challenges in conservation and chart a course for future research to tackle emerging issues and knowledge gaps. Delving into these challenges and opportunities for enhancement is crucial for devising effective strategies to safeguard animal biodiversity and foster resilient ecosystems.

5.1. Current Challenges in Preserving Animal Biodiversity

5.1.1. Habitat Loss and Fragmentation:

The primary obstacles to biodiversity conservation, namely habitat loss and fragmentation, stem from human activities like deforestation, urban expansion, and agricultural progress. As natural habitats dwindle, the populations of numerous animal species decline, rendering them more susceptible to extinction. For instance, the partitioning of tropical rainforests in the Amazon Basin imperils the survival of various species, including emblematic megafauna such as jaguars and giant otters.

5.1.2. Climate Change:

Climate change's escalating impacts, including elevated temperatures, shifting precipitation patterns, and more frequent extreme weather events, present significant risks to global animal species. Adapting to swiftly changing environmental conditions is imperative for species to avert the decline of their populations and habitats. An example of this vulnerability is evident in polar bears, whose exposure to critical hunting grounds and prey diminishes as Arctic Sea ice diminishes due to melting.

5.1.3. Invasive Species and Diseases:

The introduction of non-native animals and viruses to ecosystems can result in catastrophic consequences for native animal populations, disrupting ecological balance and contributing to the loss of biodiversity. Invasive species, exemplified by the cane toad in Australia and the brown tree snake in Guam, have inflicted harm on local animal communities and disturbed ecological processes, underscoring the urgent necessity for implementing management approaches to alleviate their impacts.

5.1.4. Overexploitation and Illegal Wildlife Trade:

The unsustainable utilization of wildlife for sustenance, medicinal purposes, and commercial activities, coupled with the illicit trade of endangered species, poses a grave menace to animal diversity. Relentless poaching and trafficking of iconic species like elephants, rhinoceroses, and pangolins persist unchecked, pushing numerous populations to the brink of extinction and undermining global endeavours for conservation.

5.2. Emerging Issues and Research Gaps for Future Exploration

5.2.1. Microplastics and Pollution:

Recent research unveils the pervasive peril of microplastic contamination to terrestrial and aquatic ecosystems, potentially impacting animal populations dwelling in the soil. Microplastics, derived from the degradation of plastic waste, have the capacity to accumulate in soil and water systems, where they could be ingested by animals, entering the food chain and potentially jeopardizing ecosystem well-being and biodiversity.

5.2.2. Microbial Interactions and Soil Health:

Despite its fundamental role in ecosystem functionality and resilience, the intricate interplay between soil-dwelling animals and microbial communities remains inadequately understood. Delving into the dynamics of microbial interactions within soil ecosystems, encompassing mutualistic relationships with flora and fauna, can unveil pivotal mechanisms governing nutrient circulation, carbon storage, and soil fertility, thereby laying the groundwork for innovative conservation tactics that leverage soil-animal interdependencies.

5.2.3. Ecological Resilience and Adaptation:

As ecosystems confront escalating environmental pressures and disturbances, comprehending the mechanisms that underlie ecological resilience and adaptability is paramount for the preservation of biodiversity. Exploring the capacity of animal populations to adjust to evolving environmental circumstances, such as alterations in habitat and climatic fluctuations, can inform conservation strategies intended to bolster ecosystem resilience and enhance species survival amidst global transformations.

5.2.4 Synthesis of Data and Interdisciplinary Collaboration:

Bridging disciplinary divides and amalgamating diverse datasets are indispensable for advancing our comprehension of soil-animal interactions and their implications for biodiversity conservation. Integrating ecological, genetic, and remote sensing methodologies can furnish comprehensive insights into ecosystem dynamics, guiding empirically grounded management decisions. Moreover, fostering interdisciplinary cooperation among scientists, policymakers, and stakeholders is vital for translating research findings into efficacious conservation policies and facilitating the dissemination of information to address pressing conservation challenges.

Tackling existing obstacles and forthcoming trends in biodiversity conservation demands a concerted endeavour to tackle emerging issues, fill research lacunae, and foster multidisciplinary collaboration. Embracing a holistic approach that integrates scientific inquiry, policy innovation, and community engagement can lay the foundation for biodiversity preservation, safeguarding animal diversity and fostering the symbiosis of soil-animal interactions for the betterment of current and future generations.

CONCLUSION

The exploration of the intricate network of interactions between soil and animals, alongside the preservation of biodiversity, has led to significant findings and implications underscoring the importance of promoting synergy between soil ecosystems and animal biodiversity. As we approach the conclusion of our journey, we reflect on these findings and their extensive implications for the sustainable management of soil ecosystems.

1. Key Insights on Animal Biodiversity Preservation

1.1. Interconnectedness of Soil-Animal Interactions:

The present chapter has unveiled the interconnectedness of soil, flora, and fauna, underscoring the pivotal role of animal biodiversity in shaping ecosystem dynamics and resilience. Ranging from symbiotic relationships among soil microorganisms, plants, and animals to the ripple effects of climate change on soil-dwelling animal communities, the complex network of interactions underscores the intricacy and vulnerability of soil ecosystems.

1.2. Conservation Mandates:

The significance of upholding animal biodiversity in soil ecosystems cannot be overstated. As ecosystems fragment and deteriorate, coupled with escalating human-induced pressures, numerous animal species face imminent extinction. Mitigating these challenges and safeguarding the biodiversity that bolsters ecosystem functionality and resilience necessitates the implementation of effective conservation strategies and the promotion of interdisciplinary cooperation.

1.3. Innovation and Adaptation:

Embracing innovation and flexibility holds paramount importance in addressing emerging challenges and bridging research gaps in biodiversity conservation. Whether through leveraging cutting-edge technologies like DNA metabarcoding or exploring innovative conservation approaches such as community-based programs and the reintroduction of keystone species, innovation presents promising prospects for enhancing our comprehension of soil-animal interactions and guiding evidence-based conservation measures.

2. Implications for Sustainable Soil Ecosystem Management

2.1. Ecosystem-Based Approaches:

To ensure the sustainable management of soil ecosystems, comprehensive, ecosystem-centric methodologies prioritizing biodiversity conservation and resilience must be adopted. Acknowledging the intrinsic value of soil-dwelling animals and their contributions to ecosystem services such as nutrient cycling, soil fertility, and carbon sequestration enables us to integrate biodiversity considerations into land management practices and policy frameworks.

2.2. Advocating for Biodiversity-conscious Practices:

Promoting the adoption of biodiversity-friendly techniques like agroecology, organic farming, and habitat rehabilitation fosters harmonious coexistence between agriculture and wildlife. These initiatives not only enhance soil quality and productivity but also establish habitat corridors and sanctuaries for soil-dwelling animal communities, thereby promoting biodiversity conservation within agricultural landscapes.

2.3. Education and Outreach:

Advocating for comprehension and recognition of soil-animal interactions, as well as the imperative of biodiversity preservation, is crucial for fostering stewardship and engagement among a diverse array of stakeholders. By enhancing our understanding of ecosystem interrelations and the benefits they provide, we can empower individuals and communities to champion conservation efforts and advocate for sustainable management of soil ecosystems.

The synergy of soil-animal interactions exemplifies a blueprint for biodiversity conservation and sustainable management of soil ecosystems. Through embracing interdisciplinary collaboration, innovation, and adaptive management strategies, we can navigate the complexities of conservation issues and chart a course for a future where humanity and the environment coexist harmoniously. As custodians of the planet, it is our shared responsibility to safeguard the richness of life that sustains us and ensure the enduring resilience and vigour of soil ecosystems for future generations.

REFERENCES

Ahmed, N., & Al-Mutairi, K. A. (2022). Earthworms effect on microbial population and soil fertility as well as their interaction with agriculture practices. *Sustainability (Basel)*, 14(13), 7803. 10.3390/su14137803

Ayres, E., Wall, D. H., & Bardgett, R. D. (2009). Mutualistic interactions and spatial gradients in a plant-microbial-invertebrate food web. *Ecology Letters*, 12(4), 490–499.

Baker, C. C., Castillo Vardaro, J. A., Doak, D. F., Pansu, J., Puissant, J., Pringle, R. M., & Tarnita, C. E. (2020). Spatial patterning of soil microbial communities created by fungus-farming termites. *Molecular Ecology*, 29(22), 4487–4501. 10.1111/mec.1558532761930

Banerjee, S., & Van Der Heijden, M. G. (2023). Soil microbiomes and one health. *Nature Reviews. Microbiology*, 21(1), 6–20. 10.1038/s41579-022-00779-w35999468

Barnett, K. L., & Facey, S. L. (2016). Grasslands, invertebrates, and precipitation: A review of the effects of climate change. *Frontiers in Plant Science*, 7, 197607. 10.3389/fpls.2016.0119627547213

Bröring, J. M., Goss-Souza, D., Baretta, C. R. D. M., Sousa, J. P., Baretta, D., Oliveira Filho, L. C. I., & Klauberg-Filho, O. (2023). Soil microbial carbon and activity along with land use and geographic gradients. *Soil Science Society of America Journal*, 87(4), 856–867. 10.1002/saj2.20545

Brosi, B. J., & Briggs, H. M. (2013). Single pollinator species losses reduce floral fidelity and plant reproductive function. *Proceedings of the National Academy of Sciences of the United States of America*, 110(32), 13044–13048. Advance online publication. 10.1073/pnas.130743811023878216

Cheik, S., Shanbhag, R. R., Harit, A., Bottinelli, N., Sukumar, R., & Jouquet, P. (2019). Linking termite feeding preferences and soil physical functioning in Southern-Indian woodlands. *Insects*, 10(1), 4. 10.3390/insects1001000430621180

Chen, M., Shao, M. A., Wei, X., Li, T., Zhang, Q., Sun, S., Yang, X., & Gan, M. (2022). Earthworm (Metaphire guillelmi) activities increase the risk of soil erosion–a simulation experiment. *Earth Surface Processes and Landforms*, 47(7), 1734–1743. 10.1002/esp.5343

Chen, M. K., Wang, S. J., Chen, W. Q., Cao, R., Cao, Q. B., Wang, P., & Li, S. H. (2019). Effects of ant nesting on soil microbial biomass carbon and quotient in tropical forest of Xishuangbanna. *Ying Yong Sheng tai xue bao=. Journal of Applied Ecology*, 30(9), 2973–2982. 10.13287/J.1001-9332.201909.03631529872

Chen, Q. L., Hu, H. W., Yan, Z. Z., Li, C. Y., Nguyen, B. A. T., Zheng, Y., Zhu, Y.-G., & He, J. Z. (2021). Termite mounds reduce soil microbial diversity by filtering rare microbial taxa. *Environmental Microbiology*, 23(5), 2659–2668. 10.1111/1462-2920.1550733817921

Chinnasamy, P., & Honap, V. U. (2023). Spatiotemporal variations in soil loss across the biodiversity hotspots of Western Ghats Region, India. *Journal of Earth System Science*, 132(2), 90. 10.1007/s12040-023-02098-x

Chown, S. L., Bergstrom, D. M., Houghton, M., Kiefer, K., Terauds, A., & Leihy, R. I. (2022). Invasive species impacts on sub-Antarctic Collembola support the Antarctic climate-diversity-invasion hypothesis. *Soil Biology & Biochemistry*, 166, 108579. 10.1016/j.soilbio.2022.108579

Cours, J., Bouget, C., Barsoum, N., Horák, J., Le Souchu, E., Leverkus, A. B., Pincebourde, S., Thorn, S., & Sallé, A. (2023). Surviving in changing forests: Abiotic disturbance legacy effects on arthropod communities of temperate forests. *Current Forestry Reports*, 9(4), 189–218. 10.1007/s40725-023-00187-0

Erlwein, A. (2022). Exploring Ecosystems Health: Effects of Increments of Biodiversity and Trophic Complexity on the Stability of a Simple Gaian Ecosystem Model. *Agro Sur*, 50(2), 13–24. 10.4206/agrosur.2022.v50n2-02

Ertiban, S. M. (2019). Soil fauna as webmasters, engineers and bioindicators in ecosystems: Implications for conservation ecology and sustainable agriculture. *American Journal of Life Sciences*, 7(1), 17–26. 10.11648/j.ajls.20190701.14

Fierer, N. (2017). Embracing the unknown: Disentangling the complexities of the soil microbiome. *Nature Reviews. Microbiology*, 15(10), 579–590. 10.1038/nrmicro.2017.8728824177

Finger, A., Smedley, M., Halley, D., Frachon, N., & Ahrends, A. (2023). Managing small populations—Using genetic data and trial translocations to help inform suitable conservation measures for the alpine blue-sowthistle (Cicerbita alpina) in Scotland. *Environmental Research. Ecology*, 2(2), 025004. 10.1088/2752-664X/acdef9

Gerlich, H. S., Holmstrup, M., Schmidt, N. M., & Høye, T. T. (2023). Phenological sensitivity of climate across taxa and local habitats in a high-Arctic arthropod community. https://doi.org/10.32942/X24590

Goncharov, A. A., Leonov, V. D., Rozanova, O. L., Semenina, E. E., Tsurikov, S. M., Uvarov, A. V., Zuev, A. G., & Tiunov, A. V. (2023). A meta-analysis suggests climate change shifts structure of regional communities of soil invertebrates. *Soil Biology & Biochemistry*, 181, 109014. 10.1016/j.soilbio.2023.109014

Görres, J. H., & Amador, J. A. (2021). The soil fauna. In *Principles and applications of soil microbiology* (pp. 191–212). Elsevier. 10.1016/B978-0-12-820202-9.00008-3

Hoare, R., Foggin, C., Alexander, S., & Chakombera, D. (2022). Combating the use of cyanide poison in ivory poaching. *African Journal of Wildlife Research*, 52(1), 90–98. 10.3957/056.052.0090

Khan, M. A., Ahmad, W., & Paul, B. (2018). Ecological impacts of termites. *Termites and Sustainable Management: Volume 1-Biology, Social Behaviour and Economic Importance*, 201-216. 10.1007/978-3-319-72110-1_10

Kiçaj, H. (2023). Ecology of soil animals (Diplopoda class, Myriapoda group). *Scientific Horizons.*, 26(5). Advance online publication. 10.48077/scihor5.2023.37

Li, S., Xie, J., & Paudel, B. (2023). Do ecological restoration projects undermine economic performance? A spatially explicit empirical study in loess plateau, China. *Remote Sensing (Basel)*, 15(12), 3035. 10.3390/rs15123035

Liu, X., Ma, D., Wu, Y., & Li, L. (2018). Effects of drought events on soil microbial community and activities in a temperate grassland. *Land Degradation & Development*, 29(11), 3886–3895.

Lu, L., Li, G., He, N., Li, H., Liu, T., Li, X., Whalen, J. K., Geisen, S., & Liu, M. (2023). Drought shifts soil nematodes to smaller size across biological scales. *Soil Biology & Biochemistry*, 184, 109099. 10.1016/j.soilbio.2023.109099

Maisnam, P., Jeffries, T. C., Szejgis, J., Bristol, D., Singh, B. K., Eldridge, D. J., Horn, S., Chieppa, J., & Nielsen, U. N. (2023). Severe prolonged drought favours stress-tolerant microbes in Australian drylands. *Microbial Ecology*, 86(4), 3097–3110. 10.1007/s00248-023-02303-w37878053

Majer, J. D., & Agosti, D. (2016). Ants visible from space influence soil properties and vegetation in steppe rangelands of Iran. *Sociobiology*, 63(4), 1063–1068. 10.13102/sociobiology.v63i4.1195

Mudrák, O., Uteseny, K., & Frouz, J. (2012). Earthworms drive succession of both plant and Collembola communities in post-mining sites. *Applied Soil Ecology*, 62, 170–177. 10.1016/j.apsoil.2012.08.004

Murguia, J. M., Ordoñez, P., Corral, L., & Navarrete-Chacón, G. (2022). *Payment for Ecosystem Services in Costa Rica: Evaluation of a Country-Wide Program*. IDB, Inter-American Development Bank. 10.18235/0004259

Murphy, J. T., Breeze, T. D., Willcox, B., Kavanagh, S., & Stout, J. C. (2022). Globalisation and pollinators: Pollinator declines are an economic threat to global food systems. *People and Nature*, 4(3), 773–785. 10.1002/pan3.10314

Raza, M. B., Bhoi, T. K., & Samal, I. (2019). Soil arthropods as a nutrient enhancer. *International Journal of Chemical Studies*, 7, 1687–1692.

Reddy, C. S., Jha, C. S., & Dadhwal, V. K. (2016). Assessment and monitoring of long-term forest cover changes (1920–2013) in Western Ghats biodiversity hotspot. *Journal of Earth System Science*, 125(1), 103–114. 10.1007/s12040-015-0645-y

Reyes, F., Sorgonà, A., Briones, M. J., Crecchio, C., & Sofo, A. (2023). Plant Growth and Root Morphology Are Affected by Earthworm-Driven (Eisenia sp.) Changes in Soil Chemico-Physical Properties: A Mesocosm Experiment with Broccoli and Faba Bean. *Journal of Soil Science and Plant Nutrition*, 23(3), 4078–4090. 10.1007/s42729-023-01325-0

Rouland-Lefèvre, C. (2011). Termites as pests of agriculture. *Biology of termites: a modern synthesis*, 499-517. 10.1007/978-90-481-3977-4_18

Sangotayo, A. O., Chellappa, J., Sekaran, U., Bansal, S., Angmo, P., Jasa, P., Kumar, S., & Iqbal, J. (2023). Long-term conservation and conventional tillage systems impact physical and biochemical soil health indicators in a corn–soybean rotation. *Soil Science Society of America Journal*, 87(5), 1056–1071. 10.1002/saj2.20563

Sawicka, B., & Egbuna, C. (2020). Pests of agricultural crops and control measures. In *Natural remedies for pest, disease and weed control* (pp. 1–16). Academic Press., 10.1016/B978-0-12-819304-4.00001-4

Sofo, A., Khanghahi, M. Y., Curci, M., Reyes, F., Briones, M. J., Sarneel, J. M., Cardinale, D., & Crecchio, C. (2023). Earthworm-driven changes in soil chemico-physical properties, soil bacterial microbiota, tree/tea litter decomposition, and plant growth in a mesocosm experiment with two plant species. *Plants*, 12(6), 1216. 10.3390/plants1206121636986903

Thakur, M. P., Künne, T., Unsicker, S. B., Biere, A., Ferlian, O., Pruschitzki, U., Thouvenot, L., Türke, M., & Eisenhauer, N. (2021). Invasive earthworms reduce chemical defense and increase herbivory and pathogen infection in native trees. *Journal of Ecology*, 109(2), 763–775. 10.1111/1365-2745.1350433664527

Valladares, F. (2023). More biodiversity to improve our health: The benefits to human well-being of favouring functional and diverse ecosystems. *Metode Science Studies Journal*, (13), 111–117.

Wang, X., Zhang, Q., Zhang, Z., Li, W., Liu, W., Xiao, N., Liu, H., Wang, L., Li, Z., Ma, J., Liu, Q., Ren, C., Yang, G., Zhong, Z., & Han, X. (2023). Decreased soil multifunctionality is associated with altered microbial network properties under precipitation reduction in a semiarid grassland. *iMeta*, 2(2), e106. 10.1002/imt2.106

Wu, J., Ayres, E., Vance, G. F., Luo, Y., & Guo, Q. (2012). Effects of elevated CO2 and N deposition on litter decomposition and soil microbial community in a Tibetan alpine meadow. *Plant and Soil*, 351(1-2), 101–112.

Yan, Z. Z., Chen, Q. L., Li, C. Y., Nguyen, B. A. T., Zhu, Y. G., He, J. Z., & Hu, H. W. (2021). Termite mound formation reduces the abundance and diversity of soil resistomes. *Environmental Microbiology*, 23(12), 7661–7670. 10.1111/1462-2920.1563134097804

Zhang, Z. Q. (2011). *Animal biodiversity: An outline of higher-level classification and survey of taxonomic richness*. Magnolia press.

KEY TERMS AND DEFINITIONS

Biodiversity Preservation: The deliberate and urgent endeavour to safeguard the rich tapestry of life forms within ecosystems, essential for maintaining ecological balance and resilience.

Conservation Practices: Proactive strategies and interventions aimed at mitigating biodiversity loss, habitat degradation, and species extinction, imperative for sustaining ecosystem services and human well-being.

Ecosystem Resilience: The capacity of ecosystems to absorb disturbances, adapt to changing conditions, and maintain essential functions and services, underscoring the necessity of conserving biodiversity and ecosystem integrity.

Mutualistic Relationships: Symbiotic partnerships between different species, characterized by reciprocal benefits that optimize resource utilization, bolstering ecosystem stability and productivity.

Soil-Animal Interactions: The intricate and mutually influential relationships between soil microorganisms and animals, vital for ecosystem stability and nutrient cycling.

Soil-Dwelling Animals: Organisms inhabiting soil habitats, including microorganisms, invertebrates, and small vertebrates, integral to soil health, fertility, and ecosystem functioning.

Sustainable Management: Holistic and adaptive approaches to resource utilization and ecosystem stewardship that prioritize ecological integrity, social equity, and economic viability, essential for ensuring the long-term health and sustainability of ecosystems and human societies.

Synergistic Mechanisms: Interactions among soil, plants, and animals that synergistically enhance ecosystem functions, resilience, and productivity, emphasizing the interconnectedness of biotic and abiotic components.

Chapter 3
Zoning of Vietnam's Forest Fund Based on Climatic Conditions and Organic Mass Growth for Environmental Improvement

Vasily Kovyazin
Saint Petersburg Mining University of Empress Catherine II, Russia

An Nguyen
http://orcid.org/0000-0002-4881-7947
Saint Petersburg Mining University of Empress Catherine II, Russia

Andrei Zubov
Saint Petersburg Mining University of Empress Catherine II, Russia

ABSTRACT

The results of research on climatic conditions in three geographical (forest vegetation) zones of Vietnam are presented. The zones were identified to assess the increase in wood organic mass in relation to changes in climatic conditions. Each so-called forest vegetation zone is characterized by certain indicators of climatic conditions, which affect the growth of wood organic mass. The relationship between the growth of wood mass, the main evaluation indicator of the forest fund, and climatic conditions is reflected based on the model of the Swedish climatologist H. Peterson. Climatic factors influencing wood growth were taken from reference books, except for solar energy. The methodology for calculating the influx of solar energy to forest areas in connection with the sun's altitude is provided in this work. The relationship between the growth of wood reserves and the climatic conditions of forest growth has been established in various forest lands, a dependency has been established, and zoning of the forest fund of Vietnam has been proposed, taking into account soil quality

DOI: 10.4018/979-8-3693-3374-7.ch003

INTRODUCTION

The Socialist Republic of Vietnam stretches in a narrow strip from west to east for 50-500 km along the coast of the South China Sea and from north to south for about 1500 km. Almost 75% of the country's area is occupied by mountains and plateaus, with heights ranging from 500 to 1000 meters. The slopes of the mountains are steep and inaccessible, and various forests grow on them, which are complex in structure, productivity, and assessment parameters (Zhang et al., 2024). The main indicator of these forests is the growth of the organic mass reserve, the magnitude of which varies depending on the climatic and soil conditions of forest growth. These forest conditions vary across the territory of Vietnam. Forest zoning has been carried out to assess the productivity of the country's forest resources and develop forest fund land monitoring.

STUDY AREA

The subject of the research is the lands of the forest fund of the Socialist Republic of Vietnam. The country is located on the eastern coast of the Indochinese Peninsula, closer to the center of Southeast Asia on maritime, land, and air routes of international communication. The territory of Vietnam is represented by land and sea. The land area covers 331,212 km^2, extending from 8°34' to 23°23' north latitude and from 102°10' to 109°24' east longitude. The northern border is with China, the western border is with Laos and Cambodia, and the eastern and southern borders are with the South China Sea and the Gulf of Thailand (Kovyazin et al., 2022). Vietnam's maritime territory covers more than 1,000,000 km^2, including 5 zones: coastal, marine, adjacent to the sea, continental shelf, and economic (Hoang et al., 2021). The coastline extends for 3260 km, not counting islands. Taking into account the maritime part of the country, its territory extends from 6°50' north latitude to 117°20' east longitude. According to UN data as of March 28, 2023, the total population of Vietnam is 99,504,886 people. Vietnam is the third most populous country in Southeast Asia (after Indonesia and the Philippines) and the fifteenth in the world (Nguyen, 2020).

Three-quarters of Vietnam's territory is occupied by mountains and hills, while lowlands make up only a quarter. Lowland and foothill areas (less than 1000 meters above sea level) occupy a larger area (Kovyazin et al., 2022). Mountainous areas with rugged terrain and high mountains (over 2000 meters) make up only 1%. The highest mountain in Vietnam is Hoang Lien Son with the peak Fansipan, which is 3143 meters high. A characteristic feature of Vietnam's natural landscape is its ancient structure, which has undergone processes of new geological formation, leading to pronounced differentiation in altitude.

The terrain of the country gently slopes from northwest to southeast. Vietnam's terrain strongly influences the structure and composition of forest stands and forest management. Different terrain and climate in the country give rise to various soil types (Nascimento et al., 2020). These factors lead to the formation of diverse natural ecosystems in different parts of the country. The structures of forest ecosystems in Vietnam vary in age, species composition, and other assessment parameters. There is a process of transition from moist tropical forests in the south to mixed forests in the northern mountainous regions. The southern tropical forests are rich in deciduous plant species such as sandalwood, oak, teak, and many others. Plains and hills provide conditions for the formation of various forest vegetation conditions and forest exploitation technologies. However, forest management is challenging in high mountainous areas, especially the harvesting of mature and over-mature stands, due to steep slopes and complex terrain.

Mountainous areas are subject to soil erosion, which poses a serious threat to forest restoration and conservation. The accuracy of determining the erosion surface by photogrammetric methods enhances the spatial resolution of satellite imagery (Kornilov et al., 2023). Machine learning methods employing neural networks have been applied to increase the speed and reliability of analysis and processing of land cover data (e.g., Khan & Sudheer, 2022; Ushakov et al., 2021). Investment in the country's mineral resource enterprises is required (Matrokhina et al., 2023). Vietnam's relief and climate create diversity in the structural, spatial (Baltyzhakova & Romanchikov, 2021), and lithological composition of forests, but also influence the development, management, protection, and sustainable use of forest resources. The method of continuous spatial data collection and deep learning using cartographic web technologies has been utilized to assess the diversity of forest composition (Kopylova & Starikov, 2021).

In the valleys of the Red River, Mekong River, and some others, agricultural crops are grown, mainly rice, due to their flat terrain (elevation ranging from 0 to 10 meters above sea level) and fertile soils (Kovyazin et al., 2023). The valleys were formed by the silt deposits from major rivers, so the clayey soils are rich in nutrients: phosphorus, potassium, and nitrogen. The total area of the valleys is almost 15 thousand square kilometers. Irrigation systems are used for rice cultivation in these areas. The capital of Vietnam, Hanoi - a political, economic, and cultural center, is located here. The Mekong Delta has a larger area (about 39 thousand square kilometers) than the Red River valley and is situated at a lower elevation. It is one of the largest deltas not only in Vietnam but also in all of Southeast Asia (Pham et al., 2023). This delta is characterized by abundant water resources and fertile soils, providing large volumes of rice production for both domestic consumption and export.

CLIMATIC AND ANTHROPOGENIC FACTORS INFLUENCING THE GROWTH OF FOREST STOCK

The development of forest ecosystems is influenced by climatic (Gunathilaka et al., 2018) and anthropogenic factors: solar energy, temperature regime, amount of precipitation, duration of the growing season, soil quality, and management practices (see table 1).

The greatest influence on the formation of wood growth is exerted by these same factors (climatic conditions, soil condition, tree species composition and properties, age of the forest stands, and forestry management practices). The growth of wood stock in forests is a crucial indicator in monitoring forest resources in terms of the following aspects. Firstly, it allows distinguishing the stability of growth, development, and restoration processes of forest ecosystems (Wiegant et al., 2022). Secondly, it helps determine the contribution of forest ecosystems to climate regulation by absorbing carbon from the atmosphere during photosynthesis and maintaining its oxygen ratio within 21%. This reduces carbon emissions into the atmosphere by industrial enterprises. Thirdly, wood stock growth is used to assess the sanitary condition of the natural ecosystem and the pace of tree growth within it. Fourthly, wood stock growth determines ecosystem services: water supply for the population, regulation of wind speed, absorption of moisture from the soil, and preservation of ecosystem biodiversity. Wood stock growth enables the determination of future possibilities for obtaining and using forest resources: wood, mushrooms, berries, medicinal raw materials, predicting their potential use for various sectors of the economy: construction, furniture production, musical instruments, paper, etc. (Rahmoune et al., 2024).

Table 1. Factors influencing the growth of forest stock (Authors' development)

Climatic	Soil	Anthropogenic factors	Composition and age of the stand	Biota
Amount of light energy	Soil type	Main felling	Species composition of plants	Presence of pests
Temperature	Agrochemical Properties	Land-use change	Biological features of the species	Plant Diseases
Rainfall	Soil fertility	Forest fires	Ecology of the species	Competition between individuals
Length of the growing season	Soil hydrology	Comprehensive forest care	Age of trees	
	Bonitet Class	Pollution of the environment		

Development of Forest Zones of Vietnam Using Bioclimatic Index H. Peterson

Growth is essential for predicting the productivity and resilience of forest ecosystems to natural and anthropogenic factors; it enables forest management and forecasts its future applications and impacts on the environment. Various methods based on the dependence of potential forest productivity on climatic factors are used to assess the influence of climatic factors on forest ecosystems. Climate refers to the long-term statistical regime of atmospheric conditions (Liu et al., 2024) characteristic of a particular territory (zone) due to its geographical location. Methods of air analysis and remote sensing of the Earth (Pashkevich et al., 2021) have been applied to assess atmospheric conditions in remote areas. A whole range of climatic factors influences the growth of wood stock in forest resources: the quantity, intensity, and wavelength of solar energy, which shapes the thermal regime, soil type, its richness, and the moisture of the territory (Schulte et al., 2023). These factors are predominant for preserving forest resources to improve the environment by enhancing the photosynthesis process in forest plants. An assessment of forest resource productivity can be conducted by applying the index developed by the Swedish climatologist H. Peterson. The authors of this study propose determining forest productivity based on H. Paterson's bioclimatic index or the CVP climatic model (C-climate; V-vegetation; P-productivity). Climatologist H. Peterson demonstrated in 1956 that the productivity of forest stands and the rates of organic mass growth depend on the climatic conditions of growth: solar radiation, air temperature, precipitation, and duration of the growing season. These same conditions also determine the formation of various soil types. The index characterizes the ability of woody plants to grow in relation to changing climatic factors (Roitsch et al., 2023). H. Peterson's proposed formula underscores the unity of factors influencing the growth of woody plants in forest ecosystems. The index value is calculated using the formula:

$$Icvp = \frac{Tv.P.G.E}{Ta.12.100} \tag{1}$$

where: I_{cvp} - H. Peterson's index; Tv - average temperature of the warmest month, °C; Ta - difference between the average temperature of the warmest and coldest months, °C; P - annual precipitation, mm; G - duration of the growing season in months; E - evapotranspiration index (evaporation reduction factor, the values of which depend on the location of the forest area). It is found by the formula:

$$E = 100 \times \frac{Rp}{Rs} \tag{2}$$

where: Rp - amount of incident solar energy at the pole; Rs - amount of incident energy in the forest vegetation zone.

The mentioned climatic factors determine the climatic conditions for tree growth, site class, and soil type. The CVP model (index) is developed for predicting the maximum potential wood growth per stock.

The authors of the article use this index to divide the territory of the Socialist Republic of Vietnam into forest vegetation zones, assess the growth and condition of woody vegetation by analyzing data obtained through satellite Earth observation, establish the degree of solar activity (Wang et al., 2022) in plant ecosystems, which can be used for analyzing the composition, condition, and prognosis of forest ecosystems. This index is frequently utilized in studies of the potential productivity of forests in various countries. Vietnam is characterized by a tropical and subtropical climate, which varies seasonally and differs from other countries at similar latitudes with cool winters and not very hot summers. Spring is warm, not very humid, with air temperatures not rising above +30°C, and the minimum average monthly temperature dropping to +21°C (table 2).

Table 2. Differences in temperature and water regimes in different zones of Vietnam (Authors' development)

Forest vegetation zone	Temperature (°C)			Precipitation (mm)			Humidity (%)		
	2010	2015	2020	2010	2015	2020	2010	2015	2020
North	24	23.5	22.5	1700	1850	1750	82	84	84
Central	24.5	25	26.5	1750	1800	1800	83	82	83
South	27.5	28	30	1800	2000	1900	78	79	81

The above information on climatic parameters indicates the following: The northern zone has the lowest air temperature and precipitation values, while the atmospheric humidity is the highest. In the southern zone, the highest values of temperature and precipitation are noted, but with reduced air humidity. Over the observation period from 2010 to 2020, there has been an increase in temperature by 2-3 degrees due to greenhouse gases and an increase in precipitation by 50-100 mm, with air humidity rising by 3-5%. Such dynamics of climatic factors will affect the process of woody vegetation growth. During the autumn-winter period (from October to April), the temperature regime significantly differs in the northern, central, and southern forest vegetation zones, with differences ranging from 5 to 8.5 degrees Celsius. In the summer months, the temperature difference decreases to 1-2°C, which is due to the similar values of solar height during this time period. Vietnam spans 1750 km in latitude. This factor, along with the presence of the massive Truong Son mountain range, determines the temperature regime of the territory.

Temperature

In the northern part of Vietnam, the average air temperature ranges from +18 to +24°C. The maximum monthly average temperature reaches +29.5°C, while the minimum average temperature drops to +12°C. In the central part of Vietnam, the temperature varies from +23 to +27°C. The maximum monthly average temperature rises to +30°C, and the minimum decreases to +16.5°C. The southern part of the country is shielded from northern winds by mountains, resulting in stable air temperatures throughout the year. In the south, temperatures fluctuate between +21 and +27.5°C. The maximum monthly average temperature reaches +29°C, while the minimum drops to +21°C (Figure 1).

Figure 1. Dynamics of average monthly air temperature in Northern, Central, and Southern Vietnam

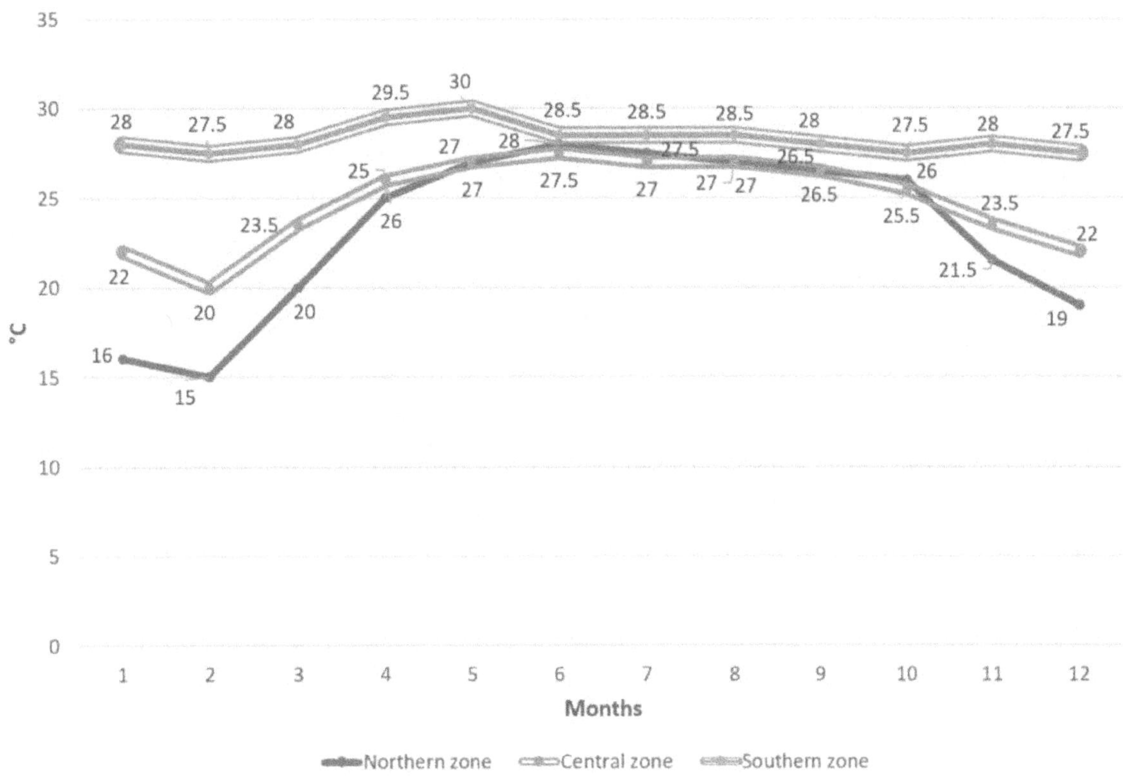

Source: Authors' development

Precipitation

Vietnam is located in the tropical climatic zone and is washed by the South China Sea, with a coastline stretching 3260 kilometers from north to south. Additionally, the presence of mountain ranges creates conditions for the development of the southeast monsoon, which brings a lot of moisture from the sea to the mainland, resulting in intense rainfall across the country. The average annual precipitation varies depending on the zone of the country from 1400 to 2500 mm. In the northern part of the country, summer rains are frequent, starting in May and lasting until October, with average annual rainfall ranging from 1400 to 1800 mm. In Central Vietnam, summers are hot and dry, with occasional heavy rains starting in August and ending in October. In the southern part of the country, the rainy season begins in May and lasts until October, with average annual precipitation ranging from 1600 to 1800 mm (figure 2).

Figure 2. Dynamics of average monthly precipitation in forest zones of Vietnam

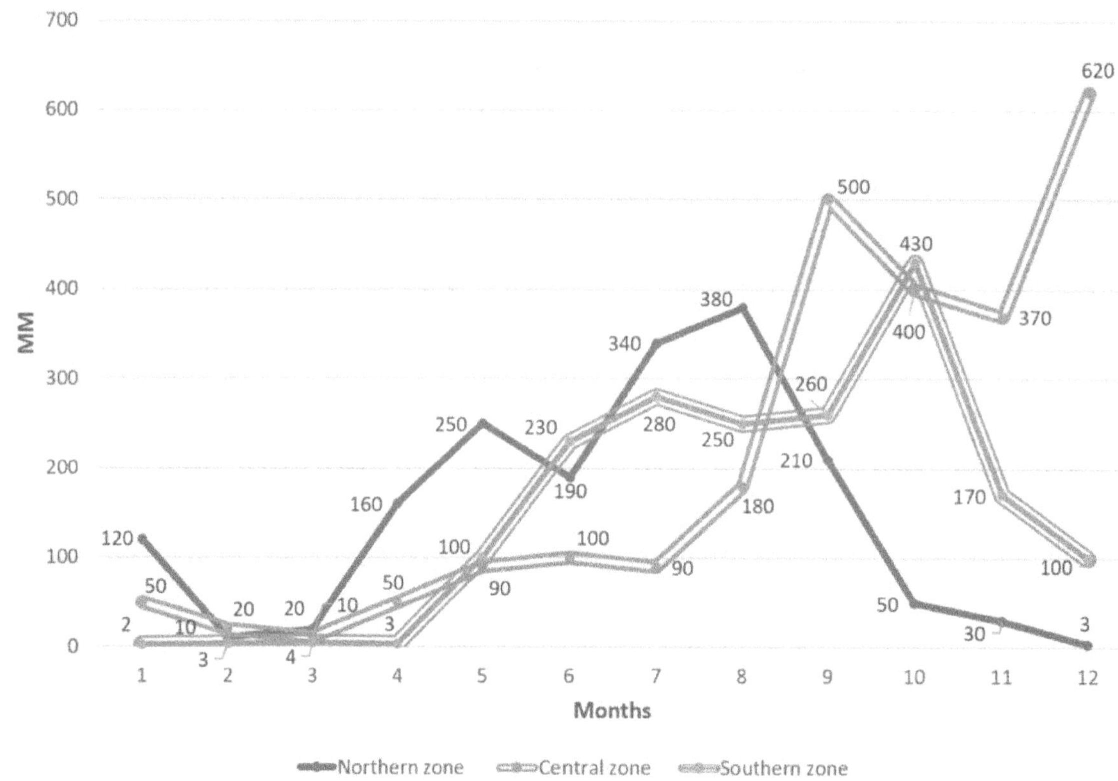

Source: Authors' development

Humidity

During the winter period (January-March), the amount of precipitation is almost the same in all zones due to low cloud cover, but starting from March, cloudiness differentiates significantly, and a substantial difference in precipitation is observed, ranging from 40 to 615 mm. Therefore, the summer is humid and warm. The air humidity in Northern Vietnam averages 82–85%, in Central Vietnam it is 84–86%, and in Southern Vietnam it ranges from 78–84% (Kovyazin et al., 2022) (figure 3). The average monthly air humidity varies significantly across each zone, with differences ranging from 3.5 to 13.5%.

Figure 3. Dynamics of average monthly air humidity in forest zones of Vietnam

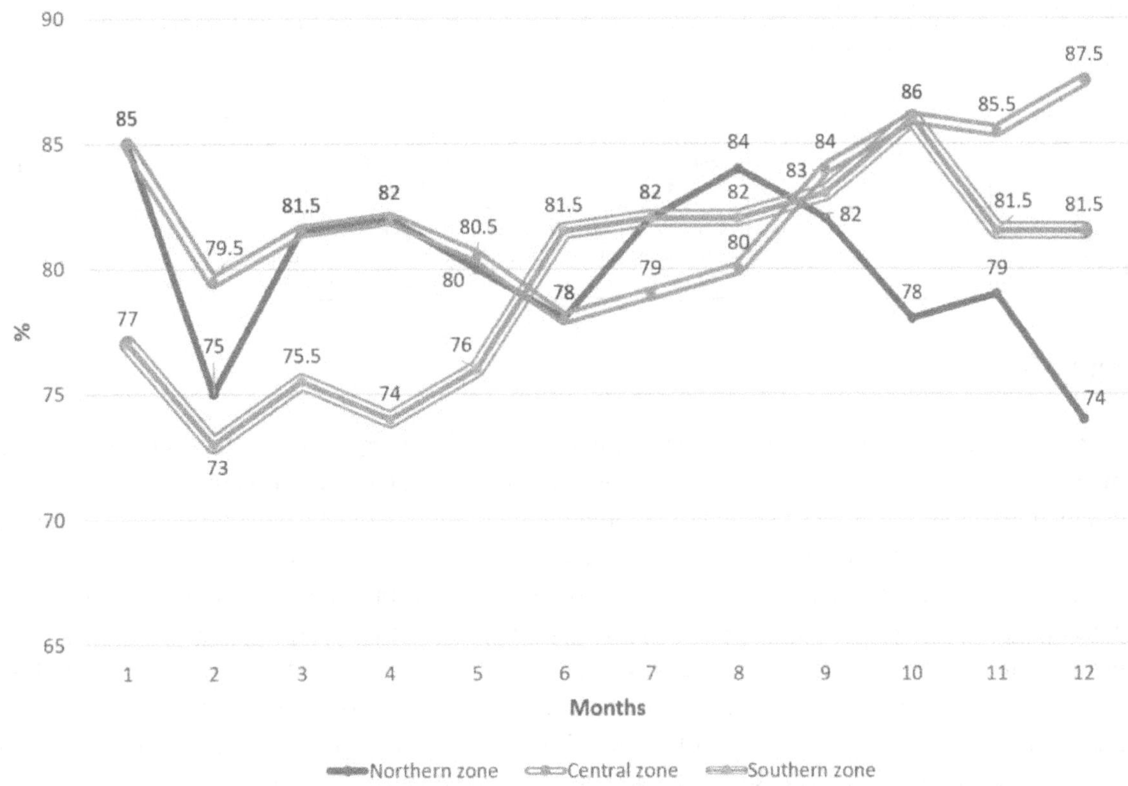

Source: Authors' development

Based on the research findings, the following conclusions can be drawn. *The northern zone* is characterized by hot and humid summers with prolonged downpours. The autumn is mild and warm. Winter is damp and cold, with temperatures dropping to +4°C at night, and even snowfall occurs in the mountains. Drought sets in during the latter half of spring. *The central zone* is characterized by temperature stability, as the Truong Son mountain range acts as a barrier to southwest winds. Rain falls in the plains during autumn, which ends with the onset of winter. Winter (December-February) is cool, with an average temperature of +20°C. The summer brings a dry season. The mountainous areas experience harsher climate conditions than the plains, with no heat, winter temperatures dropping to 0°C, and drizzly rains. *The southern zone* remains stable in terms of climatic conditions throughout the year. There are two distinct seasons based on precipitation: dry (December-March) and wet (April-November). During the dry season, only about 9% of the annual rainfall occurs. The rains are brief, followed by sunshine, leading to periods of drought during this season.

The length of the growing season also varies in different zones of Vietnam (Table 3).

Table 3. Duration of the growing season in different zones of Vietnam (Authors' development)

Forest vegetation zone	Length of the growing season (number of months)	Brief description
North	9	Vegetation usually goes through different flowering seasons throughout the year, depending on weather conditions. Plants such as rice and corn often have seasonal cycles.
Central	10	There is usually a rainy season and a dry season, which affects the growth of stands and cultivated plants. The propagation cycle of plants depends on specific geographical and climatic conditions
South	12	Plants in the southern zone can usually reproduce all year round due to the warm climate and the availability of sufficient moisture. However, there is also the impact of the rainy season and drought

Solar Radiation

The flow of solar radiation refers to the amount of radiant energy passing through a given surface per unit time. In actinometric observations, the density of radiation flux through a unit surface area is usually measured, sometimes referred to as radiation intensity or irradiance (Toming, 1977). The intensity of solar radiation at a particular point on the Earth's surface depends on many factors, all of which are difficult to account for (He et al., 2023). Even the solar constant radiation, strictly speaking, is not constant and varies within an 11-year solar activity cycle. Direct solar rays are absorbed, pass through layers of the atmosphere of varying density, scatter from clouds and dust, and reflect off the Earth's surface. In this process, electromagnetic waves of different lengths (from gamma rays to radio waves) are absorbed, scattered, and reflected differently. The percentage contribution to the total radiation when transitioning from "winter" to "summer" (direct and scattered) can fluctuate from 11/89% to 51/49%. While it is difficult to precisely determine scattered and reflected radiation, direct solar radiation depends on the duration of daylight and the angle at which rays strike the horizontal surface of the Earth.

The duration of the day depends on the declination of the Sun δ, which varies throughout the year from -ε to +ε degrees, where ε is the angle of inclination of the ecliptic to the celestial equator, approximately equal to 23°26'. The angle of inclination of the ecliptic ε, strictly speaking, also changes insignificantly over time. The value of the Sun's declination δ, on any given day can be calculated using the Julian Date (JD).

The Julian Date is a dating system that uses the number of days elapsed since the beginning of the Julian year (January 1, 4713 BCE). This dating system was developed by Julius Caesar in 45 BCE and was used for documenting historical events for many centuries (Seleshnikov, 1970). The time interval T, counted from the epoch of 1900 (January 0.5), and measured in Julian centuries containing 36,525 ephemeris days (About calendars and chronology), is calculated based on the Julian Date.

$$T = \frac{JD - 2415020.0}{36525} \quad 3$$

The refinement of the angle of inclination of the ecliptic and the declination of the Sun over the time interval T is detailed in Meyos (1988).

For the sake of simplifying the calculation of solar energy received, it is typically expressed in hours of sunshine on a surface perpendicular to the solar rays, which is proportional to the cosine of the zenith distance Z. Thus, the number of hours of direct radiation E (perpendicular) between Julian dates JD1 and JD2 can be calculated using the formula:

$$E = \sum_{JD1}^{JD2} \int_{tr}^{ts} cosZ \, dt \quad 4$$

where t_r and t_s are the hourly angles of sunrise and sunset for each Julian dates. It should be noted that the integration cosZ is carried out over time. The hourly angles of sunrise and sunset are calculated by the formula Pandul (2010):

$$cost = -tan\varphi \times tg\delta \quad 5$$

Where φ is the latitude of the location. For sunrise, t_r = 360° - arccos*t*, and for sunset, t_s = arccos*t*.

For example, on April 1, 2023, in Saint Petersburg (φ = 60°), we obtain t_r = 97.6°, t_s = 262.4°, and the duration of the day is – (360 + t_r – t_s)/15 = 13.1h.

The cosine of the zenith distance at any given time is calculated by the formula Pandul (2010):

$$cosZ = sin\varphi \times sin\delta + cos\varphi \times cos\delta \times cost \quad 6$$

In one day, on April 1, 2023, we obtain only 4.66 hours of direct solar radiation perpendicular to the Sun's beam.

The closer the Sun is to the horizon, the greater the influence of air mass - a measure of the amount of air along the line of sight when observing a celestial body through the Earth's atmosphere. It is used to calculate the loss of light intensity and luminosity in astronomy and actinometry.

Various scientists have approximated the change in the cosine of the zenith distance due to air mass. For example, Rosenberg (1963) proposed a fairly compact interpolation formula that gives satisfactory results.

$$cosZ := cosZ + 0.025 \times e^{-11 \times cosZ} \quad 7$$

In 1989, Fritz Kasten, together with Andrew Young, published refined air mass data in accordance with the 1972 standard atmosphere (Kasten & Young, 1989). Their formula yields good results for all solar angles for an atmosphere at sea level pressure (760 mm Hg) and a

temperature of 15°C. The change in the cosine of the zenith distance looks like this:

$$cosZ := cosZ + 0.052572 \, (90° - Z + 6.07995°)^{-1.6364} \quad 8$$

On a very sunny day, the intensity of solar radiation varies from 950 W/m^2 to 1220 W/m^2. The average value on a clear day is approximately 1000 W/m^2 or 3.6 MJ per hour.

The algorithm presented is close to the data provided in SNiP 23-01-99* Construction climatology (2020) [Table 8.1]. This set of rules establishes the climatic parameters used in the design of buildings, structures, heating, ventilation, air conditioning systems, water supply, urban and rural planning, and the zoning of forest areas in the Russian Federation and Vietnam.

The duration of the growing season is assessed according to the criteria of Benavides, Diodato [Benavides et al., 2009; Diodato, Bellocchi, 2020], and it is established that it begins with an average temperature of no less than 6°C and sufficient precipitation. The level of precipitation was determined using the E. De Martonne index [1926], which should be above 20 [Pardè, 1958]:

$$Idm = \frac{12 \times P}{T + 10} \quad 9$$

Where: I_{dm} - E. De Martonne index, mm/°C; P - monthly precipitation sum, mm; T - average monthly temperature, °C.

Research [Benavides et al., 2009; Diodato, Bellocchi, 2020] has shown the feasibility of replacing the insolation index E, which reflects the intensity of evaporation, with a modified indicator f [Gandullo, Serrada, 1977], which better reflects the characteristics of small territories. The f indicator is calculated by the formula [Diodato, Bellocchi, 2020]:

$$f = \frac{2500}{Psun + 1000} \quad 10$$

Where: f - modified insolation indicator, J; Psun - annual amount of sunshine hours.

The amount of direct solar radiation in different forest zones of Vietnam is shown in Table 4.

Table 4. Calculation of direct solar radiation in different zones of Vietnam (Authors' development)

Province/ zone	First Date	Second Date	Latitude of the administrative center of the zone	Direct Solar Radiation (MJ/m2)		Difference in average daily direct solar radiation (J/m2)
				Per year	Per day	
Thanh Hoa/ North	01.01.2023	31.12.2023	19°48'26"	9167	25.115	115000
Kon Tum/ Central	01.01.2023	31.12.2023	14°23'00"	9400	25.753	753000
Dong Nai/ South	01.01.2023	31.12.2023	10°57'00"	9508	26.049	1049000

Soil

Soil is an important factor in shaping the growth of wood reserves (Baldoni et al., 2023), but its productivity also depends on climatic factors. Climatic conditions influence the formation of soil types, so each forest vegetation zone is characterized by its own type of soil (Table 5). The structural-geological features of the formation of the mentioned soil types have been investigated using a remote method according to the procedure outlined in Talovina et al. (2022).

Table 5. The main types of soil in the analogous provinces of the forest vegetation zones of Vietnam (Authors' development)

Province		Soil type	Square		Brief characteristics of the soil
			ha	%	
Northern Zone (Thanh Hoa)	1	Yellow-red sand	688171	61,91	It is widely distributed in low mountains at an altitude of less than 900 m.
	2	Fluvial Deposits	165554	14,89	The uniqueness of this soil is its layering (fluvik), and it is also close to surface or underground water sources: the nature of the soil is determined by the quality of minerals coming from the upper horizon.
	3	Red-Yellow Mountain	67054	6,03	It is formed at an altitude of more than 900 m on rugged terrain, on steep slopes.
	4	Anthropo-genically disturbed	26977	2,43	It is formed in conditions of fragmented relief, with a large slope, with a disturbed layer of soils as a result of many years of irrational use in agriculture.
	5	Grey	16918	1,53	It is found in plains and mountains, characterized by good drainage and a light particle size distribution, including quartz, kaolinite, halosite and gothite.
Central Zone (Kon Tum)	1	Cinnamon- Yellow	579788	60,30	It is found in forests of a humid tropical climate, with the predominance of the processes of mineral decomposition of clay and the accumulation of iron and aluminum.
	2	Red-yellow	343228	35,70	It is found in the mountains, at altitudes of more than 900 m.
	3	Fluvial Deposits	16663	1,73	It is formed from precipitation applied by the water of rivers and streams.
Southern Zone (Dong Nai)	1	Yellow-red clay	285653	48,46	It is formed on the products of weathering of the parent rock or various soil components, and is common in humid tropical climates.
	2	Chernozem	128265	21,76	It is formed on basalt deposits or weathering products of yellow basalt soils. They are found in valleys with the presence of basalt.
	3	Grey	66809	11,33	It is found in humid tropical climates and has a light granulometric structure, contributing to the accumulation of salts.
	4	Fluvial Deposits	26030	4,42	It is formed from precipitation brought by the waters of rivers or mixed river-sea.
	5	Acid sulphate	25305	4,29	It is formed on lands containing toxic substances: iron, aluminum and sulfuric acid residues. They formed about 10,000 years ago.

From the data provided, it can be seen that in the northern forest vegetation zone, yellow-red sandy soil predominates (61.91%), in the central zone, brown-yellow soil (60.30%), and in the southern zone - yellow-red clayey soil (48.46%).

Yellow-red sandy soil is formed under the canopy of moist tropical forests. The layer of organic matter accumulation is usually thin, the organic matter content in the soil is low, and fulvic acid often predominates in the humus composition. In the B horizon of the soil, there is an accumulation of iron and aluminum oxides, creating an overall red-yellow color. The content of primary minerals is low, except for resistant minerals (quartz, kaolin). The soil is acidic, poor in humus, often dry, with a low nutrient content. It is easily washed away and degraded.

Brown-yellow soil contains a high content of organic matter, especially humus, and mineral substances such as potassium, phosphorus, and nitrogen. Humus retains moisture and provides plants with nutrients. The soils are neutral or slightly alkaline. Decomposition of substances, especially through humification, helps to increase the organic content in the soil.

Red-yellow clayey soil creates conditions for mineralization processes and the formation of organic matter. The presence of humus and mineral substances such as iron oxides gives the soil a red-yellow color, but the decomposition of organic matter and its transformation into mineral compounds affect the soil color.

The conditions for the formation of red-yellow soils are associated with a warm climate, sufficient precipitation, rich organic matter, and mineral components, as well as mineralization and transformation of organic matter. The interaction of these factors plays an important role in the formation of red-yellow soils.

Based on the data provided above, indices of factors such as climate, vegetation period, and stand productivity were calculated using the model of H. Peterson, and the obtained climatic indices of H. Peterson are presented in Table 6.

Table 6. Values of the H. Peterson index of forest vegetation zones in Vietnam (Authors' development)

Forest vegetation zone	Values of the H. Peterson index		
	Minimum	Maximum	Medium
Northern (monsoon tropical zone)	1755.25	1988.57	1839.37
Central (middle tropical zone)	2323.33	3358.44	2823.92
Southern (tropical zone)	3770.00	6820.00	5175.71

The location of forest vegetation zones in Vietnam is shown in Fig. 4. The warm and humid climate of Vietnam contributes to the development of diverse forest ecosystems. Almost a third of the country's territory is covered by tropical and subtropical forests. In the moist deciduous forests, up to 300 species of woody plants and shrubs grow. The foothill regions of Vietnam are used for growing food crops (Dao et al., 2022). Plains are occupied by agricultural crops, with bamboo and palm groves located nearby. Mangrove forests grow on coastal plains, which are flooded by sea water during high tides (Pham et al., 2019). Analyzing the data, it can be concluded that the climatic conditions of the country determine the type of plant communities, their productivity, and the growth of woody biomass. However, each part of the country is characterized by different forest conditions, which determine the species composition, productivity (site quality class), and soil formation technology (e.g., Khlyustov et al., 2022; Sanches et al., 2018).

ORGANIC MASS GROWTH OF FOREST ECOSYSTEMS

The main indicator of the forest ecosystem's role in improving the environment is the indicator of the increase in its organic mass, which characterizes the development and condition of woody plantations. The increase in organic matter varies significantly across different climatic zones, but forests perform ecological functions (Figure 5) on disturbed forest lands due to the extraction and processing of mineral resources in Russia (Bykowa et al., 2023) and in Vietnam. The forest ecosystem, through the increase in organic mass, provides the country's economy with timber, non-timber products, and performs environmental functions: it regulates the balance of carbon dioxide and oxygen in the atmosphere, absorbs dust and harmful substances, and maintains thermal, light, and water regimes.

Figure 4. The scheme of the location of forest vegetation zones in Vietnam

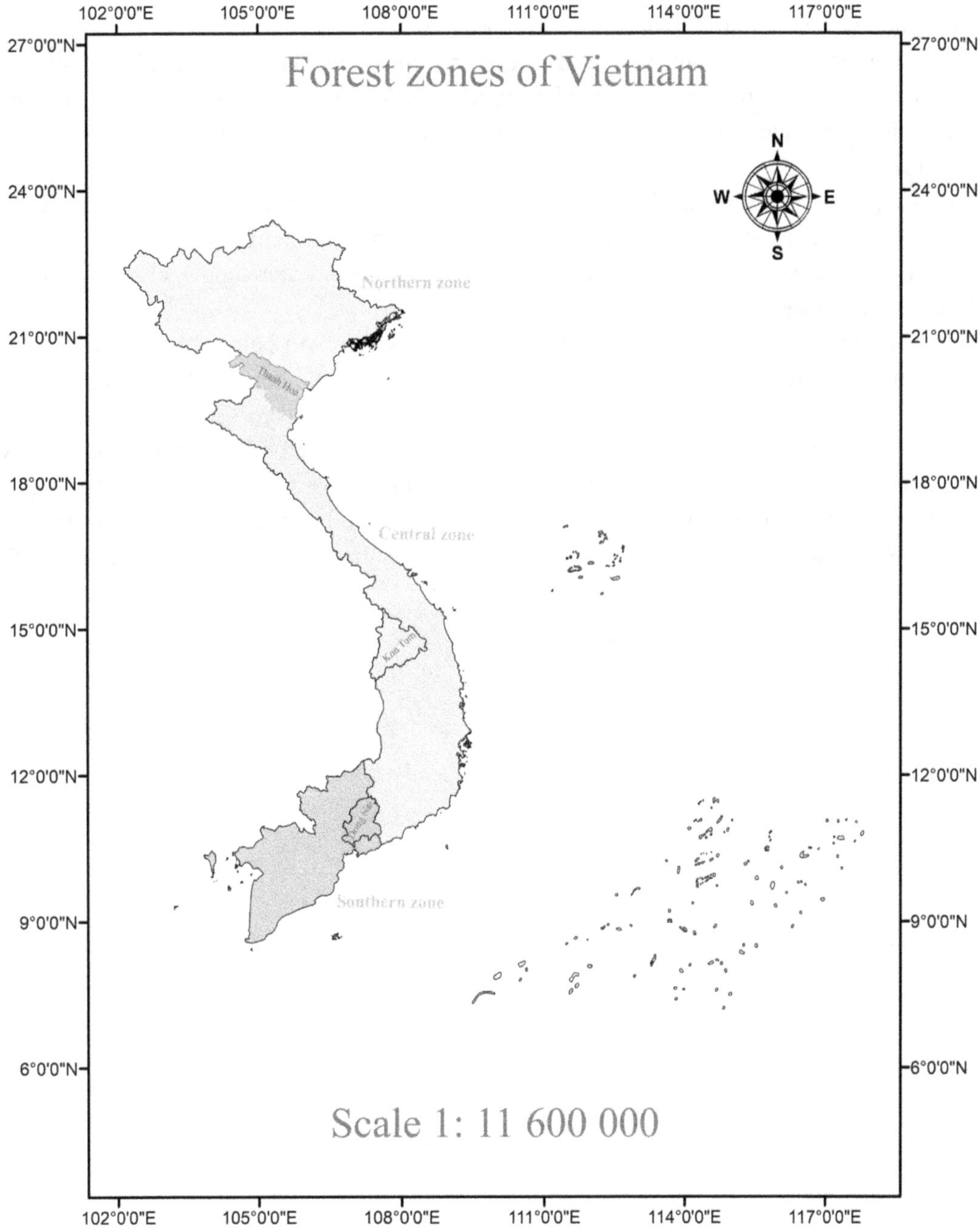

Source: Authors' development

The growth indicator represents an increase in biomass, volume, area, or number of forest resources over a certain period of time. The growth of forest stock is one of the key indicators in monitoring the forest fund and is important in several aspects:

1. Sustainable use of forest resources: Forest stock growth allows assessing the sustainability of forest growth and utilization processes. This indicator helps determine how effectively the forest is being used and regenerated after logging or other forms of use.
2. Assessing the potential contribution of forests to climate change mitigation: Trees absorb carbon from the atmosphere through photosynthesis and store it in their biomass. Forest stock growth allows estimating how much more carbon can be retained in the forest stock and reduce its emissions into the atmosphere.
3. Assessing ecosystem health: Forest stock growth is an indicator of the forest ecosystem's health. It allows evaluating the health and viability of the forest ecosystem (Skachkova & Gureva, 2023), as well as its potential to provide ecosystem services such as water supply, climate regulation, and biodiversity conservation.
4. Assessing the potential of forest resources: Forest stock growth helps assess future opportunities for using forest resources. It allows determining how much timber will be available in the future and predicting potential needs for forest resources in various sectors such as construction, furniture production, paper production, and others. Thus, the growth of forest stands plays a key role in understanding the state and management of the forest fund, as well as in forecasting its future possibilities using machine learning (Bykowa et al., 2022), and in preserving the environment through the use of biological methods, utilizing microorganisms: bacteria, fungi, and microalgae (Sozina & Danilov, 2023).

Figure 5. Forest ecosystem products

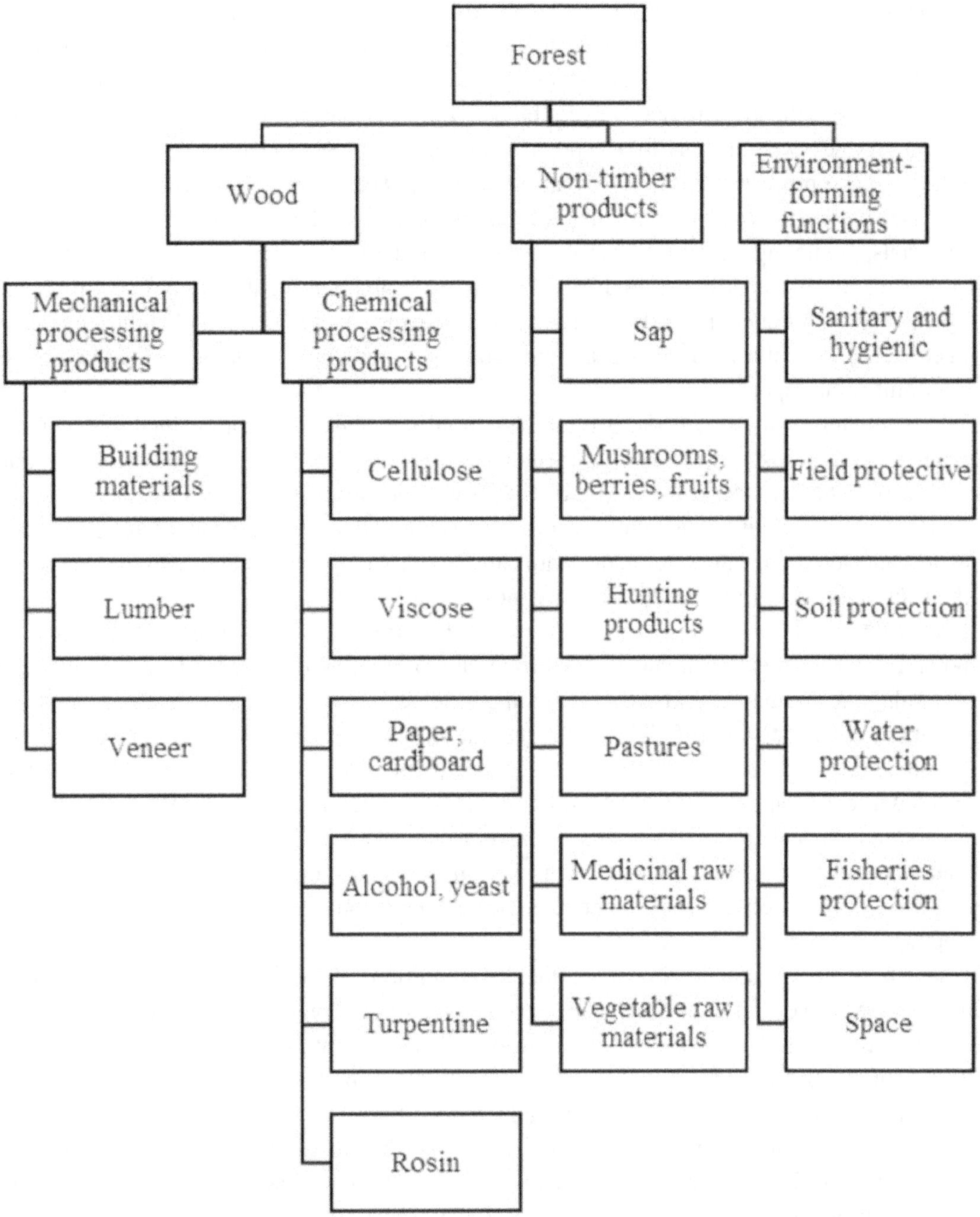

Source: Authors' development

The increase in timber stock in the forest is a key indicator in monitoring the forest fund in the following aspects. Firstly, it allows distinguishing the stability of growth, development, and regeneration processes of forest ecosystems. Secondly, it determines the contribution of the forest ecosystem to climate regulation by absorbing carbon from the atmosphere (Soeprobowati et al., 2023) during photosynthesis and maintaining the oxygen ratio within 21%. This reduces carbon emissions into the atmosphere from industrial enterprises. Thirdly, the increase in stock is used to assess the sanitary condition of the natural ecosystem and the growth rates of tree stands within it. Fourthly, the growth in stock determines ecosystem services: water supply for the population, regulation of wind speed, absorption of moisture from the soil, and preservation of ecosystem biodiversity. The increase in timber stock helps identify future opportunities for obtaining and using forest resources: timber, mushrooms, berries, medicinal raw materials, predicting their use for various sectors of the economy: construction, furniture manufacturing, musical instruments, paper production, and more (Rahmoune et al., 2024). Growth is necessary for forecasting the productivity and stability of forest ecosystems to natural and anthropogenic factors (Zhou et al., 2022); it allows managing the forest fund and predicting its future application possibilities and environmental impact.

Forestry management activities such as forest care cuttings, soil cultivation during forest culture establishment, mineral fertilizer application, and pruning of branches on prospective specimens can contribute to an increase in timber stock growth. The use of growth indicators in forestry serves several purposes:

1. Assessment of the current state of forest resources: Growth allows for evaluating the effectiveness of forest resource management and identifying problematic areas where measures are needed for the restoration or improvement of forest stands.
2. Forecasting the future development of forest resources: Growth analysis enables the prediction of potential forest development and the determination of optimal management strategies, such as determining the optimal proportion of tree felling in selective cuttings in main use areas, and selecting suitable tree species for artificial forest regeneration based on climatic and forest conditions.
3. Monitoring changes in forest resources: Regular growth measurements allow tracking the dynamics of growth and changes in the composition of stands, which is beneficial for the forest service in assessing the results of forest resource management activities.

Thus, indicators of organic mass growth play an important role in forest resource management, helping to identify effective management strategies and contributing to the sustainable and productive development of forest ecosystems.

Research by various authors has shown that the conditions of forested areas are determined by the hydrological regime of the land, which depends on the amount of precipitation in the area, the intensity of evaporation, and the granulometric composition of the soil (e.g., Fragou et al., 2020; Prasad et al., 2022). Scientists have used forest stations located in different parts of the globe to study the potential productivity of forest ecosystems, which is characterized by the annual growth increment per hectare (Table 7). To reflect the correlation between climate and the productivity of forest ecosystems, some indicator is needed. The indicator of organic mass growth is used to assess the health and growth rates of forest stands, as well as for predicting the productivity and resilience of forests to natural and anthropogenic factors.

Table 7. Biomass and primary production of major forest ecosystems (Authors' development)

Ecosystem Name	Biomass of the system, t/ha	Primary production, t/ha
Deciduous forest	100-250	10-30
Mixed forest (Humid)	500-1500	25-60
Rainforest (Coral reef)	15-50	50-120

The data provided indicate a greater biomass in the humid tropical ecosystem, and accordingly, the highest wood growth is observed there.

CONCLUSION

The research provides the geographical location and relief of Vietnam to assess the influence of climatic factors: solar energy, temperature regime, precipitation amount, duration of the growing season, and soil quality on the development and productivity of forest ecosystems. To calculate the solar energy supplied to the forest, the astronomical dependence of the Sun's height on the time of year was used, determining the Julian date. Based on this date, the illumination time and the angle of inclination of the Sun to the plane of the Earth's surface were calculated. The assessment of climatic factors was carried out using the model (index) of H. Peterson, allowing the identification of forest zones in Vietnam. The Northern, Central, and Southern zones differ in productivity, organic mass growth, and species composition of forest resources. Forest zones are also necessary for monitoring the forest fund and assessing the role of forest ecosystems in improving the environment, reflecting the characteristics of the development of plantations and forest soils and the condition of tree species. H. Peterson's climatic indices vary significantly across the forest zones of Vietnam. In the northern part, the average index value is 1839.37, in the center – 2823.92, and in the south – 5175.71. Favorable conditions for the development of forestry and the formation of organic mass growth of stands are present in the southern part of the country. Favorable environmental conditions for the external environment are formed in southern Vietnam. Climatic conditions of forest zones shape different soil types. In the northern zone, yellow-red sandy soil predominates, in the central zone – brown-yellow, and in the southern zone – yellow-red clayey soil. Climatic soil conditions determine various plant communities. The results of the research serve as the basis for land monitoring and the development of an effective forest industry strategy aimed at achieving economic goals and ensuring the sustainable development of forest ecosystems.

REFERENCES

About calendars and chronology. (n.d.) Kalendari. http://calendar.orthlib.ru/Kalendari.pdf

Baldoni, N., Francioni, M., Trozzo, L., Toderi, M., Fornasier, F., D'Ottavio, P., Corti, G., & Cocco, S. (2023). Effect of wood gasification biochar on soil physicochemical properties and enzyme activities, and on crop yield in a wheat-production system with sub-alkaline soil. *Biomass and Bioenergy*, 176, 106914. 10.1016/j.biombioe.2023.106914

Baltyzhakova, T. L., & Romanchikov, A. Y. (2021). Spatial analysis of subway passenger traffic in Saint Petersburg. *Geodesy and Cartography (Vilnius)*, 47(1), 10–20. 10.3846/gac.2021.11980

Bykowa, E., Skachkova, M., Raguzin, I., Dyachkova, I., & Boltov, M. (2022). Automation of Negative Infrastructural Externalities Assessment Methods to Determine the Cost of Land Resources Based on the Development of a "Thin Client" Model. *Sustainability (Basel)*, 14(15), 9383. 10.3390/su14159383

Bykowa, E. N., Khaykin, M., Shabaeva, Y. I., & Beloborodova, M. D. (2023). Development of methodology for economic evaluation of land plots for the extraction and processing of solid minerals. *Journal of Mining Institute*, 259, 52–67. 10.31897/PMI.2023.6

Dao, L. T. A., Nguyen, T. A., & Chandio, A. A. (2022). Climate change and its impacts on Vietnam agriculture: A macroeconomic perspective. *Ecological Informatics*, 74, 101960. 10.1016/j.ecoinf.2022.101960

Fragou, S., Kalogeropoulos, K., Stathopoulos, N., Louka, P., Srivastava, P. K., Karpouzas, S., Kalivas, D. P., & Petropoulos, G. (2020). Quantifying Land Cover Changes in a Mediterranean Environment Using Landsat TM and Support Vector Machines. *Forests*, 11(7), 750. 10.3390/f11070750

Gunathilaka, R. P. D., James, C. R. S., & Christopher, M. F. (2018). Adaptation to climate change in perennial cropping systems: Options, barriers and policy implications. *Environmental Science & Policy*, 82, 108–116. 10.1016/j.envsci.2018.01.011

He, D., Yang, R., & Cao, J. (2023). The relationship between interannual variability of winter surface solar radiation over the Southeast Asian low-latitude highlands and the circumglobal teleconnection. *Atmospheric Research*, 287, 106694. 10.1016/j.atmosres.2023.106694

Hoang, C. T., Bui, T. P., & Duong, V. H. (2021). Species biodiversity and distribution of seagrass beds in several coastal areas of central Vietnam. *Regional Studies in Marine Science*, 41, 101531. 10.1016/j.rsma.2020.101531

Kasten, F., & Young, A. T. (1989). Revised optical air mass tables and approximation formula. *Applied Optics*, 28(22), 4735–4738. 10.1364/AO.28.00473520555942

Khan, A., & Sudheer, M. (2022). Machine learning-based monitoring and modeling for spatio-temporal urban growth of Islamabad. *The Egyptian Journal of Remote Sensing and Space Sciences*, 25(2), 541–550. 10.1016/j.ejrs.2022.03.012

Khlustov, V. K., Vasenev, V. V., & Ganokhin, A. M. (2022). The importance of climatic indicators on the productivity of pine forests. *Environmental Management*, 5, 121–131.

Kopylova, N., & Starikov, I. P. (2021). Methods of displaying geospatial information using cartographic web technologies for the arctic region and the continental shelf. *Geodezia i Kartografia*, 971(5), 15–22. 10.22389/0016-7126-2021-971-5-15-22

Kornilov, Y. N., Romanchikov, A. Y., & Bogoliubova, A. A. (2023). Estimating deformation process through single shot close range photogrammetry method in Agisoft Metashape. *Geodezia I Kartografia*, 84(10), 2–11. 10.22389/0016-7126-2023-1000-10-2-11

. Kovyazin, V. F., Kozodaev, A. N., Nguyen, T. X., & Dang, T. L. A. (2022). *Monitoring of land resources of the Socialist Republic of Vietnam*. Lan Publishing House.

Kovyazin, V. F., Nguyen, T. A., & Nguyen, T. T. (2023). Monitoring the forest fund lands of Kon Tum province, Vietnam using remote sensing data of Earth. *Geodezia I Kartografia*, 84(8), 57–64. 10.2238 9/0016-7126-2023-998-8-57-64

Liu, Z., He, S., Men, W., & Sun, H. (2024). Impact of climate risk on financial stability: Cross-country evidence. *International Review of Financial Analysis*, 92, 103096. 10.1016/j.irfa.2024.103096

Matrokhina, K. V., Trofimets, V. Y., Mazakov, E. B., Makhovikov, A. B., & Khaykin, M. M. (2023). Development of methodology for scenario analysis of investment projects of enterprises of the mineral resource complex. *Journal of Mining Institute*, 259, 112–124. 10.31897/PMI.2023.3

Mees, J. (1988). *Astronomical formulas for microcalculators*. Jean Meyos.

Ministry of Construction of Russia. (2020). *SP 131.13330.2020. SNiP 23-01-99* Construction climatology*. Ministry of Construction of Russia.

Nascimento, C. M., Mendes, W. S., & Silvero, N. E. Q. (2020). Soil degradation index developed by multitemporal remote sensing images, climate variables, terrain and soil atributes. *Journal of Environmental Management*, 277, 111316. 10.1016/j.jenvman.2020.11131632980636

Nguyen, T. X. (2020). *Monitoring of land in the Socialist Republic of Vietnam for cadastral registration in the context of changing market relations* [Dissertation, Saint-Petersburg Mining University]. Saint-Petersburg Mining University.

Pandul, I. S. (2010). *Geodetic astronomy in relation to solving engineering and geodetic problems*. Politekhnika.

Pashkevich, M. A., Danilov, A. S., & Matveeva, V. A. (2021). Remote sensing of chemical anomalies in the atmosphere in influence zone of korkino open pit coal mine. *Eurasian Mining*, 35(1), 79–83. 10.17580/em.2021.01.16

Pham, T. D., Yokoya, N., Bui, D. T., Yoshino, K., & Friess, D. A. (2019). Remote Sensing Approaches for Monitoring Mangrove Species, Structure, and Biomass: Opportunities and Challenges. *Remote Sensing (Basel)*, 11(3), 230. 10.3390/rs11030230

Pham, V. T., Do, T. A. T., Tran, D. H., & Do, T. N. A. (2023). Classifying forest cover and mapping forest fire susceptibility in Dak Nong province, Vietnam utilizing remote sensing and machine learning. *Ecological Informatics*, 79, 102392. 10.1016/j.ecoinf.2023.102392

Prasad, P., Loveson, V. J., Chandra, P., & Koth, M. (2022). Evaluation and comparison of the earth observing sensors in land cover/ land use studies using machine learning algorithms. *Ecological Informatics*, 68, 101522. 10.1016/j.ecoinf.2021.101522

Rahmoune, M., Radjef, M. S., Boukherroub, T., & Carvalho, M. (2024). A new integrated cooperative game and optimization model for the allocation of forest resources. *European Journal of Operational Research*, 316(1), 329–340. 10.1016/j.ejor.2024.01.018

Roitsch, D., Abruscato, S., Lovrić, M., Lindner, M., Orazio, C., & Winkel, G. (2023). Close-to-nature forestry and intensive forestry – Two response patterns of forestry professionals towards climate change adaptation. *Forest Policy and Economics*, 154, 103035. 10.1016/j.forpol.2023.103035

Rosenberg, G. V. (1963). *Twilight. M.* Fizmatgiz.

Sanches, I. D., Feitosa, R. Q., Achanccaray, P., Montibeller, B., Luiz, A. J. B., Soares, M. D., Prudente, V. H. R., Vieira, D. C., & Maurano, L. E. P. (2018). Lem benchmark database for tropical agricultural remote sensing application. *The International Archives of the Photogrammetry, Remote Sensing and Spatial Information Sciences*, XLII-1, 387–392. 10.5194/isprs-archives-XLII-1-387-2018

Schulte, M., Jonsson, R., Eggers, J., Hammar, T., Stendahl, J., & Hansson, P. (2023). Demand-driven climate change mitigation and trade-offs from wood product substitution: The case of Swedish multi-family housing construction. *Journal of Cleaner Production*, 421, 138487. 10.1016/j.jclepro.2023.138487

Seleshnikov, S. I. (1970). *History of the calendar and chronology*. Science.

Skachkova, M., & Guryeva, O. (2023). Monitoring of the State of Saint Petersburg Green Spaces by Remote Sensing Data. *Ecology and Industry of Russia*, 27(5), 51–57. 10.18412/1816-0395-2023-5-51-57

Soeprobowati, T. R., Sularto, R. B., Hadiyanto, H., Puryono, S., Rahim, A., Jumari, J., & Gell, P. (2023). The carbon stock potential of the restored mangrove ecosystem of Pasarbanggi, Rembang, Central Java. *Marine Environmental Research*, 193, 106257. 10.1016/j.marenvres.2023.10625738086137

Sozina, I. D., & Danilov, A. S. (2023). Microbiological remediation of oil-contaminated soils. *Journal of Mining Institute*, 260, 297–312. 10.31897/PMI.2023.8

Talovina, I. V., Krikun, N. S., Yurchenko, Y. Y., & Ageev, A. S. (2022). Remote sensing techniques in the study of structural and geotectonic features of iturup island (the kuril islands). *Journal of Mining Institute*, 254, 158–172. 10.31897/PMI.2022.45

Tooming, H. G. (1977). *Solar radiation and crop formation*. Gidrometeoizdat.

Ushakov, E., Aleksandrova, T., & Romashev, A. (2021). Neural Network Modeling Methods in the Analysis of the Processing Plant's Indicators. *Advances in Intelligent Systems and Computing*, 1259, 36–45. 10.1007/978-3-030-57453-6_4

Wang, Q., Robson, T. M., Pieristè, M., Kenta, T., Zhou, W., & Kurokawa, H. (2022). Canopy structure and phenology modulate the impacts of solar radiation on C and N dynamics during litter decomposition in a temperate forest. *The Science of the Total Environment*, 820, 153185. 10.1016/j.scitotenv.2022.15318535065130

Wiegant, D., Bakx, J., Flohr, N., Oel, P., & Dewulf, A. (2022). Ecuadorian water funds' use of scale-sensitive strategies to stay on course in forest and landscape restoration governance. *Journal of Environmental Management*, 311, 0301–4797. 10.1016/j.jenvman.2022.11485035278923

Zhang, L., Qi, S., Li, P., & Zhou, P. (2024). Influence of stand and environmental factors on forest productivity of Platycladus orientalis plantations in Beijing's mountainous areas. *Ecological Indicators*, 158, 111385. 10.1016/j.ecolind.2023.111385

Zhou, P., Zhang, H., Huang, B., Ji, Y., Peng, S., & Zhou, T. (2022). Are productivity and biodiversity adequate predictors for rapid assessment of forest ecosystem services values? *Ecosystem Services*, 57, 101466. 10.1016/j.ecoser.2022.101466

Chapter 4
Improving the Use of Land Resources in Tajikistan by Optimizing the Choice of Crops

Kurbonali Partoev
National Academy of Sciences of Tajikistan, Tajikistan

Safarmadi Mirzoali
Tajik State Pedagogical University, Tajikistan

Navruz Saidaliev
Tajikistan National University, Tajikistan

ABSTRACT

Tajikistan's position in the desert zone, arid climate and mountainous terrain determined the characteristics of its soil cover. The soils mainly belong to the following types: light gray soils, typical gray soils, and mountain carbonate soils. In the Republic of Tajikistan, the total land fund or total land area is 14255.4 thousand hectares, of which 3746.0 thousand hectares are farmland. Over the past 30 years, the state of land resources in the Republic of Tajikistan has deteriorated. The share of irrigated land in relation to the area of agricultural land/arable land decreased from 14.8% to 14.4%; area of arable land per capita – from 0.16 to 0.12 ha/person. The share of agricultural land in relation to the total territory decreased by 1%. At the same time, there are natural limitations on the area of land resources of 3746.0 thousand hectares of farmland. This actualizes the need for their effective and rational use. 90% of the area of rain-fed arable land shows signs of degradation, of which 40% are highly degraded. Additionally, intensive agricultural activities on mountain slopes inevitably lead to soil erosion. They are eroded, which, as the gorges grow, leads to a reduction in the area of arable soil. Slopes up to 250 m high are widely cultivated without anti-corrosion measures.

DOI: 10.4018/979-8-3693-3374-7.ch004

INTRODUCTION

The Republic of Tajikistan is a mountainous country, 93% of whose territory is occupied by mountain ecosystems. The vertical zonation where it is possible to grow crops extends at altitudes from 350 to 4000 m above sea level. Tajikistan is rich in such important natural resources as water energy, the number of sunny days throughout the year (more than 300 sunny warm days a year), biodiversity (more than 23 thousand species of flora and fauna), etc.

The rarest species of animals and birds live here, listed in the Red Book: mountain goats, snow leopards, storks, hawks and others. The largest nature reserves in the country: Tigrovaya Balka, Ramit, Dashtidzhumsky and Sarikhosorsky reserves.

Figure 1. Rasht Valley of Tajikistan

An important energy-intensive sector in the republic is the agricultural sector, which annually provides about 24% of the GDP (Gross Domestic Product) in the republic.

The length of the territory of the republic from west to east is 700 km, from north to south 350 km. Tajikistan is a mountainous republic. The absolute altitude above sea level ranges from 300 to 7495 meters. The total area of the republic's land fund is 14.3 thousand hectares. Of these, about 4.58 million hectares are in economic use (including 3.3 million hectares of pasture land)(Figure 1).

The soil cover of the republic consists of: plain-low mountain, with gray soils; mid-mountain, with mountain browns; alpine, with high-mountain meadow-steppe, steppe, desert-steppe, zang, desert soils; nival, unformed skeletal soils in cracks of high-mountain rocks, hollows, etc. Main types of soil in Tajikistan: clayey; loamy; limestone; sandy; sandy loam and swampy.

Figure 2. Soil layer section in Vakhsh district, Tajikistan

Tajikistan's position in the desert zone, arid climate and mountainous terrain determined the characteristics of its soil cover. It is dominated by gray soils - the main zonal type (Figure 2).

Taking into account the growing population of the republic, scientists are faced with the task of finding and implementing innovative ways and methods of obtaining the maximum volume of necessary products and biological mass per unit area while maintaining and increasing the fertility and quality of land and soil resources.

In the republic, the total land fund or total land area is 14255.4 thousand hectares, of which 3746.0 thousand hectares are farmland, 673.1 thousand hectares are arable lands, 2909.8 thousand hectares are pastures, etc. . Over the past 30 years, the state of land resources in the Republic of Tajikistan has deteriorated.

The share of irrigated land in relation to the area of agricultural land/arable land decreased from 14.8% to 14.4%; area of arable land per capita – from 0.16 to 0.12 ha/person. The share of agricultural land in relation to the total territory decreased by 1% (http://www.cawater-info.net/ecoindicators/pdf/land-tj.pdf).

At the same time, there are natural limitations on the area of land resources of 3746.0 thousand hectares of farmland. This actualizes the need for their effective and rational use. 90% of the area of rain-fed arable land shows signs of degradation, of which 40% are highly degraded. Of the irrigated arable lands, 22% of the area is highly degraded, 38% shows signs of low and moderate degradation, and 40% shows no signs of degradation. Degradation of forest plantations affects about 70% of the land area, of which 22% are classified as highly degraded, 30% of the area is classified as low and moderate, 48% do not show signs of degradation.

Available estimates indicate the dependence of the agricultural sector on the state of land and soils and the need for special attention to the choice of approaches to the use of agroclimatic resources and crops (https://www.unpei.org/files/pdf/the_economic_cost_of_agricultural_and_land_degration_in_tajikistan_in_russian.pdf).

Intensive agricultural activities on mountain slopes inevitably lead to soil erosion. They are eroded, which, as the gorges grow, leads to a reduction in the area of arable soil. Slopes up to 250 m high are widely cultivated without anti-corrosion measures.

Jerusalem artichoke (Tuberous sunflower, *Helianthus tuberosus* L.) is a sweet and highly nutritious root crop, like sunflower, it belongs to the Asteraceae family, but is distinguished by the formation of tubers of various sizes, shapes and colors in the underground part. Jerusalem artichoke in Russia is called “earthen pear”, “wild sunflower”, “sun root”, in Europe “Jerusalem artichoke”, in Kazakhstan “Chinese potato”.North America is considered the homeland of Jerusalem artichoke, where it grows wild and was introduced into culture by the Indians before Europeans arrived there [Pasko, 1999; Filipov, 1970; Gibbons et.al.,1989; Klaussofer, et.al., 1998; Praz et.al., 1987].

The plant came to Europe through France, where it received the name "Jerusalem artichoke" (from the name of the Brazilian Indian tribe - Jerusalem artichoke). Jerusalem artichoke came to Russia at the beginning of the 18th century [Eikhe, 1957; Shazzo et.al, 2013; Shain, 1931; Staravoitov, et.at.2002; Sumin et.al., 2008; Swanton et. al., 2005; Titok et.al., 2018].

HISTORY OF ORIGIN TOPINAMBOUR

Topinambour, is more correct - sunflower tuberiferous (Helianthus tuberosus L.) - family Astrovye perennial plant (Asteraceae). Along with sunflower olive (Helianthus annuus), it concerns the most known and significant representatives of a sort sunflower (Helianthus). Being the near relation of sunflower, topinambour (often named also «an earthen pear») represents long-term tuberiferous a family plant Compositae, possessing powerful root system and reaching in height to 3-4 metres.

The native land of topinambour - the North America where it grows in a wild kind. For the first time root crops of an earthen pear were used in food from outside by the Brazilian Indians - of an American Indian tribe «tupinamber», using in food tubers topinambour. The name «topinambour» has been given a plant in a random way when in a XVII-th century Indians of the Brazilian tribe of Tupinambus have been brought to France. Tubers of an unknown plant which began to call a word «topinambour» - derivative of the tribe name have been found out In them.

To Europe topinambour has got in the XVII-th century beginning thanks to the traveller and pioneer Samjuelju де to Shamplenu, which have sent tubers topinambour from Canada to France and for the first time compared them on flavouring qualities with an artichoke (and most likely thanks to such comparison since then topinambour in Europe it is more known as «the Jerusalem artichoke»). Since then for more than 500 years cultivation of rather unpretentious growth to conditions topinambour was widely adopted in many countries of Europe (and together with the increased popularity at Europeans topinambour also has received also other names among which - «the Canadian potato», «jirasol», «a solar root»). In Europe and to this day from cultural grown up topinambour make fructose, spirit, liquor, wine and beer, and also tubers of this plant throughout many years are traditional components of dishes Belgian, Dutch, French, German and English cuisine . [Usanova et.al., 2003;Usanova, 1967; 2005; Usanova et.al., 1993].

Thanks to the active scientific researches of biochemical structure developed recently and treatment-and-prophylactic properties topinambour interest to this unique dietary foodstuff has substantially increased in Canada, the USA, Brazil, China, India, Japan, Russia, Belgium, Holland, Germany, Hungary etc.). And received from topinabour and irreplaceable for sick of a diabetes inulin has been officially recognised in many countries of the world by absolutely safe for the person by a food component. Topinambour in the World cultivated in more than 2.5 million hec.

TOPINAMBOUR IN TAJIKISTAN

Under the message of the doctor of Litvinov V. N (1958) topinambour to Tajikistan has been brought by migrants from the Russian Federation as early as the twentieth years of the XX-th century. Then basically it was grown up as an ornamental plant from the population.

In 1950 A.P.Saverkin has obtained the important scientific data about good growth topinambour on a number with other food cultures in a mountain zone to the Gissar valley at height about 3000 m above sea level (Litvinov, 1965). In 1953-1955 in the conditions of the Gissar valley of our republic on an experimental site special experiments on studying topinambour have been made. Those years to collective farms and state farms was роздано on 5000 kg tubers topinambour for cultivation and reproduction.

Within 1956-1958 topinambour it was grown up in the conditions of Gissar, and Vahsh valleys, in area Garma, in Leninabad area and in the Western Pamir. Then from crops topinambour in collective farms, state farms and on skilled sites it has been received on the average on 500 s a biomass and on 250 tubers about 1 hectare.

In conditions areas-Ishkashim, Rashtkalin and Shugnan about 1 hectare topinambour it has been received on 500-600 sentere green mass weights and on 150-200 sentere tubers (Litvinov, 1958; 1965).

The conducted researches by the Tadjik scientist A. Ergashev (1976) in the conditions of high mountains (village Siyohkuh of Varzab area) have shown high factor of an economic crop topinambour (K_{hoz}.) at an irrigation 0.37 - 0.41 and in conditions rein field 0.33 - 0.39 that it testifies to high intensity of photosynthesis of this plant. Our research have shown that till now in many areas of republic topinambour is grown up from the population as decorative culture (Partoev et. al., 2015).

Botanical Characteristics of Topinambour

The elevated part topinambour on appearance reminds sunflower, the downy stalk branched out on a top in length from 1,5 to 3 m; the bottom leaves of heart-shaped-jajtsevidnoj form, top prodolgovato-jajtsevidnye, large, length 10-20 sm and width 5-12 sm, on edge gear, petiolar, dark green colouring; the inflorescence - a basket, diameter 1-1,5 sm, consists from ложноязычковых yellow цветков. Flowering later, in September-October. A fruit - семянка. The root system rod, gets on depth to 2 m, forms numerous underground runaways (столоны) on which tubers are formed from 10-30 and more. Under the form tubers can be pear-shaped, веретеновидными, roundish, with strongly pronounced convex kidneys-eyes, colouring varies from white and yellow colour to rose-red and violet (mainly pulp white, and a skin zheltovato-brown). Weight of tubers from 10 to 150 g depending on a grade, a zone of cultivation and agricultural technicians. Now by selectors are deduced more than 300 grades topinambour.

Tubers topinambour contain in the structure: inulin (11-17%), fructose, фруктоолигосахариды, amino acids (to 8%) (including 8 - irreplaceable amino acids - arginin, gistidin, valin, leicin, izilecin, lisiн, triptofan, methionine and fenilalalin), karotinoids, vitamins B_1, B_2, B_3 (PP), B_6, B_9, With, pectins (to 10%), organic acids (lemon, malinovaya, apple, amber, fumarovya, fat acids (0,4%-0,7%), nitrogenous substances, cellulose (to 6%), and also rather wide set of makro у microcells (potassium, sodium, magnesium, iron, phosphorus, manganese, calcium, silicon, copper, zinc, sulphur, chrome, iodine, a pine forest, aluminium, cobalt, molybdenum, fluorine, etc.). By quantity of vitamins C, B_1 and B_2 topinambour more than in 3 times surpasses a beet, carrots and a potato [Ushakov, 1932;Ustimenko,1960;Ustimtnko-Bakumovsky, 1972; Wieczochek, 1988Yuritsina, 1988].

Topinambour as Food and Feed Crops

Now topinambour it is cultivated in many countries of the world. The earthen pear essentially surpasses a sugar beet, corn and a potato in productivity, and on food value tubers topinambour are in the lead among the majority grown up on territory of Russia of vegetables. Also topinambour often land for increase of fertility of agricultural grounds, for protection of crops against winds, and also sometimes use for decorative gardening of city parks and rest zones (Figure 3).

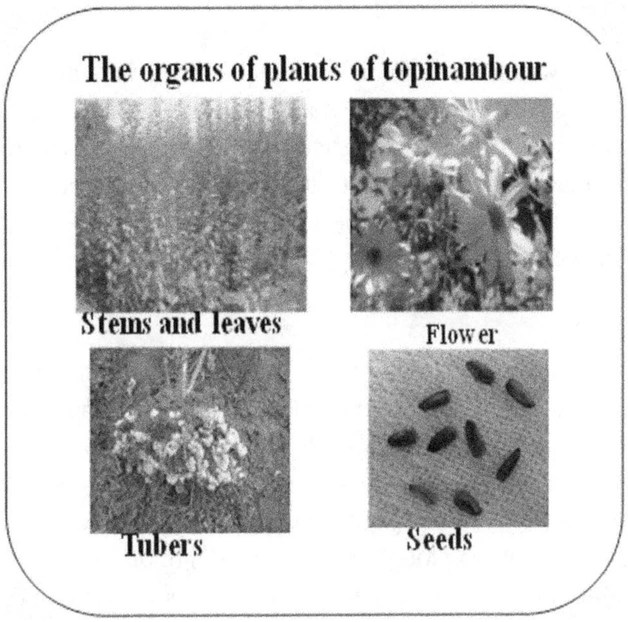

Containing in the structure a considerable quantity of pectins tubers topinambour are used in fruit candy manufacture, jelly, jam and jams for dietary and baby food. Besides, topinambour already throughout many tens years application in the bakery, meat, dairy and canning industry finds, is used in manufacture alcoholic and soft drinks, in manufacture of various biologically active additives. In pharmaceutical industry tubers topinambour are used basically as a source inulina.

But it is traditional in cookery of many countries of the world tubers topinambour already throughout many centuries find more various application. It is necessary to note, as at such various kinds of culinary processing root crops topinambour not only keep substantially the food value and weight of curative properties, and also get more sated taste and aroma in comparison with fresh tubers of this plant.

Fried and baked tubers topinambour use as a part of stuffings for pies, pies and fritters. From an earthen pear prepare also creams-soups, kissels, jam, compotes, kissels, kvass, mashed potatoes, cutlets, fritters, baked puddings, various garnishes for the second dishes.

In the USA and other countries of the world the crushed tubers topinambour from which prepare tasty dietary coffee enjoy wide popularity dried, and then.

House batch, in which preparation the powder topinambour is used (the dried up and crushed tuber topinambour) long does not harden (it it is connected by that fructose a part to an earthen pear promotes long deduction of a moisture by the baked bakery products).

It is necessary to notice what to clean a tuber topinambour because of its freakish, knotty form difficult enough (for simplification of this procedure is better to scald preliminary topinambour on a frying pan and then to clean off from it a peel thus, as usual peel young potatoes).It is recommended for preservation of the maximum advantage of a tuber topinambour only carefully to clear it of the soil rests under flowing water by means of a rigid brush and to use this root crop in the culinary purposes directly with a peel representing really a well so necessary organism to person of silicon, iron and other useful substances.

RECOMMENDATIONS ABOUT CULTIVATION OF TOPINAMBOUR

Rotation. In connection with that, topinambour is a perennial plant and in one place can successfully grow within 8-10 years it usually grow up out of a crop rotation. But, for its normal growth and development by its predecessors can be almost all agricultural crops. On soils where such plants as a lucerne, bean and a number of forage crops productivity topinambour will be more than were grown up when grain, vegetable and olive cultures (flax, sesame, sunflower, safflower and others) will be predecessors. Because topinambour concerns to family complex, it is desirable not to plant it after sunflower crops. Besides, topinambour unlike other cultures gives a good harvest green weights both on salted (pH=8-10) and in sour soils (pH=4-6).

Time of planting. Good terms of planting for topinambour are autumn (November, December). Also it is possible to plant it in February and in March. Landing terms also depends on district and a landing zone. In mountain areas at height more than 2000 m above sea level good term of planting is April. Tubers topinambour if put towards the winter (November, December) it is necessary to plant of 10-12 sm on depth, and at landing in the spring on depth 6-8 sm. If the long-term alfa-alfa is the predecessor topinambour, it is desirable to plough the earth on depth of 30-35 sm since autumn, and landing to spend in the spring that till the spring a part the rests of root systems of a alfa-alfa e well decayed. And at landing topinambour in the spring on the earths where other cultures were cultivated better to spend superficial processing early in the spring on depth 16-18 see In a mountain zone it is desirable to spend landing topinambour when the soil on depth of 10-12 sm heats up till 5-7 sm. As a whole in mountain areas topinambour it is possible to plant during the period from March, 15th till May, 15th. Depth of planting tubers in a mountain zone should be 6-8 sm, and at late terms of landing depth of landing can be 10-12 sm.

Seed material, schemes of planting and fertilizer entering at planting. As a planting material tubers with weight of 15-30 gram are recommended

Large tubers it is possible to cut, leave in each part on 2-3 eyes and to plant. The expense of a planting material on 1 hectare can be 700 1200 kg. The planting scheme topinambour should be 70 x 35-40 sm. Tubers topinambour it is possible to put with the help potato planting machinery or in the manual. Also it is possible to cut at first a furrow a cultivator, then manually to put in the middle of a furrow of tubers, also to bring fertilizers and then by means of a tractor and piers to level numbers. It considerably accelerates landing terms. At landing it is possible to bring all recommended norms phosphoric (100 kg/hectares in the form of operating substances) and potash (50 kg/hectares) fertilizers. Also it is recommended to bring 10-15 t/hectares of organic fertilizers at landing topinambour.

Interrow processings and fertilizer of plants. If planting topinambour has been spent in November or December, it is necessary to spend in the beginning March one cultivation processing of crops. This operation helps in destruction of weeds and preparationof soils and accelerations of shoots of plants.

This processing can be spent by means of cultivators, pets or to the manual. At carrying out of the given processing it is necessary to watch, that depth of processing was 12-14 sm and on distance of 10-12 sm from the crest centre.

The second processing of numbers can be spent already after occurrence of shoots when plants reach stalk heights 20-30sm and have more than 10-15 unit of leaves.

Simultaneously it is possible to combine with carrying out of this processing of entering of mineral fertilizers from calculation of nitric fertilizers of 100 kg/hectares in the form of operating substance (ammoniac saltpeter or urea). Also with carrying out of this processing it is necessary to spend and окучивание numbers.

Application of fertilizers. When topinambour it is grown up on poorer soils, it is desirable to bring on 1 hectare of 30-40 tons of organic fertilizers, 80-100 kg/hec. nitric fertilizers, 120-140 kg/hec. of phosphoric and 40-50 kg/hectares of potash fertilizers. Topinambour also it is very sympathetic to entering of ashes into soil and microcells.

Watering of topinambour. In mountain areas, annual norms of deposits makes 500-700 mm topinambour can give a good harvest without watering. And in valley areas it is necessary to water topinambour for vegetation of 5-7 times.

Experiences show that length рядков optimum should make 60-80 m. very important that the first vegetative watering is spent slowly and rational way.

Topinamboury it is necessary within 7-10 days once it is necessary to water. For vegetation it is necessary topinamboury to water 5-7 times. The expense of water during vegetation on 1 hectare makes approximately 1500-2500 $м^3$.

After the first hay crop (at height of 15-20 sm) entering of nitric fertilizers in number of 40-50 kg/hectares. After application of fertilizers it is necessary to water. Watering stop 10-12 days prior to harvesting.

Struggle against weed plants, illnesses and insects. Topinambour basically in the vegetation beginning can be muffled weed plants. Therefore it is recommended carrying out in the beginning of vegetation by carrying out once interrow processing. It can be spent a tractor, pets or to the manual. From illnesses basically there is a fungoid illness-mealy dew, and from wreckers it is possible to meet in soil worm, which leave in tubers a plant. Tubers also in a storage time can be amazed wet or dry by decay.

Harvesting and its storages. Biological weight of stalks, leaves topinambour where heat suffices clean in the beginning of July and in the end of October. This weight can be fed an animal, or will dry or use for manufacturing of a silo or сенажа. And tubers topinambour clean in October and keep in warehouses in soil, or in sand before their realization.

The temperature of storage of tubers topinambour should be - 0,5^{0}C or + 1^{0}C at humidity of air of 80-90%. The height tubers should be 0,5-1,0 m and from above a small layer of earth, grasses, sand, sawdust etc.

THE TOPINAMBOUR AS MEDICINE PLANT

The tuber topinambour is rich with carbohydrates 77% from which are necessary on a share polisakharid inulina, turning at long storage of an earthen pear in fructose. Thanks to the high maintenance of polymer of fructose inulina the tuber topinambour already throughout long time is really an irreplaceable and major component of a food allowance of the people sick of a diabetes. It is connected by that natural fructose, on 95% from which consists inulin, is capable unlike glucose not only to get into cages without participation of a hormone of insulin (and accordingly to provide cages with necessary energy), but also is high-grade "assistant" to glucose in the various exchange processes proceeding in a human body. Besides, molecules inulina, somewhat promote penetration into cages of a human body of glucose (for this reason containing in topinamboure inulin often name vegetative analogue of insulin).

The albuminous structure of an earthen pear is characterised by the high maintenance irreplaceable (not synthesised in a human body) amino acids among which the in the lead position is occupied with amino acid аргинин. Promoting escalating of muscular weight and increase of endurance amino acid аргинин plays an important role in development of hormones of growth, improves erectile function (raises a potentiality) and promotes immunity strengthening.

A part of tubers topinambour amino acid methionine appreciably strengthens clearing action on a human body of other biochemical components topinambour - insulin and клетчатки, plays an important role in natural synthesis of insulin, and also, as well as amino acid аргинин, interferes with development fatty infiltration to a liver.

Amino acid lysine with which root crops topinambour are rich, takes part in process кроветворения, natural development of the collagen, natural synthesis of enzymes, hormones and antibodies, and also promotes effective mastering necessary for bones and calcium joints.

As a part of an earthen pear also there are various organic acids (lemon, malonovмaya, apple, amber, fumarovya), in many respects defining flavouring qualities topinambour, rendering anti-oxidant and the bactericidal action, interfering adjournment in joints of salts of uric acid, and essentially improving digestion process (organic acids topinambour raise secretor function of salivary glands, strengthen allocation of bile and pancreatic juice, and also make active motor function of intestines).

On the average more than 8% of weight of a tuber topinambour are necessary on pectin's (differently - the pertinacious substances often named «soluble cellulose»). Possessing knitting and properties pectin's adsorb on the surface and deduce from a human body the various ballast substances promoting development of an atherosclerosis and gallstones of illness, salt of heavy metals, radionuclide, toxins of a various origin. Besides, pectin's, also as well as containing in topinamboure inulin, promote improvement intestinal and support normal balance of useful intestinal microflora.

Contains also in a tuber topinambour and rough insoluble cellulose, making active motorno-evakutornuju function of intestines and being excellent highly effective "cleaner" of intestines from every possible harmful substances. Insoluble, also as well as containing in an earthen pear inulin reduces sugar level in blood, interferes with development of locks and warns formation of stones in a gall bladder, and also substantially reduces risk of development of a dysbacteriosis, colitis, an enteritis, a chronic gastritis, diseases of a pancreas, an atherosclerosis, adiposity, allergic diseases, a diabetes of 2nd type, a hemorrhoids, oncological diseases of a thick gut, a mammary gland.

A part of root crops topinambour vitamin C renders anti-inflammatory, antiallergic, restorative and spasmolytic action, and also is the important participant of natural synthesis of collagen - the squirrel of the connecting fabric representing a structural basis of a skin, nails, hair, bones, vessels and various parts of articulate connections.

Containing in tubers topinambour group B vitamins take part in carbohydrate, vodno-salt, albuminous, fatty exchanges, in regulation of functions of digestive, nervous, muscular and cardiovascular systems, play an important role in haemoglobin synthesis, development of sexual hormones and synthesis of the hormones developed by adrenal glands, thyroid and pancreatic glands. Also as well as the vitamin A formed in a human body from containing in tubers topinambour carotinoids, group B vitamins promote strengthening of immunity and improvement of a condition of a skin, hair, nails and organs of vision.

Tubers topinambour, besides, are perfectly balanced on macro- and to microelement structure. In the maintenance of phosphorus, potassium, iron, silicon and zinc topinambour essentially surpasses such root crops as carrots, a beet and a potato. Potassium containing in an earthen pear plays an important role in regulation of a vodno-salt exchange, participates in transfer of nervous impulses to muscles, speeds up work of a cardiac muscle (myocardium).

Silicon (which maintenance in solid topinambour reaches 8%) is the powerful catalyst of fatty and albuminous exchanges, takes part in natural synthesis of various enzymes and hormones, and also a key role in high-grade mastering by an organism by the person more than 70 vitamins and mineral substances. This major microcell which is a part topinambour, also substantially stimulates process of natural development by a human body fibers of collagen and promoting that to strengthening of walls of blood vessels, cages of a skin, bone and to a fabric, improvement of a condition of nails and hair. Deficiency of silicon in a human body often leads to development of an atherosclerosis, a diabetes, a hypertension, diseases of a skin, teeth, eyes and joints, остепорозу, to a craw, a dysbacteriosis and urolithic illnesses, and also causes infringement of hormonal balance, a sleeplessness, chronic weariness and depression.

Containing in a tuber topinambour iron is the main participant of process of development by a fiber organism blood of the haemoglobin "responsible" for transport of oxygen from lungs in all cages of a human body. It is necessary to note, a part topinambour zinc, silicon, potassium, magnesium, manganese, phosphorus and chrome play an important role in natural development of insulin by a pancreas. And anti-inflammatory property of an earthen pear is substantially caused by presence at its structure of zinc and manganese.

Topinambour - irreplaceable foodstuff for sick of a diabetes and adiposity. Thanks to the high maintenance inulina a tuber topinambour and products made from it (the powder, broths, infusions, syrups) long since and traditionally are high on the list in a food allowance of 1st diabetes 1st and 2nd type. At a diabetes of 1st type (a diabetes) containing in topinamboure inulin (consisting on 95% from fructose) gets into cages of a human body without participation of a hormone of insulin, it is high-grade replacing glucose in exchange processes and appreciably satisfying instead of glucose power cellular hunger. Besides, fragments of molecules inulin somewhat facilitate penetration into cages and the glucose.

Not split by gastric juice inulin and containing in a tuber topinambour cellulose to adsorb a significant amount of glucose formed as a result of food intake, interfering its superfluous quantity in blood. Thanks to the above described complex action containing in topinamboure inulina and cellulose proof and stable decrease the glucose maintenance in blood at 1st type sick of a diabetes, not accompanied characteristic for a diabetes by sharp fluctuations of level of sugar in blood within days is observed, and accordingly, the daily requirement sick of a diabetes in insulin entered in the form of injections essentially decreases. In turn, stable decrease in level of glucose stimulates insulin development by special cages of a pancreas (natural synthesis of insulin also substantially is promoted a part of a tuber topinambour vitamins B_1 and B_3, carotinoids, by amino acids and methionine, zinc, silicon, potassium, phosphorus, magnesium, manganese, chrome) in blood. Containing in topinamboure pectins, inulin and insoluble connect and deduce from an organism harmful by-products of the metabolism broken as a result of a diabetes (acetone, etc.) . At a diabetes of 2nd type (diabetes) introduction in a food allowance topinambour and products on its basis promotes: to reduction pancreatic hormoneto strengthening sensitivity of cages to insulin), to increase of efficiency of mastering by glucose cages, reduction glucose from intestines in blood, stimulation of natural development by an insulin pancreas, to formation increase in a liver glycogen. And also the regular use of products on a basis topinambour promotes considerable improvement of a fatty exchange and decrease in superfluous weight of a body at 2nd type sick of a diabetes (also at sick of a diabetes at the long use in food of an earthen pear essentially improve properties of blood, the maintenance in cholesterol blood decreases, work of a liver and other bodies of digestive system improves, synthesis of hormones of a bark of adrenal glands, hormones of sexual glands and a thyroid gland is normalised).

By clinical supervision it has been revealed that at the regular use rich inulin and tubers topinambour and products on their basis at 1st diabetes 1st or 2nd type such complications, as diabetic, diabetic retinopathy, serious vascular diseases of the bottom extremities much less often develop.

Advantage topinambour at diseases of digestive system. Inulin and pectin's with which the tuber topinambour is rich, promote normalisation of balance of useful intestinal microflora (for this reason topinambour traditionally in national medicine it is used as rather effective auxiliary means as a part of preventive maintenance and dysbacteriosis treatment). The same substances in a complex with containing in tubers topinambour organic acids and insoluble cellulose essentially make active intestines, promoting elimination of locks and effective clarification to a human body from ballast and toxic substances.

Rendering anti-inflammatory and action on mucous membranes of the gastro enteric path, well influencing for work and the pancreas blood supply, rendering and gepattopprotctornya action (thanks to present at structure inuliny, to amino acids аргинину and to methionine), a tuber topinambour and products on its basis are useful at such diseases, as a gastritis, an enteritis, stomach ulcer of a stomach and a duodenal gut, a dyskinesia biliary dyskinesia ways, biliousness illness, a chronic cholecystitis, a sharp and chronic hepatitis of a various origin, a fatty dystrophy of a liver, a cirrhosis. Eliminating intestinal gripes and various dispirited frustration (diarrheas, locks, a heartburn) the tuber topinambour promotes increase of stability of bodies to digestion to a bacterial and virus infection, and also interferes with penetration into bodies of a gastroenteric path of various parasites.

The regular use in food of an earthen pear, and also products on its basis (a syrup and chips from topinambour, a powder, broths and infusions from topinambour, etc.), sating a human body necessary vitamins, macro- both microcells and biologically active substances, is rather effective in preventive maintenance and treatment of many illnesses of the person.

Cardiovascular diseases (a hypertension, an atherosclerosis, an ischemic heart trouble, a tachycardia, a heart attack, a stroke, thromboses, a thrombophlebitis, etc.)

In a tuber topinambour the complex of substances well influencing for work of heart and blood vessels (among such substances - inulin, potassium, magnesium, calcium, phosphorus, vitamin C and group B vitamins) contains. Thanking these substances at the regular use topinambour and products containing it promote improvement properties of blood, interfere with formation of blood clots, strengthen and raise elasticity of walls of blood vessels, reduce arterial pressure at a hypertension and stop angiospasms, prevent development of inflammatory processes in cardiovascular system, and also promote decrease in blood of level of cholesterol, preventing thereby formation on walls of vessels of the cholesteric plaques provoking development of an atherosclerosis.

Anaemia (anaemia). Containing in a tuber topinambour iron, manganese, sulphur, zinc, chrome, silicon, copper, cobalt, amino acid and also some vitamins of group B play an important role in synthesis of fiber of blood of the haemoglobin which is carrying out transportation of oxygen on all human body.

Diseases of the oporno-impellent device (a gout, or the gouty arthritis often named in the people «adjournment of salts», an arthritis (including revmotoids and a polyarthritis), art roses, an osteochondrosis, an osteoporosis, etc.). The potassium which excellent source is the earthen pear, promotes restoration of a normal vodno-salt exchange in a human body, interfering with adjournment of salts of uric acid in articulate connections. Present at a tuber topinambour silicon, zinc, copper, sulphur, carotid and vitamin C, amino acids lizine, methionine and treonin are participants of natural development by a collagen human body - the fiber of a connecting fabric which is a structural basis synovial of a cover of a joint, and bone fabrics, sinews and other components of articulate connection. A part topinambour phosphorus and calcium, also as well as silicon, are necessary for high-grade formation of a bone fabric, and containing in topinambour amino acid lizin is necessary for correct mastering by a phosphorus and calcium human body.

Diseases and traumatic damages of a skin (eczemaa, seboriya, vitiligo, ulcers, wounds, burns, ugrevaya a rash, etc.). At treatment of dermatological diseases and skin traumas it is useful to combine the regular use in food possessing and anti-inflammatory properties topinambour with its local external application (in a kind pulp from the crushed tubers).

Diseases of organs of vision (a cataract, a glaucoma, diabetic retinopaty, molecular dystrophies, "night blindness", conuctiva, etc.). In a tuber topinambour contain rather useful to eyes carotinouds, vitamin C, magnesium, phosphorus, chrome, zinc, manganese, copper, and also vitamins B_1, B_2 B_3, B_6, responsible for harmonious work of nervous system of an eye.

Diseases of respiratory organs (including accompanied by sharp cough and hoarseness of a voice) (OPVI, bronchial allergy an asthma, a tuberculosis of lungs, a bronchitis, a laryngitis, etc.)

And also introduction in a food allowance topinambour and products containing it can bring notable advantage:

In treatment of infringements erectile functions at men (the tuber topinambour is rich promoting increase of a potentiality by zinc and amino acid lizin).

At decrease workavaibility and endurance, at a syndrome of chronic weariness, at dream infringements (the use in food topinambour is useful for the people often testing physical and psychoemotional loadings, and also it is recommended to those who suffers a sleeplessness. It is connected by that topinambour is rich and the phosphorus promoting increase of endurance and physical strength, and also as a part of an earthen pear there is having sedative an effect a magnesium and group B vitamins, and possessing antidepressant property amino acid, having in a complex rather salutary influence on a functional condition of nervous system).

Cosmetic properties of tubers topinambour. Masks from the wiped tubers topinambour apply for a long time in the cosmetic purposes at flabby and wrinkled, and also at the angry and inflamed face skin. Masks from an earthen pear - also excellent means (the tuber structure topinambour includes a wide set of the substances stimulating natural development of collagen, providing elasticity and elasticity of a skin (among such substances - vitamin C, carotinoids, silicon, copper, zinc, sulphur, and also some containing in topinamboure amino acids). The zinc which excellent source is the earthen pear, possesses anti-inflammatory property and normalises secretion of sebaceous glands of a skin, preventing occurrence ugrevoy rashes).

Syrup from topinambour. Syrup structure: tubers topinambour (70%), water (30%). The raw materials for manufacture of a syrup offered on our site from topinambour are grown up in non-polluting areas without use of any unhealthy person of chemical fertilizers. The syrup from topinambour is made at the temperature which is not exceeding 55°C that allows to keep in the course of manufacture of this product in its structure the maximum quantity of vitamins, macro- and microcells and other substances necessary for a human body.

From topinambour which is useful for using both to adults, and children, it is recommended to use a syrup as natural any dishes and drinks. Possessing low glicemimcaly an index (GI = 13-15) a syrup topinambour - an irreplaceable product in a food allowance sick of a diabetes.

The recipe of preparation of juice from tubers of topinambour. Well washed up under flowing water and carefully cleared tubers topinambour to pass through a juice extractor, a meat grinder or on a small grater. Of the received weight to squeeze out juice and to filter it through a gauze. The juice prepared thus topinambour is necessary for storing in the refrigerator.

At a diabetes it is recommended to enter into a food allowance fresh or boiled tubers topinambour (them at this disease 3 times a day 15 minutes prior to meal use on 1 piece), or regularly to use in food in a warm kind broth of an earthen pear - on 100 ml 3 times a day (for preparation of such broth 3-4 tubers topinambour cook on slow fire in 0,5 waters within 20 minutes). It is possible to use also a sick diabetes in food and infusion from tubers topinambour (for preparation of such infusion 4 items of a spoon of the tuber crushed on a grater fill in 1l of boiled water and insist in a thermos within 3 hours).

At a heartburn, a stomach ulcer and a duodenal gut, a gastritis with the raised acidity, gastric pains, intestinal gripes, locks it is recommended to accept on 2 items to a spoon 3 times a day 15 minutes prior to meal juice topinambour, or to use 3 times a day 15 minutes prior to meal infusion of an earthen pear (1 part a spoon of the crushed tuber topinambour to fill in with 1 glass of boiled water and to insist 1-2 hours).

At a hypertension, an anaemia (anaemia), a headache, a pancreatitis, a cholecystitis broth from crude or dried tubers topinambour will benefit. For preparation of such broth on 1l of water it is necessary to take 5 crude root crops of an earthen pear or 3 items of a spoon of the crushed tubers topinambour. Broth from fresh tubers is prepared by 12-15 minutes, and crushed dried up - 30 minutes. To use such broth 3 times a week are recommended on 1 litre during the day.

External application of juice from tubers topinambour (in the form of the compresses imposed of 2 times a day) in a combination to its internal use (on 100 ml 3 times a day) promotes treatment of a polyarthritis and pigmentation restoration at such disease, as витилиго. At an atherosclerosis, a thrombophlebitis, a tachycardia, an ischemic heart trouble and other cardiovascular diseases, and also at a tuberculosis and diseases (urolithic illness, a cystitis, a pyelonephritis, etc.) it is useful to eat systems daily 1-2 fresh crude tubers topinambour and also to drink within 20 days 3-4 times a day 15 minutes prior to meal infusion from an earthen pear (for preparation of such infusion 2 items of a spoon of the crushed dried up tubers of an earthen pear it is necessary to fill in 0,5 l of boiled water and to insist 8-10 hours).

Containing in the structure a considerable quantity of teeth of silicon necessary for health, calcium and phosphorus topinambour is recommended to be drunk juice to children and adults for caries preventive maintenance. Infusion of tubers topinambour it is useful to rinse an oral cavity at a toothache and an inflammation of gums.

Fresh juice from tubers topinambour, dissolved half water, is recommended to drink daily 2-3 times a day on an empty stomach a sick tuberculosis, a bronchial asthma, a bronchitis and other diseases of respiratory organs.

At a cold will bring simplification burying in a juice nose topinambour (4 times a day till 11-12 a thaw).

Sick adiposity it is useful to drink in the morning (1 hour prior to meal) and in the evening (in 1 hour after meal) on 50 g expressed juice of an earthen pear within not less than 15 days. To the people, suffering excess weight (and also at such diseases as a diabetes, a cystitis, an anaemia leicoz, a pyelonephritis) is recommended also 3 times a day to drink on 1 glass (between food intakes) tea from a powder topinambour which 2 items of a spoon of a powder topinambour on 0,5 l of boiled water prepare from calculation.

Will bring notable anaesthetizing effect at diseases of a backbone and joints (an osteochondrosis, an arthritis, an art rose), and also at stretching's, crises and other traumas of the oporno-impellent device a bath with broth from the fresh or dried up tubers topinambour (for preparation of such bath it is necessary to put 1 kg in 8 l of cold water fresh or 150 g the dried up tubers topinambour to finish to boiling and to cook on weak fire within 30 minutes. Then broth to filter and pour out in a heat bath - temperature of 39-40 degrees). Time of reception of a bath should make not less than 15 minutes. As a part of treatment of diseases of the oporno-impellent device such bath is recommended to be accepted daily within 18-20 days.

Culinary recipes on a basis of topinambour. The fresh or dried up tubers topinambour to cut, place in glasswares, to fill in with cool water and to put in a warm place. In 4-5 days kvass tasty and useful to health can be poured on bottles and to place in storage in the refrigerator.

Coffee from the dried up tubers topinambour. Tubers of an earthen pear carefully to wash up, small to cut, dry up within several days on air, and then still to dry on slow fire in an oven. Before coffee preparation the dried up tubers topinambour to roast to a golden shade on a frying pan (without oil addition), and then to crush in a coffee grinder. To boil on weak fire crushed topinambour some minutes, then to insist under a cover - and useful vitamin coffee it is ready to the use.

Boiled topinambour. Carefully to clear and wash up root crops of an earthen pear, to fill in with the hot added some salt water, to add for preservation of colour of a vegetable of a few vinegar (15 g on 1 l of water) and to cook to readiness. Before giving on a table boiled topinambour to water with sour cream, mayonnaise or the kindled butter.

Fermented of topinambour. The cleared and washed up tubers topinambour to cut on thin circles, densely to lay in the glass or enameled ware, to fill in with a cold brine (on 1 l of water - 1,5-2 items of a spoon of salt) and to leave under oppression in a dark warm premise for 2-3 days, then to transfer ware with topinambourом to a cool place. To the use in food квашеный topinambour it will be ready in 2-2,5 weeks. Slices квашеной an earthen pear can be added in vinaigrette or salad, and also it is possible to submit as a garnish.

Fried of topinambour. The cleared and washed up tubers topinambour to douse boiled water, to dry a paper napkin, to cut straws and to roast on vegetable oil, as a potato.

Topinambour in hot fan. For this purpose to shake up 1 egg and to add to it 2 items of a spoon of a flour, 2 items of a spoon of sour cream, salt and black ground pepper. Then cleared and boiled to readiness topinambour to dip in кляр and to fry in a considerable quantity of hot vegetable oil. To submit topinambur-fri is better with sour cream and greens.

Winter salad with topinambour. The cleared tubers of an earthen pear to rub on a large grater, to add grated apples on a large grater, sauerkraut and the onions cut by half rings. Salad to fill with salt and vegetable oil, to season with fennel and parsley greens.

Spring salad with topinambour. The cleared tuber topinambour and 5 pieces of a radish to rub on a large grater and to add there small cut green onions, young garlic, sorrel leaves (or spinach), a sheet celery and parsley. To salt to taste, to season with vegetable oil and lemon juice.

Vinaigrette with topinambour. Components: 400 gboiled (or crude) a tuber topinambour, 200 g pickles, 200 g boiled carrots, 200 g a boiled beet, 100 g sauerkraut.

Cleared topinambour, carrots and a beet to cut straws, to add the sauerkraut, small cut onions and pickles. All to mix, fill with vegetable oil or mayonnaise, to season vinaigrette to taste with salt and sugar.

Baked pudding from topinambour. To wash up and clear tubers of an earthen pear, to boil them in slightly added some salt water. Then to cut boiled topinambour circles, to lay out in the form greased with fat for roasting, to fill in in regular intervals from above with sour cream, to strew grated cheese, small fennel and parsley greens and to bake to readiness in an oven. To submit a baked pudding from topinambour hot, having strewed before giving on a table small green onions.

Cream soup from an earthen pear. Washed up and cleared topinambour to cut with thin circles. Then small to cut onions and to roast it a little together with the cut earthen pear. Then in a boiling chicken or vegetable broth to lay out the browned onions with topinambour and to cook on average fire - 15-20 minutes. For thickening soup-mashed potatoes it is necessary to prepare refuelling. For this purpose a flour small amount to dissolve with a broth and to brown 4-5 minutes on slow fire on creamy or vegetable oil. In boiling soup to add the browned flour, dense cream or sour cream. Then soup to shake up in a blender and to boil even 5 minutes. Ready cream soup from topinambour to fill before giving on a table stem greens.

Fritters from topinambour with carrots. Components: 500 g crude topinambour, 500 g crude carrots, 2 eggs, 2 items of a spoon of a flour, vegetable oil, salt.

To rub on a small grater cleared both washed up topinambour and carrots, to add there 2 eggs, to salt and mix. In нагретое on a frying pan vegetable oil to spread a spoon the received weight and to fry fritters to readiness from two parties.

Berry juice with a syrup topinambour. The fresh or frozen berries to shake up in a blender together with the syrup added to taste topinambour and to dissolve with water to the necessary concentration. In a ready berry berry juice it is possible to add cinnamon, vanilla or mint leaves.

Fruit cocktail with a syrup topinambour. In a blender to shake up 1 banana, 2 kivy, 1 tangerine, a few the different fresh or frozen berries and 0,5 l of water. To add to taste a syrup topinambour, cinnamon, vanilla or mint.

Plov from topinambour. For preparation oshi plow from topinambour in carrots place use tubers topinambour.

Contra-indications to the use of topinambour. Individual intolerance. In some cases the use in tuber food topinambour in a crude kind can cause meteorizm. For the purpose of the prevention strengthened gazering in intestines topinambour it is necessary to use in food, having subjected to preliminary thermal processing (that is in a boiled, baked or stewed kind) or in a combination to the spices interfering gazering (for example, a coriander and caraway seeds). Power value 100 g a crude tuber topinambour - 57,3 kcal.

The Topinambour as Bioenergy Crops

Topinambour in the future can using as wood materials and biogas. One hec. topinambour can providing more than 4500m^3 biogas (tabl.1).

Table 1. Energy potacial of topinambour plant

Topinambour	Biogas	Biomass from one hec.	Biogas from one hec.	Number family can provide biogas
1 MT	50 m^3	100 MT/ hec.	4500 m^3	12 family/ year

Also from one mt/tubers of topinambour can providing more than 80 liter of bioetanol.

Botanical Characteristics of Jerusalem Artichoke

Jerusalem artichoke (*Helianthus tuberosus* L.) is a perennial plant of the Asteraceae family. Jerusalem artichoke (earthen pear) is a tuberous plant of the Asteraceae family, which has a powerful root system and reaches a height of 3-4 meters.

- Main botanical characteristics of Jerusalem artichoke:
- The stem is erect, 2 to 4 m long;

The leaves are heart-ovate, large, 10-20 cm long and 5-12 cm wide, serrated along the edge, petiolate, dark green in color;

- Inflorescence - a basket, 1-1.5 cm in diameter, consists of false-ligulate yellow flowers;
- Fruit - achene;
- The root system is taprooted, penetrates to a depth of 2 m, forms numerous shoots (stolons), on which 10-30 or more tubers are formed;
- Tubers - pear-shaped, spindle-shaped, round, with convex buds-eyes, with colors - white, yellow, rose-red and purple, pulp - white, skin - yellowish-brown). The mass of tubers is from 10 to 150 g. Breeders around the world have bred more than 500 varieties of Jerusalem artichoke [Shazzo et.al., 2013].

The homeland of Jerusalem artichoke is North America. Jerusalem artichoke tubers were first used as food by Brazilian Indians long before the discovery of North America by Europeans.

The name "Jerusalem Jerusalem artichoke" was given to the plant by chance when the Indians of the Brazilian Tupinambus tribe were brought to France in the 17th century. Tubers of an unknown plant were found in their possession, which began to be called the word "Jerusalem artichoke" - a derivative of the name of the tribe [Yaroshevich et.al., 2010; Shazzo et.al., 2013].

Jerusalem artichoke came to Europe at the beginning of the 17th century thanks to the traveler and discoverer Samuel de Champlain, who sent Jerusalem artichoke tubers from Canada to France and for the first time compared them by taste.

Since then, for more than 500 years, the cultivation of Jerusalem artichoke, which is very unpretentious to growing conditions, has become widespread (2.5 million hectares and more than 300 varieties) in many countries around the world with different names: "Canadian potato", "jirasol", "sun root"), "sanchok", "kartoshkai homkhurak", "kartoshkagul", etc.) [Partoev et.al., 2015].

Chemical Composition of Jerusalem Artichoke Tubers

Jerusalem artichoke tubers contain: inulin (11-17%), fructose, fructooligosaccharides, amino acids (up to 8%) (including 8 essential amino acids - arginine, histidine, valine, leucine, isoleucine, lysine, tryptophan, methionine and phenylalanine), carotenoids, vitamins B_1, B_2, B_3 (PP), B_6, B_9, C, pectin's (up to 10%), organic acids (citric, malonic, malic, succinic, fumaric), fatty acids (0.4%-0.7%), nitrogenous substances, fiber (up to 6%), as well as a very wide range macro- and microelements (potassium, sodium, magnesium, iron, phosphorus, manganese, calcium, silicon, copper, zinc, sulfur, chromium, iodine, boron, aluminum, cobalt, molybdenum, fluorine, etc.) [Shazzo et.al., 2000;2013].

In recent years, many countries around the world have been searching for and studying plants that can be used as raw materials for the production of bioethanol (biofuel). For these purposes, some grains, cruciferous crops and highly productive wild plants have already found practical application.

In our opinion, such plants that meet the above criteria in the conditions of Tajikistan include Jerusalem artichoke (earthen pear).

The earthen pear was brought to Tajikistan in the 20s of the last century by settlers and was used as an ornamental plant [Litvinov, 1958]. Plantings of earthen pear could be found near residential buildings, in front gardens, orchards, and less often in vegetable gardens. In 1950 A.P. Saverkin, testing various varieties of forage crops in the highlands of the Gissar Range, reported good development of the earthen pear at an altitude of about 3000 m above sea level [Litvinov, 1958].

In order to obtain feed mass in 1953-1955. In Tajikistan, in special experimental plots, the propagation of earthen pear planting material was carried out and 500 centners of tubers were transferred to collective and state farms of the republic for sowing.

Figure 4. The young farmer Vaisiddin Sharipov happy with high yield of Jerusalem artichoke

Dushanbe, 2018

In 1956-1958 earthen pear was cultivated in the natural climatic conditions of the Gissar and Vakhsh valleys, in the Garm group of districts, in the Leninabad region and in the Western Pamirs. In the production crops of collective farms, as well as in experimental plots, the earthen pear on average yielded 500 centners of green mass, and 250 centners of tubers per hectare.

In high mountain conditions, on collective farms in the Ishkashim, Roshtkala and Shugnan districts, it was established that the earthen pear is a valuable forage crop, capable of producing 500-600 centners of green mass and 150-200 centners of tubers per hectare [Litvinov, 1965].

In world practice, interest in Jerusalem artichoke has either increased or waned over different periods, which was determined by the activity of studying and promoting the potential capabilities of this crop and the high productivity of the above-ground mass and tubers. In the first half of the 20th century, Jerusalem artichoke was studied for use as a forage crop.

However, extensive biochemical studies in recent decades, starting from the 80s of the 20th century, have shown that, along with the nutritional benefits, Jerusalem artichoke has high prospects for use as a food, medicinal and industrial (bioenergy) crop [Yaroshevich et.al., 2010;2014].

Jerusalem artichoke is an unpretentious, perennial, highly productive plant, resistant to adverse conditions, effectively using solar energy for growth and development.

In the conditions of Tajikistan, the earthen pear can be planted in all valley-foothill, mountain and highland areas in places with rugged terrain on the slopes of ravines, in ravines, in floodplains, that is, in places where other crops cannot grow [Litvinov,1965; Partoev et.al., 2015](Figure 4) .

The high ecological plasticity of Jerusalem artichoke plants is evidenced by the fact that in the autumn period of the year they are able to tolerate frosts down to –7…–8°C. Tubers can freeze and thaw repeatedly without losing their viability. Undug tubers remain viable with a snow cover of 0.2...1.0 m, even when the soil freezes and the air temperature drops to –35°C and below [Pasko, 1999, 2003; Yaroshevich et.al., 2010].

It should be especially emphasized that the natural polyploid level of Jerusalem artichoke and its ability for vegetative propagation create great prerequisites for the selection of varieties with a high level of heterosis, including those with undamped heterosis. In addition, species of different ploidy of the genus Helianthus are used by researchers in interspecific hybridization to create economically valuable hybrids [Pasko, 1999; Kochnev et.al., 2002; Yaroshevich,2014]. For example, highly productive hybrids of Jerusalem artichoke with sunflower have been obtained – Jerusalem artichokes (varieties "Novost VIR", "Vostorg", etc.) [Pasko, 1999]. Two Russian sunflower varieties ("Progress" and "Novinka") were bred by academician G.V. Pustovoit based on hybrids of crossing *H. annuus* and *H. tuberosus* [Pasko, 1999; Yaroshevich, 2014]. The earthen pear is capable of producing very high yields of more than 1000 c/ha of green mass and 500 c/ha of tubers. The average yield of earthen pear is usually about 400-500 c/ha of green mass and 250 c/ha of tubers [Litvinov, 1958].

In Tajikistan, the yield of green mass of earthen pear under irrigated conditions reaches 1000 c/ha, and tubers - 400-450 c/ha; on rainfed lands, 300 c/ha of green mass and 120 c/ha of tubers [Litvinov,1965].

In the valley regions of Tajikistan, by the end of July, plant growth reaches a height of more than 2.5 m. The yield of green mass at this time is at least 550 c/ha. If the tops are mowed at this time, then the earthen pear will again produce the same crop of tops by autumn [Litvinov,1965].

This biological feature of the earthen pear makes it possible to lay silage from its green mass in two periods: summer and autumn.

What is important here is that the timing of ensiling pear tops coincides with the timing of mass summer and autumn ensiling of other silage crops, which makes it possible to prepare their mixtures [Litvinov,1965; Kasymov, 1975].

Jerusalem artichoke (*Helianthus tuberosus* L.) is a high-yielding agricultural crop. In the conditions of the Non-Black Earth Region of the Russian Federation, its yield of green mass can reach 60.0 t/ha, and tubers - 40.0 t/ha, provide a yield of 7.5-10.0 t/ha of feed units and 6.0-6.8 c/ha of digestible protein. Jerusalem artichoke is distinguished by high nutritional qualities due to the presence of valuable chemical components in it.Tubers contain 18-22% sugars, up to 2.5% protein, vitamins B and C. The mineral composition of ash elements contains up to 6% phosphorus, over 5% iron, which makes the tubers a particularly valuable food for young animals. Green mass contains up to 20-25% dry matter. It includes a carbohydrate complex, a significant proportion of which is occupied by a special type of carbohydrate - inulin, which is processed in the body of animals into easily digestible fructose, and also contains complete protein (which is represented by 16 amino acids, including 8 essential), as well as vitamins and fiber [Golubev et.al., 1995; Reingart et.al., 2008; Tretyakov, 2009; Shazzo et.al., 2013; Safarzoda et.al.,2014; 2015; Khalifaev et.al., 2013; 2014; 2015; Jahoni, 2010; Kondratenko, 1988; 2008; Novikova et.al., 2017].

Agroecological environmental factors have a large amplitude in different conditions of the valley and mountain zones of Tajikistan, and these factors have different effects on the growth and development of various crops. In particular, the authors [Sardorov 1977; Partoev et.al., 2015; Saidaliev, 2018; Saidaliev et.al., 2015] inform that between such characteristics of Jerusalem artichoke as stem height, number of leaves, number of tubers and productivity there is a direct average correlation depending on the cultivation zone.

Knowledge of the influence of various agroecological environmental factors on a number of polygenic traits of Jerusalem artichoke is of great scientific and practical importance in the process of breeding new promising genotypes of Jerusalem artichoke in the future.

The founder of peace and unity, Leader of the nation, President of the Republic of Tajikistan Emomali Rahmon in 2011 set the task for the country's scientists: "Scientists of Tajikistan must develop ways and methods for producing bioethanol (bio gasoline) based on the use of Jerusalem artichoke (earthen pear). The Academy of Sciences, together with the Ministry of Energy and Industry, local executive government authorities must develop alternative sources of energy use and prepare recommendations for their use in the national economy of the country."

THE STUDY RESULTS OF TOPINAMBOUR

In Conditions of Tajikistan

Considering useful properties topinambour us there was an object in view - to study feature of growth and development topinambour in the conditions of Gissar and Rasht valleys of Tajikistan on watering and without watering, as food and fodder culture (at height of 820-2700 m above sea level).

The following entered into a problem of experiences:

· To Choose different variants of entering of mineral fertilizers under topinambour,
· To Spend planting to spend accounts of approach of the basic growth phases and development topinambour (shoots, branching, бутонизации, flowerings, formations of tubers),
· To Define dynamics of formation of a biomass in the basic phases of development a plant,
· To carry out the biochemical analysis of tubers and elevated weight of plants,
· To spend the description of a plant during vegetation,
· The Account of productivity and efficiency of plants.

Material and Technique of a Bookmark of Experience

As material for carrying out of our researches tubers of a grade "the Interest", grown up in the conditions of Javanese area served. Tubers of the given grade topinambour are got by us at Public organization (NGO) «Is advisory - the Information Network» ("PUSSYCATS").

Experiences on studying of feature of growth and development topinambour have been put in the conditions of Gissar and Rasht valleys of Republic Tajikistan.

On the basis of skilled a site of Institute of botany, physiology and genetics of plants of AN PT located in Dushanbe experience with following variants has been put:

- Planting topinambour without watering (rein fields);
- Planting with watering carrying out.

At landing of tubers topinambour used following norms of entering of mineral fertilizers on a site without watering:

1. Without fertilizers (control),
2. Entering NPK (50:50:25kg/hectare) in the form of operating substances,
3. Entering NPK (75:75:50 kg/hectare) in the form of operating substances,
4. Entering NPK (100:100:75 kg/hectare) in the form of operating substances,
5. Entering organic fertilizers (a liquid from a water solution a plant ejhornija-2 from calculation of 5 litres on 10 metres).

Frequency of experiences triple and рендомизированная. On a site with watering put experience with following schemes of landing: the landing Scheme: 70x20sm; 70x30sm; 70x40sm.

Also here on this skilled site liquid organic fertilizer (a liquid of a water plant - ejhornija-2 from calculation of 5 litres on 10 running metre is introduced. Ploughing skilled a site have spent on April, 5th by means of wheel tractor MTZ-82 on depth 25 see For soil preparation have spent at first ploughing and have every other day spent processing fields and the furrow by means of a cultivator KNS - 4,2 cut. Рядки for landing topinambour have spent on depth of 10-15 sm under the scheme of 70 sm between numbers. The tubers topinambour have been planted on a skilled site manually in cut ready for landing of numbers. Workers on buckets took tubers topinambour and manually on one piece put in the middle of the cut furrows, and then also have manually covered tubers with soil . Tubers were the covered soil approximately 5-7 sm.

Technique of a Bookmark of Experiences in the Rasht Valley

Experience is put on April, 25th, 2012 in the conditions of village “Sasybulak” Muksu” of Jirgital area. Experience consisted of two parts: a landing site topinambour without watering (богарный the earths, the earth sandy at height of 2000 metres above sea level) and landings to a site with watering carrying out (at height of 2200 metres above sea level), and also on богаре (at height of 2700 metres above sea level). The planting scheme is 50x20sm.

THE STUDY RESULTS IN GISSAR VALLEY

As have shown researches growth and developments topinambour pass in the Gissar valley normally in spite of the fact that landing has been spent rather late.

From figure 5 it is visible that in the beginning of June on height of plants between variants of experience of an especial difference it is not observed. However, since начла July the height of plants in a variant with watering amplifies also it is observed till the end of vegetation. The height of plants in variants in a shade is much less, than at variants without a shade - without watering and in a variant with watering. Thus, the tree shade negatively affects on growth and development of plants topinambour, and watering promotes considerable strengthening of growth of plants almost during all vegetative period topinambour. In variants in a shade - without watering and without a shade, without watering because of lacks of a moisture of soil, growth of plants proceeds to grow to the beginning of August, and then it does not increase and stabilised at one level. In a variant with watering height and growth of plants during the period August - October increases also it proceeds before the vegetation termination. Thus the soil moisture and waterings play an important role in accumulation and dynamics of growth of plants topinambour.

Figure 5. Dynamics of growth of plants topinambour in the conditions of the Gissar valley

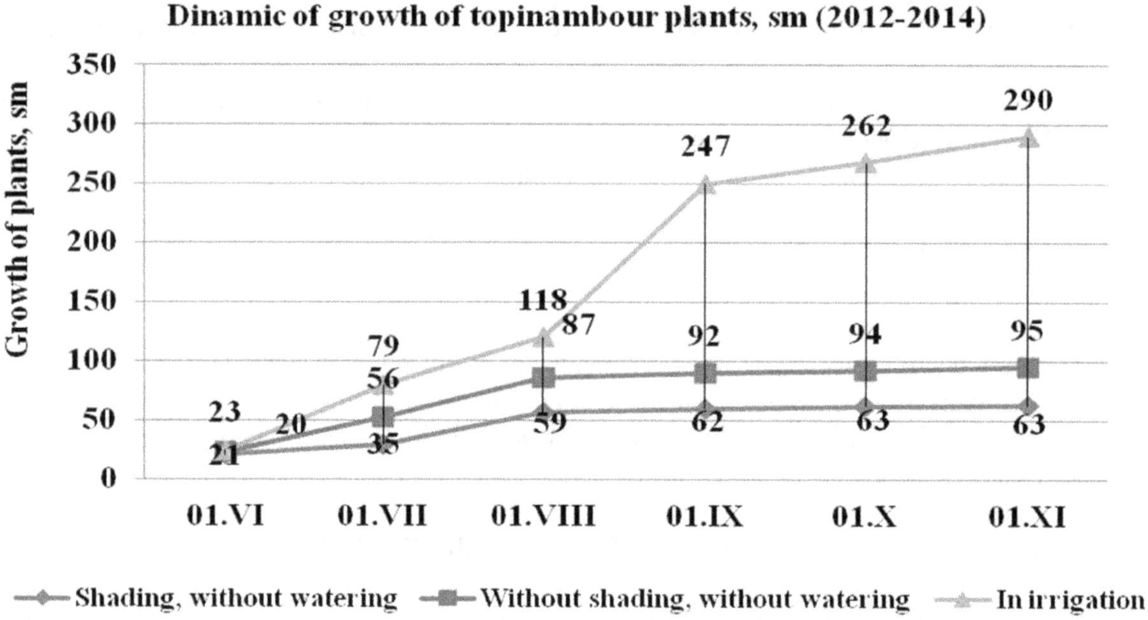

From figure 6 it is visible that in June by quantity of leaves on one plant between variants of experience of an especial difference it is not observed. However, since начла August of quantity of leaves at plants in a variant with watering increases intensively and it proceeds till the end of vegetation. The quantity of leaves in variants in a shade is much less, than at variants without a shade - without watering and with watering. Thus, the tree shade also negatively affects on formation of quantity of leaves at plants topinambour, and carrying out of waterings promotes considerable strengthening of process of formation of leaves at plants. In variants in a shade - without watering and without a shade - without watering because of lacks of a moisture of soil the quantity of leaves on plants proceeds to increase prior to the beginning of August, and then does not increase quantities of leaves and keeps at one level. In a variant with watering of quantity of leaves in July - October increases also it proceeds before the vegetation termination. Thus, waterings have played an important role increase in quantity of leaves topinambour.

Figure 6. Dynamics of increasing of number leafs of topinambour in the conditions of the Gissar valley

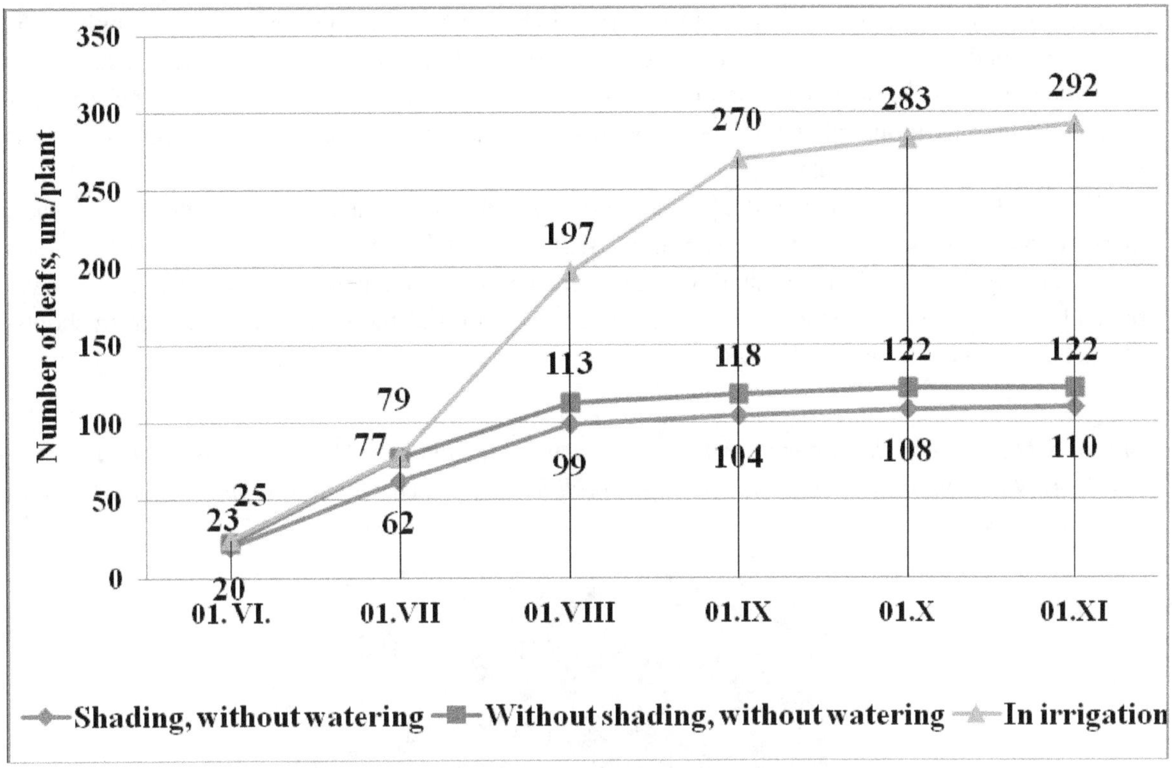

Figure 7. Yield of biomass of topinambour with fertilyzers NPK-II

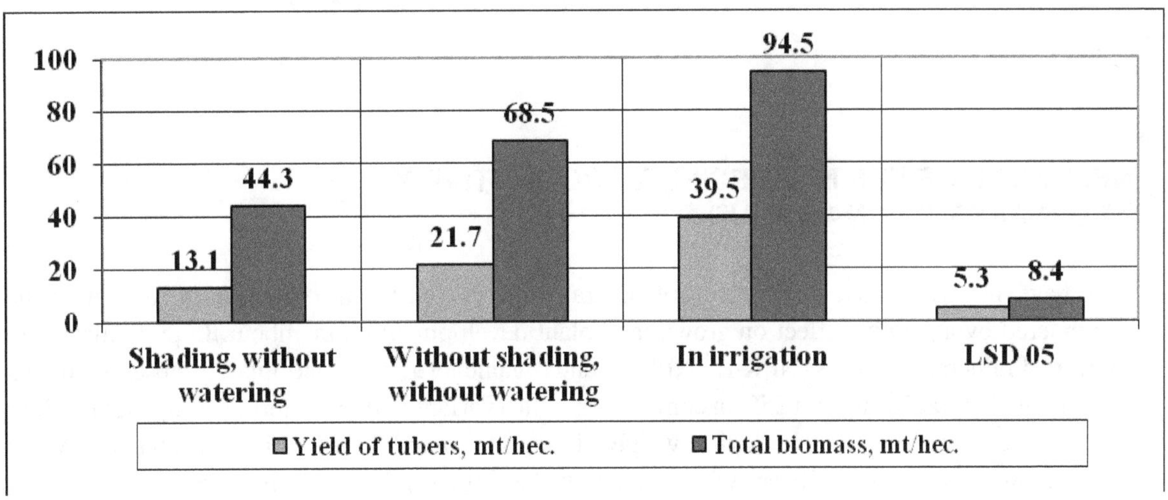

From figure 7 it is visible that on the basis of the general biomass between variants the essential difference is observed. In a variant in a shade - without watering the poor harvest on the general biomass, than in variants without a shade - without watering and in a variant with watering (accordingly on 52% and 97.9%) and accordingly and on a crop of tubers (on 70.0 and 184.2%) is observed. Thus, the tree shade negatively influences formation of the general biomass topinambour. In a watering variant the general biomass at topinambour increases by 30.2%, and a crop of tubers on 72.9% than in a variant without a shade - without watering.

Thus, watering plays an important role in growth and development strengthening and efficiency increase topinambour in the conditions of the Gissar valley of Tajikistan.

Apparently from fig. 3 topinambour well developed and accumulated a good harvest of a biomass in the end of vegetation to the beginning of November. Duration of the vegetative period from shoots of plants has made more than 165 days (Figure 8).

Figure 8.The head of the innovative project academician Ahmedov H.M. and post-graduate student Saidaliev N.Kh. on a skilled site look through conditions of growth and development a plant topinambour, Tajikistan

INFLUENCE OF FERTILIZERS FOR PRODUCTIVITY OF BIOMASS TOPINAMBOUR

As have shown our researches entering of mineral fertilizers (NPK) and organic fertilizers - ejhornija-2 are rendered by a positive effect on growth and plant development topinambour. In particular at joint entering of mineral fertilizers (NPK-I), (NPK-II) and organic fertilizer - ejhornija-2 is observed increase in height of plants at 56 sm or at 25% in comparison with a variant without entering of mineral fertilizers.

In particular on the general biological weight of plants topinambour essential influence has rendered joint entering of mineral fertilizers with organic fertilizer (ejhornija-2) that is visible from mentioned below figure 9.

Figure 9. Influence of fertilizer to yield of topinambour, mt/hec

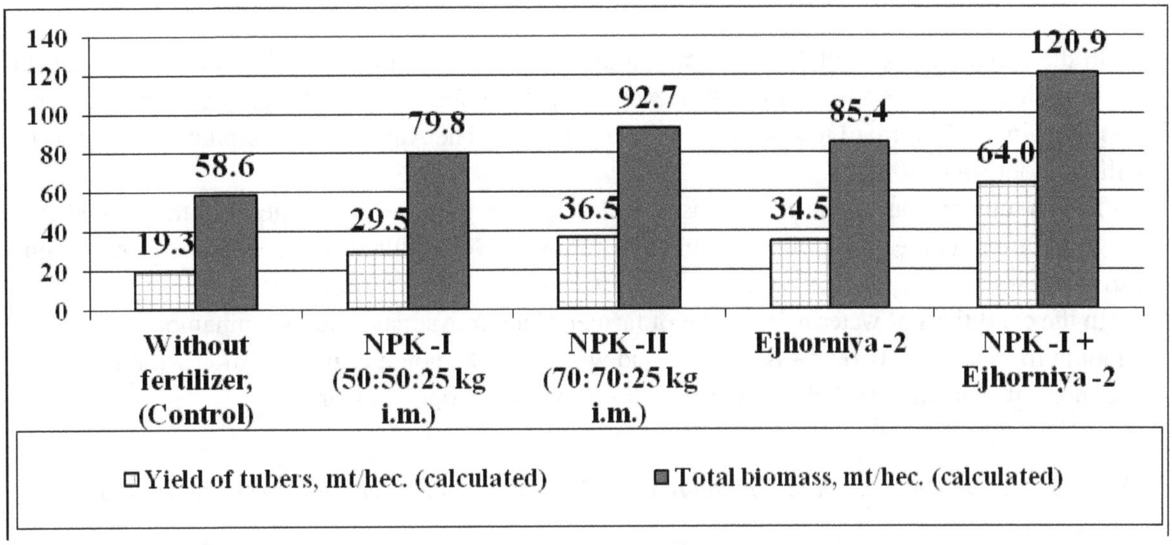

Apparently from figure 9. Joint entering of organo-mineral fertilizers (NPK-I + ejhornija-2) leads to substantial growth of the general biomass topinambour in comparison with variants without entering of mineral fertilizers (on 161.2%) and with entering of mineral fertilizers NPK-I (on 94.9%), NPK-II (on 64,7%) and ejhornija-2 (on 77.8%) separately.

Also joint entering NPK-I + ejhornija-2 leads to substantial growth of a crop of tubers topinambour in comparison with variants without entering of mineral fertilizers (in 5,1 times) and with entering of mineral fertilizers NPK-I (in 3.4 times), NPK-II (in 2,5 times) and ejhornija-2 (in 2,7 times). Thus, the most effective variant considers joint entering of organo-mineral fertilizers (NPK-I + ejhornija-2) under topinambourом at watering of plants. In this variant it is observed increases in productivity of the general biomass at 161.2% and productivity of tubers more in five times than in control (without application of fertilizers).

At joint entering natural - mineral fertilizers (NPK (50:50:25 kg i.m.) +Эйхорния-2) almost twice increases weights of tubers from the general biomass of plants, than in a control variant. Thus also it is observed reduction of weight of roots, leaves and stalks, than in control. Almost such change in a parity of formation of bodies of plants topinambour from the general biomass is observed and at entering of organic fertilizer ejhornija-2 in the pure state. Thus, stimulating action of organic fertilizer ejhornija-2 on formations of a crop of tubers from the general biomass of plants topinambour is observed that has great value in increase in an exit of tubers from area unit.

GROWTH AND DEVELOPMENT TOPINAMBOUR IN THE RASHT VALLEY LAYAKHSH (JIRGITAL) DISTRICT

In the conditions of Jirgital area landing topinambour was spent on April, 25th, 2012 in Layakhsh district at height of 2100 metres above sea level. Soils of this site according to soil a cartogram concerned to mountain карбонатным to soils and in them pH = 6,8. The Amount of precipitation in a year there will drop out about 400 mm.

Tubers topinambour in the conditions of watering have spent on приусадебном the ground area of the local farmer of the given village Mahkamova Idibeka. Soils skilled sites concerned group mountain carbonate soils.

In the conditions of watering on a site of farmer Mahramova mass shoots topinambour it is noted in a month from landing as year was too cold and with a small amount of precipitation within May - July, than normal years that has led to a delay of shoots of tubers topinambour.

Figure 10. Development a plant topinambour in Jirgital area at height of 2700 metre altitude

We had been spent on September, 20th, 2012 survey of a skilled site without watering (on not waterin to the earth) and on watering at height of 2100 metres on a sea level in the Lahshsky file of Jirgital area (Figure 10).

On rein field plants of topinambour despite shortage of a moisture, nevertheless had growth of plants to 60-80 sm, blossomed and formed till 7-12 piece of tubers. And in the conditions of plant watering reached height of 260 sm and intensively blossomed.

Plants topinambour well developed as well in the conditions of watering at height of 2700 metres above sea level and their growth has reached heights of 200 sm (Figure 11).

Figure 11. Dynamics of grows of plant topinambour

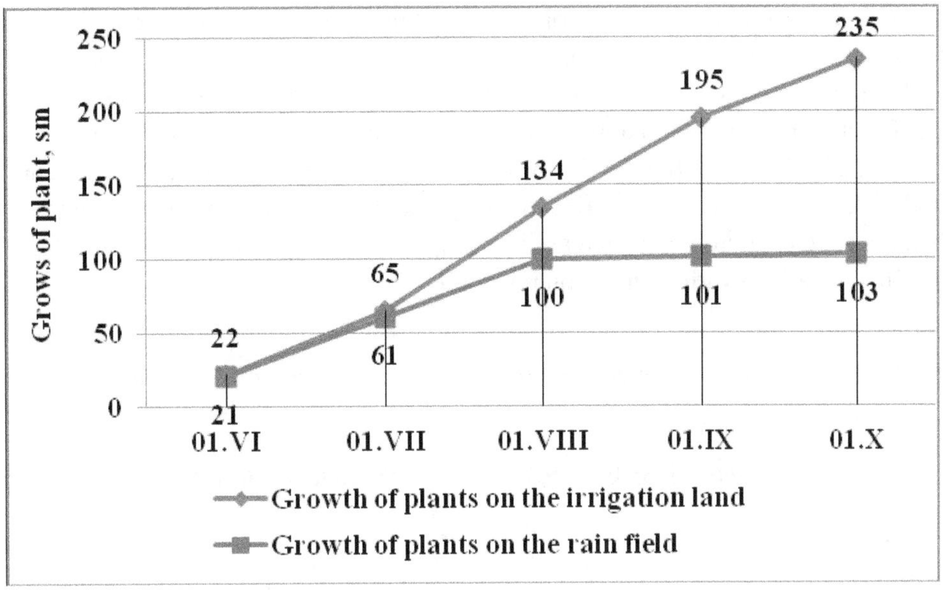

As show the data figure 11 at watering dynamics of growth of plants topinambour since the beginning of July quickly it grows also proceeds till the end of vegetation. In a variant without watering since the beginning of August dynamics of growth of plants is stabilised and till the end of vegetation almost remains at the same level.

Figure 12. Dynamics of formation of number leaves of topinambour, Jirgital district

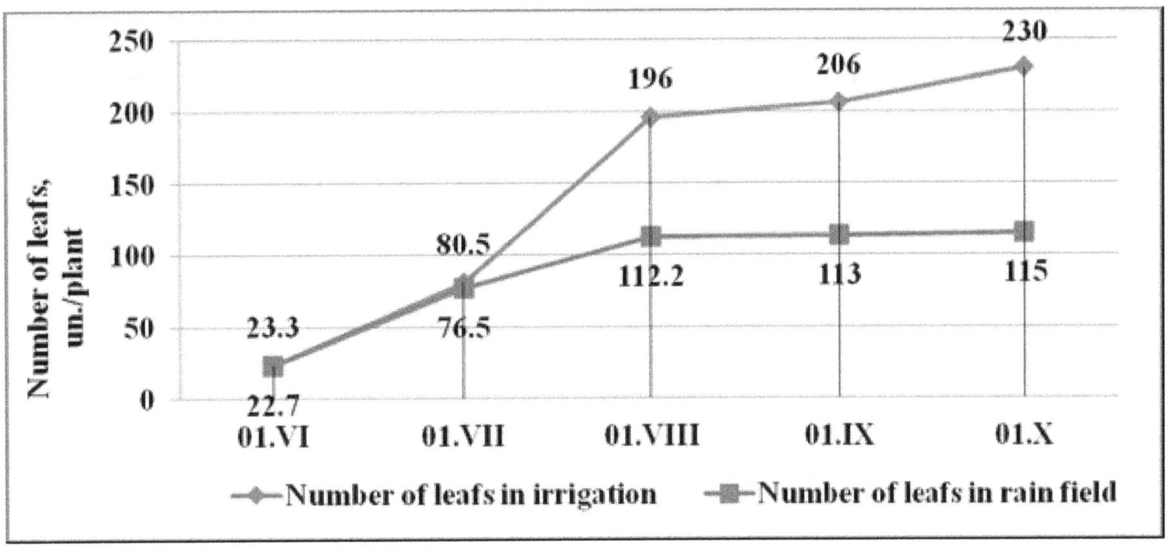

From figure 12 it is visible that the quantity of leaves at watering since the beginning of June prior to the beginning of October increases by one plant, and at landing on богаре increase the quantity of leaves is observed prior to the beginning of July, then this indicator is stabilised at one level till the end of vegetation. Hence in the conditions of a mountain zone at cultivation topinambour on богаре it is possible from the beginning of August an elevated part topinambour to mow down for use on a forage of cattle or for other purposes, and at watering of it it is possible to make for two months later.

From figure 13 it is visible that on the average the biomass crop topinambour on rain field the irrigation land makes 48.4 t/hec. that testifies to utility of cultivation topinambour on rain field. Also it is established that watering promotes increase in a crop of a biomass topinambour on 29.1 t/hectares (or on 60.1%), in comparison with planting on rain land.

Figure 13. A biomass crop topinambour on rain field and in irrigation land, mt/hec (calculated)

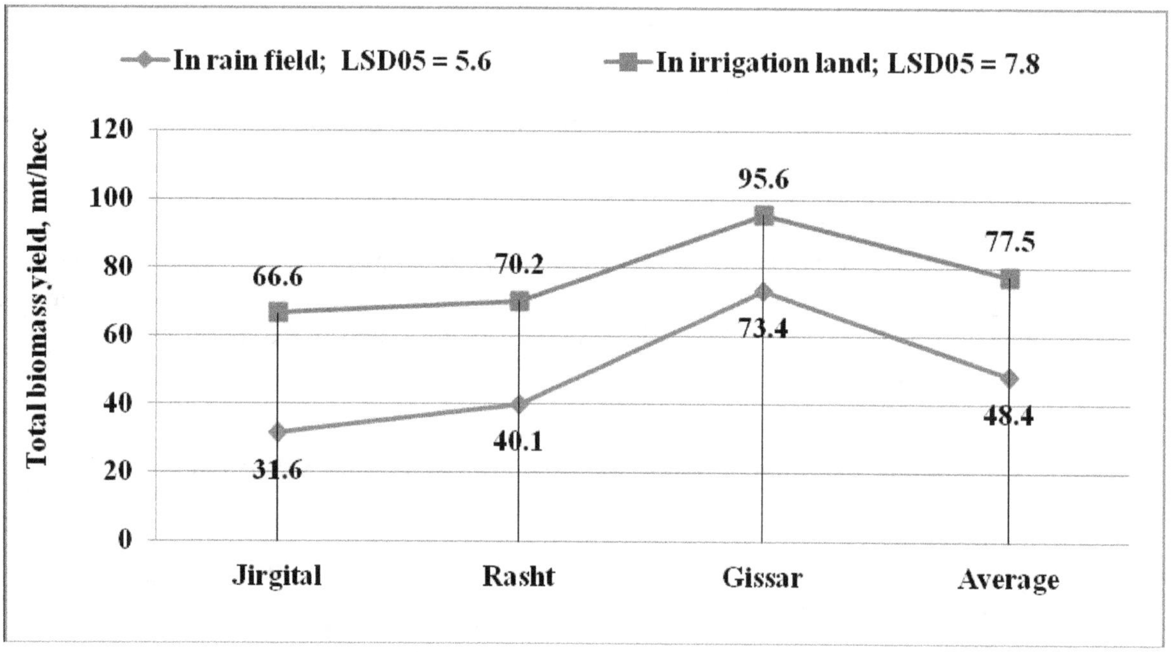

Figure 14. Yield of tubers in rain and irrigation lands, mt/hec (calculated)

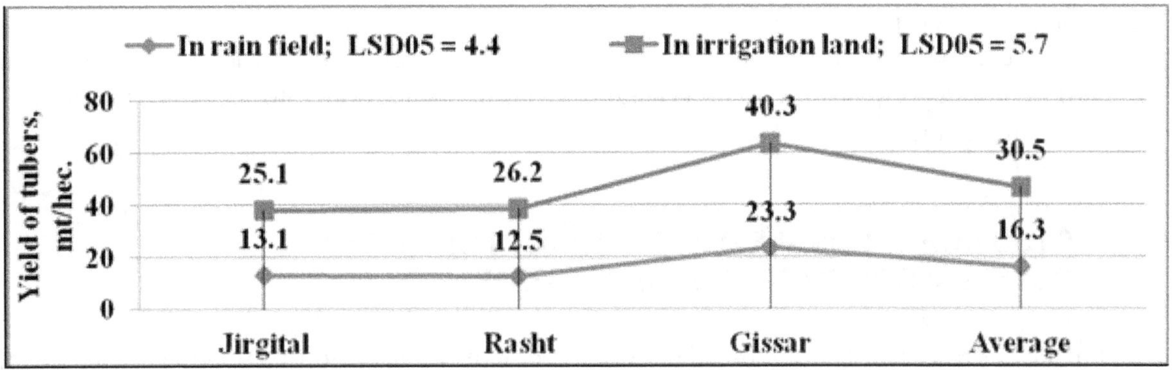

From figure 14 it is visible that on the average the crop of tubers topinambour on rain lands the earths makes 16.3 t/hectares that testifies to a high indicator of efficiency topinambour on богарных the earths. It is established that watering promotes increase in a crop of tubers topinambour on 14.2 t/hectares (or on 87.1%), in comparison with landing on богаре. By us also it is established that on the average on all zone of cultivation topinambour on watering and on богаре the parity of bodies of plants (parts) from lump essentially does not change (Figure 15).

Figure 15. A parity of bodies of plants topinambour under different conditions of cultivation (In % from the general biomass of plants)

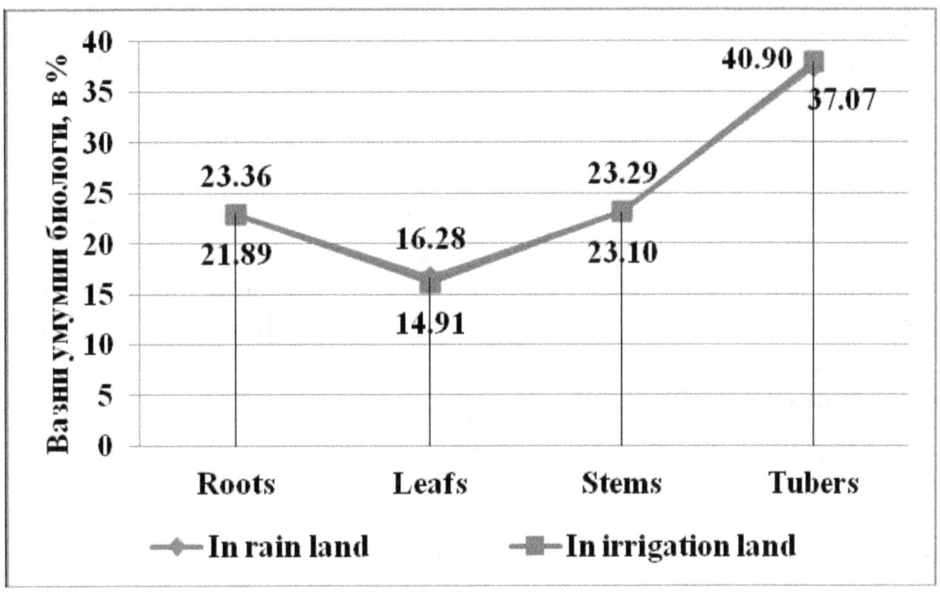

As show the data figure 15 approximately from the general biomass of plants topinambour different parts or bodies make following quantity: roots approximately 23-25%, leaves - 14 - 16%, stalks – 22-23% and tubers - about 38- 39%.

Thus, the cultivation variant topinambour on the irrigation earths has not led to essential change of parities of parts of plants from the general biomass in comparison with a cultivation variant on богарных the earths.

THE STUDY OF JERUSALEM ARTICHOKE IN DIFFERENT AGROSYSTEM CONDITIONS OF TAJIKISTAN

Purpose of research. The purpose of our work is to study the characteristics of growth and development, the biological productivity of plants, the yield of total biomass and tubers of Jerusalem artichoke in the valley and in the mountainous zone of Tajikistan and to assess the prospects of using Jerusalem artichoke as a food product, as animal feed and as a raw material for biofuel production[Pashchenko et. al., 1999; Akhmedov et.al., 2003; Ergashev et.al., 2018; Sadriddinov, 2018].

Materials and Research Methods

Scientific work to study the characteristics of growth and development, as well as the manifestation of a number of morphological polygenic traits of various Jerusalem artichoke genotypes, was carried out during 2016-2020. in various agro-ecological conditions of the Republic of Tajikistan: Vose - 460 meters. above sea level (a.s.l.), Vakhsh - 600 m.a.s.l., and in the city of Dushanbe - 840 m.a.s.l., Muminabad - 1200 m.a.s.l. m., Rasht - 1800 m.a.s.l., Lyakhsh -2000 m.a.s.l.,Vakhdat city (Kanask) - 2560 m above sea level (Figure16).

In these areas, the sum of effective temperatures ranges from 895 to 376^{0}0C. The amount of precipitation in the form of rain and snow throughout the year ranges from 300 to 800 mm. The soils in the areas where the experiments were carried out are poor in humus content (0.5-2.0%) in the arable horizon. Here the soil type is mainly gray soils and carbonate soils.

The material for our research was elite and varietal seed tubers (I-II seed reproduction) of various genotypes/varieties of Jerusalem artichoke (*Helianthus tuberosus* L.).

We obtained the source materials from the collection of the Institute of Botany, Physiology and Plant Genetics of the National Academy of Sciences of Tajikistan (IBFGR NAST), the All-Russian Research Institute of Plant Growing (VIR, Maikop Experimental Station, Republic of Adygea) and the Kuban Agrarian University (Russia). The number of studied Jerusalem artichoke varieties in these areas was 16-20 pieces. Jerusalem artichoke varieties were grown on the basis of generally accepted agricultural technology for each agro-ecological zone. Depending on the altitude above sea level, tubers of Jerusalem artichoke genotypes/cultivars were planted during March–April according to a planting pattern of 70 x 35 cm. Jerusalem artichoke varieties were planted in quadruplicate.

Figure 16. The area and heights above sea level where the research is conducted plant of an artichoke (2016-2020)

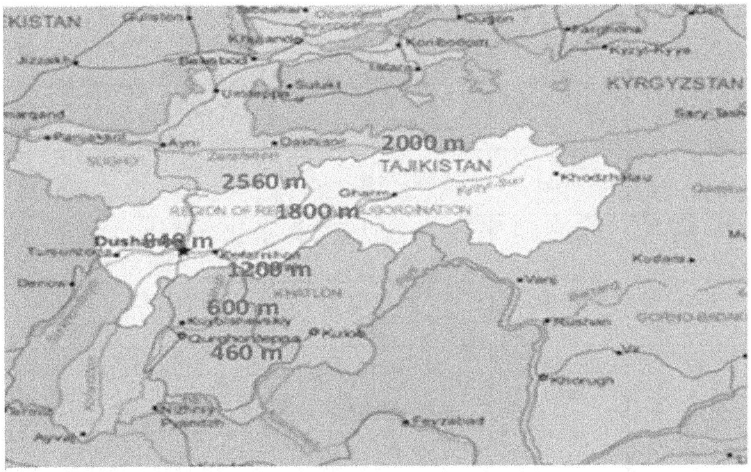

The total number of plants from each genotype/variety was 80 plants. During the growing season of genotypes/varieties of Jerusalem artichoke, the following agrotechnical work was carried out: application of mineral fertilizers ($N_{100}P_{150}K_{80}$ kg/ha), two inter-row treatments (manually), cultivation of rows, hilling of rows and 5-7 times of vegetation irrigation. The standard variety of Jerusalem artichoke was the "Interes" variety (Russian selection). During the growing season of Jerusalem artichoke, the following phenological surveys and observations were carried out: counting seedlings, plant height in different phases of plant development, number of leaves, number of stems, root mass, number of tubers, tuber mass, total plant biomass.

Information on the average daily air temperature, the sum of effective temperatures (over 10^0C) and the amount of precipitation was taken from the State Institution "Meteorological Station" in the city of Dushanbe, Republic of Tajikistan. Statistical data processing was carried out according to [Dospekhov, 1985] using the Microsoft Excel 2007 computer program.

Results and Discussion

In the various ecological zones where our research was carried out, there is a significant influence of such agroecological conditions of the area as altitude above sea level and the sum of effective temperatures on the formation of various useful characteristics of Jerusalem artichoke (Table 2).

From Table 2 it can be seen that polygenic characteristics of Jerusalem artichoke, such as the mass of stems, leaves and roots, the mass of tubers and the total biomass of plants, vary greatly depending on the vertical zonation of location and sea level. During the growing season of plants, with increasing altitude above sea level, a proportional decrease in the sum of effective temperatures (over 10^0C) is observed. The largest amount of the sum of effective temperatures is observed in the southern part of the republic at an altitude of 460 m above sea level in the Vaseysky district (3760^0C), and the smallest number of effective temperatures is observed in the highlands of Central Tajikistan, in the city Vahdat in the Kanask section at an altitude of 2560 m above sea level (895^0C).

Table 2. Area, altitude, temperatures and efficiency of a sun artichoke (an average for 2016-2020)

Area and altitude, m	Height above sea level, m	Soil type	Effective temperature during the year, °C	Mass of stems, leaves and roots, g/ plant	Mass of tubers, g/ plant	Total biomass, g/ plant
Vose, 460 m	460	Light gray soil	3760	2817±1,4	1575±1,6	4392±1,7
Vakhsh, 600 m	600	Light gray soil	3455	2190±1,3	1500±1,8	3690±1,4
Dushanbe, 840 m	840	Typical gray soil	2610	2040±1,6	855±1,7	2895±1,5
Muminabad, 1200 m	1200	Light gray soil	2280	1100±1,4	590±1,5	1690±1,7
Rasht, 1800 m	1800	Carbonate soils	1300	1000±1,2	450±1,1	1450±1,9
Lyakhsh, 2000 m	2000	Carbonate soils	1210	810±1,5	370±1,8	1180±1,3
Kanask, 2560 м	2560	Carbonate soils	895	520±1,1	250±1,3	770±1,4
Average	-		-	**1496,7**	**798,6**	**2295,3**
SSD_{05}	-		-	**8,2**	**9,1**	**10,4**

Scientists from the Institutes of Botany, Physiology and Plant Genetics of the National Academy of Sciences of Tajikistan during 2005-2017. As a result of using the selection method among the Jerusalem artichoke variety "Interest", a new Jerusalem artichoke variety "Sarvat" ("Wealth") was developed (Figure 17).

The new variety of Jerusalem artichoke, unlike the mother variety "Interes", has more even and smooth tubers, which are convenient for canning tubers. This variety was zoned throughout the republic in 2017. Under production conditions, its yield exceeds other varieties by 25-30%.

The yield of tubers of the Sarvat variety is 45-50 t/ha, and the total biological mass is 60-70 t/hec. Jerusalem artichoke tubers are a good remedy for reducing sugar levels in human blood. The variety is grown on rocky and infertile soils. (*The authors of the variety are Partoev K., Akhmedov Kh.M., Mirzoev N.R., Saydaliev N.Kh. and Yasinov Sh.M.).*

Figure 17. The tubers of variety «Sarvat» ("Wealth") (left) and of variety «Interest» (right), 2017

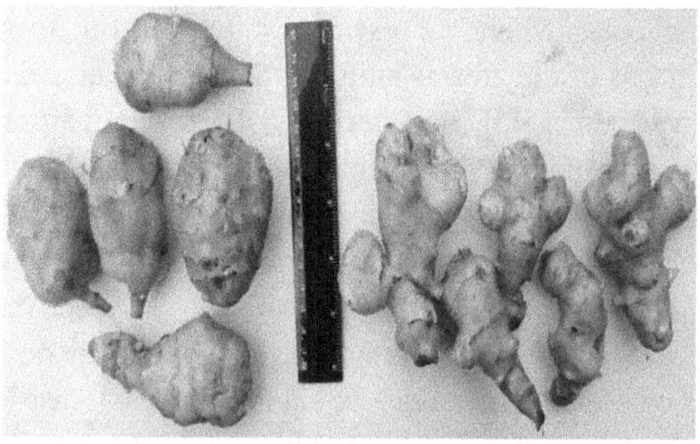

The producrivity of Jerusalem artichoke varieties given in table 3.

Table 3. Productivity of Jerusalem artichoke varieties (2016-2020)

Variety	Mass of stems, t/hec	Mass of roots, t/hec	Mass of tubers, t/hec	Total biomass, t/hec
“Sarvat:	31	53	144	228
“Interes”	22	23	95	140
Difference, %	41	130	52	62,9

Figure 18. Reasearch plot for studing of collection varietyes of Jerusalem artichoke, Dushnbe, 840 m altitude, 2018

Our experiments have shown (Figure 18) that the relationship between such indicators as the sum of effective temperatures, the mass of stems and leaves, the mass of tubers and the total biomass of Jerusalem artichoke plants is strongly positive (Figures 19,20,21).

Figure 19. Correlation between the sum of effective temperatures, mass of stalks and leaves of sun artichoke

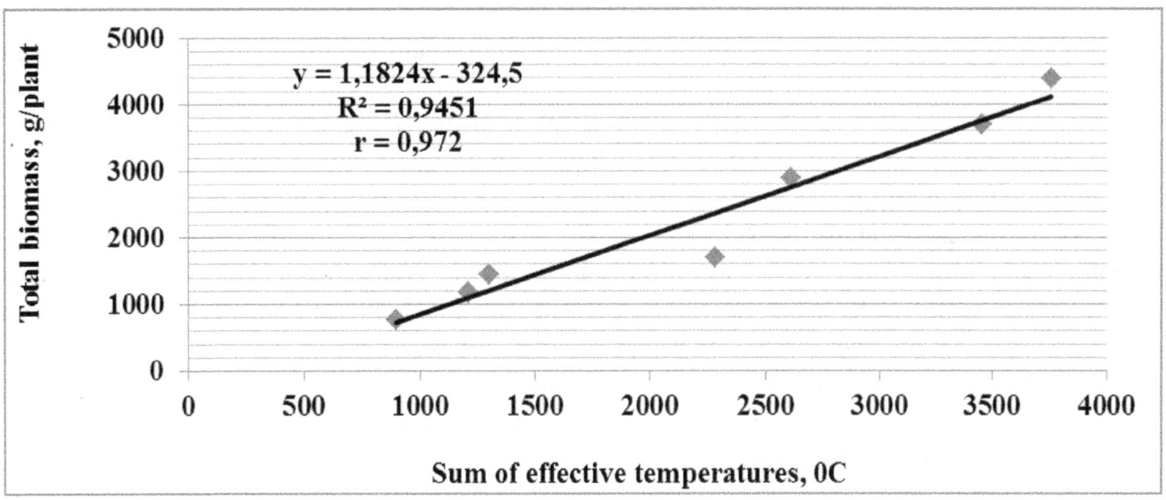

Figure 20. Correlation between the mass of tubers and the general biomass of sun artichoke

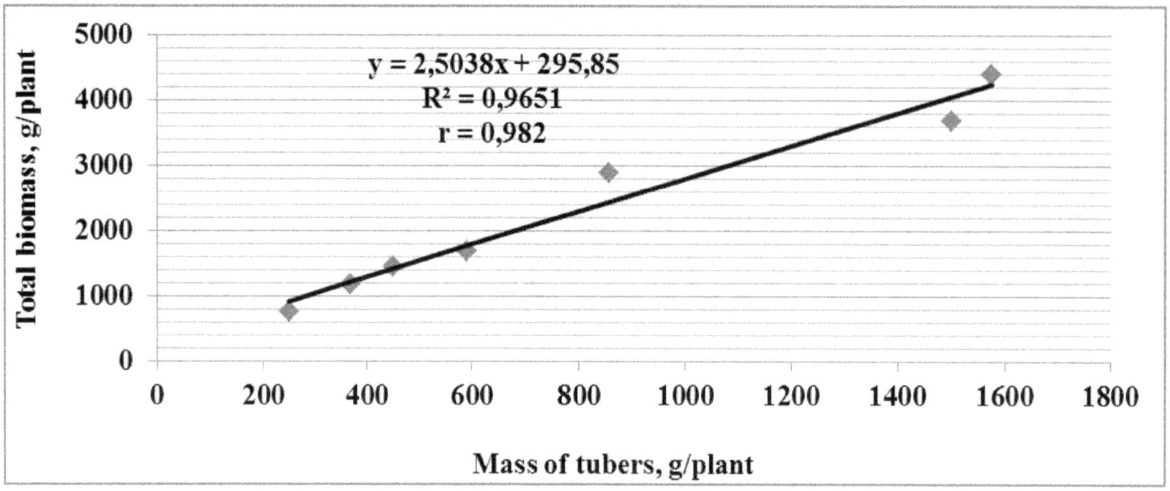

Figure 21. Correlation between the mass of stems, leaves, roots and the general biomass of plants of sun artichoke

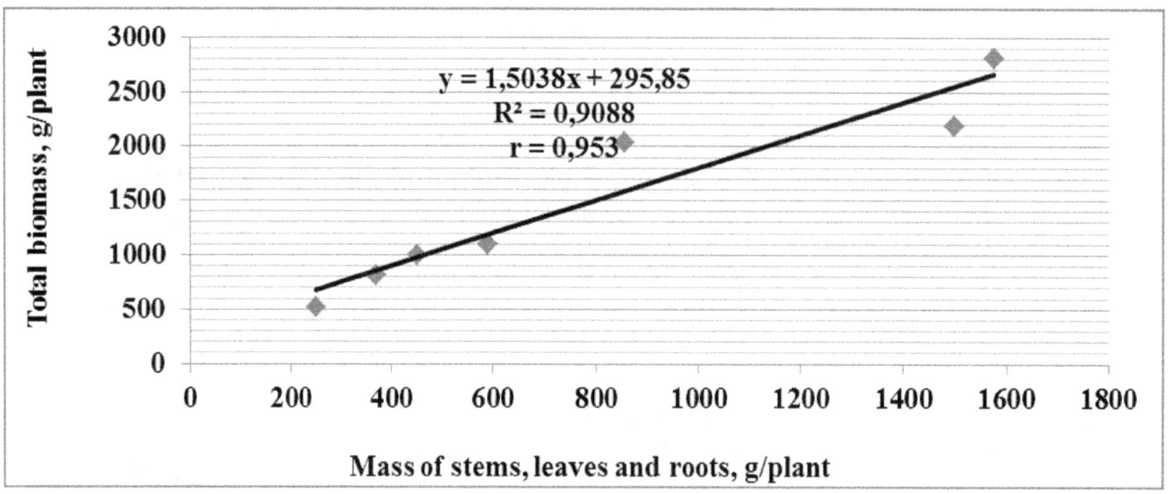

As can be seen from Figures 19, 20 and 21, with an increase in the sum of effective temperatures, a significant increase in the total biomass of Jerusalem artichoke is observed and the correlation between these indicators is r = 0.972. The correlation between the traits weight of stems, leaves, roots and total biomass was r = 0.953 and between the weight of tubers and total biomass was r = 0.982.

Thus, in different environmental conditions of Tajikistan, the main influencing factor for the formation of Jerusalem artichoke productivity is the sum of effective temperatures, which varies significantly depending on the height above sea level.

As follows from Figure 14, the yield of beetles of Jerusalem artichoke varieties at altitudes from 460 m to 840 m above sea level is 63.0 and 34.2 t/ha, respectively. The yield of Jerusalem artichoke at altitudes of 1200 to 2000 m decreases from 23.6 to 14.6 t/ha, and at an altitude of 2560 m above sea level - 10.0/ha.

Thus, as the altitude increases from 470 m to 2550 m, the yield of Jerusalem artichoke tubers decreases from 63.0 t/ha to 10.0 t/ha (Figure 22).

The yield of Jerusalem artichoke tubers at an altitude of 2560 m is 6.3 times higher than at an altitude of 460 m above sea level.

Figure 22. A harvest of tubers and the general biomass of a sun artichok depending on height above sea level (an average for 2016-2019)

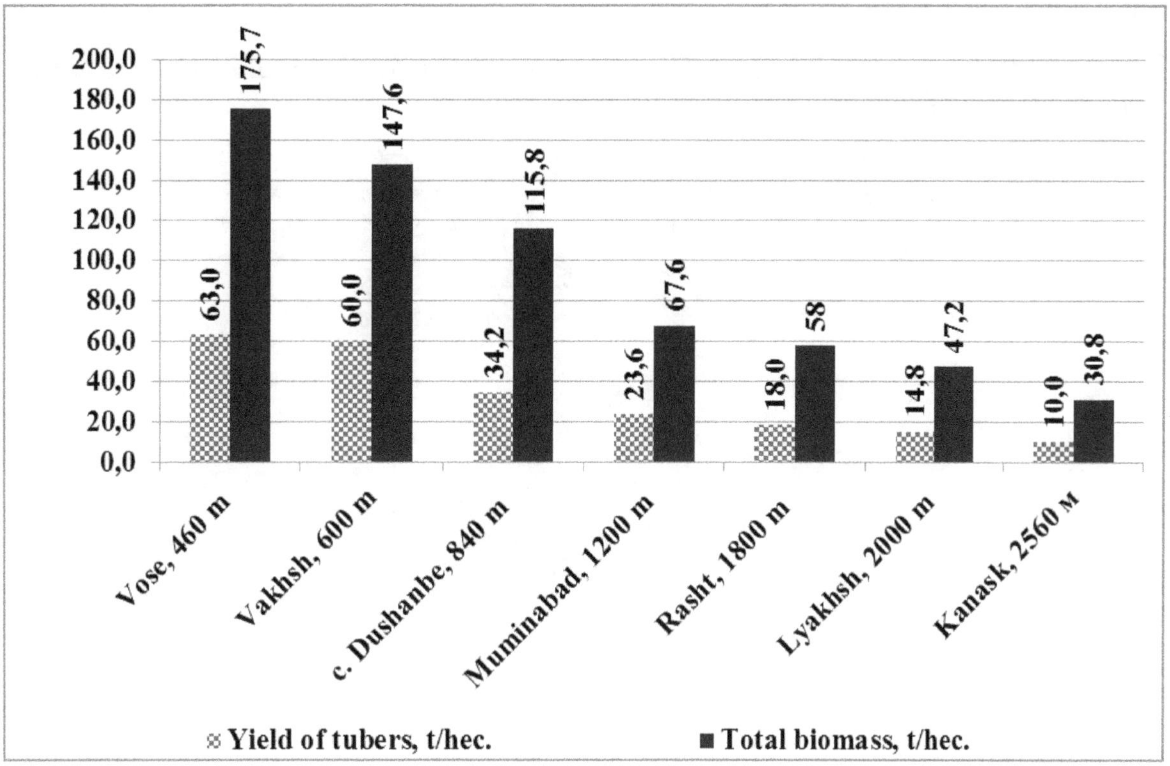

At an altitude of 2560 m above sea level, due to the low level of the sum of effective temperatures, a significant decrease in the yield of Jerusalem artichoke varieties is observed, compared to the cultivation of Jerusalem artichoke at altitudes of 460-840 m above sea level. On average, the tuber yield of Jerusalem artichoke varieties at various levels from altitude above sea level is 31.9 t/ha, which shows the high potential of Jerusalem artichoke plants for tuber yield in the conditions of the Republic of Tajikistan as an important food product.

In the future, Jerusalem artichoke can play an important role in ensuring food security in different agro-ecological conditions of Tajikistan. Analysis of the yield of the total biomass of Jerusalem artichoke also shows a positive effect of altitude above sea level on this polygenic trait of Jerusalem artichoke. For example, if the total biomass of Jerusalem artichoke varieties at altitudes of 460 m and 600 m above sea level, respectively, is 175.7 and 147.6 t/ha, then this figure at altitudes of 2000 and 2560 m, respectively, is 47, 2 and 30.8 t/ha ha or 3.7 and 4.8 times less than at altitudes of 460 and 600 m above sea level.

In terms of the yield of total biomass, Jerusalem artichoke varieties at altitudes of 840, 1200 and 1800 m above sea level also differ significantly from each other (the difference is in the range of 9.6-10.8 t/ha or 16.8 - 22.9%).

We have also studied methods for obtaining various products from the roots, tubers and leaves of Jerusalem artichoke plants, which are shown in the figures below (Figure 23 and 24).

Scientists of the Institute of Botany, Plant Physiology and Genetics of the National Academy of Sciences of Tajikistan during 2010-2022 managed to obtain a new variety of Jerusalem artichoke in the conditions of Tajikistan "Sarvat" and a patent for methods of growing Jerusalem artichoke from roots and stolons, which are given in the text below.

Figure 23. Different agriculture materials (tea, beverages, cellulose, alcohol, disele oil and others manufactured from of Jerusalem artichoke, 2016-2020)

Figure 24. Different agriculture food production manufactured from of Jerusalem artichoke, 2016-2020

CONCLUSION

By planting Jerusalem artichoke in the future, it is possible to increase the yield of biological mass of plants per unit area by two to three times and thereby help strengthen the forage base for livestock farming and food security in the republic. On the other hand, through widespread sowing of Jerusalem artichoke in various soil types, it is possible to significantly increase the economic return of land resources of the Republic of Tajikistan in the future.

Analysis of the yield of the total biomass of Jerusalem artichoke also shows a positive effect of altitude above sea level on this polygenic trait of Jerusalem artichoke. For example, if the total biomass of Jerusalem artichoke varieties at altitudes of 460 m and 600 m above sea level, respectively, is 175.7 and 147.6 t/ha, then this figure at altitudes of 2000 and 2560 m, respectively, is 47, 2 and 30.8 t/ha ha or 3.7 and 4.8 times less than at altitudes of 460 and 600 m above sea level.

In terms of the yield of total biomass, Jerusalem artichoke varieties at altitudes of 840, 1200 and 1800 m above sea level also differ significantly from each other (the difference is in the range of 9.6-10.8 t/ha or 16.8 - 22.9%).

At cultivation topinambour on watering essential strengthening of dynamics of growth of plants, formations of quantity and the area of leaves, biological efficiency and productivity in comparison with cultivation on богарных the earths is observed.

In conditions in a shade and without watering carrying out essentially ингибируются growth and developments of plants, efficiency and productivity of plants, than in the conditions of cultivation without a shade without watering and on watering.

Joint entering of organo-mineral fertilizers (NPK-I +ejhornija-2) on watering to lead to increase in the general biological weight topinambour in comparison with variants without entering of mineral fertilizers (on 161.2%), with entering of mineral fertilizers NPK-1 (on 94.9%), NPK-II (on 64.7%) and ejhornija-2 (on 77.8%) separately.

Also joint entering NPK-I + ejhornija-2 leads to substantial growth of a crop of tubers topinambour in comparison with variants without entering of mineral fertilizers (in 5.1раз) and with entering of mineral fertilizers NPK-I (in 3.4 times), NPK-II (in 2.5 times) and ejhornija-2 (in 2.7 times). Thus, the most effective variant considers joint entering of organo-mineral fertilizers (NPK-I + ejhornija-2) under topinambourом at watering of plants. In this variant it is observed increases in productivity of the general biomass at 161.2% and productivity of tubers more in five times, than in control (without application of fertilizers).

Stimulating action of organic fertilizer ejhornija-2 on formation of a crop of tubers from the general biomass of plants topinambour on watering that has great value in increase in an exit of tubers from area unit is observed. At joint entering органо - mineral fertilizers (NPK (50:50:25 kg i.m.) + ejhornija-2) almost twice increases weights of tubers from the general biomass of plants, than in a control variant.

Different parts of plants topinambour from the general biomass are distributed by a following parity: roots of 23-25%, leaves - 14 - 16%, stalks - 22 23% and tubers - 38 39%. The cultivation variant topinambour on the irrigation earths has not led to essential change of parities of parts of plants from the general biomass in comparison with a cultivation variant on богарных the earths.

The general crop of a biomass topinambour on the irrigation earths of the Gissar and Rasht valley fluctuates within 66.6 - 95.6 t/hectares, and on богарных the earths - 31.6 - 73.4 t/hectares. On the average the biomass crop topinambour on watering makes 77.5 t/hectares, and on богарных the earths - 48.4 t/hectares that testifies to efficiency and utility of cultivation topinambour on irrigation and on rein field of our republic.

Watering promote increase in a crop of a biomass topinambour on 29.1 t/hectares (or on 60.1%) a crop of tubers on 14.2 t/hectares (or on 87.1%) in comparison with cultivation without watering.

The weight and tubers topinambour with the big hunting are eaten by animals, rabbits and birds and topinambour the animal industries and receptions of a biomass for synthesis of biogases and bioetanol in the conditions of our republic in the future are highly effective culture for strengthening of a forage reserve.

It should be noted that, taking into account the growing population of the republic, scientists are faced with the task of finding and implementing innovative ways and methods of obtaining the maximum volume of necessary products and biological mass per unit area while maintaining and increasing the fertility and quality of land and soil resources.

In different environmental conditions of Tajikistan, the average yield of the total biomass of Jerusalem artichoke varieties when grown at altitudes from 460 m to 2560 m above sea level is 91.8 t/ha, which can be an important bioresource for supporting the feed supply in livestock farming, as well as as a product food (tubers) for the population in the future.

Thus, it can be noted that the most suitable zones for cultivating and obtaining high yields of Jerusalem artichoke in the conditions of Tajikistan are areas located at altitudes from 460 to 1200 m above sea level, with a total effective temperature during the growing season of plants from 2280 to 3760°C and relatively with a low humus content (0.5-2.0%) in the arable horizon. In the future, Tajik farmers, by planting Jerusalem artichoke, can obtain a high mass of this crop, which is so important for strengthening the feed supply in livestock farming, as well as food for the population.

Figure 25. The patent about new methods of plantinf sun artichoke and certificate of variety of "Sarvat"

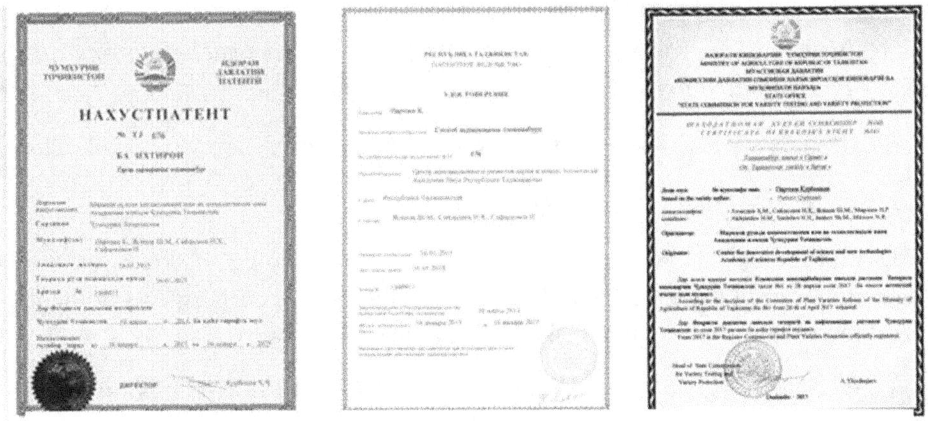
НАХУСТПАТЕНТ

REFERENCES

Akhmedov, Kh. M., & Karimov, Kh. S. (2003). *The possibility for produce and using the biological of gaze in Tajikistan.*

Dospekhov, B.A. (1985). *Technique of field experiment.* M.-L.

Eikhe, E. P. (1957). *Jerusalem artichoke or earthen pear.* M.-L., Publishing House of the USSR Academy of Sciences.

Eikhe, E. P. (1957). *Jerusalem artichoke or earthen pear.* M.L.

Ergashev, A. (1976). Intensity and dynamics of formation of photosynthesis products in Jerusalem artichoke. *Physiol. and biochem. cult. rast.*, *3*(8), 299-303.

Ergashev, A., Partoev, K., & Safarmadi, M. (2019). *Productivity of Jerusalem artichoke varieties.* International scientific and practical conference Current tasks of efficient use of land in the context of innovative development of the agricultural sector in the context of global climate change, Dushanbe.

Filippov, M. I. (1970). Jerusalem artichoke – a valuable forage crop. *Abstracts. V symposium for new silage plants. Part 1.* L.: Kolos.

Gibbons, W. R. (1989). Batch and continuous solid-phase fermentation of Jerusalem artichoke tubers. *Journal of Fermentation Technology*, 67(4), 258–265.

Golubev, V.N., Volkova, I.V., Kushalakov, Kh.M. (1995). *Jerusalem artichoke, properties, composition, processing methods, applications.* M.-L.

Jahoni, A. (2010). *The medicins and food plants.* Dushanbe, Adib.

Kasymov, J. K. (1975). *Agricultural crops of Tajikistan.*

Khalifaev, D. R., Khalifaev, D. R., & Popov, D. M. (2015). *Receiving and standardization of tincture of homeopathic matrix from fresh tubers of Jerusalem artichoke.* Modern aspects of the use of plant raw materials and raw materials of natural origin in medicine, Moscow.

Khalifaev, D. R., Safarzoda, R. Sh., & Gulmamadova, G. G. (2013). *About the technology for obtaining dry extract from the Helianthus tuberosus plant.* The contribution of medical sciences to practical healthcare, Dushanbe.

Khalifaev, D. R., Safarzoda, R. Sh., & Popov, D. M. (2014). *Amino acid composition of grass Jerusalem artichoke cultivated in Tajikistan.* Pharmacy development and registration of medicines, Moscow.

Klausshofer, H., Abraham, B., & Leichtfried, G. (1998). Hydrolaseaktivität bei Topinambur und Zuckerind. *Zuckerind*, 113(3), 209–215.

Kochnev, N.K., & Kolinicheva, M.V. (2002). *Jerusalem artichoke - bioenergy culture of the XXI century.* M.-L.

Kondakov, A. M. (1988). Alternative energy sources. *Geography at school.* M. *Pedagogy*, 185.

Kondratenko V.V. (2008). Complex processing of Jerusalem artichoke into functional food products. *Innovative technologies in the field of refrigerated storage and food processing*, (132-133).

Litvinov, V. N. (1958). *Earthen pear in the Gissar Valley of the Tajik SSR.* [Dissertation... Ph.D. agricult. Sci., Stalinabad].

Litvinov, V. N. (1965). *Forage crops of Tajikistan*. Irfon.

Novikova, L.Yu, Novikova, S.D., & Kiru, E.V. (2017). Manifestation of economically valuable traits in potato varieties (Solanum L.) under climate change on the European territory of Russia. *Agricultural biology*, *52*(1), 75-83.

Partoev, K., Saydaliyev, N.H., Yasinov, Sh.M. & Sadriddinov S. (2015).Correlation communication between signs of a sun artichoke. *News of the Orenburg state agricultural university*, *6*(56), 36-37.

Pashchenko L.P. (1999). Rational aspects in processing Jerusalem artichoke. *Storage and processing of agricultural raw materials,7*, 13-17.

Pasko, N. M. (1999). Jerusalem artichoke - biotechnical potential for food, medicinal, technical, feed and environmental purposes. *Sat. works "Innovative technologies and products"*. Novosibirsk.

Pasko, N. M. (2003). Selection and seed production of Jerusalem artichoke. Sat. scientific works, Moscow.

Praznik, W., & Beck, R. H. (1987). Inulin composition during growth of tubers of Helianthus tuberosus. *Agricultural and Biological Chemistry*, 51(6), 1593–1599. 10.1271/bbb1961.51.1593

Reingart, E.S., Kochnev, N.K., Ponamarev, A.G., & Zvyagintsev, P.S. (2008). Prospects for the use of Jerusalem artichoke for the production of bioethanol. *Achievement of science and technology of the RPK*, *1*, 38-40.

Safarzoda, R. Sh., Khalifaev, D. R., & Popov, D. M. (2014). *Amino acid composition of grass Jerusalem artichoke cultivated in Tajikistan*. Pharmacy development and registration of medicines, Moscow.

Safarzoda, R. Sh., Khalifaev, D. R., & Popov, D. M. (2015). *Amino acid composition of grass Jerusalem artichoke cultivated in Tajikistan.* Pharmacy development and registration of medicines, Moscow, 4.

Saidaliev H., Partoev, K., Yasinov, S. (2015). *Biopotensial of topinambour in conditions of Tajikistan*. Izvestya Orenburg Government agrarian University.

Saidaliev, H. H. Kh. (2018). *Productivity of topinabure depending with cultivations in Gissar and Rasht valyes of Tajikistan* [Dissertation, Dushanbe].

Sardorov, M. N. (1997). *Productivity and photosynthetic activity of combined crops of alfalfa with cereal crops in the conditions of Central Tajikistan* [Dissertation, Dushanbe].

Shain, S.S. (1931). *Ground pear is a fodder crop*. M.-L.

Shazzo, R. I., Kaishev, V. G., Gish, R. A., & Cornea, R. I. (2013). *Jerusalem Artichoke:biology, agricultural cultivation, a place in the ecosystem, processing technology (yesterday, today, tomorrow).* Krasnodar.

Shazzo, R.I., Kasyanov, G.I. (2000). Functional foods. M.-L.

Starovoitov, V.I., Starovoitova, M.B., Voronov, N.V. (2002).Cultures of the XXI century. *Technology and equipment for the village*, *8*, 9-13.

Swanton, C. J., & Hamill, A. S. (2005) *Jerusalem artichoke*. BUKVI. www.bukvi.ru/stat/stat 394.html

Titok, V. V., Rupasova, Zh. A., Kuptsov, N. S., & Popov, E. G. Duba,r D.A., Pashkevich, P.A., & Veevnik, A.A. (2018). *Jerusalem artichoke in Belarus*. Minsk.

Tretyakov, V. F. (2009) *Bioethanol - state and prospects in energy and petrochemicals*. Materials II int. Sci-Tech. conf. Alternative sources of raw materials and fuel, Minsk.

Usanova, Z. I. (1967).Photoperiodism of Jerusalem artichoke. *Proceedings of the Tula State University. agricultural experimental station*. 1,Tula: Priokskoe book. publishing house, 161-181.

Usanova, Z. I. (2005). Programming the yield of earthen pear. *Current problems of agricultural science and practice*. Tver.

Usanova, Z. I., & Ivanov, V. A. (1993). Features of the cultivation of Jerusalem artichoke in the north-eastern part of the Central region of Russia. *Abstracts of reports V interregion. intersectoral. scientific-prod. conf.* Tver.

Usanova, Z.I., Oserbaev, A.K. (2003). Production process of the Skorospelka earthen pear in the northern part of the Central region of Russia. *Non-traditional natural resources and innovative technologies and products*. M: RANS.

Ushakov, T. V. (1932). Earthen pear in Ingushetia. Ordzhonikidze: Publishing house. *Heart (British Cardiac Society)*, 57.

Ustimenko, G.V.(1960). *Earthen pear. M.: State*. Publishing house of agricultural literature.

Wieczochek, A. (1988). Helianthus tuberosus as a potential energy source. *Acta aliment.Hol*, 14(2), 115–122.

Yaroshevich, M. I. (2010). Jerusalem artichoke is a promising multi-purpose crop. *Proceedings of BSU*, 4(2), 1–12.

Yaroshevich, M. I. (2014). *Jerusalem artichoke is a promising multi-purpose crop*.

Yuritsina, T.N. (1998). Jerusalem artichoke is an amazing plant. *Potatoes and vegetables*, *3*, 17-21.

Chapter 5
Unveiling Urban Oasis: Insights From OSM Data on Green Spaces

Munir Ahmad
http://orcid.org/0000-0003-4836-6151
Survey of Pakistan, Pakistan

Froilan Delute Mobo
http://orcid.org/0000-0002-4531-8106
Philippine Merchant Marine Academy, Philippines

ABSTRACT

OpenStreetMap data is a valuable resource for analyzing urban green spaces globally. This study reviews extensive literature showcasing diverse applications of OSM data in understanding green infrastructure distribution, accessibility, and characteristics. Researchers utilize various methodologies including spatial analysis, feature extraction, classification, network analysis, temporal analysis, and integration with other datasets to quantify green infrastructure and assess urban green space quality. OSM data offers substantial opportunities with global coverage, open access, rich attribute information, timeliness, and customizability, enabling nuanced analyses and comparative studies. However, limitations such as data quality variability, inconsistent tagging, and biases in coverage exist, necessitating careful consideration and validation methods. Despite challenges, judicious utilization of OSM data can advance understanding of urban green spaces, aiding evidence-based decision-making in urban planning and environmental management.

INTRODUCTION

Green spaces within an urban region play a crucial part in developing environmental sustainability, public health, and quality of life in cities around the globe. Often referred to as the "the lungs" of urban areas because they bring numerous (and very diverse) positive effects to both people and the surrounding environment. Cities' freshness is raised by green urban spaces. From mitigating air and noise pollution to

DOI: 10.4018/979-8-3693-3374-7.ch005

reducing urban heat island effects, urban green spaces contribute significantly to improving the overall ecological balance of cities (Anguluri & Narayanan, 2017; Badiu et al., 2016; Ramaiah & Avtar, 2019).

Further, urban green spaces contribute to recreational activities for the local residents with opportunities for physical team exercises and also become a meeting point for relaxing and socializing (Dos Santos et al., 2019; Kothencz et al., 2017). Research has demonstrated that having green spaces in one's vicinity positively affects their mental health (Nutsford et al., 2013), helps them to deal with stress, and promotes stronger community bonds. For one, the urban parks are not just aesthetic spaces but also provide habitats for many different types of plants and animals which maintain the biodiversity and thrive urban ecological processes (Threlfall et al., 2017).

Consequently, city planners and policymakers are currently turning their focus to the sustainable extension of green spaces as a core issue. However, it is essential to have accurate and up-to-date spatial data for effective management of such places to well and systematically plan these areas. This is where OpenStreetMap (OSM) data emerges as a valuable resource. OSM functions as a collaborative mapping initiative, offering an openly accessible dataset that serves as a valuable resource for businesses (Wiki, 2023). OSM data provides a valuable resource for analyzing and understanding the distribution, characteristics, and significance of green spaces within urban environments. OSM plays a pivotal role in advancing sustainable development across economic, environmental, and societal domains (Ahmad & Ali, 2023).

To this effect, the primary objective of this chapter is to bridge the gap between the valuable resource of OSM data and its application in urban green spaces. To achieve this goal, the chapter is structured logically, moving from general concepts to more specific and targeted applications. The initial sections provide essential background information, emphasizing the value of both OSM data and green spaces independently. Subsequently, the chapter narrows its focus to demonstrate how OSM and green spaces can be effectively integrated to increase mutual benefits.

URBAN GREEN SPACES

Urban green spaces are all areas in the cities that are predominantly greenery covered with vegetation of parks and gardens, green belts, urban forests, and other natural or semi-natural areas. This type of space is either created on purpose or stands preserved because it offers the residents ecological, social, and economic benefits.

Environmental Benefits

In terms of ecological and environmental advantages, urban green spaces play a pivotal role in supporting the urban environment through various means, including improving air quality, regulating temperatures, managing stormwater, and preserving biodiversity. Industry and traffic have evolved to be the major pollutants in urban ecosystems; trees and plants can remove a considerable number of various pollutants, such as CO2 and NO2, by naturally purifying the atmosphere. By sequestering CO2, urban green spaces act as carbon sinks, helping to mitigate the negative effects of climate change by reducing the amount of greenhouse gases in the atmosphere. Tree branches and leaves can absorb NO2 through stomata on the leaf surface. This helps in mitigating air pollution besides enhancing human health and the environment as well. Therefore, it also improves the entire experience of urban life (Mwendwa & Giliba, 2012).

Urban spaces have higher temperatures than the outskirts and temperature is usually intensified by human activities which include industrialization, transportation, and the building of highways. Land cover like concrete, asphalt, and buildings absorb and retain the heat that builds up elevated urban temperatures, especially in densely constructed areas that have less vegetation to provide shade. On the other hand, trees and shrubs give out natural shade through their canopies which bring down the amount of solar radiation reaching the ground. Decreasing temperatures in cities can provide significant benefits to the livability of urban areas The green infrastructures and urban greening projects can be leveraged to allow cities to cope with climate change, which in turn will contribute to shaping a more sustainable and friendly urban environment (Jaganmohan et al., 2016; Oliveira et al., 2011; Sugawara et al., 2016; Yang et al., 2017).

Urban communities often face many challenges when it comes to stormwater management because roads, sidewalks, and buildings that are present are all impenetrable to rainwater and, as a result, rainwater cannot infiltrate the ground (Berland et al., 2017; Ibrahim et al., 2020; Prudencio & Null, 2018). That causes excessive overlanding of water, which can overwhelm drainage systems and lead to flooding situations during periods of heavy rain. Green spaces like parks, gardens, and green belts, play an unquestionable role in reducing these challenges as they work as natural stormwater management systems. Components of green nature, such as trees, grass, and even shrubs, use their roots and leaves to absorb the rainwater. This mechanism allows water to infiltrate into the soil, reducing the volume of runoff that flows over impermeable surfaces.

Urban green spaces represent a significant conservation area for numerous herbaceous and animal species and, to some extent maintain the biodiversity of a given urban area (Aronson et al., 2017; Kong et al., 2010; Threlfall et al., 2017). Green spaces cover a range of habitats, from forests to wetlands, meadows, and riparian areas. This kind of landscape sets the stage for diverse ecosystems, which provide a home to many plant and animal species. The urban pockets of greenery (such as parks, gardens, and cyclic motion routes) fulfill diverse ecological functions (food, housing, breeding ground, and migration) for urban wildlife. The green spaces usually provide a conducive environment for pollination and other processes of ecology which include: seed dispersal, nutrient cycling, and soil formation. The reserves create functional space for nature, and together with preserving the habitat of native species, support the urban ecosystem resilience.

Social Benefits

The social benefits of urban green spaces include recreation & leisure, community gatherings, mental health improvement, and educational opportunities (Baur, 2018; Wajchman-Świtalska et al., 2023). Recreational and leisure areas provided by green areas and parks are exceedingly influential in maintaining urban inhabitants' mental and physical well-being. Public parks and green spaces are places where people can indulge in several enjoyable activities such as walking, jogging, biking, playing games, etc. Doing such sports leads to individuals' health improvement since cardiovascular health and strength are improved, weight is managed through planning meals and finally, the obesity and inactive lifestyle is reduced. Parks are community spaces where groups of people with different kinds of opinions get together and exchange thoughts. green space around us not only disclosed productivity but also brought an overall sense of well-being and happiness that ultimately made the urban environment more pleasant and habitable.

The green spaces, play as the vital meeting points for the communities, where the community residents can congregate and socialize at the pleasant venues to either hold ceremonies and celebrations or just entertain themselves with different activities (Bell, 2005; de Kleyn et al., 2020). Public green spaces enable everyone, irrespective of their age, origin, or preferences, to come together, come and socialize. Whether it is a park in the neighborhood, a community garden, or a plaza of the community these are places where people encounter each other and become associated. Moreover, green spaces are usually the centers of a number of activities in the city including concerts, events, fairs, and community gatherings. Those events connect people and give cultural diversity when people integrate and form a harmonious community alike. Furthermore, green spaces can be a hub for the community where they engage and partake in civic activities. Community members can join in environmental activities like pouring clean-up events, gardening projects, or advocacy campaigns to better and protect the shared spaces which represent a sense of unity and agency for our residents.

Access to green spaces contributes to better mental health and the enjoyment of life (Astell-Burt & Feng, 2019; Nutsford et al., 2013). There is a specific aura of psychological relaxation and physical detachment that can be achieved by engaging outdoors which leads to lowering the level of stress hormones like cortisol. Green spaces are sanctuaries that can help to deliver to people the quietude they need to shut out the pressures of the city lifestyle as they move away from the hustle and bustle and rejuvenate. Local and public green spaces promote physical activity with numerous people enjoying walking, cycling, and jogging in parks, which are also known for relieving stress and improving mood. The action of exercise initiates the production of the neurotransmitters called endorphins, which are associated with feelings of happiness and mental serenity, and therefore mental health is functioning properly.

Through urban greening, interactive learning activities can be put in place so as to creatively exhibit nature, ecology, and sustainability. Educational institutes and community organizations have established environmental education programs, which play an important role in providing guided tours for educational purposes. These programs provide a broad knowledge ranging from plant identification to animal observation, ecosystem dynamics, and environmental conservation. There are some urban green spaces that feature nature interpretation centers or visitor centers which contain educational exhibits, interactive displays, and a system of education. These facilities assist in educating locals concerning the kinds of ecosystems, biodiversity, and environmental problems existing within the community. Educating through activities such as learning how to plant trees, making things from nature, discovering birds around the area, and gardening in green areas helps people in practical ways of raising their curiosity, thinking critically, and environmental stewardship skills.

Economic Benefits

Similarly, the economic benefits of urban green spaces include property value enhancement, tourism and recreation revenue, and healthcare cost reduction. Developing green spaces around and growing plants has been shown to relate to good property value and economic development (Mwendwa & Giliba, 2012). Green areas are an essential component appearing in the scenes of the prevailing neighborhoods so that they enhance their visual appeal and contribute to the creation of the sensation of beauty and serenity. Properties beside the parks, natural sceneries, or green areas tend to be more expensive than others owing to their picturesque views and environment of nature. The residents get to enjoy a high-quality life through close access to the green spaces that offer opportunities for recreation, relaxation, and social interaction. Properties near parks or recreational venues are considered more desirable thus, they are likely to increase demand and property values. Neatly preserved green spaces also contribute to the attractiveness of commercial districts and thus the shopping malls, inviting many to go and do their shopping there. Businesses existing alongside green spaces tend to attract more pedestrians and thus augment their sales as the economy around such an area is boosted (Biao et al., 2012; McCord et al., 2014).

Green spaces are usually the main components of the tourism industry allowing longer stays for tourists thus having positive economic effects for local communities (Lestari & Pratika, 2020; Muliasari et al., 2021). Properly cared parks, botanical centers, and natural reserves attract tourists and visitors who search for recreation amid nature and cultural activity. Moreover, these green spaces serve as reflection points for sightseeing, picnicking, bird-watching, and photography. When tourism and recreation happen in these green spaces, it increases the economic activity of the neighborhood around. To the local business community, tourists spend money on a variety of services and leisure time, such as accommodations, restaurants, cafés, and souvenir shops. Furthermore, green spaces have different kinds of activities and events going on all year round, and that attract big crowds of visitors which generates business revenue. Such activities are dances, food fests, art displays, and outdoor markets in parks and public squares.

Green spaces lead to better public health outcomes by contributing to the reduction of medical expenses related to sedentary lifestyles and stress disorders (Zhang et al., 2021). Green spaces allow people to get involved with outdoor recreation and exercises like walks, runs, cycling, and sports. The constant energy burning without increasing weight reduces the risk of obesity, cardiovascular diseases, diabetes, and other chronic disorders, which helps healthcare budgets by reducing the expenditure on these disorders. Green spaces serve as venues for socializing, community events, and group activities, fostering social connections and support networks. Strong social ties have been linked to better mental and physical health outcomes, reducing the likelihood of isolation, loneliness, and associated healthcare needs.

OPENSTREETMAP (OSM)

OpenStreetMap (OSM) is a Wiki mapping project for which users from all over the world are able to generate, edit as well as share their geospatial information. It is a freely available and freely editable platform, which allows any person to create maps for different possible applications – travel navigation, planning city, disaster response, environment conservation, etc. OSM uses the volunteer force, sometimes called "mappers", to boost the map quality and to improve its properties. This volunteer group consists of people with different origins and locations and what they have in common is the fact that they are either professional cartographers or other kinds of people who have got an interest in the mapping. The map editing tools from OSM are convenient and user-friendly, meaning that volunteers can directly add and edit map features in their web browser as if they are drawing in a drawing application (Bertolotto et al., 2020).

The most fundamental operation called iD Editor possesses an easy-to-use interface that helps users generate all the components of the maps such as roads, buildings, landmarks, and so on. Moreover, OSM and the available editing tools, including JOSM (Java OpenStreetMap Editor) and Potlatch, with their advanced editing features and offline editing function provide the advanced capability of map editing to the community of mappers. Such tools are preferred by senior mappers or those who already have mapped for some projects. OpenStreetMap is based on accessibility, which implies that anyone can become a mapmaker from any part of the world, with any kind of background, and cartographic expertise. The OSM community pays great support to the new mappers by providing online forums, mailing letters, and through social media. Advanced mappers provide help and guidance to beginners showing their different mapmaking methods, advice on appropriate practices, and the tag patterns used in OSM.

OSM enables the availability of map data to the general public without fees, without charging for access, and does not impose any restriction for its use. People, businesses, and organizations can use it freely since they won't have to purchase a license or make a monthly contribution to obtain the mapping data. The map data in the OpenStreetMap is licensed under the Open Database License (ODbL) which is an open-source license permitting it to be redistributed without cost and for other purposes as long as the original author of the data is credited. ODbL license establishes freedom for both commercial and non-commercial use of the map data. It thus implies that the OSM platform enables businesses to use their data for developing apps like routing, navigation, and location-based services. On the other hand, people and nongovernmental organizations can use it for research, advocacy, and community programs. Despite the Open Database License, it necessitates the provision of proper attribution to OpenStreetMap and all additions made to the data must be licensed under the same Open Database License. This ensures transparency and accountability in the use and distribution of OSM data.

The OSM layer includes almost all objects of a geographical character such as roads, buildings, landmarks, natural features (e.g., rivers, mountains, and forests) as well as attractions that may be of interest to tourists and visitors (e.g., businesses, amenities, and tourist attractions). OSM strives to cover the entire world from urban-populated areas to remote rural areas or even wilderness areas. This worldwide jurisdiction ensures that OSM can be applicable and significant in all areas that users may want to check out, regardless of their geographic location. OSM is continuously updated and improved by its community of contributors, ensuring that the map data remains current and accurate. Users can, for instance, add new features, edit the map data, and verify the actuality of the information depicted on the map in order to keep the map up-to-date with any possible changes in the real environment (Anderson et al., 2019).

The tagging system in OSM allows users to describe map features with specific attributes and metadata, providing a flexible and customizable framework for categorizing geographic data. OSM signifies each map element as a "node", "way", or "relation" and users can carry out the tagging in order to provide some contextual information (descriptive information) about the characteristic of the map features. For instance, a road segment may be labeled as "road_type=highways" to denote that it is a primary road. OSM tagging system is very dynamic and enables the users to create custom tags and add more specifics for the map objects, such as their attributes. This is the most important advantage, which can display a multitude of geographic aspects, from transport networks, land-use patterns, amenities, POIs, and administrative boundaries. While users can make their own tags, OSM maintains a mechanism for a uniform tagging approach to ensure consistency and interoperability of map data. In addition to other considerations, these conventions enable proper tagging, thus increasing consistency across different mapping projects and regions. The interoperability of the Tagging system in OSM gives it the potential to serve virtually all the needs of different users, such as navigation, urban planning, environmental monitoring, disaster response, and tourism. It stays on the top list for users' customization of geographical data as they meet the needs of the specific location and tastes.

Maintaining the accuracy and reliability of map data is essential for the usability and credibility of OSM. OSM provides validation tools and plugins that allow users to check and correct mapping errors, discrepancies, mistakes, inconsistencies, and omissions in the map data. These tools help users in the observation and identification of tagging conventions, detection of geometry errors, and also verification of the completeness and correctness of the map features. The OSM community plays a crucial role in reviewing and verifying the accuracy of map data. A board of experienced mappers and validators is responsible for checking the changesets that come from the users; they give feedback, solve disputes, and undertake the necessary measures to make sure that there are no mistakes in the mapping data. OSM runs artificial intelligence algorithms to suppress prevalent mapping abnormalities. These algorithms trace the suspicious changesets, identify potential vandalism or data vandalism, and pinpoint areas that require review or verification by the community of the system. OSM encourages users to cite reliable data sources and references from other users when they either add a new feature or edit an existing feature. Through referencing trustworthy sources, contributors can amplify the credibility and accuracy of their input and make sure that the data on the map is derived from verifiable information.

OSM data is identifiable as a very versatile piece of architecture, which can be put to work in a large number of mapping programs, travel navigation systems, and GIS. OSM platform has APIs that allow apps to interact with its data automatically. These APIs make it possible to fetch map data, perform spatial queries, and update the map database in real time. Developers may create custom maps using OSM data for a wide variety of use cases. Through APIs, developers have the opportunity not only to customize map styles for their purposes but also to add layers on top of a base map and to integrate interactive features such as markers, tooltips, and overlays. OSM data can be used to develop routing services that provide directions and navigation assistance to users. Developers can implement routing algorithms that calculate optimal routes for various modes of transportation, such as walking, cycling, driving, and public transit, based on OSM's road network and transportation infrastructure data. OpenStreetMap data helps geospatial studies by providing a wide array of spatial data suitable for spatial computing and spatial science projects. OSM developers can perform spatial queries, and analysis of patterns and trends to derive deeper insights about urban mobility, land use, environmental conditions, and other geographical phenomena.

OSM is also a vital tool for humanitarian agencies and disaster management operations and has brought the huge potential of volunteer mapping to help relief efforts in crisis-affected areas (Solís et al., 2021). An important role for volunteer mappers is mapmaking efforts on behalf of people who are affected by natural disasters, conflicts, or humanitarian crises. The volunteers utilize satellite images, aerial images, and other remote sensing data to map infrastructure, settlements, elements of hazards, etc on the ground. The mapping work of humanitarian efforts focuses on mapping the damaged buildings in the natural disaster zone, the road networks, the evacuation routes, the water sources, the medical facilities, and other key features that can play important roles in emergency responses. These maps, produced through humanitarian mapping efforts, confer valuable geospatial awareness and nodal context to humanitarian responders. The humanitarian mapping project also bridges gaps between the volunteers, humanitarian organizations, government agencies, and the local people using community engagement and collaboration. Through the common effort of mapping the crisis areas, volunteers directly contribute to the saving of lives, reducing vulnerabilities, and building resilience in regions frequently affected by the disaster.

OSM produces a result of its communal members themselves by pursuing projects that serve the purpose of boosting the geo-coverage, accuracy, and resolution in a particular place. The map schemes of these projects can either be directed towards filling voids on the map, updating old data, or even adding specialized levels of info to the map-dependent on their priorities.

OPENSTREETMAP FOR URBAN GREEN SPACES

Numerous investigations have explored the utilization of OpenStreetMap (OSM) data to facilitate the analysis of green infrastructure in various contexts. For instance, Kumar & Vuilliomenet (2021) employed OSM data to quantify the presence of green infrastructure across 184 European cities within the framework of the CCM coverage. Additionally, O. Klimanova et al. (2021) utilized OSM to compile an inventory of ecosystems actively supporting green infrastructure. Moreover, Ma et al. (2023) assessed the accessibility of green spaces to green infrastructure in China by extracting green space data from the OSM platform.

In studies by O. A. Klimanova & Illarionova (2020) and Arthur & Hack (2022), OpenStreetMap data was utilized to derive land use categories, identifying and classifying agricultural lands, green areas, as well as landmarks such as streets, parking lots, and public parks. Furthermore, Venter et al. (2020) employed OSM data to identify and delineate parks based on their geometries and attribute descriptions, encompassing designations like parks, playgrounds, and protected areas.

Ludwig et al., (2021) specifically focused on analyzing land use parcels obtained from OSM data in the city of Dresden, Germany. Texier et al., (2018) discovered that the OSM dataset closely resembled official cadaster-based approaches like UrbIS in Brussels when analyzing urban green spaces. It is important to note that OSM data may not comprehensively represent municipal commonages or city green belts on the outskirts of smaller cities. Despite this limitation, Venter et al., (2020) have often found OSM data to be the most accessible option, particularly when seeking national coverage. The literature on the utilization of OpenStreetMap data for analyzing green infrastructure and green spaces is extensive, with Table 1 providing a summary of selected studies, including details such as the country of focus, the scope of analysis, and specific tags or keys employed.

Table 1. Selected studies on OSM for green spaces

Tags/Keys	Values	Country	Analysis	Reference
Natural	wood, tree_row, tree, scrub, heath, grassland, fell, shrub	Europe	Correlation between green infrastructure and the cultural vibrancy	(Kumar & Vuilliomenet, 2021)
Landuse	allotments, farmland, farmyard, flowerbed, forest, meadow, orchard, vineyard, cemetery, grass, greenhouse_horticulture, plant_nursery, recreation_ground, normali_green			(Kumar & Vuilliomenet, 2021)
Leisure	garden, nature_reserve, park			(Kumar & Vuilliomenet, 2021)
Landuse	farmland	Russia	Quality of green infrastructure	(O. Klimanova et al., 2021)
Vegetation	wood			(O. A. Klimanova & Illarionova, 2020)
Landuse	forest, orchard, grass, meadow, village_green, recreation_ground, garden, park, allotments, cemetery			(O. A. Klimanova & Illarionova, 2020)
Not Mentioned	park, playground, and protected area	South Africa	Distribution of green spaces	(Venter et al., 2020).

continued on following page

Table 1. Continued

Tags/Keys	Values	Country	Analysis	Reference
Land use	forest; recreation ground, grass, meadow, orchard, greenfield, village green, and vineyard	Belgium	Spatial dimensions of urban green spaces	(Texier et al., 2018)
Leisure	common, dog park, garden; golf course; nature reserve; park and pitch			(Texier et al., 2018)
Natural	scrub; fell; grassland; heath; moor and wood.			(Texier et al., 2018)

As indicated in Table 1, the studies referenced encompassed a diverse array of green space tags, offering valuable insights into the analysis of green infrastructure and its multifaceted applications across diverse regions.

Methodologies for Analyzing Urban Green Spaces

Analyzing urban green spaces using OpenStreetMap data involves several methodologies that leverage the attributes and spatial information available in the dataset.

Spatial Analysis

Spatial analysis involves studying geographic data and exploiting its use to identify spatial patterns, connections, and dynamics. In urban areas, OSM data can be used for spatial analysis to study spatial distribution, density, and accessibility of green spaces within a city or region (Mete & Yomralioglu, 2023; Spangler et al., 2023). Some common spatial analysis techniques include:

Proximity Analysis: This involves measuring the distance between urban green spaces and other features or locations, such as residential areas, schools, or public transportation hubs. The proximity analysis method can help identify the hard-to-reach places to support better urban planning and development strategies.

Spatial Clustering: Spatial clustering methods group together the features that are of a similar type or location, depending on their adjacency. In the context of urban green spaces, it can highlight the tendencies in the distribution of green space, which can be the neighborhood parks, or green corridors, and help to visually understand urban green network structure.

Spatial Interpolation: The spatial pattern interpolation methods involve predicting values to be filled in cases where either data is missing or data is not available. In the general context of urban green spaces, spatial interpolation helps in the estimation of uncovered or insufficient green space data in those particular areas of urbanites. The use of spatial interpolation is beneficial in mapping green space distribution across the city's local authorities to make the right decisions toward the city's greening.

Feature Extraction

Feature Extraction focuses on identifying and extracting specific features or characteristics from geospatial datasets, such as OSM data. The extraction of the features from the OSM data of urban green spaces involves the detection and extraction of data related to parks, gardens, forests, and semi-natured areas. Feature extraction techniques may include:

Querying OSM Data: Researchers can query OSM data, based on a set of predefined tags or attributes, associated with green spaces: for instance, "leisure=park" or "landuse=forest" to find the specified features.

Extracting Spatial Features: Having identified relevant features, a set of spatial features such as boundaries, centroids, or area measurement values can be extracted from OSM data that can be used for further analysis. This helps researchers in assessing the magnitudes of urban green spaces, as well as the areas, and spatial coverage of those spaces within a city or region.

Classification and Mapping

Classification and mapping operate on categorizing and visualizing urban green spaces based on their characteristics and spatial attributes. When it comes to OSM data, classifying and mapping techniques can be integrated to discriminate between different species of green spaces and map their location area. Some common techniques include:

Machine Learning: The machine-learning algorithms can be trained to divide OSM data into various categories of features, like parks, playgrounds, sports centers, or natural preservations, by using their properties and spatial attributes. This helps in the automated identification of urban green areas by city perspectives at a larger scale (Carranza et al., 2022).

Image Classification: Image classification techniques can be incorporated with satellite/aerial imagery data combined with OSM data to classify terrain types and urban green spaces. This provides a visual representation of green space distribution and helps identify areas of interest for further analysis or intervention (Ludwig et al., 2021; Zurbaran et al., 2019).

Network Analysis

Network analysis entails studying the connectivity and accessibility of green spaces that are networked within a city or in a region by examining the network of streets, highways, and other means used for transportation. As a result of network analysis from OSM data, pedestrian and bicycle routing links and green space connections between a lot of green spaces can be identified.

Pedestrian and Cycling Networks: OSM data containing information such as the location of footpaths, cycle lanes, and other walking/cycling facilities is important. By employing this data, the researchers can determine possible routes for people on foot or bikes to connect them to different urban green spaces and facilitate their movements between parks, trails, and recreational areas (Cohen & Dalyot, 2021).

Identifying Barriers and Obstacles: The network analysis tool can also help to expose areas that are prone to road blockages like highways, busy streets, or obstacles with physical barriers like rivers or railways. Awareness of these hurdles is crucial to ensure connecting and visiting of greenspaces as well as efficient urban mobility.

Temporal Analysis

Temporal analysis involves the monitoring of changes in urban green spaces using dynamic data from past OSM datasets. With this methodology, researchers can be in a better position to track the development of green lands, discover patterns, and realize why certain factors lead to the creation of large, separated, and high-quality green spaces.

Tracking Changes: Temporal analysis can help to demonstrate movement in growth or reduction of urban green spaces over a period of time, and settlement patterns or biochemical cycles within such areas. Such knowledge is important in manning and planning urbanization, evaluating the performances of conservation and proactively identifying the areas at risk of environmental degradation.

Impact Assessment: Both temporal analysis and spatial analysis may be employed in order to determine the impact of urbanization, environmental policymaking, or natural disasters on available and accessible green space. Through the analysis of various data such as historical OSM data with other datasets, researchers may reveal if these catchments have significantly affected the number and quality of urban green space.

Integration With Other Datasets

Integration with other datasets involves combining OSM data with additional sources of information, such as satellite imagery, census data, or environmental data, to enhance the analysis of urban green spaces.

Satellite Imagery: Overlaying OSM information with satellite images gives a chance to analyze vegetation coverage, track environmental changes, and spot places with environmental significance for urban settings.

Census Data: Integration of demographic data in the OSM data helps researchers calculate the equity of green space spread, search for communities that use green spaces least, and in detail review the disparities in access to green areas depending on social, economic, and other factors.

Environmental Data: Combining OSM data with environmental datasets allows researchers to analyze the ecological characteristics of urban green spaces, such as biodiversity, habitat quality, and ecosystem services. This information can inform conservation efforts and urban planning decisions aimed at preserving and enhancing green space ecosystems.

Opportunities for Analyzing Urban Green Spaces

OSM data offers many opportunities for analyzing urban green spaces, including its global coverage, accessibility, and rich attribute information as illustrated below.

Global Coverage

OSM covers different regions on a global scale which means that it is relevant for the cities, countries, and geographic areas around the world (Joseph et al., 2023; Minaei, 2020; Zhang et al., 2015). Having global coverage placed OSM as a valuable tool for analyzing different regions, and facilitating comparative studies and assessments of the green spaces in any urban area around the globe.

Free and Open Access

Sharing of OSM data is neither charged nor subjected to any conditions that hinder the public from accessing it. This open access drives large-scale usage and inclusion of people, groups, and different global systems to contribute to OSM datasets. The outcome of this is that OSM enjoys bigger and varying communities that produce better, updated, and more detailed geographic data.

Rich Attribute Information

OSM is rich in attribute information and helps to access more specific information about urban green spaces, explaining its characteristics and features more deeply. This includes information about the types of green spaces (e.g. parks, gardens, forests) their sizes, amenities (e.g. playgrounds, sporting facilities), and accessibility features (e.g. footpaths, entrances). Such comprehensive multi-attribute information allows experts to conduct nuanced analyses to assess and explore the nature of these spaces.

Timeliness

OSM information is revised constantly by contributors, and thus it has always remained current and fully represents changes and advancement on the ground. This continuous updating ensures that the information on urban green spaces available in OSM remains relatively up-to-date compared to traditional databases which have a longer and less frequent update cycle. An abode of current information makes it easier for researchers to record their analysis and thus improve the relevance and accuracy of the reports.

Customizability

OSM data can be customized and tailored for specific research analysis and therefore, researchers can present any kind of data they want related to their objectives. This extradition allows researchers to focus on particular aspects of urban green spaces, such as specific types of green infrastructure or particular geographic areas. Researchers can modify OSM data according to their research needs which helps them to conduct detailed and meaningful studies that can be referenced directly to their research questions and objectives.

Limitations in Analyzing Urban Green Spaces

Researchers should be mindful of its limitations, particularly regarding data quality, and consistency as elaborated below.

Data Quality

Although updates and improvements take place on OSM datasets, their quality can vary depending on factors such as the region and the level of contribution (Yeboah et al., 2021). In certain regions, data may be not available or limited with patches on it due to the small number of mappers and less active mapping community. This variability in data quality can pose challenges to the accuracy and reliability of the data for researchers relying on OSM data for their analyses.

Inconsistent Tagging

OSM classifies urban green spaces based on user-generated tags. However, the tagging of features in OSM can be variable across different regions and between different contributors, though. Such variability can prove a difficult challenge when attempting to do a consistent analysis of urban green spaces because the same types of features may be defined using different terms or following different classification criteria. Therefore, it becomes a necessity for researchers to carefully consider tagging features and in the case of the multiplicity of conventions conduct a possible standard harmonization.

Lack of Standardization

Unlike authoritative datasets sanctioned by the government, OSM does not operate under standardized protocols and regulatory guidelines for data approval. The absence of a uniform approach can hamper the reliability and uniformity values of the data as various contributors may follow completely different steps in data mapping and attribution activities. Thus, OSM information can only be trusted within some quality standards, and must not be fully relied upon as it often does not abide by standardized rules.

Limited Detail

While OSM supplies the location and basic characteristics of urban green spaces, it might not have the exact information that is essentially required for some kinds of analysis. For instance, currently, OSM has no information about vegetation types, ecological characteristics, or detailed boundary information for urban green spaces which are just initialized as recreation areas. Through the lack of many crucial parameters, some crucial analyses may find constraints in information availability.

Bias and Coverage Gaps

OSM data may exhibit biases in coverage, with certain areas or types of urban green spaces being overrepresented or underrepresented. For example, urban areas with more active OSM communities may have more detailed and up-to-date data compared to rural areas or regions with fewer contributors. Additionally, there may be gaps in coverage for certain types of urban green spaces or features, which can affect the comprehensiveness of analyses conducted using OSM data. Researchers need to be aware of these biases and coverage gaps when using OSM data and consider alternative data sources or validation methods to mitigate potential biases.

CONCLUSION

OpenStreetMap data presents a valuable resource for analyzing urban green spaces across various regions globally. The extensive literature reviewed in this study demonstrates the diverse applications of OSM data in understanding the distribution, accessibility, and characteristics of green infrastructure. From quantifying green infrastructure presence to assessing the quality and spatial dimensions of urban green spaces, researchers have employed OSM data in a range of methodologies, including spatial analysis, feature extraction, classification, network analysis, temporal analysis, and integration with other datasets.

The opportunities presented by OSM data are substantial. Its global coverage, free and open access, rich attribute information, timeliness, and customizability empower researchers to conduct nuanced analyses tailored to their specific research objectives. OSM data enables comparative studies across diverse geographic regions and facilitates collaboration among global communities to contribute to and update geographic datasets continually.

However, it is essential to acknowledge the limitations and challenges associated with OSM data. Variability in data quality, inconsistent tagging, lack of standardization, limited detail, and biases in coverage can affect the reliability and comprehensiveness of analyses. Researchers must exercise caution and consider alternative data sources or validation methods to address these limitations effectively.

While OSM data offers significant opportunities for analyzing urban green spaces, researchers should approach its utilization judiciously, considering both its strengths and limitations. By doing so, they can harness the power of OSM data to advance our understanding of urban green spaces and support evidence-based decision-making in urban planning and environmental management efforts.

Future research could focus on improving the quality and consistency of OSM data, particularly in regions with limited coverage or sparse contributions. Collaborative efforts to engage local communities and increase mapping activities could help address data quality issues and enhance the reliability of OSM datasets. Developing standardized tagging conventions for urban green spaces within OSM could facilitate more consistent and reliable analyses. Future studies could explore the feasibility of establishing common tagging guidelines to ensure uniform classification and attribution of green space features across different regions and contributors.

REFERENCES

Ahmad, M., & Ali, A. (2023). Mapping the Future of Sustainable Development Through Cloud-Based Solutions: A Case Study of OpenStreetMap. In *Promoting Sustainable Management Through Technological Innovation* (pp. 153–176). IGI Global. 10.4018/978-1-6684-9979-5.ch011

Anderson, J., Sarkar, D., & Palen, L. (2019). Corporate editors in the evolving landscape of OpenStreetMap. *ISPRS International Journal of Geo-Information*, 8(5), 232. 10.3390/ijgi8050232

Anguluri, R., & Narayanan, P. (2017). Role of green space in urban planning: Outlook towards smart cities. *Urban Forestry & Urban Greening*, 25, 58–65. 10.1016/j.ufug.2017.04.007

Aronson, M. F. J., Lepczyk, C. A., Evans, K. L., Goddard, M. A., Lerman, S. B., MacIvor, J. S., Nilon, C. H., & Vargo, T. (2017). Biodiversity in the city: key challenges for urban green space management. In *Frontiers in Ecology and the Environment, 15*(4). 10.1002/fee.1480

Arthur, N., & Hack, J. (2022). A multiple scale, function, and type approach to determine and improve Green Infrastructure of urban watersheds. *Urban Forestry & Urban Greening*, 68, 127459. 10.1016/j.ufug.2022.127459

Astell-Burt, T., & Feng, X. (2019). Association of Urban Green Space with Mental Health and General Health among Adults in Australia. *JAMA Network Open*, 2(7), e198209. 10.1001/jamanetworkopen.2019.820931348510

Badiu, D. L., & Ioj, . (2016). Is urban green space per capita a valuable target to achieve cities' sustainability goals? Romania as a case study. *Ecological Indicators*, 70, 53–66. 10.1016/j.ecolind.2016.05.044

Baur, J. W. R. (2018). Urban green spaces, recreation and spiritual experiences. *Leisure/ Loisir, 42*(2). 10.1080/14927713.2018.1449131

Bell, S. (2005). Nature for people: The importance of green spaces to communities in the east midlands of England. In *Wild Urban Woodlands*. New Perspectives for Urban Forestry. 10.1007/3-540-26859-6_5

Berland, A., Shiflett, S. A., Shuster, W. D., Garmestani, A. S., Goddard, H. C., Herrmann, D. L., & Hopton, M. E. (2017). The role of trees in urban stormwater management. In *Landscape and Urban Planning, 162*. 10.1016/j.landurbplan.2017.02.017

Bertolotto, M., McArdle, G., & Schoen-Phelan, B. (2020). Volunteered and crowdsourced geographic information: The openstreetmap project. *Journal of Spatial Information Science*, 20(20). 10.5311/JOSIS.2020.20.659

Biao, Z., Gaodi, X., Bin, X., & Canqiang, Z. (2012). The Effects of Public Green Spaces on Residential Property Value in Beijing. *Journal of Resources and Ecology*, 3(3), 243–252. 10.5814/j.issn.1674-764x.2012.03.007

Carranza, J. P., Piumetto, M. A., Lucca, C. M., & Da Silva, E. (2022). Mass appraisal as affordable public policy: Open data and machine learning for mapping urban land values. *Land Use Policy*, 119, 106211. 10.1016/j.landusepol.2022.106211

Cohen, A., & Dalyot, S. (2021). Route planning for blind pedestrians using OpenStreetMap. *Environment and Planning. B, Urban Analytics and City Science*, 48(6), 1511–1526. 10.1177/2399808320933907

de Kleyn, L., Mumaw, L., & Corney, H. (2020). From green spaces to vital places: Connection and expression in urban greening. *The Australian Geographer*, 51(2), 205–219. 10.1080/00049182.2019.1686195

Dos Santos, T. B., Do Nascimento, A. P. B., & De Mora Regis, M. (2019). Green areas and quality of life: Use and environment perception of an urban park in São Paulo city, Brazil. *Revista de Gestao Ambiental e Sustentabilidade*, 8(2). 10.5585/geas.v8i2.1316

Ibrahim, A., Bartsch, K., & Sharifi, E. (2020). Green infrastructure needs green governance: Lessons from Australia's largest integrated stormwater management project, the River Torrens Linear Park. *Journal of Cleaner Production*, 261, 121202. 10.1016/j.jclepro.2020.121202

Jaganmohan, M., Knapp, S., Buchmann, C. M., & Schwarz, N. (2016). The Bigger, the Better? The Influence of Urban Green Space Design on Cooling Effects for Residential Areas. *Journal of Environmental Quality*, 45(1), 134–145. 10.2134/jeq2015.01.006226828169

Joseph, J., van Dijk, M., & Krisztin, T. (2023). Do Large Estates Benefit Smallholder Neighbours? Evidence from Malawi. *The Journal of Development Studies*, 59(4), 570–592. 10.1080/00220388.2022.2147831

Klimanova, O., Illarionova, O., Grunewald, K., & Bukvareva, E. (2021). Green infrastructure, urbanization, and ecosystem services: The main challenges for Russia's largest cities. *Land (Basel)*, 10(12), 1292. 10.3390/land10121292

Klimanova, O. A., & Illarionova, O. I. (2020). Green infrastructure indicators for urban planning: Applying the integrated approach for Russian largest cities. *Geography, Environment. Sustainability*, 13(1), 251–259.

Kong, F., Yin, H., Nakagoshi, N., & Zong, Y. (2010). Urban green space network development for biodiversity conservation: Identification based on graph theory and gravity modeling. *Landscape and Urban Planning*, 95(1–2), 16–27. 10.1016/j.landurbplan.2009.11.001

Kothencz, G., Kolcsár, R., Cabrera-Barona, P., & Szilassi, P. (2017). Urban green space perception and its contribution to well-being. *International Journal of Environmental Research and Public Health*, 14(7), 766. 10.3390/ijerph1407076628704969

Kumar, V., & Vuilliomenet, A. (2021). Urban nature: Does green infrastructure relate to the cultural and creative vitality of European cities? *Sustainability (Basel)*, 13(14), 8052. 10.3390/su13148052

Le Texier, M., Schiel, K., & Caruso, G. (2018). The provision of urban green space and its accessibility: Spatial data effects in Brussels. *PLoS One*, 13(10), e0204684. 10.1371/journal.pone.020468430332449

Lestari, N. P., & Pratika, Y. (2020). Tourist perception of the intention to visit green open space. *Jurnal Inovasi Ekonomi*, 5(01). 10.22219/jiko.v0i0.11018

Ludwig, C., Hecht, R., Lautenbach, S., Schorcht, M., & Zipf, A. (2021). Mapping Public Urban Green Spaces Based on OpenStreetMap and Sentinel-2 Imagery Using Belief Functions. *ISPRS International Journal of Geo-Information*, 10(4), 251. 10.3390/ijgi10040251

Ma, Y., Brindley, P., & Lange, E. (2023). From Modelling and Analysis of Accessibility of Urban Green Space to Green Infrastructure Planning: Guangzhou as a Case Study. In *Adaptive Urban Transformation: Urban Landscape Dynamics, Regional Design and Territorial Governance in the Pearl River Delta, China* (pp. 249–266). Springer.

McCord, J., McCord, M., McCluskey, W., Davis, P. T., McIlhatton, D., & Haran, M. (2014). Effect of public green space on residential property values in Belfast metropolitan area. *Journal of Financial Management of Property and Construction*, 19(2), 117–137. 10.1108/JFMPC-04-2013-0008

Mete, M. O., & Yomralioglu, T. (2023). A Hybrid Approach for Mass Valuation of Residential Properties through Geographic Information Systems and Machine Learning Integration. *Geographical Analysis*, 55(4), 535–559. 10.1111/gean.12350

Minaei, M. (2020). Evolution, density and completeness of OpenStreetMap road networks in developing countries: The case of Iran. *Applied Geography (Sevenoaks, England)*, 119, 102246. 10.1016/j.apgeog.2020.102246

Muliasari, I. G. A. D., Suartika, G. A. M., & Saputra, K. E. (2021). Conversion of Public Green Open Spaces to Facilitate the Tourist Industry. *IOP Conference Series. Earth and Environmental Science*, 903(1), 012011. 10.1088/1755-1315/903/1/012011

Mwendwa, P., & Giliba, R. A. (2012). Benefits and challenges of Urban green spaces. *Zhongguo Renkou Ziyuan Yu Huanjing*, 10(1), 73–79. 10.1080/10042857.2012.10685062

Nutsford, D., Pearson, A. L., & Kingham, S. (2013). An ecological study investigating the association between access to urban green space and mental health. *Public Health*, 127(11), 1005–1011. 10.1016/j.puhe.2013.08.01624262442

Oliveira, S., Andrade, H., & Vaz, T. (2011). The cooling effect of green spaces as a contribution to the mitigation of urban heat: A case study in Lisbon. *Building and Environment*, 46(11), 2186–2194. 10.1016/j.buildenv.2011.04.034

Prudencio, L., & Null, S. E. (2018). Stormwater management and ecosystem services: A review. In *Environmental Research Letters, 13*(3). 10.1088/1748-9326/aaa81a

Ramaiah, M., & Avtar, R. (2019). Urban Green Spaces and Their Need in Cities of Rapidly Urbanizing India: A Review. In *Urban Science, 3*(3). 10.3390/urbansci3030094

Solís, P., Anderson, J., & Rajagopalan, S. (2021). Open geospatial tools for humanitarian data creation, analysis, and learning through the global lens of YouthMappers. *Journal of Geographical Systems*, 23(4), 599–625. 10.1007/s10109-020-00339-x

Spangler, K. R., Brochu, P., Nori-Sarma, A., Milechin, D., Rickles, M., Davis, B., Dukes, K. A., & Lane, K. J. (2023). Calculating access to parks and other polygonal resources: A description of open-source methodologies. *Spatial and Spatio-temporal Epidemiology*, 47, 100606. 10.1016/j.sste.2023.10060638042531

Sugawara, H., Shimizu, S., Takahashi, H., Hagiwara, S., Narita, K., Mikami, T., & Hirano, T. (2016). Thermal Influence of a Large Green Space on a Hot Urban Environment. *Journal of Environmental Quality*, 45(1), 125–133. 10.2134/jeq2015.01.004926828168

Threlfall, C. G., Mata, L., Mackie, J. A., Hahs, A. K., Stork, N. E., Williams, N. S. G., & Livesley, S. J. (2017). Increasing biodiversity in urban green spaces through simple vegetation interventions. *Journal of Applied Ecology*, 54(6), 1874–1883. 10.1111/1365-2664.12876

Venter, Z. S., Shackleton, C. M., Van Staden, F., Selomane, O., & Masterson, V. A. (2020). Green Apartheid: Urban green infrastructure remains unequally distributed across income and race geographies in South Africa. *Landscape and Urban Planning*, 203, 103889. 10.1016/j.landurbplan.2020.103889

Wajchman-Świtalska, S., Grabowska-Chenczke, O., & Woźniak, M. (2023). Exploring the Patterns of Recreational Activity among Urban Green Spaces in Poland. *Sustainability (Basel)*, 15(6), 5425. 10.3390/su15065425

Wang, J., Rienow, A., David, M., & Albert, C. (2022). Green infrastructure connectivity analysis across spatiotemporal scales: A transferable approach in the Ruhr Metropolitan Area, Germany. *The Science of the Total Environment*, 813, 152463. 10.1016/j.scitotenv.2021.15246334952053

Wiki, O. (2023). *About OpenStreetMap*. https://wiki.openstreetmap.org/wiki/About_OpenStreetMap

Yang, C., He, X., Wang, R., Yan, F., Yu, L., Bu, K., Yang, J., Chang, L., & Zhang, S. (2017). The effect of urban green spaces on the urban thermal environment and its seasonal variations. *Forests*, 8(5), 153. 10.3390/f8050153

Yeboah, G., de Albuquerque, J. P., Troilo, R., Tregonning, G., Perera, S., Shifat Ahmed, S. A. K., Ajisola, M., Alam, O., Aujla, N., Azam, S. I., Azeem, K., Bakibinga, P., Chen, Y. F., Choudhury, N. N., Diggle, P. J., Fayehun, O., Gill, P., Griffiths, F., Harris, B., & Yusuf, R. (2021). Analysis of openstreetmap data quality at different stages of a participatory mapping process: Evidence from slums in Africa and Asia. *ISPRS International Journal of Geo-Information*, 10(4), 265. 10.3390/ijgi10040265

Zhang, Y., Li, X., Wang, A., Bao, T., & Tian, S. (2015). Density and diversity of OpenStreetMap road networks in China. *Journal of Urban Management*, 4(2), 135–146. 10.1016/j.jum.2015.10.001

Zhang, Y., Wu, Q., Wu, L., & Li, Y. (2021). Measuring community green inequity: A fine-scale assessment of beijing urban area. *Land (Basel)*, 10(11), 1197. 10.3390/land10111197

Zurbaran, M. A., Wightman, P., & Brovelli, M. A. (2019). A MACHINE LEARNING PIPELINE ARTICULATING SATELLITE IMAGERY and OPENSTREETMAP for ROAD DETECTION. *International Archives of the Photogrammetry, Remote Sensing and Spatial Information Sciences - ISPRS Archives, 42*(4/W14). 10.5194/isprs-archives-XLII-4-W14-255-2019

KEY TERMS

Community: The term "community" in the context of mapmaking and urban planning means a group of people with at least one connecting interest, objective, or geographical object. Mapping communities usually consist of volunteers as well as activists and researchers who team up with the community to collect and analyze geographic data, come up with solutions for issues faced in the neighborhood or region, and push for a change for the good in their locality or region.

Mapping: Mapping is the process of creating visual representations of geographic data on a map. It involves collecting, organizing, and displaying spatial information in a way that is understandable and useful for decision-making. Mapping can be done using various tools and technologies, including geographic information systems (GIS), satellite imagery, and GPS devices.

OpenStreetMap (OSM): OpenStreetMap is a free and open-source interactive mapping project with the goal of creating a free and editable map of the world. It depends on input provided by volunteers all over the world who are equipped with GPS tools and aerial imagery among other sources. OSM data can be used for various purposes which range from navigation, urban planning, emergency response, grassroots community mapping, and environmental analysis.

Spatial Analysis: Spatial analysis is a set of techniques that are applied in order to analyze geographic data, with a goal to reveal spatial relationships, patterns, and processes. It consists of a study of the spatial distribution of features, the detection of linear or cluster trends, and forming predictions on the spatial processes. Spatial data analysis tools include buffer analysis, hot spots analysis, kriging, and spatial regression.

Urban Green Spaces: Urban green spaces generally refer to the areas of the cities covered with vegetation, such as parks, gardens, forests, and other natural and semi-natural spaces. These spaces have environmental, social, and recreational amenities for urban residents to enjoy improved air quality, more biodiversity, opportunities for physical activity, relaxation, and close social relationships.

Chapter 6
Soil Color Characteristics From Unmanned Aerial Vehicle (UAV) Images and Possible Applications

Mikhail Yurievich Kataev
http://orcid.org/0000-0002-7710-5463
Department of Control Systems, Tomsk State University of Control Systems and Radioelectronics, Russia

Evgeny Yurievich Kartashov
http://orcid.org/0000-0003-3056-4555
Tomsk State University of Control Systems and Radioelectronics, Russia

Igor Borisovich Sorokin
Agrochemical Service Station, Russia

ABSTRACT

The development of the production of unmanned aerial vehicles (UAVs), their reduction in cost and mass production, and the ability to carry various measuring devices have led to an expansion of the capabilities of remote sensing of agricultural fields. The development of methods for monitoring agricultural fields has led to the emergence of several areas of application of optical devices, which are mostly aimed at studying vegetation. The characteristics of the soil are studied much less frequently, since most of the year it is covered with snow or vegetation, and only in the spring or after plowing does the soil become available for study using unmanned aerial vehicles. In this regard, the authors do not consider ground-based methods for studying soil, since they are well developed and widely known. Therefore, it is necessary to systematically and comprehensively study the application of UAV remote sensing of soil systems that are feasible for analysis. In this article, we review the available literature and present the results of our own research in the field of using UAVs to study the color characteristics of soil. The presented results will be useful for improving the understanding of UAV applications for monitoring soil characteristics.

DOI: 10.4018/979-8-3693-3374-7.ch006

INTRODUCTION

Lack of sustainability in agriculture is caused by environmental conditions, loss of soil productivity, loss of biodiversity and volume, and deteriorating water quality. Today, food production is one of the biggest challenges facing humanity and requires a multidisciplinary study of the impact of environmental problems, climate change, agricultural production limits, soil degradation and deteriorating water quality. In any case, knowledge of the state of the soil of agricultural enterprises is the most important attribute that allows you to count on an understandable solution to the problems of food production.

Currently, the direction associated with the digital transformation of agriculture is acquiring a high rate of development. Transformation refers to the introduction of digital technology approaches into typical agricultural processes, which are designed to create new business processes. The resulting digital agriculture after transformation is a sphere of agricultural activity that is based on such elements as: "smart farming", "smart greenhouse", "smart field", "smart herd", "smart greenhouse", "smart warehouse", etc. The use of such "smart" systems should take the management of agricultural enterprises to a new level, where the degree of uncertainty from natural changes, technical and human factors will be minimal. This requires organizing the collection, processing and analysis of a variety of data of different types and at different times, which will optimize agricultural business processes, increase the productivity, profitability and sustainability of agricultural enterprises.

The transition to "Digital Agriculture" means the massive use of information systems containing a large amount of data on all information, technological, technical and human resources that ensure the functioning of each individual farm. The most important thing is that all "smart" systems of this approach must interact with each other in order to achieve maximum efficiency of each activity. For example, "Smart Farm" and "Smart Greenhouse" are associated with the "Smart Field" direction, where, in fact, information about the state of vegetation, soil, air and water should be automatically collected by a variety of technical means (in the near future robotic).

The purpose of the research is to identify color patterns of different types of soils to assess the conditions for identifying types using RGB images.

To achieve this goal, the following tasks were solved:

1) identify agricultural tasks where there is a need to monitor the soil,
2) identify the features of solving problems of soil identification from images obtained using UAVs based on color characteristics,
3) consider the features of color models necessary to understand the color of the soil,
4) find the features of soil image formation necessary for its confident segmentation by images,
5) apply the identified features to specific images of agricultural fields in order to identify soil areas based on color characteristics.

The object of research is the soil type and its RGB images obtained using a UAV.

Research methods - collection of statistical information about brightness variations in RGB spectral channels for different types of soils from images obtained using UAVs.

Keywords: image, color, soil, characteristics, digital camera, UAV

REMOTE SENSING IN PRECISION AGRICULTURE

The precision agriculture approach provides tools and technologies to determine variability in soil and plant health and yields under real-world conditions. This approach assumes spatiotemporal differentiation in the conduct of the agricultural process. Solving this problem involves obtaining and processing a large amount of data collected from ground-based and remote instruments.

Spectral images obtained through remote sensing are used for mapping soil conditions, identifying and classifying crop species, detecting crop (plant) stress, monitoring weeds and crop diseases, moisture, soil temperature, yield mapping, etc. When using spectral Remote sensing images for an agricultural production process must take into account the geometric accuracy of measurements, instrument features and the state of the soil and atmosphere, etc. There are four types of resolutions in remote sensing that must be taken into account when analyzing images. These are spatial, temporal, spectral and radiometric. Among these types of resolution, spatial and spectral resolution make important contributions to the ability to extract the desired information from an image.

Geometric accuracy. When obtaining images with spectral instruments (camera, spectrometer, etc.) installed on an unmanned aerial vehicle (UAV), aircraft or satellite, various factors are inevitably affected (for example, changes in the position in time and space of the platform, topographic relief, etc.). All these factors lead to geometrically distorted objects in the image (distortion of the shape of the object and its position relative to reality). This kind of distortion is corrected using special mathematical approaches quite accurately. Once the images are converted, they can be used for precision agriculture applications.

Spatial resolution refers to the space occupied by a pixel and determines the size of the smallest objects in an image. For an image with high spatial resolution, small objects in the image can be detected and vice versa, with low spatial resolution, most objects are difficult to distinguish. The pixel size is determined by the distance between the device and the Earth's surface, viewing angles and field of view of the device. Images obtained by manned aircraft have a higher spatial resolution (tens of centimeters) than satellite images (from units to hundreds of meters).

However, images obtained using a UAV are not repeatable over time, due to the weak stable position of the device (for example, a photographic digital camera). This results in a change in spatial resolution from one image to another, obtained by repeating the same flight path. Comparative analysis of such images requires mathematical transformations. In this regard, satellite images are significantly more stable over time.

Spectral resolution refers to the ability of an instrument to distinguish individual wavelength intervals in the electromagnetic spectrum. The smaller the distinguishable wavelength range, the better the spectral resolution. Lower spectral resolution and the number of spectral channels make it possible to provide more detailed information about the spectral properties of the studied objects on the Earth's surface. This important property is used in agriculture to calculate, for example, vegetation, water, soil, snow index, etc.

Temporal resolution refers to the frequency with which images are measured at the same geographic location. This parameter is important for obtaining images on the basis of which it is possible to construct spatiotemporal index images and evaluate, for example, the vegetation cycle of crops. Processing a sequence of such images makes it possible to calculate the timing of critical growth stages (for example, harvest time) of crops, find potential problem areas (crop diseases, etc.) and formulate precise management strategies in time and space.

Radiometric resolution is related to the technical characteristics of the instrument to distinguish changes in the measured signal. The higher the radiometric resolution, the more sensitive it is to detect small differences in the measured information, the more accurately it is possible to estimate a particular parameter necessary for studying the condition of crops, soil, etc.

The spectral information in an image captured by instruments on board satellites or unmanned aerial vehicles depends on various factors, such as absorption and scattering in the atmosphere, instrument illumination geometry, and instrument calibration, which change over time (Figure 1). The spectral values recorded in the image are proportional to the reflectance of various objects on the Earth's surface. They depend on the viewing geometry of the instrument at the time the images are taken, the location of the Sun, certain weather conditions, etc.

The purpose of spectral image processing is to detect changes in the state of crops and soil, which are detected as a result of changes in surface reflectivity over a period of time (several measurements). The efficiency and accuracy of finding changes is related to the tasks of calibration and correction of measured images. One of the adjustment steps is the process of radiometric calibration and correction. The images, which are radiometrically calibrated and corrected, make it possible to detect temporary changes in the condition of crops and track critical periods of crop growth and soil conditions (before plowing, after plowing, after rain, etc.). Correction helps determine where, when and in what intensity changes occurred.

Atmosphere

Figure 1. Influence of the atmosphere on radiation measured by remote instruments

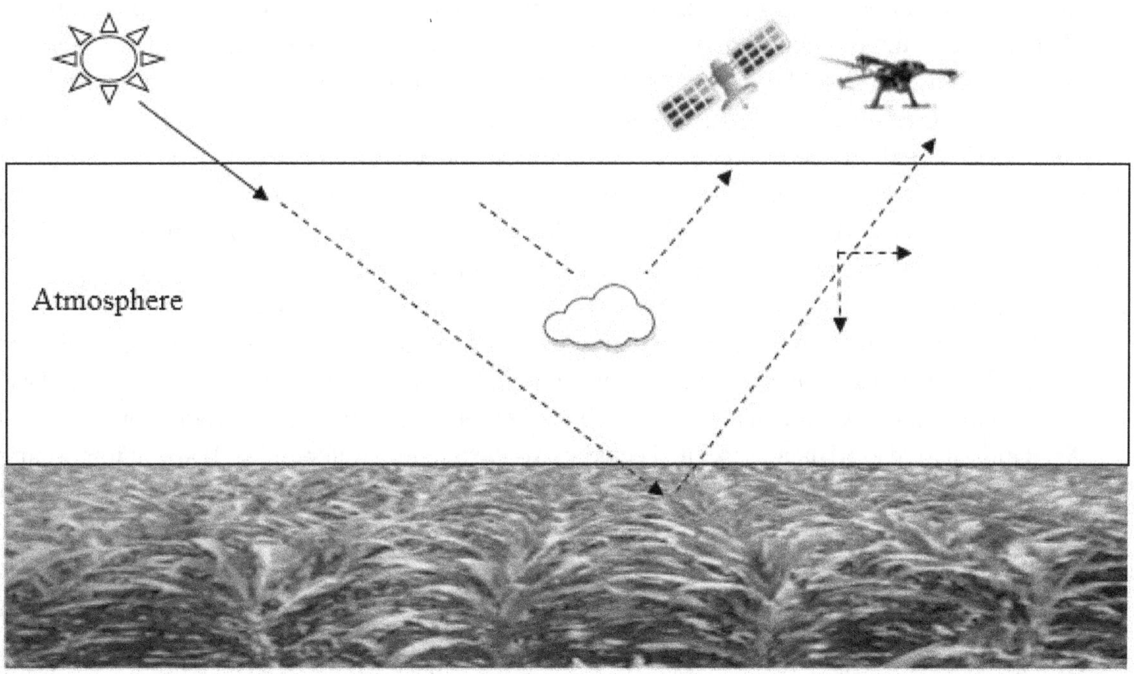

Note that for satellite images, radiometric calibration and correction is a routine, standardized operation. However, such workflows do not exist for UAV imagery. To date, most studies based on UAV images use their own image pre-processing approaches (more gear-specific) or simply use raw values.

Since reflectivity measurements are sensitive to environmental conditions: cloudy or clear skies, windy and calm periods, sun zenith angle (time of day and day of year), field measurements should be carried out under the same conditions, during sunny noon with clear skies during calm periods (wind <5 m/sec).

The capabilities of remote sensing devices (RGB image, multispectral and hyperspectral devices, temperature measurement, fluorescence, 1D, 2D, 3D laser devices, etc.) for solving agricultural problems are shown in Table-1.

Table 1. Capabilities of remote sensing instruments for solving agricultural problems

	plant leaf	plant	Area of field	Field	ecosystem
Healthy vegetatio	UAV	UAV	UAV, satellite	UAV, satellite	satellite
Diseased vegetation	UAV	UAV	UAV, satellite	UAV, satellite к	satellite
Vegetation with disease symptoms	UAV	UAV	UAV, satellite	UAV, satellite	satellite
Soil			UAV, satellite	UAV, satellite	satellite

Measurements should no longer be carried out on the scale of one hectare, but at the level of one square meter, then it will be possible to adjust agricultural technologies to increase productivity even in low-yielding and “variegated” areas in terms of fertility, while reducing costs. Naturally, the abundance of digital data requires the introduction of new principles of scientific validity of decisions made with elements of artificial intelligence.

Each agricultural field is a formation heterogeneous in shape, soil type and topography, which must be taken into account to obtain the maximum possible efficiency of use. The digitalization of agriculture is to obtain from each field the maximum, but economically feasible yield of a given quality product when weather and resource conditions change. The primary role in the digitalization process is to fill the sphere of agricultural production not only with technical and telecommunication devices, but also with appropriate information support. The basis for planning the results of agricultural work is knowledge of the parameters of soil fertility in each section of the field. In addition, land plots may be temporarily removed from crop rotation; there are abandoned plots, the control of which is also important. These issues are dealt with by the agrochemical service of the Ministry of Agriculture of Russia, whose specialists have already carried out the “digitization” of all agricultural lands with indicators of soil fertility and the ecological and toxicological state of soils, but in integrating this information with other data and sources of information, we see the path to the digitalization of all agricultural production.

Agricultural fields and their associated ecosystems (vegetation, water and soil) are an important part of terrestrial ecosystems, and spatio-temporal soil monitoring is an important basis for successfully managing production objectives. The traditional way of studying soil characteristics is mainly based on field research inside an agricultural enterprise or with the involvement of specialized agrochemical organizations. This method of obtaining soil characteristics is important and accurate, but it is point-wise (about one point per hectare) and does not affect the entire field area.

With the development of science and technology for monitoring using UAVs, it becomes possible to fly over large areas and obtain high spatiotemporal resolution images. Computing tools and corresponding software make it possible to process large sets of images, forming mosaics that are possible for viewing, accumulation and analysis. Satellite remote sensing methods can also play some help in this work, but they have a lower spatial resolution (paid 0.5-1 meter and free 5-10 meters). In addition, satellite images have limitations associated with the presence of clouds, as well as the time of passage of satellites over a given agricultural field (repeated visits), which vary from 2 to 15 days.

In this regard, the use of unmanned aerial vehicles is more flexible, easier to manage and use regularly, even in the presence of low clouds. UAVs with on-board measurement sensors and in combination with geo-spatial positioning technology make it possible to obtain images in remote areas. Geographically referenced images allow the use of geographic information systems, which expands the capabilities of data obtained from UAVs.

The combination of UAV technology and image processing techniques can take full advantage of the data obtained in this way and provide new insights for many scientific research and practical applications. Among the surface measurement devices that can be installed on UAVs are: high-resolution RGB digital cameras, thermal infrared sensors, laser devices (LiDAR), multispectral sensors, hyperspectral sensors, radars and other sensors. Among this set of sensors, the most common is the digital camera, which produces high-resolution RGB images that can reach 1-10 cm. The images captured by digital cameras provide rich data for analysis and application in managing production tasks in the agricultural field.

Areas of application of RGB images for agricultural fields are related to soil quality and processes that arise from human influence and natural conditions on the field. One such process is soil erosion, which is a pressing issue for agricultural sustainability. Under certain human and environmental conditions in an agricultural field, accelerated rates of soil erosion may occur and need to be monitored. Another task is related to assessing soil quality, when annual monitoring of the field area is necessary, since the application of existing agricultural methods often exceeds the rate of natural soil regeneration.

It is very important to develop software that allows you to monitor the condition of agricultural fields using technical vision technology in monitoring mode. The fields clearly represent three types of surface: soil, vegetation and artificial structures. Thus, identifying areas of only soil is, in one way or another, associated with finding all types of surface. To do this, it is proposed to use unmanned aerial vehicles to obtain images of agricultural fields from the moment of plowing to harvesting and carry out image processing. Maintaining a database of measured and processed values for all measurements with geographic reference, as well as the necessary accompanying information (meteorology, satellite measurements). The use of UAVs and metadata from satellites to obtain images of fields and the use of a software package for analyzing the state of fields, monitoring soil areas, classifying plant types and determining their vegetation state.

It can be assumed in advance that the software product being developed should have the following characteristics:

1. Independence from a specific UAV model, since most UAVs have a digital camera and a communication module as standard (as well as the ability to connect additional modules or replace existing ones),
2. Use of communication types GSM, satellite communication, Wi-Fi to transmit information flows and receive control commands,
3. Presentation of analysis results (output data) in a form understandable to users without special education,
4. In addition to determining the vegetative state of plants, determining the type of plants and soil,
5. Linking to GIS systems,
6. User notification system about the current state of plants and soil,
7. Support of the full cycle of work (assistance in operation and technical support),
8. Assessment of the required resources for monitoring agricultural areas,

9. Providing recommendations for solving problems that have arisen in agricultural areas (an example of such situations: violation of natural favorable conditions for crops, detection of weeds, soil leaks, wet soil, etc.),
10. Implementation of comprehensive monitoring with/without human participation (automation),
11. Monitoring the harvest according to the proposed timing and comparative control of the soil condition before plant germination and after harvesting (assessment of degradation),
12. Provision of computing power (for processing data from objects),
13. Processing and output of statistical data for reporting,
14. Obtaining data for assessing productivity and other soil-related indicators.

SOFTWARE ARCHITECTURE

A possible software architecture is characterized by the diagram presented in Figure 2. We receive data as input - an array of images (or one image) taken from a UAV. Next we work on each image. We present the image as an RGB texture. Let's save it to the image database. Then we will subject the image to pre-processing. Pre-processing includes light correction and dividing each image into a certain number of blocks to increase the accuracy of the final results. Then, in the texture and color processing block, we obtain a number of special indicators that characterize the object of the image. After completing the above steps, you need to assign a specific color to each surface class and mark the surface in the image with this color. The next step is to combine the existing images into an orthomosaic.

Then we will calculate the area in m2 of a certain class of plants and soil, as well as their percentage. In order to determine in the field which crops have sprouted and which plants are weeds and interfere with germination, we compare the results in the “Analysis” block with templates from the template database. We also use the templates to determine their stage of growth. The output is a report file with an orthomosaic (or a single image), the type of plant and its condition, as well as the area occupied in the image, the type and color characteristics of the soil.

As a result of shooting with a regular digital camera, images are saved in JPG format. This format provides image compression with little loss of quality. However, most camera manufacturers have the ability to save images in RAW format, which significantly improves the quality of the original images. The RAW format allows you to change the illumination and improve the histogram immediately after the UAV has flown over. The main disadvantage of this format is that the image size increases, which leads to a decrease in the number of images. As already mentioned, the original images are RGB images. But this is not the only color model. Color models provide different methods for defining colors (Figure 2). In each model, colors are defined by using specific color components.

Figure 2. Typical scheme of the program

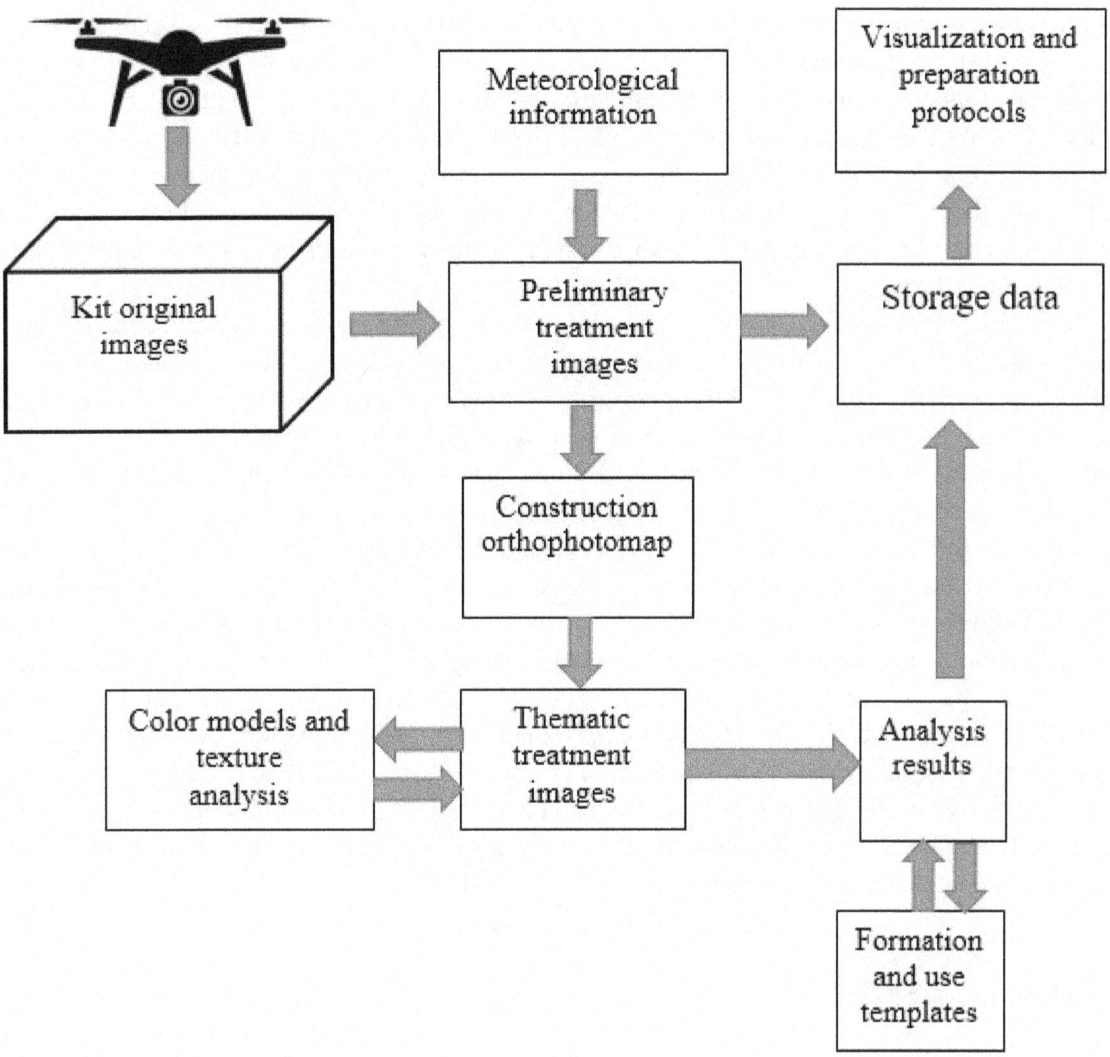

DEFINITION OF COLOR SYSTEM

To measure and represent color information, it is first necessary to understand its fundamental physical and psychological properties. Color is the result of the interaction of light, an object, and an observer (or recording device). When interacting with an object, the light is modified in such a way that the recording device (for example, like the human vision system) perceives the modified light as a certain color. For color to exist as such, all three of these elements must be present. In fact, color is a phenomenon caused by the human vision apparatus's perception of light.

The basis of the mathematical description of color in colorimetry is the experimentally established fact that any color, subject to certain conditions, can be represented as a mixture (sum) of certain quantities of three linearly independent colors, i.e. colors, each of which cannot be represented as a sum any quantities of the other two colors. There are infinitely many groups (systems) of linearly independent colors, but only a few of them are used in colorimetry. The three selected linearly independent colors are called primary colors. These colors define a color coordinate system or color scheme - a set of primary colors used to produce all the others. Then the three numbers describing a given color are the quantities of primary colors in a mixture whose color is visually indistinguishable from a given color - the color coordinate of a given color.

Being related to a standard observer under certain unchanging conditions, standard color mixing data and colorimetric color systems built on them actually describe only the physical aspect of color, without taking into account changes in the color perception of the eye when observing conditions change and for other reasons. The representation of color using a color coordinate system should reflect the properties of human color vision. Therefore, it is assumed that the basis of all color schemes is the so-called physiological color system.

The properties of color vision are taken into account in colorimetry based on the results of experiments with color mixing. Such experiments perform visual equalization of pure spectral colors (that is, colors corresponding to monochromatic light of different wavelengths) with mixtures of the three primary colors. When plotting graphically the dependences of the quantities of primary colors on the wavelength, functions of the wavelength are obtained, called color addition curves or simply addition curves.

Color schemes can be divided into two groups: schemes representing color from emitted and reflected light. We see objects because they either emit light or reflect reflected light. In the first case, objects acquire the color of the radiation they emit, and in the second, their color is determined by the color of the light falling on them and the color that they reflect. An example of a radiating object is a monitor screen, and a reflective object is paper with paint applied to it.

RGB COLOR SYSTEM

In fact, the basis of all color schemes is a system whose addition curves have been determined experimentally. Its primary colors are pure spectral colors corresponding to monochromatic radiation with wavelengths of 700.0 nm (red), 546.1 nm (green) and 435.8 nm (blue). This system, adopted by the International Commission on Illumination (CIE) in 1931, was called the international colorimetric system CIE RGB or simply RGB (R - RED, G - GREEN and B - BLUE). The RGB system is additive. In such systems, color is obtained by adding primary colors. In this case, the absence of all colors is black, and the presence of all colors is white. The additive color system works with emitted light, such as from a computer monitor.

CMYK COLOR SYSTEM

Colored, non-luminous objects absorb part of the white light spectrum that illuminates them and reflect the remaining radiation. Depending on the region of the spectrum in which absorption occurs, objects reflect different colors (are colored in them). Colors that use white light by subtracting certain parts of the spectrum from it are called subtractive. To describe them, the subtractive CMY model is used (C - CYAN, M - MAGENTA, Y - YELLOW). In this model, primary colors are formed by subtracting the primary additive colors of the RGB model from white. It is clear that in this case there will also be three main subtractive colors: cyan (white minus red), magenta (white minus green), yellow (white minus blue).

The CMY color system was widely known long before computers were used to create graphics. Its primary colors: cyan, magenta and yellow are, in fact, the heirs of the three primary colors of painting (blue, red and yellow). The change in shade of the first two is due to the difference in the chemical composition of artistic inks from printing inks. Both artistic and printing inks cannot produce very many shades. To improve the quality of the print, black has been added to the main printing inks (and to the model). It was she who added the last letter to the name of the CMYK model (the black component is shortened to the letter K, since this ink is the key ink in the color printing process).

HSB COLOR MODEL

The RGB and CMYK color systems are based on limitations imposed by the hardware (in the case of RGB, these are monitors, scanners, etc., in the case of CMYK, these are printing inks). A more intuitive way to describe color is to represent it in terms of hue (H - HUE), saturation (S - SATURATION) and brightness (B - BRIGHTNESS) - the HSB color system. Tone is a specific shade of color that is different from others: red, green, blue, etc. Saturation of a color describes its relative intensity (or purity). By reducing the saturation of, for example, red, we make it more pastel, closer to gray. The brightness (illuminance or intensity) of a color indicates the amount of black tint added to the color, making it darker.

The HSB system has an important advantage over other systems: it is more consistent with the nature of color and is in good agreement with the human color perception model. Many shades can be quickly and conveniently obtained in HSB, then converted to RGB or CMYK, modifying in the latter case if the color has been distorted. Therefore, the HSB system is often used when the user selects a color. The HSB model color space uses a cylindrical coordinate system, and the set of valid colors is represented by a hexagonal cone placed on top. The base of the cone represents bright colors and corresponds to $V = 1$. However, the colors of the $V = 1$ base do not have the same perceived intensity. Hue (H) is measured by the angle measured around the vertical axis OV. In this case, the red color corresponds to an angle of 0°, the green color corresponds to an angle of 120°, etc. Colors that complement each other to white are opposite each other, that is, their tones differ by 180°.

LAB COLOR SYSTEM

The abbreviation LAB usually refers to the CIE L*a*b* color space, which is currently an international standard. In the CIE L*a*b system, the L coordinate means lightness (ranging from 0 to 100), and the a,b coordinates mean the position between green-magenta and blue-yellow. Figure 3 shows sections of the CIE L*a*b* color body for two lightness values.

Figure 3. Color slices in Lab space

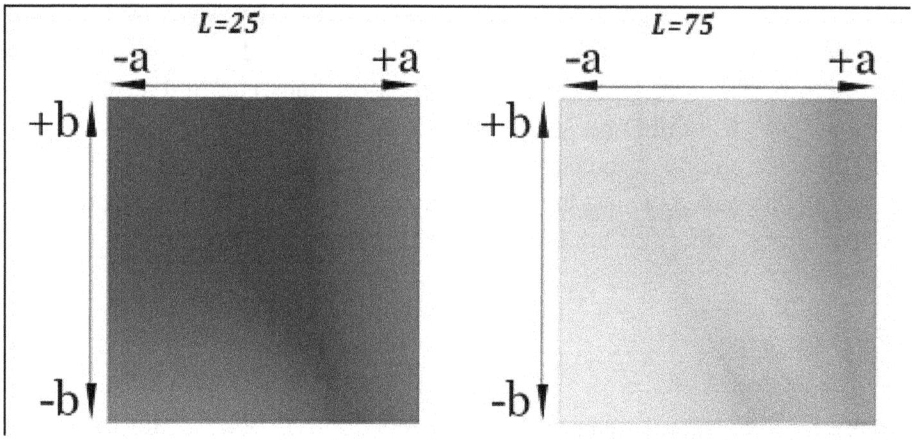

INFORMATION SYSTEMS RELATED TO AGRICULTURAL LAND

Information system (IS) "Electronic Atlas of Agricultural Lands" [https://sovzond.ru/services/gis/ogv/federal/azsn/] is intended to provide state authorities, local governments, legal entities and individuals with up-to-date information about lands agricultural purposes, their areas, condition, degree of soil degradation, reclamation status. The IS is designed to view satellite imagery data and thematic maps built on their basis (for example, the NDVI vegetation index) highlighting different levels of detail: types of land, types of agricultural crops, reclaimed land, degraded agricultural land, etc.

Another information system for agricultural lands [https://efis.mcx.ru/efis] is closed to free access and contains information about the location, condition and actual use of land plots throughout the Russian Federation. Associated with this system is another open program - the Web information system "ERFIS ZSN Remote Sensing Analyst" [http://pro-vega.ru/maps/efis.shtml?lang=rus] is being developed by the Space Research Institute of the Russian Academy of Sciences.

The Central Information and Analytical System of the Russian Ministry of Agriculture [http://mcx.ru/analytics/infosystems] is important for the agricultural industry. The system was developed in order to form a unified analytical base for monitoring the state of the agro-industrial complex and supporting management decision-making by the leadership of the Ministry. The objective of this project is the creation and implementation of "digital services" related to the activities of the Ministry of Agriculture of the Russian Federation. This information and analytical system allows you to obtain a single digital space for solving agricultural problems.

By order of Rosselkhoznadzor, the Russian company Scanex developed the GIS "Demetra" [https://www.scanex.ru/thematic/projects/rosselkhoznadzor-demetra/]. The program is designed to increase the efficiency of obtaining information on the condition of agricultural land, ensuring the relevance, reliability and effectiveness of state control in the field of use of agricultural land, correlating factual information with data from the State Land and Property Cadastre.

The WISE Soil Database for Crop Simulation Models [https://www.isric.org] contains information on soil parameters such as: calcium, carbon, cation exchange capacity, electrical conductivity, nitrogen, organic matter, soil density, soil classification, depth soils, soil horizons/profile, soil acidity, soil salinity, soil texture and soil holding capacity. WISE is being built as part of the World Inventory of Soil Emission Potentials project, which collects soil profiles around the world in a single format and which can be used as input to build simulation models of crop growth.

The Harmonized World Soil Database [https://www.fao.org/soils-portal/soil-survey/soil-maps-and-databases/harmonized-world-soil-database-v12/en/] contains the main areas: landscape, vegetation cover and soil quality. The information is collected on all soils of the world and is a raster database with a spatial step of 30 arc seconds (about 1 km). The resulting raster database consists of 21,600 rows and 43,200 columns that are associated with soil property data: organic carbon, pH, water storage capacity, soil depth, soil cation exchange capacity, clay, total exchangeable nutrients, lime and gypsum content, percent sodium metabolism, salinity, textural grade and granulometry.

The interactive SoilGrids data-driven map system, [https://soilgrids.org/#!/?layer=ORCDRC_M_sl2_250m&vector=1] is designed to provide soil properties and classes using mathematical models with a spatial resolution of 250 m. Numerous data from all over the world, which are consistent with each other on the following parameters: bulk density (fine-grained), clay mass fraction, coarse volumetric fragments, silt mass fraction, sand mass fraction, cation exchange capacity of the soil, organic carbon content in the soil (fine-grained fraction), PH soil in H2O and soil PH in KCl.

The site contains useful information about sources of information about soils [https://gcmd.gsfc.nasa.gov/KeywordSearch/Keywords.do?Portal=GCMD&KeywordPath=Parameters%7cland+surface%7csoilS&MetadataType=0&lbnode=gcmd3] and their parameters: calcium, carbon, cation exchange capacity, denitrification rate, electrical conductivity, heavy metals, hydraulic conductivity, magnesium, nitrogen, phosphorus, potassium, soil density, soil classification, soil moisture, soil plasticity, soil porosity, soil texture, thermal conductivity, etc. By selecting a section, for example, "calcium" - you can see a list of references in the search direction, including links to databases.

The GLDAS system [https://ldas.gsfc.nasa.gov/gldas/soils] is based on soil data collected around the world and includes parameters such as sand, silt, clay, water content, porosity, etc. Spatial resolution The map is 10 km and there are two layers: 0-30 cm and 30-100 cm. More detailed maps are converted to more spatially large 25 and 100 km for some parameters: soil texture, and another model uses a different texture class. Therefore, LDAS includes a procedure to calculate a texture class based on the sand, silt, and clay fractions in a given grid cell for the model's soil texture classification scheme. Currently, top layer texture information is used in all soil layers, clay, etc.

An analysis of existing information systems in Russia and the world suggests that the existing information systems being developed in the Ministry of Agriculture do not have a sufficient set of parameters that would cover all agricultural areas. In addition, the collected information is not systematized, and numerical models useful for agricultural workers are not built on its basis. In addition, the level of personnel training in not all regions allows them to use even the information that is collected and provided to them. World soil systems cover the entire territory of the globe, but are coarse in spatial resolution and have a different number of parameters, which makes them difficult to use in practice. In addition, there are regional local conditions and farming characteristics.

Therefore, we have decided to develop our own information system, which will include data from well-known world databases, satellite and meteorological information, as well as the results of ground measurements "Tomskaya Agrochemical Service Station" [http://agrohim.tomsk.ru]. We believe that it is the integrated use and automated analysis of such a data set that will make it possible to obtain a practically useful information system for agricultural workers.

RELATIONSHIP OF PLANTS WITH SOIL CONDITION

In fact, precision agriculture means taking into account all factors and collecting the necessary data to facilitate the effective and rational management of plant growth processes, taking into account the agrochemical specifics and topography of each field. The use of information systems that have the ability to accumulate, process and analyze spatiotemporal data based on the latest achievements in the field of computer and information technology. The adoption of agro-technological solutions taking into account global positioning systems (GPS, GLONASS) and geographic information technologies (GIS) will allow their implementation at a higher, scientifically based level. The use of Earth remote sensing data from space will make it possible to make decisions when seeing information throughout the entire farm territory.

The yield is influenced by: the agrochemical condition of the soil, the terrain, the shape of the fields, the presence of forest and hydro-reclamation, the quality of seeds, technologies and timing of the start and completion of work, methods and means of protecting plants from diseases and pests, weather conditions and other aspects. Note that of all the listed aspects, the agroecological state of the soil, which is associated with the following parameters, is of particular importance: 1) Basic agrochemical indicators: soil acidity, soil organic matter, nutrient content, hydrolytic acidity, sum of absorbed bases, etc., 2) Microelements: cobalt, manganese, copper, molybdenum, zinc, nickel, boron, etc., 3) Toxic elements: cadmium, lead, chromium, mercury, arsenic, benzopyrene, pesticide residues, radionuclide activity, 4) Particle size distribution, physical properties of soils and soil microbiology: pathogenic bacteria, pathogens of fungal diseases and viral infections of plants, as well as the spread of pests.

One way to monitor the condition of the soil is visually, when it is believed that the blacker the color of the soil, the more fertile it is. This also includes the aspect of soil texture, which can give clues to the condition of the soil. Another method is associated with the analysis of chemical and bacterial components of soil in agrochemical services around the world. The third method is associated with the use of satellite technologies, when the brightness properties of spectral channels are used for analysis, which make it possible to estimate the reflective properties of the earth's surface. Reflectance depends on the number and area of surface types that fall within the satellite pixel.

Remote sensing methods play an important role in surveying the earth's surface and are a significant source of information about the state of land resources. The physical properties of the measured reflected solar and outgoing infrared radiation make it possible to record the phenology (growth) of plants throughout the growing season, as well as changes in chlorophyll content or structural changes of plants. Satellite images have long been used to study crops, their health, and monitor agricultural practices.

Back in the 1970s, it was discovered that various combinations of satellite spectral channels are sensitive indicators of the presence and condition of green vegetation. Based on the spectral channels of the visible and infrared range, it is possible to measure the spectral index of vegetation. The NDVI vegetation index is a widely used and easily calculated proxy for vegetation from satellite multispectral imagery. It should be noted that the condition of the soil and vegetation are very closely interrelated. The vegetation index is an indicator of the relative abundance and activity of green vegetation in a satellite image pixel, including leaf area index, chlorophyll content, green biomass, and absorbed photosynthetically active radiation. Thus, if there is no open soil in a satellite image pixel, its condition can be determined by long-term measurements of the vegetation index, which reflects the influence of soil condition, precipitation, chemical and textural structure.

Figure 4 shows changes in the values of the green vegetation index NDVI over a period of 8 years for four satellite pixels in the Tomsk region.

Figure 4. Values of the NDVI vegetation index of green vegetation (MODIS satellite radiometer) for a period of eight years for four satellite pixels in the Tomsk region

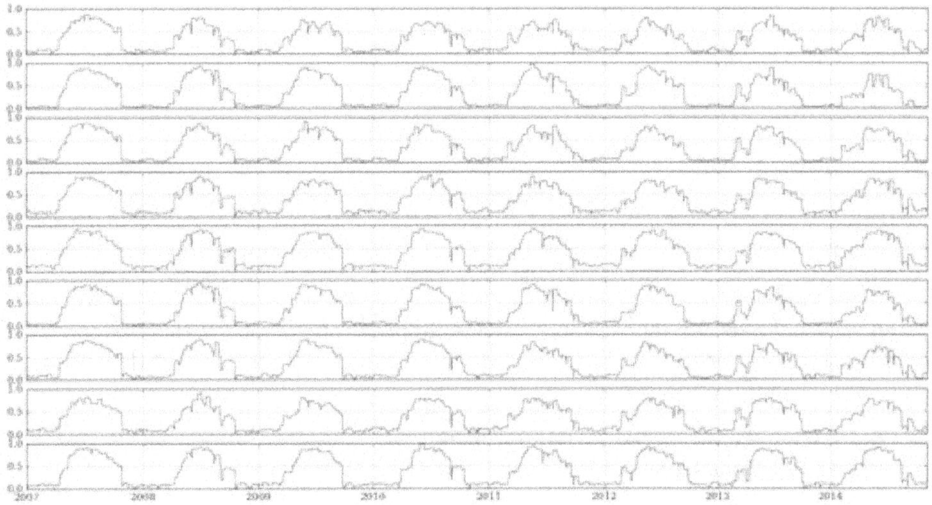

From Figure 4 it can be seen that the behavior of vegetation in four pixels differs not only in one year, but also across all years. Naturally, this behavior depends on the type of vegetation and soil that falls into the pixel, as well as changes in weather (temperature, humidity, number of sunny days, etc.).

Figure 5 shows the reflectance of water, dry soil and green vegetation.

Figure 5. Reflectance coefficients of water, dry soil, and green vegetation

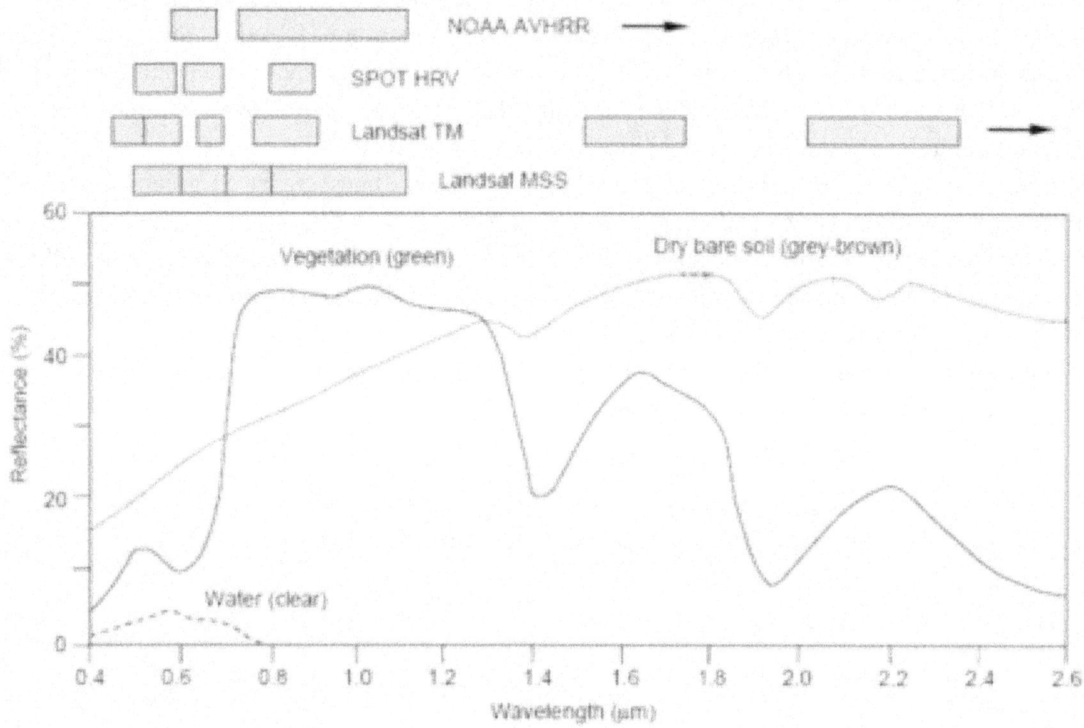

Figure 5 shows that in the range of 0.4-0.6 µm, the reflection of soil exceeds the reflection of vegetation and water. In the range of 0.6-1.3 µm, the soil reflection is less than the vegetation reflection, and for wavelengths greater than 1.3 µm, the soil reflection exceeds the vegetation reflection. In the reflection curves of vegetation and soil for wavelengths of 1.4 and 1.9 µm, dips are visible, which are associated with water absorption. Knowledge of this property allows one to determine the amount of water in soil and vegetation.

In Fig. Figure 6 shows the coefficients of dry and wet soil and shows that the reflection of dry soil is more than twice that of wet soil.

Figure 6. Dry and wet soil coefficients

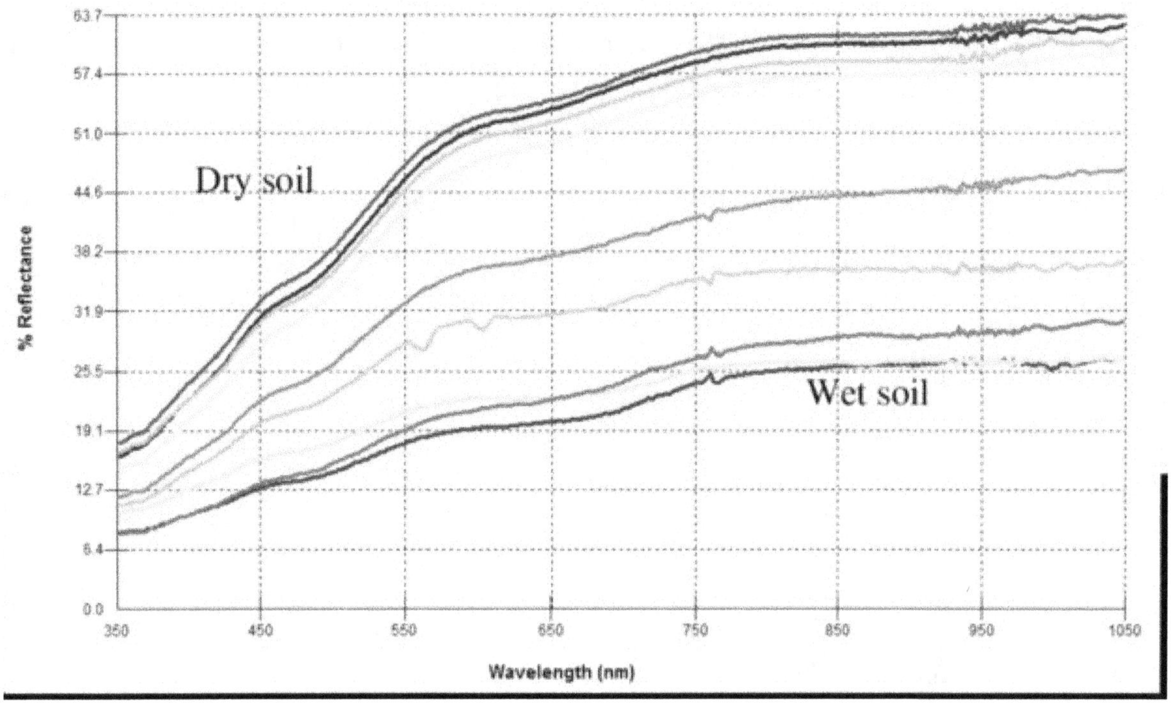

In Fig. Figure 7 shows the reflectance coefficients of healthy vegetation and under stress. The figure shows that the reflection of healthy vegetation is almost twice as large as that of vegetation under stress for the wavelength range 0.7-1.05 μm.

Wavelength, nm

Figure 7. Reflection coefficients of healthy vegetation and under stress

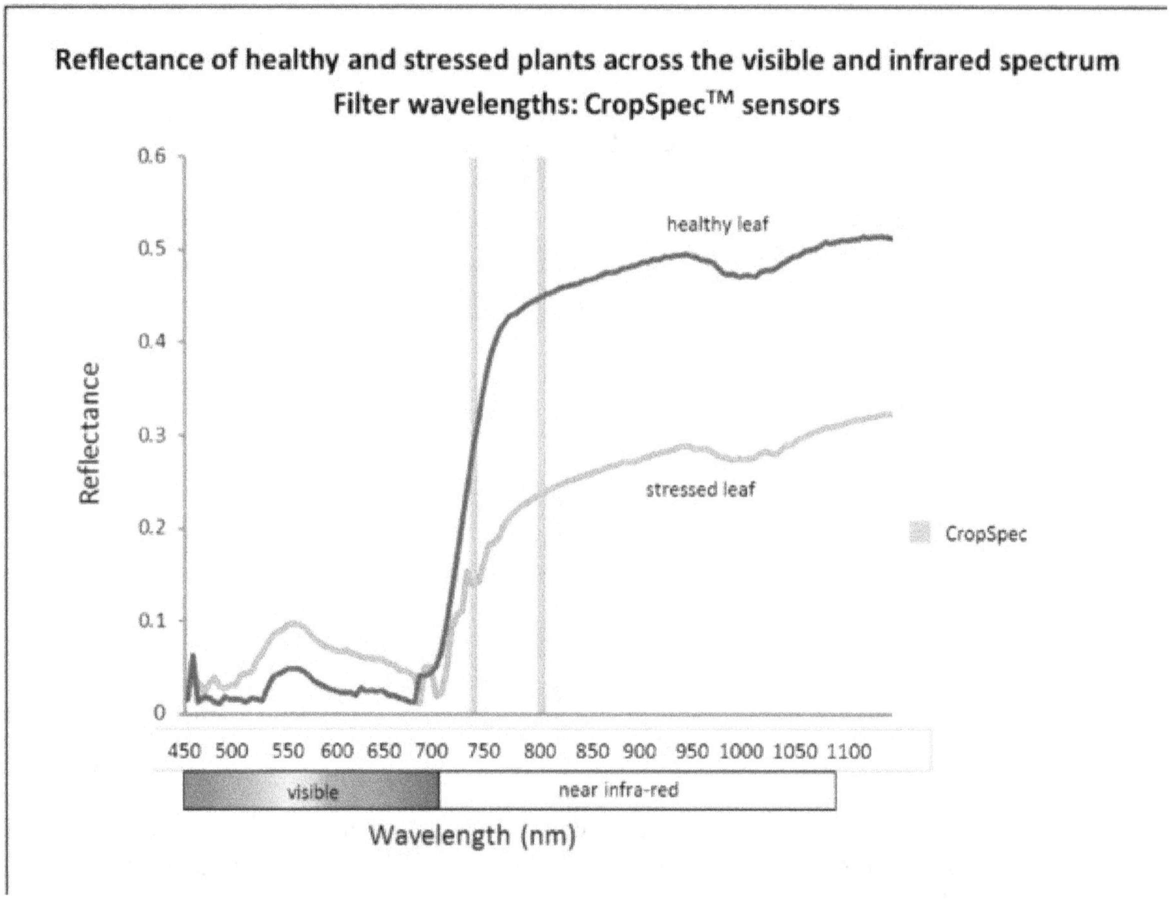

A green, healthy leaf of a plant (Figures 5 and 7) has typical spectral features that differ in the functions of the three main optical spectral regions - visible, near-infrared and infrared. In the visible region of the spectrum (400-700 nm), light absorption by pigments dominates the leaf's reflectance spectrum and results in generally lower reflectances. There are two main light absorption bands for plants: blue (450 nm) and red (670 nm) due to the absorption of the two main leaf pigments, chlorophyll A and B, which account for 65% of the total leaf pigments of higher plants. Other leaf pigments also have an important influence on reflectance in the visible spectrum.

In the near-infrared region of the spectrum (700-1300 nm), the structure of the leaves determines the optical properties, when the absorption is very low and the reflectivity becomes high. Since leaves are nourished not only by the air (humidity, carbon dioxide, methane), but to some extent by the transfer of various substances from the soil, it becomes possible to record this on the basis of spectral measurements.

Soil properties vary over time and space and are determined by internal factors (soil-forming factors - chemical composition, structure, etc.), as well as external factors (agricultural cultivation methods, land reclamation, etc.). There is a very complex relationship between yield and soil, associated with the physical and chemical properties of the soil, as well as natural atmospheric factors.

OVERVIEW OF MAPPING EQUIPMENT

Unmanned aerial vehicle (UAV), an artificial mobile object (aircraft), usually reusable, without a crew (human pilot) on board and capable of independently purposefully moving in the air to perform various functions autonomously (with the help of its own control system). programs) or through remote control (carried out by a human operator from a stationary or mobile control panel). There are automatic UAVs that operate in accordance with the programs embedded in their on-board computer. For mapping purposes, the following characteristics of a UAV are noted: 1) UAVs can be aircraft or helicopter types. 2) According to the control method, they are divided into automatic and semi-automatic. 3) The payload must be sufficient for a digital automatic camera and a spectral camera. 4) UAV models must fly at low altitudes - airspace class G, altitude up to 4.5 km in uninhabited areas. 5) Commercial availability.

When mapping, in addition to the need to obtain high-quality images, the task of correctly positioning these images in space is added. To solve the problem of spatial orientation, a small-sized GPS receiver with an antenna is installed on board the UAV. Thus, after the flight of the unmanned aerial vehicle, a set of photographs and telemetry data are generated, which include the coordinates of the photographing center, as well as roll, yaw and heading angles. Russian manufacturers use receivers of signals from satellite positioning systems of the combined GLONASS/GPS type.

Typically, the accuracy of GLONASS/GPS navigation and the features of UAV automatic control systems make it possible to achieve the following parameters when flying along an aerial photography route: – transverse displacement from the route axis – ± 10 m, – keeping the UAV at a given height – ± 15 m, – the distance from the designed photography center to the camera shutter release point is ± 5 m, – change in the UAV roll angle along the route between two images – 10°, – change in pitch angle on the route between two images – 6°.

For mapping purposes, two types of cameras are most often installed on board UAVs: conventional aerial photography and spectral. To obtain conventional RGB images, as a rule, digital cameras such as Canon-S110, SX260, Sony-QX1,DSC-RX100 A7R, A7, A7S, NEX-6, NEX-5R, NEX-5T, A5100 and Panasonic-GH3 are used . These cameras have a number of advantages: low cost, ease of replacement during a "hard landing". Before use, they must be calibrated in laboratory conditions to improve shooting accuracy. When shooting, special attention should be paid to the best possible sharpness of the frame, if the camera has the ability to use the RAW format.

As a result of shooting with an aerial camera you can get: – orthomosaic, – digital terrain models, –RGB channels for use in calculating various indices. Spectral cameras have greater accuracy and detail of information. With their help, you can obtain images of the territory in different zones of the electromagnetic radiation spectrum. The use of multispectral maps makes it possible to assess the quality of soils, identify weeds at the germination stage, and analyze the growth conditions of certain crops and soil.

The main indices calculated using spectral cameras are presented in Table 2.

Table 2. Vegetation indices

Name	Abbreviation	Formula	Comment
Normalized Difference Vegetation Index	NDVI	NDVI = (NIR - RED) / (NIR + RED)	NIR – reflection in the near infrared region of the spectrum RED – reflection in the red region of the spectrum.
Enhanced Normalized Difference Vegetation Index	ENVI	ENVI = 2.5 * ((NIR - RED) / (NIR + 6 * RED - 7.5 * BLUE + 1))	BLUE – reflection in the blue region of the spectrum
Green Normalized Difference Vegetation Index	GNDVI	GNDVI = (NIR - GREEN)/(NIR + GREEN)	GREEN – reflection in the green region of the spectrum

Research shows that the main analysis of vegetation in images is based on NDVI index readings. However, in addition to this index, scientists have calculated a large number of other indices, calculated from images of a regular RGB camera. Obtaining reliable results using conventional cameras could significantly reduce the cost of equipping UAVs for agricultural purposes

PRE-TREATMENT METHODS

Pre-processing of input data, in our case – images from a UAV, is necessary to reduce errors and inaccuracies in obtaining analysis results. Image acquisition (during the flight of the UAV) is carried out at different times, at which the angular position of the Sun changes, shadows appear associated with cloud cover or objects located along the edge of the field (as a rule, these are forest plantations). A variety of techniques are used to remove these color artifacts from an image. One technique involves using the effect of the bidirectional reflectance distribution function (BRDF). Another technique is based on the analysis of changes in reflection from the earth's surface (for example, based on Lambert's law of reflection).

THEMATIC IMAGE PROCESSING

Many researchers have put a lot of effort into studying how to detect greenness in a crop image. In addition to analysis using the NDVI index, other indices that can be used based on RGB images have been studied and based on the publications (Gulakov et al., 2011; Franlenko, 2014; Andreev et al., 1984; Haralik, 1979; Potapov, 2003; Rasmussen et al., 2016; Aasen et al., 2015; Brede et al., 2015; Honkavaara, 2013, 2014; Schaepman-Strub et al., 2006; Colomina and Molina, 2014; Christiansen et al., 2017; Smith and Milton, 1999; Burkart et al., 2018; Laliberte et al., 2011; Lelong et al., 2008; Tymchuk, 2018).

Typical methods are based on visible spectral index such as Excess Green Index (ExG), Excess Green Minus Excess Red Index (ExGR), Vegetative Index (VEG), Color Vegetation Extraction Index (CIVE), Combined Index (COM), etc. All of these methods solve the problem of greenness identification, based on the assumption that plants show a clear high degree of greenness, and the background (soil) only barely

For example, (Mercado-Luna,. 2010) used the RGB color space to determine nitrogen N deficiency in tomato seedlings. Many scientists have used similar calculations for the RGB color space to assess the health of plants and soil. Color indices were obtained from the normalized colors of the entire image. The indices used for evaluation are listed and defined in Table 3. The normalized colors r, g and b in the RGB color space are defined as follows:

r = R/(R+G+B),

g = G/(R+G+B),

b = B/(R+G+B),

where R, G and B are the color components of the input cropped image.

Table 3. RGB-based color indices

Name	Abbreviation	Formula
Normalized green red difference index	NGRDI	(g - r) / (g + r)
Kawashima index	IKAW	(R - B) / (R + B)
Excess red vegetation index	ExR	1.4r – g
Excess blue vegetation index	ExB	1.4b – r
Excess green vegetation index	ExG	2g - r – b
Excess green minus Excess red	ExGR	ExG – ExR
Green-red vegetation index	GRVI	(G - R)/ (G + R)
Red green ratio index	RGRI	R / G
Green leaf index	GLI	(2G - R - B) / (2G + R + B)
Visible atmospherically resistance index	VARI	(G - R) / (G + R - B)
Principal component analysis index	IPCA	0.994|R - B| + 0.961|G - B| + 0.914|G - R|
Normalized Difference Vegetation Index	NDVI	(NIR - r) / (NIR + r)
Green-Normalized Difference Vegetation Index	GNDVI	(NIR - g) / (NIR + g)
Vegetation Index	VEG	g / (r0.667 b0.333)
Vegetation extraction Index	CIVE	0.441r - 0.881g + 0.385b + 18.78745
Combined index	COM	0.25ExG + 0.30ExGR + 0.33CIVE + 0.12VEG
Grenness0	Grenness0	G/(R+G+B)
Grenness1	Grenness1	2G — R — B
Grenness2	Grenness2	2G + B - 2R
Grenness3	Grenness3	G — 0.39R — 0.61B

The indices were analyzed based on images of beet beds obtained during a UAV flight (Figure 8).

Figure 8. Original image of beet beds (UAV height ten meters)

From Figure 8 it is clear that in addition to vegetation, the soil can be very well detected and, therefore, its condition can be determined.

At the first stage, the greenness indices Grenness0, Grenness1, Grenness2, Grenness3 were calculated. A summary table of index values relative to belonging (earth, green leaves, yellow leaves) is shown in Table 4.

Table 4. Color index values

Название	**Grenness0**	**Grenness1**	**Grenness2**	**Grenness3**
soil	0.295 — 0.345	>-20	<56.5	-0.12-0.102
shadow	0.22 - 0.295	-	-	-
green beet stalks	>0.22	-	-15.5 - 0	-0.184-(-0.12)
Yellow leaves	0.345-0.375	-	>92	-
Green leaves	>0.375	>25	>92	>0.102

Figure 9 and Table 4 clearly show that the Grenness1 index has no difficulty in separating shadow areas (darker areas) from ground areas (light background), or yellow leaves from green leaves. Therefore, for use in the module it is proposed to take the Grenness0 index.

Figure 9. Processing using indexes1 (a – original image and b – index values1)

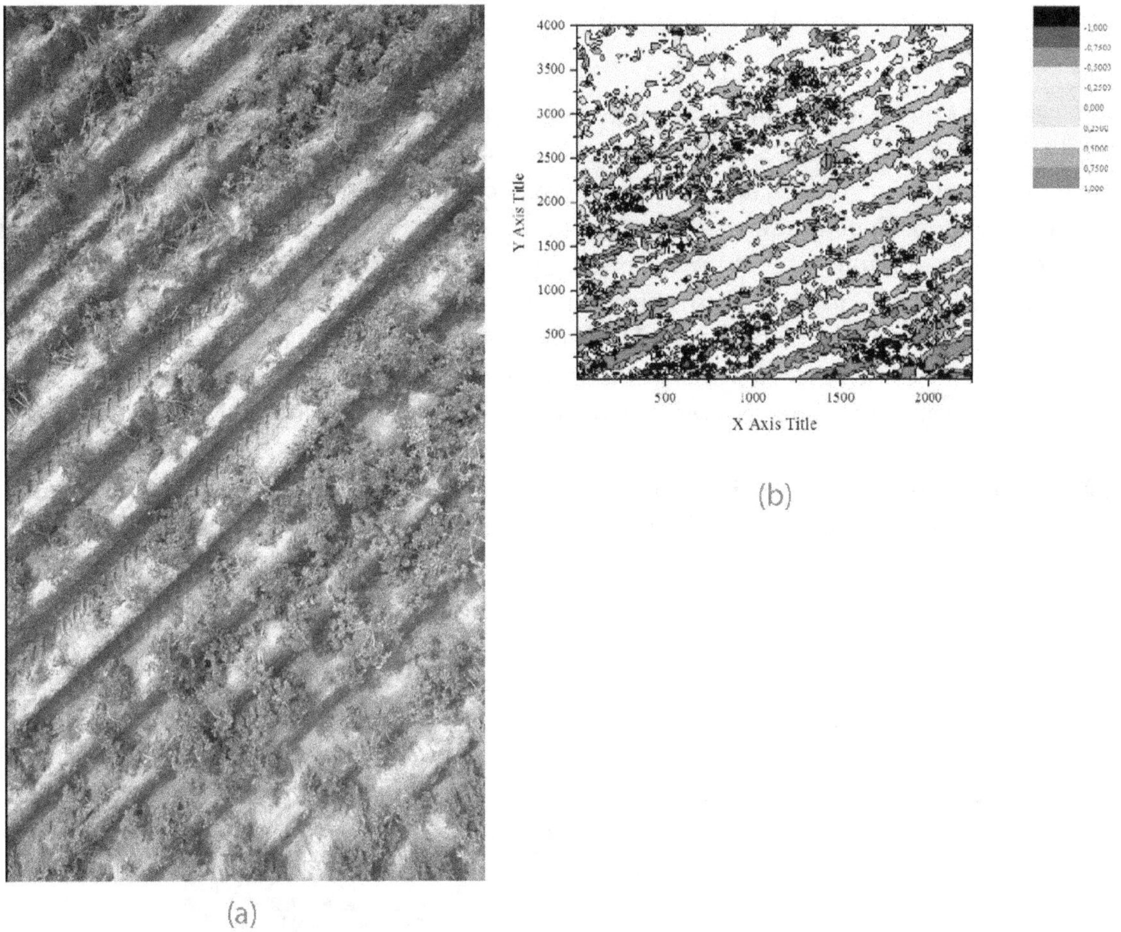

Currently, there are many different methods for image clustering, however, in narrow subject areas there are still problems that relate to the quality of clustering. The main problem associated with image clustering arises from the significant overlap of various features (for example, color), which in turn does not allow one to clearly distinguish objects in the image and perform correct clustering for them. In order to overcome this problem, the proposed method increases the number of features by using different color spaces.

Threshold methods are based on the use of threshold values, which allow you to divide images into several parts. These methods have simple implementation. The threshold method includes the direct difference method, which is calculated using the formulas:

$$F = \begin{cases} 1, \textit{если} \ |R_{m,n} - R| \le P \textit{И} \ |G_{m,n} - G| \le P \textit{И} \ |B_{m,n} - B| \le P \\ 0, \textit{если} \ |R_{m,n} - R| > P \textit{И} \ |G_{m,n} - G| > P \textit{И} \ |B_{m,n} - B| > P' \end{cases}$$

$$\begin{bmatrix} r_{m,n} \\ g_{m,n} \\ b_{m,n} \end{bmatrix} = \begin{bmatrix} R_{m,n} \\ G_{m,n} \\ B_{m,n} \end{bmatrix} \times F,$$

Where F is a multiplier, R_(m,n) is the red component of the (m, n)th pixel of the segmented image, R is the red component of the desired color, P is the threshold value, G_(m,n) is the green component of (m, n)-th pixel of the segmented image, G – green component of the desired color, B_(m,n) – blue component of the (m, n)-th pixel of the segmented image, B – blue component of the desired color, r_(m,n) – red component (m, n)th pixel of the segmented image, g_(m,n) – green component of the (m, n)th pixel of the segmented image, b_(m,n) – blue component of the (m, n)th pixel of the segmented image.

The UAV flies around a field of size x by y, at a given height (H) and at a fixed speed (V). Images are taken with a specified time step in such a way as to ensure the desired overlap of images. For flights, a sunny day and slightly windy weather is chosen to minimize possible displacement of the UAV by air masses. The field is photographed when the Sun is at its highest. As a result of the measurements, a set of intersecting images is obtained. Before processing them, they are stitched into one image (mosaic) and the perspective of the resulting image is adjusted. Figure 10 shows an example of the resulting image.

Figure 10. Example of a mosaic of an agricultural field

In Figure 10 you can see different types of surfaces ("field1", "field2" and "forest"). An important task of clustering this type of image is to calculate surface masks so that a specific surface can be isolated and its properties can be studied. The proposed technique allows you to calculate a surface type mask and use it to select the surface itself. If you select several areas in the field where the ground is visible (on the left in the image) and find the average characteristics of each color, this will allow you to build the threshold algorithm presented above. The result of this algorithm is shown below. We conducted experiments with the pixel size of these areas. To test the algorithm, we selected a separate field, which is shown in Figure 11.

Figure 11. Template for the surface "field1"

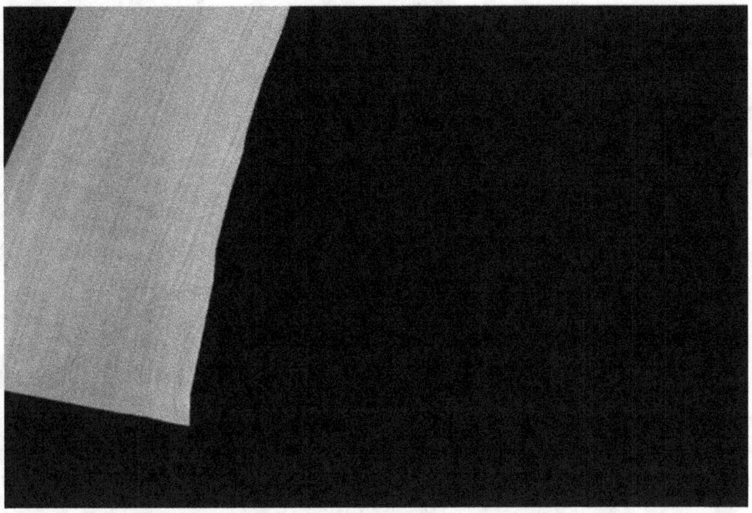

To test the selected surfaces, different sliding window sizes were used. The result of selecting the "field1" surface with a sliding window measuring 100 by 100 is shown in Image 12.

Figure 12. The result of selecting the surface "field1" with a window of 100 by 100 pixels

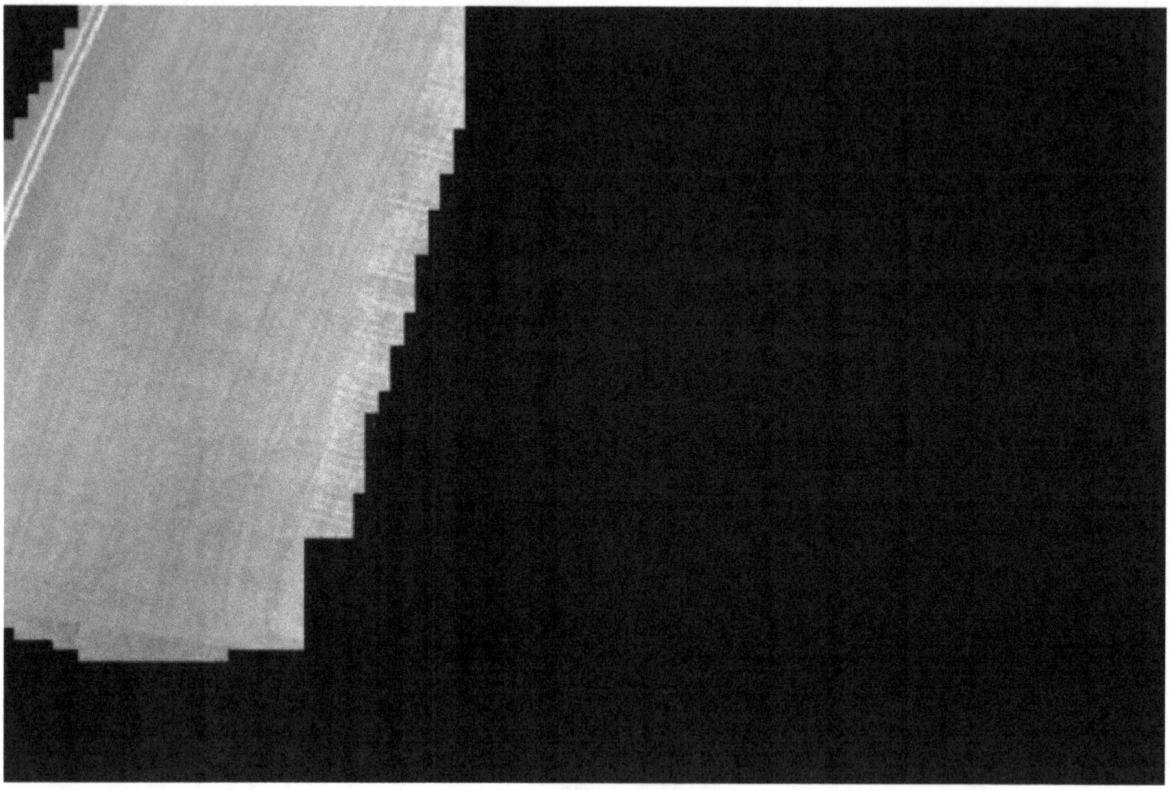

The result of selecting the "field1" surface with a sliding window measuring 50 by 50 is shown in image 13.

Figure 13. Result of selecting the "field1" surface with a 50 by 50 pixel window

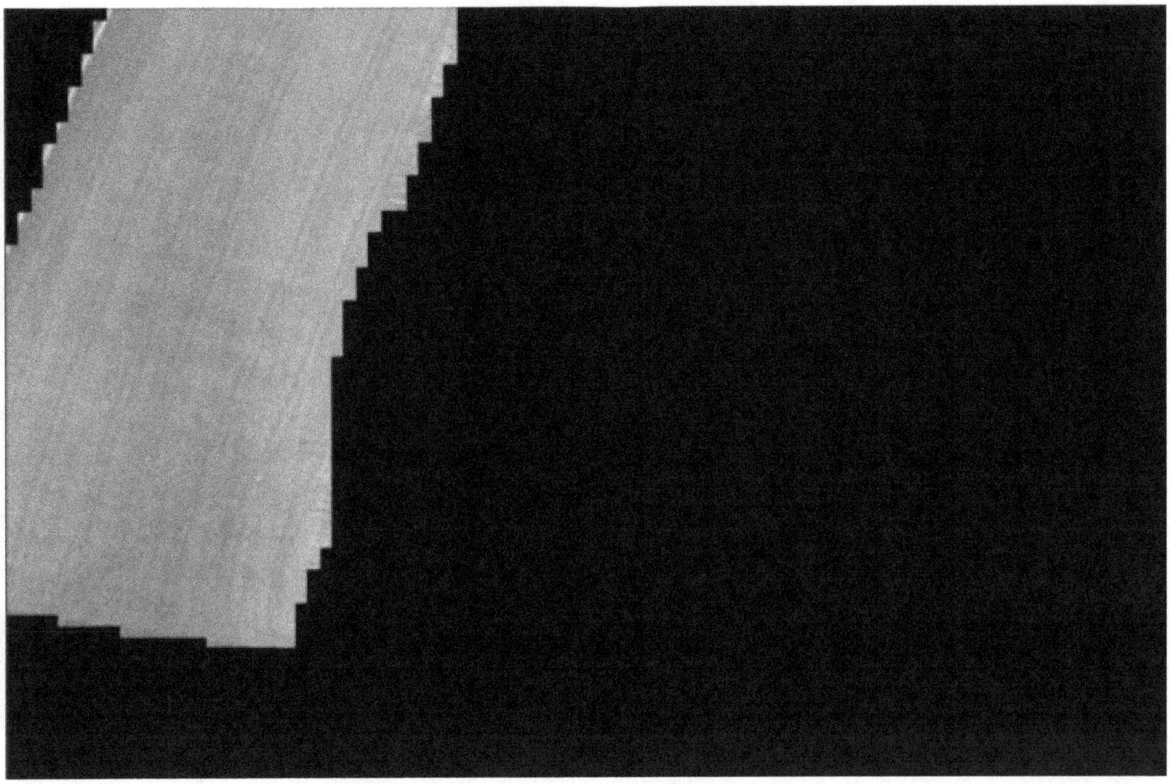

The result of selecting the "field1" surface with a sliding window measuring 25 by 25 is shown in image 14.

Figure 14. The result of selecting the surface "field1" with a window of 25 by 25 pixels

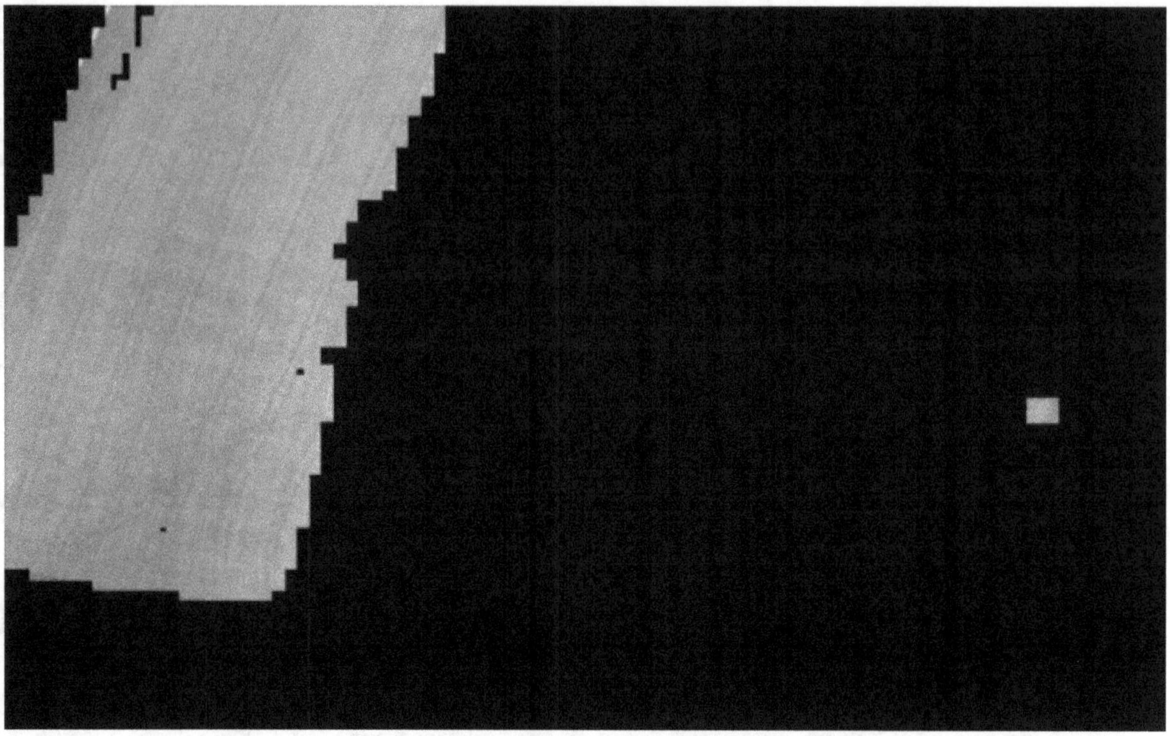

Metrics for the quality of highlighting the "field1" surface for three different window sizes are presented in Table 5.

Table 5. Metrics of quality of surface selection "field1"

Metrics	Sliding window size 100 x 100	Sliding window size 50 x 50	Sliding window size 25 x 25
Accuracy	85%	97%	98%
Completeness	99%	98%	96%
F-measure	92%	97%	97%

Table 5 shows that high scores for all three metrics were obtained for a window measuring 50 by 50 pixels.

CONCLUSION

The article discusses various aspects of the use of unmanned aerial vehicles and digital cameras installed on them. The use of digital cameras involves an image processing process, which consists of preliminary and thematic processing. At the pre-processing stage, corrective actions are performed to bring all images obtained during a UAV flight over an agricultural field to the same brightness. As part of thematic processing, work is carried out on segmentation (selecting homogeneous areas based on some characteristics) of the image with the task of identifying plants and soil. The results of processing real images of UAV flight and highlighting vegetation and soil are presented.

REFERENCES

Aasen, H., Burkart, A., Bolten, A., & Bareth, G. (2015). Generating 3D hyperspectral information with lightweight UAV snapshot cameras for vegetation monitoring: From camera calibration to quality assurance. *ISPRS Journal of Photogrammetry and Remote Sensing*, 108, 245–259. 10.1016/j.isprsjprs.2015.08.002

Andreev, G. A., Bazarsky, O. V., Glauberman, A. S., Kolesnikov, A. I., & Korzhik, Yu. V. (1984). Analysis and synthesis of random spatial textures. *Foreign Radio Electronics*, (2), 3–33.

Brede, B., Suomalainen, J., Bartholomeus, H., & Herold, M. (2015). Influence of solar zenith angle on the enhanced vegetation index of a Guyanese rainforest. *Remote Sensing Letters*, 6(12), 972–981. 10.1080/2150704X.2015.1089362

Burkart, A., Hecht, V. L., Kraska, T., & Rascher, U. (2018). Phenological analysis of unmanned aerial vehicle based time series of barley imagery with high temporal resolution. *Precision Agriculture*, 19(1), 134–146. 10.1007/s11119-017-9504-y

Christiansen, M., Laursen, M., Jørgensen, R., Skovsen, S., & Gislum, R. (2017). Designing and Testing a UAV Mapping System for Agricultural Field Surveying. *Sensors (Basel)*, 17(12), 2703–2715. 10.3390/s17122703291687 83

Colomina, I., & Molina, P. (2014). Unmanned aerial systems for photogrammetry and remote sensing: A review. *ISPRS Journal of Photogrammetry and Remote Sensing*, 92, 79–97. 10.1016/j.isprsjprs.2014.02.013

Fralenko, V.P. (2014) Methods for texture analysis of images, processing of Earth remote sensing data. *Software systems: theory and applications, 4*(22), 19–39.

Gulakov, V.K., Trubakov, A.O., Ogurtsov, S.N. (2011) Informative significance of texture characteristics based on the adjacency matrix of image pixel brightness levels. *Bulletin of Bryansk State Technical University, 2*(30), 81.

Haralik, R. M. (1979). Statistical and structural approaches to the description of textures. *TIIER*, 67(5), 98–120.

Honkavaara, E., Markelin, L., Hakala, T., & Peltoniemi, J. (2014). The Metrology of Directional, Spectral Reflectance Factor Measurements Based on Area Format Imaging by UAVs. *Photogrammetrie, Fernerkundung, Geoinformation*, 14, 175–188. 10.1127/1432-8364/2014/0218

Honkavaara, E., Saari, H., Kaivosoja, J., Pölönen, I., Hakala, T., Litkey, P., Makynen, J., & Pesonen, L. (2013). Processing and Assessment of Spectrometric, Stereoscopic Imagery Collected Using a Lightweight UAV Spectral Camera for Precision Agriculture. *Remote Sensing (Basel)*, 5(10), 5006–5039. 10.3390/rs5105006

Laliberte, A. S., Goforth, M. A., Steele, C. M., & Rango, A. (2011). Multispectral remote sensing from unmanned aircraft: Image processing workflows and applications for rangeland environments. *Remote Sensing (Basel)*, 3(11), 2529–2551. 10.3390/rs3112529

Lelong, C. C. D., Burger, P., Jubelin, G., Roux, B., Labbe, S., & Baret, F. (2008). Assessment of unmanned aerial vehicles imagery for quantitative monitoring of wheat crop in small plots. *Sensors (Basel)*, 8(5), 3557–3585. 10.3390/s805355727879893

Potapov, A. A. (2003). New information technologies based on probabilistic texture and fractal features in radar detection of low-contrast targets. *Radio Engineering and Electronics*, 48(9), 1101–1119.

Rasmussen, J., Ntakos, G., Nielsen, J., Svensgaard, J., Poulsen, R. N., & Christensen, S. (2016). Are vegetation indices derived from consumer-grade cameras mounted on UAVs sufficiently reliable for assessing experimental plots? *European Journal of Agronomy*, 74, 75–92. 10.1016/j.eja.2015.11.026

Schaepman-Strub, G., Schaepman, M. E., Painter, T. H., Dangel, S., & Martonchik, J. V. (2006). Reflectance quantities in optical remote sensing—Definitions and case studies. *Remote Sensing of Environment*, 103(1), 27–42. 10.1016/j.rse.2006.03.002

Smith, G. M., & Milton, E. J. (1999). The use of the empirical line method to calibrate remotely sensed data to reflectance. *International Journal of Remote Sensing*, 20(13), 2653–2662. 10.1080/014311699211994

Tymchuk, A. I. (2018). On texture features in the problem of segmentation of aerial photographs based on brightness dependence matrices. *Cybernetics and Programming*, (6), 31–39.

Chapter 7
Spatio-Temporal Variability of Soil and Methodological Approaches to Its Assessment by the Example of the Leningrad Region

Olga Anatolievna Pasko
http://orcid.org/0000-0001-8322-834X
National Open Institute of St. Petersburg, Russia

Alexander V. Zakharchenko
Institute of Problems of Northern Development of the Federal State Budgetary Institution of Science, Russia

Vladimir N. Goryachkin
Center for Agrochemical Service "Leningradsky" Russian Federation, Russia

ABSTRACT

In recent decades, there has been a steady trend in Russia towards a decrease in the fertility of arable soils, which is recorded by changes in their agrochemical, agrophysical, and biological indicators. In addition, most regions, in particular the Leningrad region, are characterized by the presence of limiting climatic factors that limit the possibility of obtaining high yields, despite favorable natural conditions for the development of agricultural production. Knowledge of the properties of arable land makes it possible to optimize the nutrition of agricultural crops as much as possible, taking into account the necessary and sufficient doses of fertilizers, regardless of suitable conditions.

DOI: 10.4018/979-8-3693-3374-7.ch007

INTRODUCTION

Intensification of agricultural production has led to an increase in crop yields, on the other hand, a violation of the balance of nutrients in the soils of agricultural lands (Zharnikov, Larionov and al., 2019) a decrease in the environmental sustainability of agricultural landscapes (Methodenkatalog ..., 2024; Pasko, Staurskaya and al., 2022) Preservation and improvement of soil fertility require carrying out a complex of agrotechnical, agrochemical, meliorative, erosion control, phytosanitary and other measures (Zakharchenko and Pasko, 2022; Sorokin, Kulizhsky and al., 2023) and a deep understanding of its condition and fertility (European Soil Data Center, 2024), dynamics of agrochemical, agrophysical and biological indicators based on data (Zakharchenko, Pasko et al., 2021).

Optimization of the use of soils and agricultural lands occurs by substantiating the level of basic fertilizers in soils (Pasko and Zakharchenko, 2020a), ensuring a positive balance of humus, macro- and micro-elements; introducing resource-saving environmentally sound technologies, increasing crop yields, improving the quality of their products, preserving and/or increasing soil fertility and improvement of technological properties of land plots (Pasko, Zakharchenko, et al., 2019).

The effectiveness of various soil fertility strategies can only be assessed by monitoring changes in indicators that reflect the state of soils to provide ecosystem services or by assessing the range of loads that disrupt soil functions. Soil monitoring consists of on-site surveys, which are increasingly complemented by digital measurement methods (e.g., Earth observation (Kataev, Kartashov, 2023a), remote sensing (Kataev, Kartashov, 2023b), and new data processing tools (e.g., big data, geostatistics (Tokareva, Majid et al., 2020). The widespread practice of applying fertilizers in doses averaged over large areas leads to an excess or shortage of fertilizers and, accordingly, to economic and environmental costs. Reduction of fertilizer costs is possible if they are applied differentially, taking into account the spatial variability of soils.

Therefore, knowledge of the spatial variability of agricultural soils, assessed by generally accepted indicators, makes it possible to minimize their application doses in a precision farming system (Bucci, Bentivoglio et al., 2018; Zhang, 2015).

On the other hand, smart farming technologies are becoming more widespread in modern agriculture, helping to optimize agricultural and livestock production and minimize waste and costs (Joint Research Centre, 2014). Precision farming is a technological and data-based approach to agricultural management that observes, measures and analyzes the needs of individual fields and crops.

STATE LAND MONITORING

Information on the dynamics of the main agrochemical indicators of soil condition allows us to identify the main trends in changes in their fertility and create predictive models for the development of the most effective measures for the use of land and soil resources in different countries (FOEN, 2017; BAFU, 2017; Bispo et al., 2020; Borrelli et al., 2022; Martin, I., et al., 2022). The study of soil processes is part of geosystem monitoring and is carried out mainly by remote methods in combination with ground-based studies (European Soil Data Center, 2024). The use of the results of long-term aerospace and ground-based studies of the dynamics of humidification, salinization, soil erosion and a number of other processes allows us to obtain new relevant scientific information (Komarov, 2018).

Extensive and long-term monitoring is probably the most realistic approach to obtaining objective information about differences between ecosystems and changes within them (Steinhoff-Knopp et al., 2020) and human exposure to them (Bloem, Schouten, et al., 2005). The assessment of the spatial variability of agricultural land fertility serves as the basis for calculating the necessary and sufficient doses of fertilizers (Schimmel and Amelung, 2022; Vonk et al., 2020). The preservation and enhancement of soil fertility has become a strategic task for most countries (Veerman et al., 2020; Lehmann et al., 2020, Lessmann et al., 2021). The focus of special attention of soil scientists, agrochemists and ecologists is the effect of fertilizers on the fertility of arable soils (Hirt et al., 2021; Lavallee et al., 2020; Prout et al., 2020) and crop yield De Vries et al., 2021; Grilli et al., 2021), conservation of soil biodiversity and prevention of soil degradation (Horn et al., 2022); combating the development of negative processes, in particular erosion of arable land (Panagos et al., 2020, 2021; Prasuhn, 2020; Kuhwald et al., 2022), acidification of arable land, contributing to crop loss (Zhu et al., 2020) and many more etc.

Among the urgent problems, experts note the need to standardize assessments of ecosystem services related to soil (Paul et al., 2020), to develop a quantitative assessment of soil functions – its potential and condition (Vogel et al., 2020), In Russia, the need for state monitoring of land is established by the Land Code of the Russian Federation (Article 67) (The Land Code, 2001). Land monitoring is a system of observations (surveys, surveys and surveys) of the state of the land resources of the Russian Federation for the timely identification of changes, their assessment, prediction, prevention and elimination of the consequences of negative processes. Control of changes in soil fertility, identification of the dynamics of their agrochemical composition, nutrient availability, etc. They are carried out by specialized state institutions (Agrochemical Service Centers or agrochemical stations of the Ministry of Agriculture of the Russian Federation) (Agrochemical Service of Russia, 2024).

Scientifically based data on the qualitative state of the lands and the changes taking place in them make it possible to prevent and eliminate the consequences of negative processes: droughts, excess moisture, environmental pollution by industrial emissions of pollutants, etc. Land monitoring is carried out on the basis of systematic and comprehensive observations, and, like land management and the state cadastre of immovable property, is a "tool" for land management.

The following tasks are being solved:

a) timely identification of changes in the state of lands, assessment of these changes, forecast and development of recommendations for agricultural enterprises to prevent and eliminate the consequences of negative processes;
b) information support of activities related to the maintenance of the state land cadastre, the implementation of state land control over the rational use and protection of land, other functions in the field of land management, as well as land management;
c) providing citizens with information about the agrochemical condition of the land.

ANALYSIS OF THE AGROECOLOGICAL CONDITION OF THE LANDS

The value of land as the main means of agricultural production is determined by its fertility – the ability to meet the needs of plants for nutrients, air, water, heat and to ensure the formation of yields of agricultural crops of good quality. The optimal form of soil fertility monitoring is a periodically repeated comprehensive agrochemical survey of the entire area of agricultural land, and it is assigned to the centers and stations of the State Agrochemical Service. In recent years, in the Leningrad Region, as in other regions of Russia, as a result of the unfavorable economic situation, there has been an increase in the rate of soil degradation, which may become irreversible in the future (Gavrilova and Shchukin, 2019). The greatest damage to soils is caused by the processes of overgrowth of agricultural lands with shrubs and small forests, flooding and re-humidification (Timofeeva, 2020).

The purpose of the study is to analyze the spatial and temporal variability of agrochemical indicators that determine the fertility of agricultural land.

To achieve this goal, the ***following tasks*** were solved: 1. To study the agrochemical indicators of the lands of the survey areas: pH, mobile forms of phosphorus and potassium, hydrolytic and metabolic acidity, the amount of exchange bases, humus content. 2. To assess the spatial heterogeneity of physico-chemical and agrochemical indicators using classical mathematical and statistical methods of analysis. 3. To identify the temporal variability of the studied lands over a period of ~ 50 years.

Practical significance. The results obtained can be used in the development of design and technological documentation on the use of chemicals to optimize agrochemical, physico-chemical and biological properties of the soil to increase crop yields.

OBJECTS AND METHODS OF RESEARCH

The Leningrad Region is located in the Northwestern Federal District of the Russian Federation and borders the European Union. The total length of its face on land is 2,440 km; the territory is 8,390.8 thousand hectares, including the area of the territory covered by the Ladoga and Onega Lakes – 1081.3 thousand hectares (Investment portal of the Lenin-grad region, 2024). The territory is divided into 17 municipal districts and one urban district. The distribution of the land fund of the Leningrad Region by land categories is shown in Fig. 1.

The structure of the land fund of the Leningrad region is dominated by the forest fund, which makes up 56.68% of the total area. The share of lands of settlements accounts for 2.84%, of which 93.8 thousand hectares are the area of cities and urban-type settlements, 144.3 thousand hectares are rural settlements. Industrial, transport and other special purpose lands occupy an area of 387.5 thousand hectares, of which the bulk falls on defense and security lands (294.1 thousand hectares). The lands of the water fund occupy an area of 1081.3 thousand hectares, including 1006.8 thousand hectares - a mirror of the Ladoga and Onega lakes. Agricultural lands are the most problematic in their use, accounting for 20.28%.

Figure 1. Distribution of the territory of the Leningrad region by land categories (area, thousand hectares)

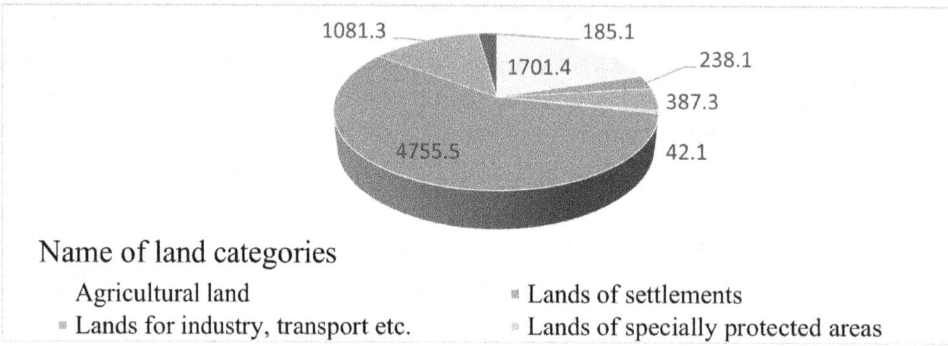

Agricultural lands are the main means of production in agriculture, according to which they have a special legal regime and are subject to protection aimed at preserving the area, preventing the development of negative environmental processes and increasing fertility. In the Leningrad Region, the area of agricultural land as of January 1, 2022 is 1701.3 thousand hectares. This category includes lands provided to various agricultural enterprises and organizations (partnerships and societies, cooperatives, state and municipal unitary enterprises, research institutions); also includes land plots provided to citizens for farming, personal subsidiary farming, gardening, and country farming, horticulture, animal husbandry, haymaking and grazing of farm animals (Federal Law "On the transfer of lands, 2004 Federal Law of 12/21/2004 N 172).

In agricultural lands, agricultural lands are of particular interest, i.e. lands that are directly used for agricultural production (Figure 2). Arable land, fallow land, hayfields, pastures and perennial plantations are classified as agricultural land, while non-agricultural land includes lands under water, including swamps, forest areas and lands under forest plantations, building lands, lands under roads, disturbed lands, other lands (ravines, sands, etc.). As of 1 In January 2022, the area of agricultural land in all land categories amounted to 798.5 thousand hectares, or 9.5% of the total land fund of the Lenin-Gradskaya region. The main share of agricultural land is concentrated in the category of agricultural land (616 thousand hectares or 77.14%) (Agricultural portal, 2024). Significant areas (108.7 thousand hectares) are located on the territory of the lands of settlements (mainly within the boundaries of rural settlements), which means which is 13.61% of the total area of this category. The land reserve contains 29.4 thousand hectares of farmland. In the structure of agricultural land, arable land (434 thousand hectares) accounts for 54.35%. Natural forage lands (hayfields and pastures) occupy 320 thousand hectares (40.08%), perennial fruit plantations – 44.5 thousand hectares (5.57%).

Figure 2. Photos of agricultural land in the Leningrad region: A – Arable land; B – Rapeseed crops

A: https://bigenc.ru/c/leningradskaia-oblast-khoziaistvo-sel-skoe-khoziaistvo-1d4b3f ;
B: https://agroprom.lenobl.ru/ru/news/27522

The Leningrad Region is located in the temperate latitudes of the northern hemisphere, in the forest zone, at the junction of the taiga subzones and mixed forests, the territory of the Leningrad region occupies 85.9 thousand square kilometers. The climate of the region is Atlantic-continental. Marine air masses cause a relatively mild winter with frequent thaws and moderately warm, sometimes cool summers. The total solar radiation averages 77 Kcal/sq.m. cm per year. The sum of temperatures above +10 degrees Celsius varies by year in the range of 1400 to 1900 degrees Celsius. According to the amount of precipitation, the Leningrad region belongs to the zone of sufficient moisture (550-650 mm per year) The greatest amount of precipitation falls in summer and autumn. The main feature of the climate is the variability of the weather, due to the frequent change of air masses, which, depending on the area of formation, are divided into marine, continental and Arctic.

The peculiarity of the Leningrad region is the heterogeneity of weather conditions in the territory, due to the large extent of the region from west to east, the diversity of the landscape and the proximity of large reservoirs (Gulf of Finland, Ladoga and Onega Lakes). Adverse factors in the region are meteorological phenomena: strong winds, including squalls and tornadoes, heat, short-term intense downpours and prolonged rains, thunderstorms, hail, forest fires, drought and floods.

According to the data for the period 1991-2020, the average annual air temperature is 6.3 °C, in the western regions 4.8-6 °C, in the eastern regions 3.7-4.8 °C. The average number of days per year with an average daily temperature above 0°C is 246 days. The average date of transition of the average daily air temperature through 0 °C towards positive values in spring is March 22, towards negative values in autumn – November 23. The warmest month is July with an average temperature from 17.1°C (Vinnytsia) to 18.5°C. Annual precipitation ranges from 632 mm (Lomonosov) to 796 mm (Lodeynoye Pole). Precipitation falls unevenly throughout the year: most of it (51%) falls during the warm period (May-September). The maximum precipitation is usually in August (86.4 mm), and the minimum is in March (35.4 mm) (Decree of the Governor of the Leningrad Region, 2021). The Baltic Sea and has a dense, well-developed river network. There are also 1800 lakes in the region.

CHARACTERISTICS OF THE SOIL COVER

The division of the territory of the Leningrad Region into soil areas and subdistricts is based on the prevailing types and subtypes of soils, mechanical composition, soil-forming rocks, as well as the depth of carbonates, the nature of the relief, the spread of swamps and swampy soils. According to them, the territory of the Leningrad Region is divided into seven natural agricultural areas (Pestryakov, 1973) (Fig.3).

Figure 3. Natural zoning of soil zones of the Leningrad region (I – Karelian isthmus, II – Prinevskaya lowland, III – Ordovician plateau, IV – southwestern plain, V - Mginsko–Tosnenskaya plain, VI - Ladoga–Volkhovskaya plain, VII - Eastern hilly upland

(according to Pestryakov, 1973)

I. The Karelian Isthmus includes the Vyborg and Priozersky districts, as well as the northern part of the Vsevolozhsky district. From a soil-geomorphological point of view, it is an area with three pronounced geomorphological elements: in the north – a crystalline massif, in the center – a moraine upland, on the coast – a terraced plain with water–glacial and glacial deposits. The soil–forming rocks are glacial moraine sandy loam deposits, water-glacial sandy loams and sands, lake-glacial sands, sandy loams and loams, ribbon clays and binomial deposits. The indigenous forests are mainly coniferous. The main body of the soil cover is represented by a wide variety of podzolic soils – from superficially podzolic and typically podzolic to sod (humus)-podzolic.

Signs of the development of the gluing process are characteristic of most of the territory of the Karelian Isthmus. In drainless depressions and depressions, mineral soils of the podzolic-marsh type are formed, more often marsh soils (peat-loamy, peat-marsh, soils of upper marshes predominate). The soils of this zone are characterized by poverty of exchange bases and high acidity, but the delayed mineralization of organic matter in conditions of periodic waterlogging contributes to a higher content of humus in them, compared with soils of other regions.

II. The Prinevskaya lowland occupies a small but well-developed agricultural territory, which stretches from Lake Ladoga to the Gulf of Finland and along the coast of the Neva Bay and covers part of the territory of Vsevolozhsky, Lomonosov and Gatchina districts. Glacial and water-glacial deposits of sandy, sandy loam and loamy composition, as well as reduced relief, combined with constant seasonal waterlogging, mainly contribute to the development of podzolic-marsh type soils – sod-podzolic gleevate, surface-gleevate, sometimes peaty illuvial-humus podzols. On elevated landforms, surface-podzolic soils are formed. Peat-swamp lowland and upland soils used for peat mining are also widespread. This area belongs to the area of intensive vegetable growing. Agricultural use of soils is associated with their active drainage, fertilization and liming.

III. The Ordovician plateau is a relatively elevated plain with a slight slope in the south and south–east direction, composed of Ordovician limestones, which sometimes act as soil-forming rocks, but in some places are overlain by quaternary glacial and water-glacial deposits with the participation of dolomitized limestones. The Ordovician plateau occupies the territory of Gatchina, Volosovsky, Kingisepp and Lomonosov districts. The soils are rich in calcium and magnesium carbonates, which contributes to the development of sod-carbonate type soils. Sod-carbonate typical, leached and podzolic soils are formed here, as well as sod-podzolic soils on carbonate deposits. Mineral waterlogged soils predominate in the southern and southeastern parts of the district.

IV. The southwestern district occupies a vast territory of the region and borders the Ordovician plateau. This is the territory of Luga, part of the Slantsevsky, Lomonosovsky and Gatchina districts. The area is characterized by an increased amount of positive temperatures, relatively flat terrain and carbonation-free soil-forming rocks. The area is characterized by a wide combination of soils of podzolic, podzolic-marsh and marsh types. Vast territories are represented by a subtype of sod-podzolic soils. Favorable climatic conditions, sandy loam and light loamy mechanical composition of soils favor their active use in agriculture. Large areas of soil here have long-term development, significant areas are planned for development under arable land.

V. The Mginsko-Tosnenskaya plain occupies the Tosnensky district. The flatness of the territory and ribbon loamy soil-forming rocks (less often lake-glacial sands and sandy loams) under conditions of excess precipitation caused waterlogging of soils. Sod-podzolic soils of normal moisture are much less common. On negative and flat relief elements, podzolic-swamp soils develop, which are widely represented in this area, and drainage depressions are occupied by swamps. From an agricultural point of view, this territory has been little developed, but the presence of huge flat massifs of mineral podzolic-marsh soils can serve as a fund for priority land development through land reclamation.

VI. The Ladoga-Volkhov plain occupies a vast territory of the Kirishsky, Volkhovsky and partially Lodeynopolsky and Tikhvin administrative districts. In the northern part, the Ladoga sandy lowland stands out, in the southwest there is a wide-spreading plateau, and in the west the Ordovician plateau, composed of limestones, continues. The Pashsko-Syassky watershed, composed of fluvioglacial deposits, stands out for this area, in the southeast there is a sandy-marsh lowland, and the entire main central part of the district is an extensive plain composed of lake sediments of loamy composition. Together with the typical sod-carbonate, leached and podzolic soils that develop in the end sections of the Ordovician plateau, podzolic soils are widely developed here, which are characterized by increased development of the turf process and the formation of an accumulative horizon.

The flatness of the territory and the presence of a large number of negative landforms contribute to the waterlogging of the territory and the development of soils of podzolic-marsh and marsh types. Peat-marsh soils of lowland and especially upland bogs are evenly found throughout the region. The

Ladoga-Volkhov plain is not characterized by high development of the territory, but it is the most promising for the expansion of arable land due to the development of massifs with podzolic-marsh and other soils that necessarily require intensive drainage.

VII. The eastern hilly upland. The extremely heterogeneous terrain of the territory is characterized by alternating hills and ridges with closed depressions and small valleys. This creates a rather complex landscape of the area. Because of this, the development of the area is low, and the main areas are occupied by forests and swamps. Typically podzolic soils are formed under forests on sandy and loamy sediments, binomial sediments are often found. Reduced landforms are occupied by swamp and podzolic-swamp soils, and sod-alluvial soils are developed in the floodplains of rivers. There are few arable soils, they are concentrated mainly on flat hilltops and small elevated plains. Recently, the development of new lands has intensified here, but the difficult terrain and soil poverty limit the possibility of their widespread involvement for agricultural use.

So, in the structure of the soil cover of the Leningrad region, 6 main types of soils are distinguished: podzolic, marsh-podzolic, sod-gley, peat-marsh, sod-carbonate, sod-alluvial. According to the soil-geographical zoning, the Leningrad Region is included in the zone of the central taiga-forest bioclimatic region and is located on the border and the southern subzones. Vegetation belongs to the taiga zone, to the subzones of the middle (in the north of the region) and southern taiga. There is a transition from coniferous forests to mixed forests in the south of the region.

The soils of the Leningrad region are predominantly sod-podzolic, and therefore have an acidic reaction of the medium. Despite the poverty of the soils of agricultural lands in the Leningrad region, they are an important production base and an indispensable source of feed for the agro-industrial complex (Timofeeva, 2020; Federal Law, 2024; Pestryakov, 1973). Agricultural lands are considered to be the most valuable, and the issues of their assessment are prioritized (The Land Code, 2021). It is known that nutrition, growth, development and crop yields are affected by soil pH. For most cultivated plants, the optimal soil acidity is in the range from 6.0 to 7.0. Macronutrients (nitrogen, phosphorus, potassium) are highly soluble and more accessible to plants in soils with an increased pH value; trace elements (boron, iron, manganese, copper, etc.) – with a reduced pH value.

At a soil pH above 7, trace elements turn into insoluble compounds and become inaccessible to plants (Agrochemical Service, 2024). The difficult natural conditions of the region impose special requirements on the production of agricultural products. In order to obtain high yields, a comprehensive and objective account of natural and climatic factors is necessary.

OBJECTS AND METHODS OF RESEARCH

The object of the study is agricultural lands of the Leningrad Region (Table 1). The subject of the study is agrochemical indicators established by the Leningradsky Agrochemical Service Center and other state agrochemical institutions during agroecological monitoring of lands.

Table 1. Spatial characteristics of research objects (by Sales of products, 2023)

Symbol of the region	Region	Area, ha		
		of the region	of agricultural land	Of arable land
LD	Leningrad Region	8 390 800	177 800	46 461
	Districts			
V	Volosovsky	273 000	16 276	11 053
Vh	Volkhovsky	5 12400	2000	2500
G	Gatchinsky	289 181	38 615	8572
K	Kingiseppsky	290 800	3 583	1 090
L	Lomonosovsky	200000	9101	3 633
Lh	Luzhsky	600600	28 995	3685
P	Priozersky	3 59700	4100	1156
Th	Tikhvinskiy	7 018 00	120	177 800
T	Tosnensky	3 655 00	4850	1400

The research complex includes chemical-analytical, hierarchical cluster and probabilistic-statistical methods. The Garmin GPSMAP 78s GPS navigator was used for soil sampling, and the data obtained was processed by BaseCamp software. The survey was conducted in accordance with the "Methodological guidelines for conducting comprehensive monitoring of soil fertility of agricultural lands in 2003" (Methodological guidelines, 2003). The territories of the former farms "Kirovsky", "New Life", "Coastal", "Kotelsky" and "Kingisepp" and others were surveyed. The cartographic basis was copies of the land management plans of these farms on a scale of 1:10,000.

An agrochemical analysis of various land plots, i.e. land plots systematically used to produce agricultural products, such as arable land, hay fields, pastures, has been carried out. Arable land is an agricultural land that is systematically cultivated and used for crops, including crops of perennial grasses, as well as clean pairs. Haymaking – agricultural land used for mowing; pasture: – for grazing animals. After a reconnaissance inspection of agricultural lands subject to agrochemical inspection, a grid of elementary plots of a specified size is applied to the cartographic scheme of soil sampling.

The selection of combined soil samples was carried out on elementary sites with a reed drill to the depth of the humus (arable) horizon, on hayfields and pastures – to the depth of the humus horizon, an arbitrary-shaped section of the earth's surface with homogeneous soil conditions was taken as an elementary site; one elementary site corresponds to one soil sample selected along the route. The route is laid in the center of each elementary section along the elongated side. Each combined sample is made up of 40 point samples and has a mass of at least 400 g.

The choice of land characteristics for agrochemical and agroecological monitoring of agricultural lands is determined by the traditions and requirements of the Rosselkhoznadzor (Methodological..., 2003). The Leningradsky Agrochemical Service Center conducted an agrochemical analysis of the soils of 224 fields in the following districts: Volosovsky; Volkhovsky; Gatchinsky; Kingisepp; Lomonosovsky; Lugsky; Priozersky; Tikhvin and Tosensky (Table 1). The following were taken into account: the area and type of land, the nature of vegetation, type and granulometric composition soil, soil pH, humus content, mobile forms of potassium and phosphorus. The determination of agrochemical parameters was carried out using standard chemical analytical methods.

The acidity of the salt extract pH was determined according to the State Standard of the USSR "Preparation of salt extract, 1985". The essence of the method is the extraction of exchangeable cations, nitrates and mobile sulfur from the soil with a solution of potassium chloride with a concentration of 1 mol/dm (1 N) at a ratio of soil to solution of 1:2.5 and potentiometric determination of pH using a glass electrode. When determining pH in samples of organic soil horizons, the extract is prepared at a soil to solution ratio of 1:25.

The total error of the method in determining pH is 0.1 pH units.

The content of *mobile phosphorus and potassium* was determined by the Kirsanov method as modified by TsINAO (Soils. Determination of mobile compounds, 1991). The method is based on the extraction of mobile compounds of these ions from the soil with a solution of hydrochloric acid of molar concentration. The content of P2O5 and K2O is determined directly from the previously constructed calibration graph (Soils. General requirements, 1991). Permissible relative deviations from the certified value of the standard sample for confidence probability P = 0.95 when determining mobile phosphorus 15-20%; potassium – 10 – 15%.

The method for determining the content *of organic matter* by the Tyurin method in the QINAO modification is based on the oxidation of organic matter with a solution of potassium bicarbonate in sulfuric acid and subsequent determination of trivalent chromium equivalent to the content of organic matter on a photoelectrocolorimeter. The maximum values of the relative error of the analysis results for a two-way positive probability P = 0.95 are as a percentage (rel.): 20 – with a mass fraction of organic matter up to 3%; 15 - St. 3 to 5%; 10 - St. 5 to 15 (Soils. Methods for determination of organic matter, 1991).

Hydrolytic acidity is determined by the Cappen method (Soils. Determination of hydrolytic acid, 1991), based on soil treatment with a solution of sodium acetic acid concentration of 1 mol/dm with a soil-to-solution ratio of 1:2.5. The limiting value of the relative error of the analysis results for a two-way confidence probability of 0.95 is 12%.

To identify the content of the *sum of absorbed bases*, the Cappen method (Soils. Determination of base absorption sum, 1988). It is based on the reaction of absorbed soil bases with hydrochloric acid and subsequent titration with sodium hydroxide of the acid residue that did not react. The limit value of the total relative error of the method with a two-way confidence probability P = 0.95 is, %: 20 – for the sum of absorbed bases up to 5 mmol / 100 g of soil and 15 – over 5 mmol / 100 g of soil. The analyses of the samples take into account the types of agricultural land, the type and granulometric composition of the soil and the nature of vegetation. The International Classification was used to describe soils (Soil Atlas, 2005; Gerasimova, 2019; Minasny, McBratney et al., 2010).

Statistical analysis of the sample and correlation analysis were carried out using Microsoft Excel and Statistica 8 programs, cluster analysis was carried out using Orange 3.29.3 software. The relationships between soil properties were revealed by information and logical analysis. The measure of similarity between objects is established using the Spearman correlation coefficient. Euclidean distance and the Word method were used for clustering (Zakharchenko, Moskovchenko et al., 2022). The result of the analysis is a system of nested sample partitions (clusters) forming a "Tree", where the core is the entire sample, and the branches and leaves are the smallest groups of data. Clusters are a combination of samples (observation points) with minimizing the variance of object parameters within point combinations relative to the distance between them (Zakharchenko, Pasko et al., 2023).

To characterize the lands, data on agrochemical indicators (pH, P_2O_5, K_2O, humus, S,ha – plot areas) were used for 9 districts in the amount of 577 observations. A dispersion analysis of 221 observations was carried out to study the influence of factors in 8 districts that affect the indicators: districts, vegetation, soils, granulometric composition, agronomic land use. In six districts (46 observations), changes in agrochemical parameters (pH, P_2O_5, K_2O, humus, S, ha) over ~50 years of their use from 1968 to 2020 were studied by parametric and nonparametric methods.

RESULTS

During the study, it was found that the soil cover of the areas of study is heterogeneous and significantly differs in its natural properties, in addition, farming systems also differ. This affects an increase in the values of the standard deviation and the confidence interval (Table 2). The pH values of the aqueous extract of soils belong to neutral or alkaline conditions (pH > 6). Two districts are distinguished (Kingisepp (one farm, conditionally K1) and Lomonosov (three farms, conditionally L1 – L3). 123 elementary plots with a total area of 43 thousand hectares are characterized by a very high content of phosphorus and potassium, but low values of the sum of exchange bases and high values of hydrolytic acidity. A base saturation level of less than 50 indicates the need for soil liming. These sites are characterized by an average or high content of organic carbon from 3.9% to 6.7%.

Table 2. Differences in agrochemical indicators for monitoring agricultural lands (pH, P2O5 mg mg/kg, K2O mg/kg, organic carbon content (humus) %, sum of exchange bases SO, mg/kg, Hydrolytic acidity mg-eq./100 g of soil, Area S thousand hectares) between administrative districts

District	N	pH±Δ	P_2O_5 мг * ±Δ	K_2O ±Δ	Humus ±Δ	SO±Δ	GK±Δ	S
K1	69	5,2±0,28	242,4±119,69	57,2±16,4	3,9±0,56	2,0±0,4	13,8±3,5	13,6
L1	13	6,6±0,19	1000,0±0,00	249,6±83,5	5,4±1,03	1,1±0,3	30,5±7,0	29,4
L2	17	6,1±0,33	669,5±169,52	270,5±67,3	5,7±0,59	1,8±0,5	27,4±7,7	4,7
L3	24	6,2±0,34	463,2±214,61	145,3±72,0	6,7±4,97	1,1±0,5	33,8±8,0	20,9
P1	6	4,9±0,76	354,2±208,91	76,8±166,6	5,7±0,47	7,5±6,4	4,9±3,0	17,7
L5	43	5,4±0,35	125,5±49,28	69,5±22,8	3,1±1,32	12,8±5,5	2,1±0,4	44,0
T1	383	5,5±0,10	394,2±54,38	80,1±13,9	5,9±0,42	14,5±0,8	2,6±0,2	4,2
L4	21	5,7±0,5	180,3±81,1	164,468,2	23,110,0	2,31,3	21,2±8,7	23,8
All Grps	555	5,5±0,10	379,7±47,63	89,0±13,0	5,5±0,41	11,5±1,1	6,7±1,6	9,4

*Values of P2O5 mg greater than 1000 (the limit of determination) are replaced by a value of 1000, which reduces the total variance or even reduces it to 0.

According to measurements of agrochemical properties of soils in 432 observation points on an area of 56.7 thousand hectares, the soils of one farm in Tosnensky (T1), Lomonosovsky (L5) and Priozersky (P1) districts do not need liming. The aqueous extraction of soils has an acidic or neutral pH of the medium; the phosphorus and potassium content is high, but low values are observed for hydrolytic acidity and high values for the sum of the exchange bases. At the same time, the degree of saturation is high, and liming is required. Due to the high phosphorus content and very high content of organic matter in the soil (> 20%), the Lomonosov district has an L4 farm with a total area of 23.8 thousand hectares. They are also characterized by low values of exchange bases, high potassium content, hydrolytic acidity with a degree of saturation of bases <50, which indicates the need for their liming.

According to the pH value of the water extract > 6, three farms of the Lomonosov district (L1, L2, L3) are distinguished, differing in high values from all others at $p < 0.05$ (Fig. 4a). The soils of the farms of Kingisepp (K1) and P1 (Priozersky) districts differ in relatively low pH values (4.9-5.2) relative to other districts. The soils of the farms of Lomonosov (L5) and Tosnensky (T1) districts have pH values close to 5.5, which is considered optimal for most crops. The phosphorus content in the soils of agricultural lands is highest in the farms of L1 and L2 Lomonosov districts, it significantly differs from other districts at $p < 0.05$. The lowest values of P2O5 content are observed in farms K1 of Kingisepp and L5 of Lomonosov districts (at $p < 0.05$). The remaining areas are characterized by an average content of mobile phosphorus (Fig. 4b).

The distribution of agrochemical indicators by five factors is considered: districts (District), vegetation of agricultural crops (Vegetation), soil types (Soil), granulometric composition (Granulometric), types of agrochemical lands (Land). Based on the analysis of variance, the relationship of these factors with agrochemical parameters (pH, P_2O_5, K_2O, Humus) at $p < 0.05$ was established (Table 3).

Figure 4. Changes in agrochemical parameters: pH (a), P2O5 mg/kg (b); K2O, mg/kg (c); Humus (Gum), % (d); sum of exchange bases SO, mg/kg (e); hydrolytic acid, mg-eq. / 100 g of soil (f) by farms in the study areas: K – Kingisepp, L – Lomonosovsky; T – Tosnensky, P – Priozersky

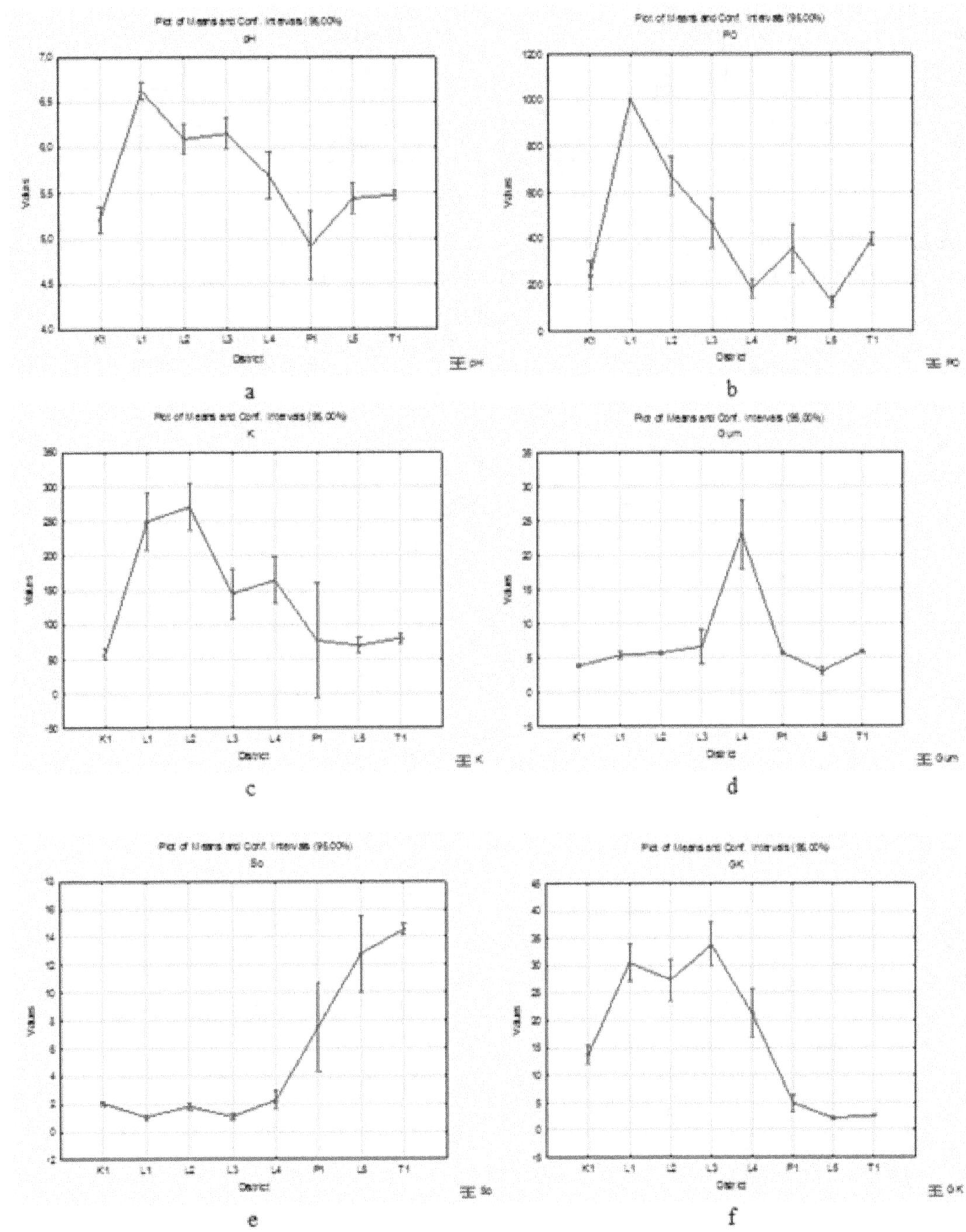

The number in the index of the district indicates the internal division into subdistricts.

Table 3. Likelihood ratio criterion for district factors, type of land use based on agrochemical parameters pH, P_2O_5, K_2O, Humus

Factor	Wilks	F	Effect df	Error df	p
Intercept	0,05	959,8604	4	204,0000	0,000
District	0,08	25,7711	28	736,9547	0,000
Land	0,66	3,8073	24	712,8810	0,000

Figure 5. Agrochemical parameters pH (a), P_2O_5 mg/kg (b); K_2O, mg/kg (c); Humus (Gum), % (d) on farms of Volosovsky (V1, V2), Kingisepp (K1), Lomonosov (L1 – L4) and Priozersky (P1) districts

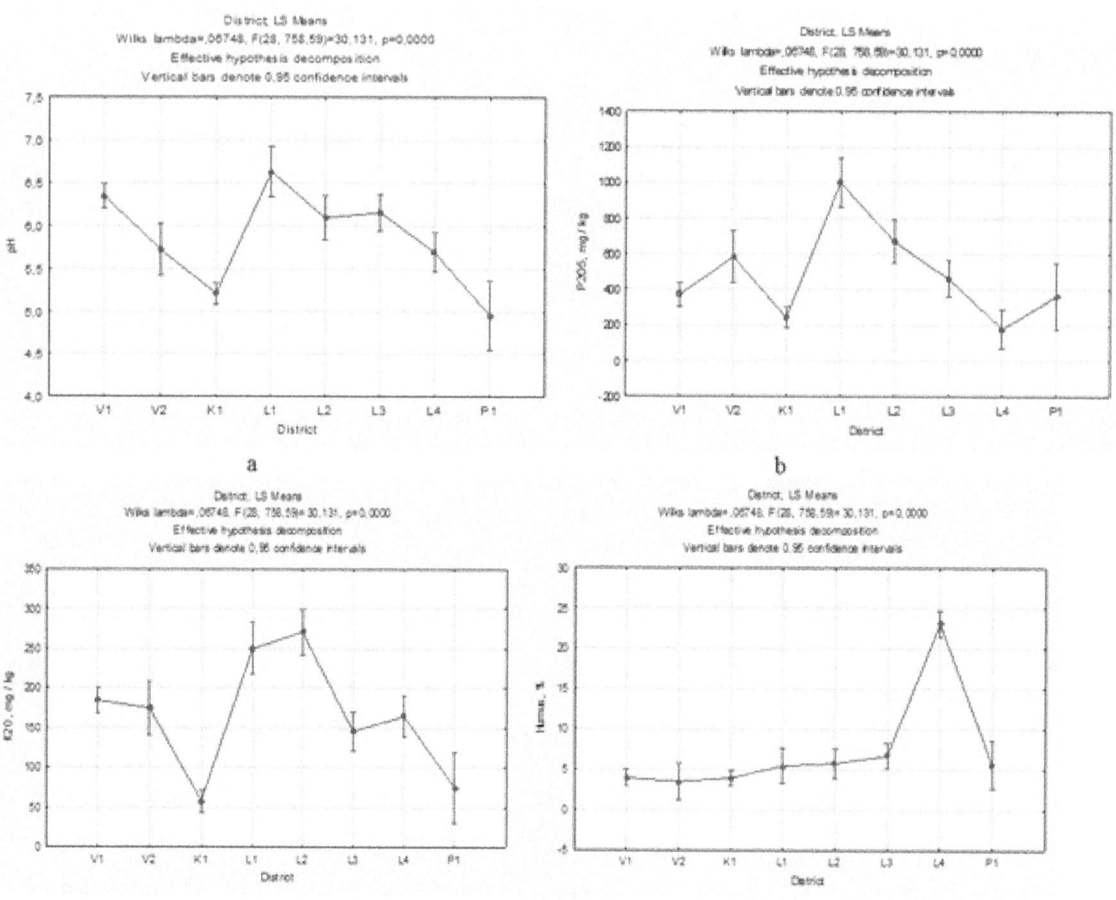

The greatest importance of Wilks is noted for the agricultural use of land. The combined effect of the factors is present at a significance level of $p < 0.05$. The areas of study are not equivalent in terms of agrochemical indicators and we have divided them into groups: the first group includes factors of geographical location (District) (Fig. 5), the second - the type of land use (Land) (Fig.6).

The soils of the Volosovsky (V1) and Lomonosovsky (L1) farms differ in high pH values (alkaline extracts). Farms L2 and L3 of the Lomonosov district are also characterized by pH values > 6. The most acidic soils are found in farms K1 of the Kingisepp and P1 of the Priozersky districts, differing from the rest with $p < 0.05$. The soils of farms with slightly acidic water extracts of soils contain little mobile phosphorus (Kingisepp (K1), Lomonosov (L4) and Priozersky (P1) districts. The maximum content of phosphorus content is observed in soils with alkaline conditions (farms L1 and L2 of the Lomonosov district). High potassium levels were found in farms of Lomonosov (L1, L2) Priozersk (P1) districts with high P2O5 content; low values were associated with acidic conditions of water extraction of soils (farms K1 of Kingisepp and P1 of Priozersk districts). A very high content of organic carbon is characteristic of the L4 farm of the Lomonosov district, which is probably dominated by reclaimed peat bogs.

Significant changes in agrochemical indicators are observed on lands that differ in agricultural use (Fig. 6).

Figure 6. Changes in agrochemical parameters pH (a), P_2O_5 mg/kg (b); K_2O, mg/kg (c); Humus (Gum), % (d) depending on the type of land use: arable land (AL), arable drained land (ALDR), hay (HM), haymaking drained(HMDR), pasture (PSDR), PSTR (drained pasture), PSTB (pasture) Arable land (AL) is extractedfrom the soil by alkaline aqueous extract at $p < 0.05$

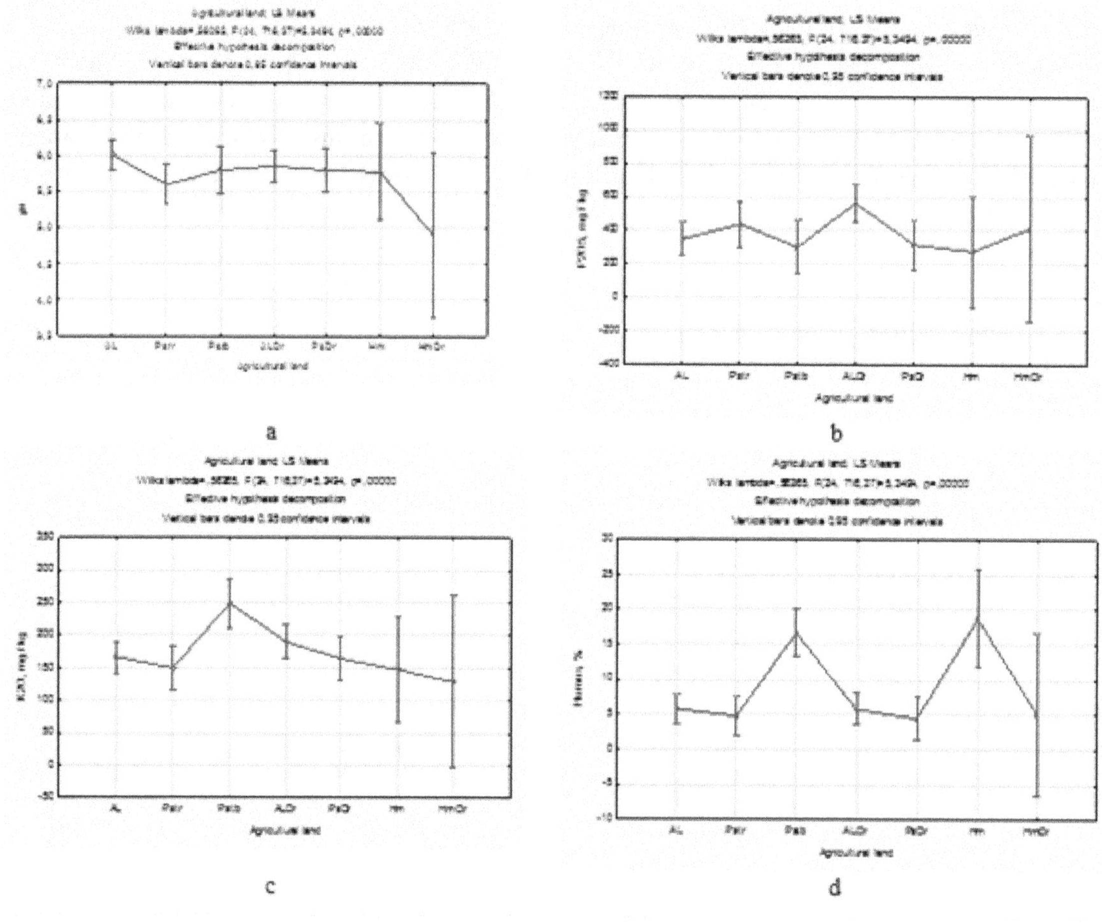

The soil of the drained hayfield (HMDR) has low average values. There is also a very high range of readings, which makes conclusions on this land grading incorrect, moreover, the effect of high error values is observed for all agrochemical indicators. According to the content of P_2O_5, the soil of arable land drained at $p < 0.05$ is allocated. For potassium, the highest content values are observed in the soils of pastures. According to the content of humus, the highest average values (>15) are distinguished in the gradations of meadows and pastures, for the rest, the average values vary within 5%.

The second group of factors is determined by the properties of soil types, granulometric composition and agricultural crops. Agrochemical indicators depend on the following factors: soil type, granulometric composition, vegetation character during the survey and their combined effect (Intercept) (Table 4) at $p < 0.05$.

Table 4. Wilks likelihood ratio criterion for Soil type and granulometric composition factors

Factor	Wilks	F	Effect df	Error df	p
Intercept	0,09	516,8875	4	200,0000	0,00
Vegetation	0,45	7,4312	24	698,9267	0,00
Soil type	0,52	6,0596	24	698,9267	0,00
Granulometric	0,59	5,6687	20	664,2748	0,00

The least influence of agrochemical factors is observed in relation to the nature of vegetation during the survey (Wilks = 0.45), the granulometric composition (Wilks = 0.59) has the greatest influence, followed by soil type (Wilks = 0.52) at $p < 0.05$. The combined action of factors also has a high level of importance.

The type of soil affects all agrochemical parameters (Fig. 7).

Figure 7. Agrochemical parameters pH (a), P2O5 mg/kg (b); K2O, mg/kg (c); Humus (Gum), % (d) depending on the type of soil: podzolic peeled (PP), sod-carbonate leached (SCL), sod-carbonate podzolic (SCP), podzolic turf, carbonate (SPC), sod-podzolic, fine-grained (SPFG), turf-podzolic, medium-grained (SPMPG) and sod-podzolic, shallow- and shallow-podzolic (SPS)

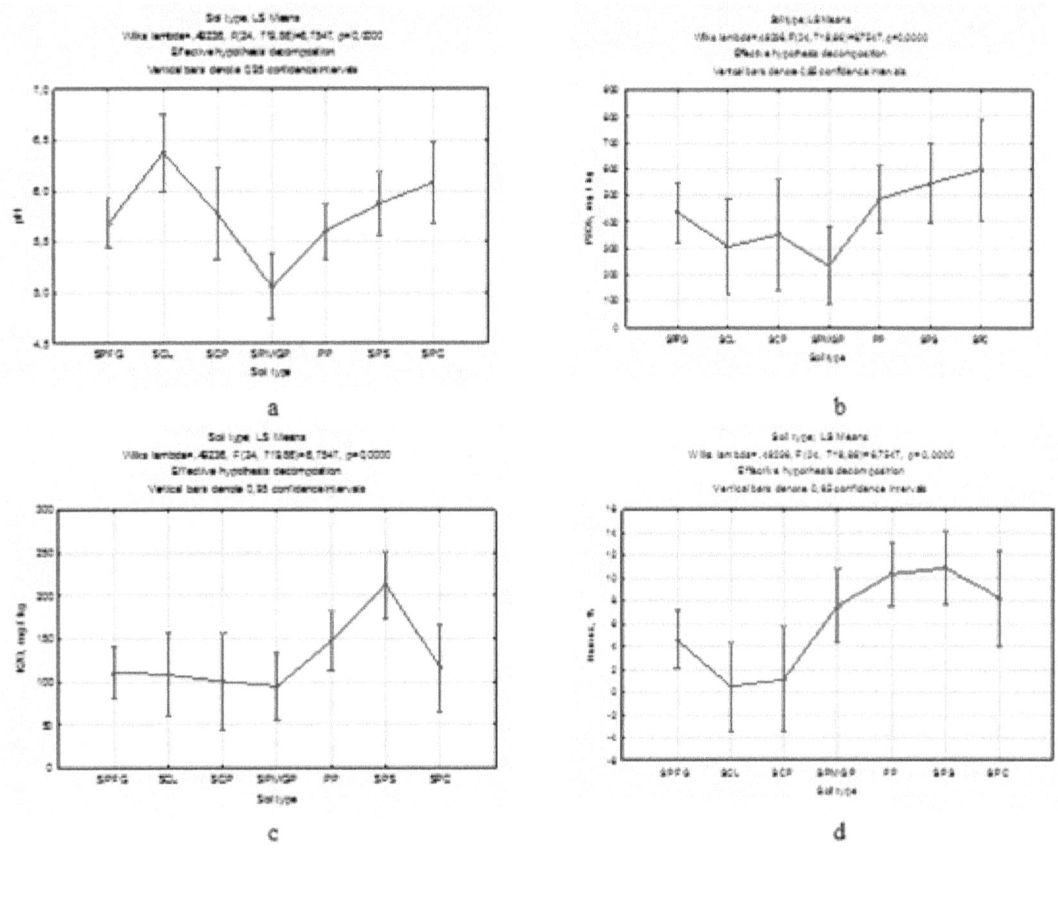

The most alkaline (> 6) conditions are observed in sod-carbonate leached (SCL) and sod-podzolized, carbonate (SPC) soils with a high content of carbonates, and sod-carbonate leached soils (SCL) are distinguished with a high level of significance $p < 0.05$. The most acidic (pH~5) soils are dry-podzolic, medium-grain soils (SPMGP). The remaining types have a slightly acidic or neutral pH of the aqueous extract.

The highest phosphorus content is found in soils: podzolic peeled (PP); sod-podzolic, shallow and shallow podzolic (SPS), sod-podzolic, carbonate (SPC), and the lowest - sod-carbonate leached (SCL) and sod–podzolic, medium-grain (SPMGP) (fig. 4b). Due to the high K2O content, shallow and shallow podzolic (SPS) are distinguished at $p < 0.05$ (Fig. 4c).

Very high values of humus content (Humus > 8%) are typical for soils: podzolic peeled (PP), shallow and shallow podzolic (SPS), and low – sod-carbonate leached (SCL) anderny podzolic, carbonate (SCP). Agrochemical parameters depend on the granulometric composition (Fig. 8).

Figure 8. Dependence of agrochemical parameters pH (a), P2O5 mg/kg (b); K2O, mg/kg (c); Humus (Gum), % (d) on the granulometric composition of the soil Loamy soils have a neutral or slightly acidic aqueous extract from the soil

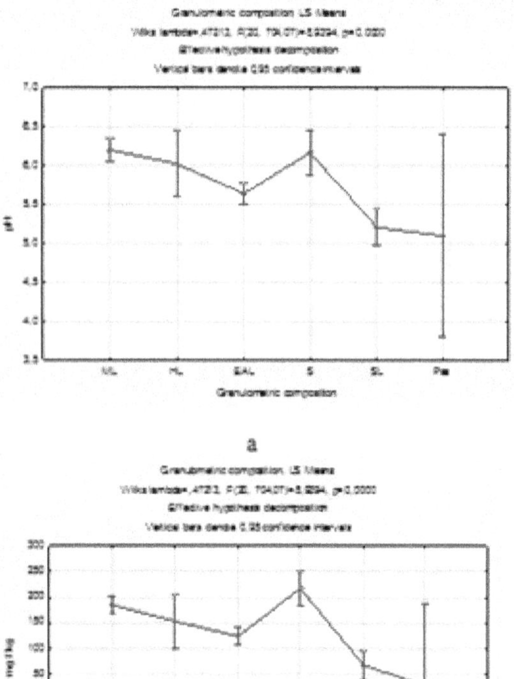

a

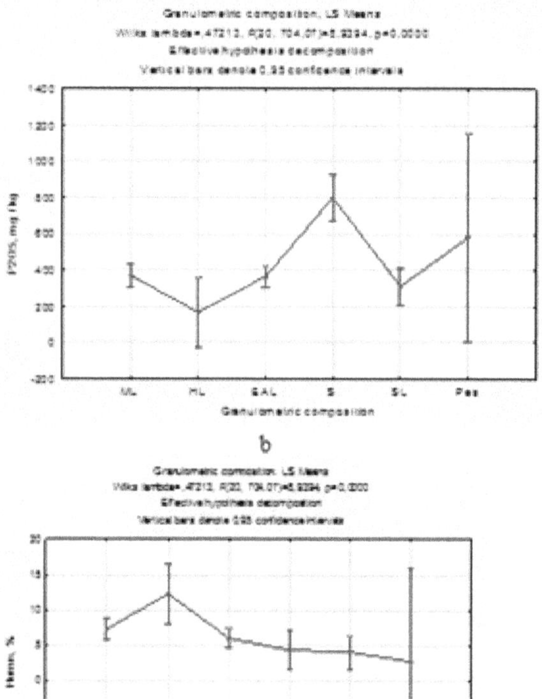

b

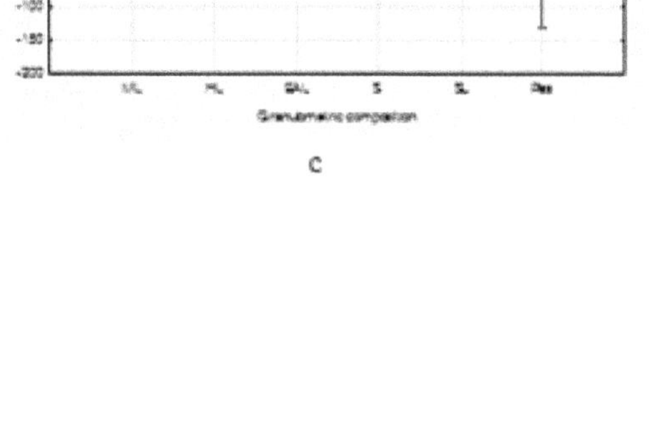

c

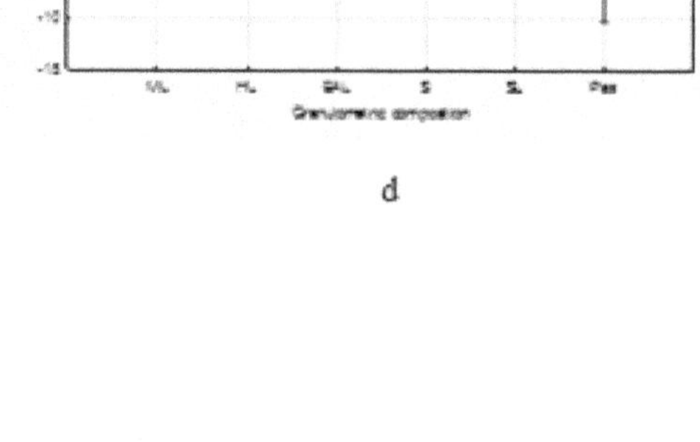

d

Light loamy and sandy soils have an acidic environment. Sandy soils are distinguished by a high range of values for all agrochemical indicators. The highest content of P_2O_5 was found in loamy soils, and low ones are typical for heavy clay (HL), light loamy (EaLm) and sandy loam (SL) soils. According to the potassium content, loams (S) are isolated at $p < 0.05$. The highest humus content (Humus > 10%) is present in heavy loamy soils (HL).

Agricultural vegetation depends on agrochemical parameters (Fig. 9). The maximum pH values are typical for arable soils (Al) and soils with stubble (Lys), the minimum values are for soils after growing potatoes and vegetables (P&V), annuals (Hrbs) and perennial grasses (PerH). The maximum content of P_2O_5 was also noted for arable land (Al) and soils with stubble (LYS), and the minimum – for soils from mowed fields (MF) after annual and perennial grasses (Hrbs). Soils with stubble (LYS) and arable land have the highest values of this indicator and significantly differ from other gradations at $p < 0.05$. A relatively high K_2O content was noted for soils with stubble (LYS), on arable land (AL), on mowed fields (MF) and after growing potatoes and vegetables (P&V).

It is especially distinguished by low values of less than 100 mg/kg at $p < 0.05$ of the soil after annual and perennial Hrbs grasses. According to the humus content, all gradations are in equal conditions, high values of soil are allocated after cereals (Cer), differing at $p < 0.05$ from the LYS and Al gradations. There are correlations between the used agrochemical indicators (Table 5). An increase in phosphorus content in soils positively correlates with humus content ($r = 0.57$). This is due to the fact that phosphorus is contained in humus substances (Zubkova and Karpachevsky, 2001). The sum of the exchange bases increases with an increase in the humus content ($r = 0.40$). A pattern is noted – the lower the amount of exchange bases, the higher the hydrolytic acidity ($r = -0.68$). The humus content contributes to an increase in the absorption capacity of the soil (Karpachevsky, 2005), which positively affects the amount of exchange bases ($r = 0.40$). There are general patterns of an increase in the phosphorus content in the soil with an increase in the pH value ($r = 0.54$), since the solubility of phosphorus decreases on acidic soils and increases on soils with a pH > 6.

Figure 9. Agrochemical parameters pH (a), P_2O_5 mg/kg (b); K2O, mg/kg (c); Humus (Gum), % (d) depending on vegetation (crops) LYS – stubble, AL – arable land; MF– last year's stubble; P&V – potatoes and vegetables; Hrbs – herbs; PerH, Cer – cereals Low K_2O values in soils are characteristic of Hrbs hayfields, after annual and perennial grasses (PerH) and cereals (Cer)

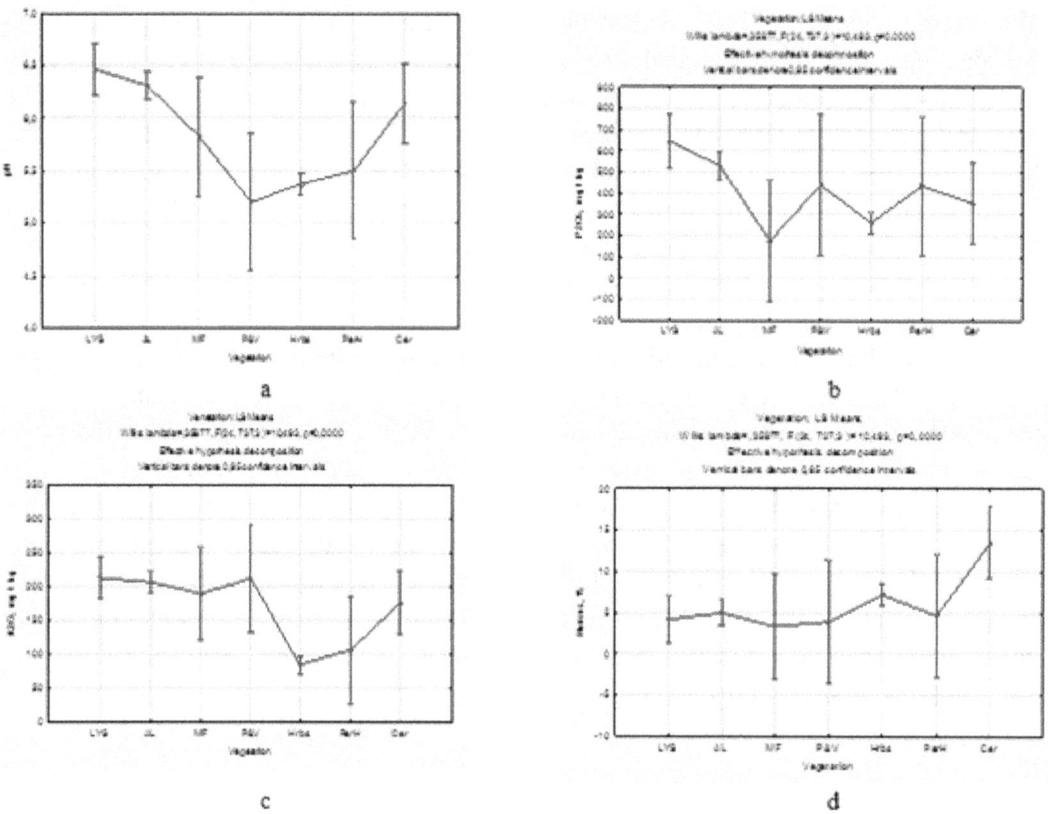

Table 5. Spearman correlation coefficient between agrochemical parameters: pH, $P2O_5$ mg/kg, K_2O mg/kg, organic carbon content (humus) %, sum of the exchanged bases SO, mg/kg, hydrolytic acidity mg-eq./100 g of soil. The most significant coefficients are marked in bold

Variable	Spearman Rank Order Correlations (district_600) MD pairwise deleted Marked correlations are significant at p <, 05000					
	pH	P_2O_5	K_2O	Humus	So	GK
pH	1,00	**0,54**	0,15	0,36	0,15	-0,20
P_2O_5		1,00	0,11	**0,57**	0,18	-0,06
K_2O			1,00	-0,05	-0,13	0,17
Humus					0,40	-0,05

continued on following page

Table 5. Continued

Variable	Spearman Rank Order Correlations (district_600) MD pairwise deleted Marked correlations are significant at p <, 05000					
	pH	P_2O_5	K_2O	Humus	So	GK
So					1,00	**-0,68**
GK						1,00

Hydroxonium H_3O+ occupies a larger volume in the soil absorbing complex (PPK) relative to Ca^{+2}, Mg^{+2}. Its introduction into the soil absorbing complex leads to a change in the amount of absorbed bases on a logarithmic scale, reflected in the negative relationship between the amount of bases and hydrolytic acid (Fig. 10).

Figure 10. The power dependence between hydrolytic acidity and the sum of the exchange bases

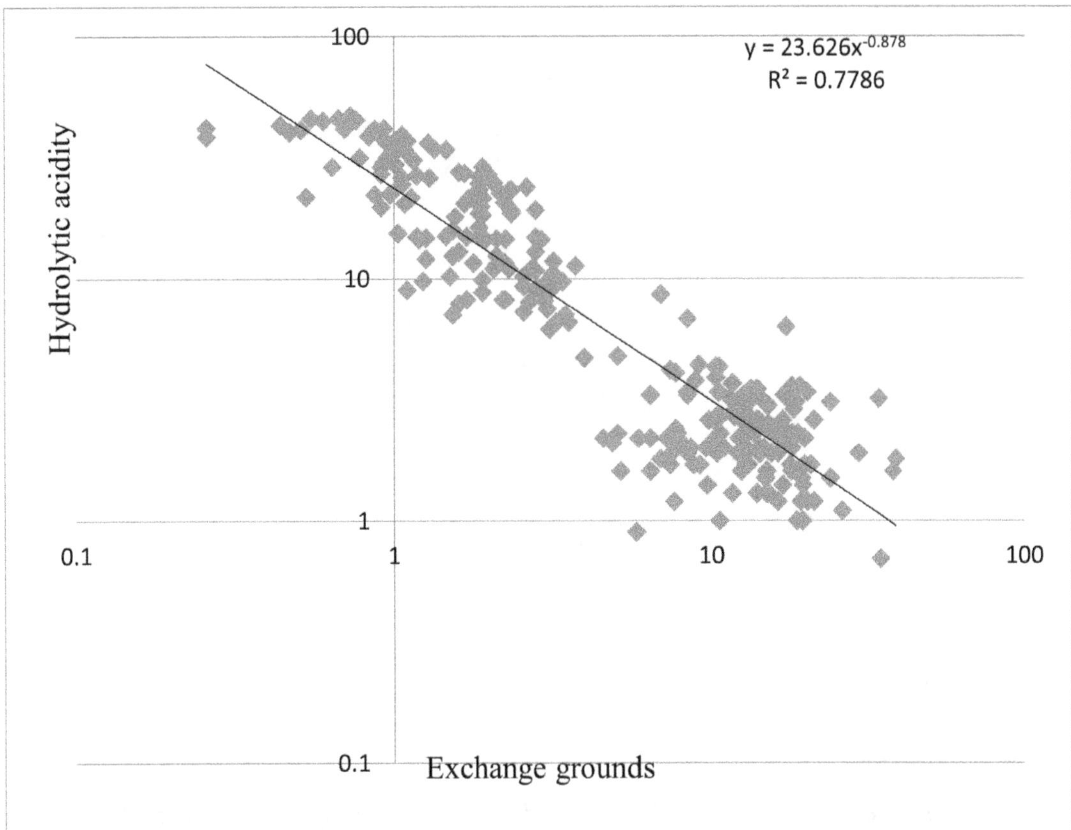

The power dependence between hydrolytic acidity and the sum of the exchange bases is described by the power dependence $Y = 23{,}626x^{-0.878}$Y with a tight coupling R2 = 0.82, regardless of the type of soil. Two groups of points are highlighted on the graph: 1) high value of hydrolytic acidity and low amount of bases; 2) low hydrolytic acidity and high base amounts. The first group includes acidic sod-podzolic soils with varieties, the second – neutral and alkaline sod-carbonate and soils of varying degrees of salinity. Given the practically functional relationship between the indicators in this region, one of them can be excluded from analytics. In other regions, for example, in the farms of the Perm Territory, such a high degree of interconnection is not observed (r = -0.47) (Mudrykh, 2021). Hierarchical cluster analysis revealed the structure of observations, which are divided into two large groups: C1-C2 and C3-C6 (Fig. 11).

To identify the characteristic properties of the clusters, statistical characteristics were determined for each of them (Table 6). The largest by area (15.7 thousand hectares) cluster C1 is distinguished by a slightly acidic aqueous extract, low phosphorus and potassium contents, a low amount of exchange bases and a low content of organic matter (Humus).

Figure 11. Cluster analysis of agrochemical parameters: pH, phosphorus content (P_2O_5), potassium content K_2O, sum of exchange bases (AAB), hydrolytic acidity (HA)

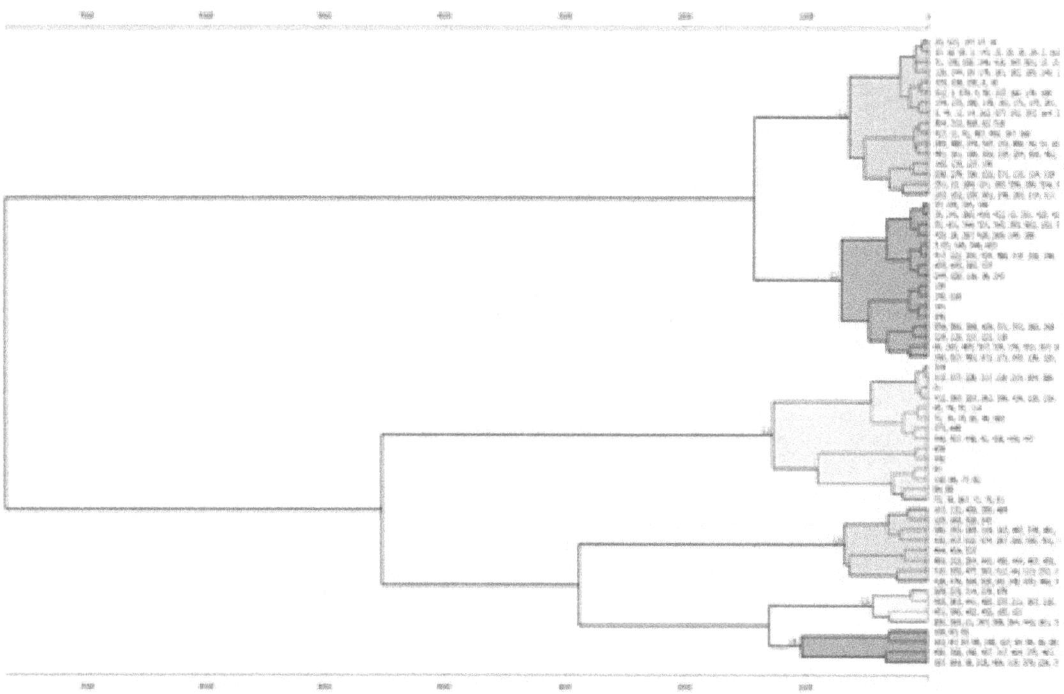

Table 6. Results of clustering of agrochemical indicators: pH, phosphorus content (P_2O_5), potassium content K_2O, sum of exchange bases (AAB), hydrolytic acidity (HA) on an area of 47.8 thousand hectares for 577 observations

Cluster	S, th. ha	N, pcs	pH	P_2O_5 mg/kg	K_2O, mg/kg	AAB, mg/kg	HA, mg-eq./kg	Humus, %
C1	15,7	198	5,2±0,1	115±11	72±10	10,0±1,8	5,7±1,6	5,5±1,6
C2	9,0	103	5,4±0,2	236±16	91±24	10,5±2,3	6,5±3,4	6,4±2,2
C3	5,3	53	6,1±0,3	994±45	186±31	10,2±4,0	14,0±7,3	7,4±2,2

continued on following page

Table 6. Continued

Cluster	S, th. ha	N, pcs	pH	P_2O_5, mg/kg	K_2O, mg/kg	AAB, mg/kg	HA, mg-eq./kg	Humus, %
C4	6,9	118	5,6±0,2	388±19	81±19	12,5±2,4	6,4±3,9	5,9±1,0
C5	5,2	34	5,9±0,2	788±23	66±43	14,0±3,7	5,7±5,8	6,9±1,2
C6	5,7	71	5,8±0,3	600±13	106±90	12,4±4,7	9,5±8,4	6,3±1,2

The C2 cluster, which occupies a smaller area (9 thousand hectares), is similar in properties to C1, but differs from it by increased hydrolytic acidity, higher content of phosphorus and organic matter. In the second group, clusters of C3 and C5 (5.3 and 5.2 thousand hectares) are distinguished by a smaller area, but very high phosphorus content, hydrolytic acidity and organic matter, and they differ in potassium content and pH shifted to the alkaline region in C3. C6 is similar to cluster C3 in properties, differing from it by a lower humus content. Cluster C4, being located in the second group, tends to the first group in terms of pH, phosphorus content and hydrolytic acidity, but differs from it in a high content of metabolic bases.

DYNAMICS OF AGROCHEMICAL INDICATORS

Of particular interest is the change in the dynamics of agrochemical indicators over time (Belenkov and Tyumakov, 2003). We have analyzed the dynamics of the main indicators of fertility of agricultural lands in the Leningrad region from 1968 to the present day (2020-2020). The nonparametric Wilcoxen criterion showed that significant differences at $p < 0.05$ are observed for phosphorus and humus (Table 7).

The phosphorus content has decreased by more than 3 times over 50 years from 1187±459 to 206±51 in Volosovsky district (V) (Table 8). During this time, the content of humus in the soil decreased by 0.3%. In the area of g, the decrease reached 0.4%. The pH values of the aqueous solution and the potassium content have changed slightly. Temporary changes in agrochemical parameters were studied in the Kingisepp, Gatchina and Volosovsky districts (Fig. 12).

Table 7. Comparison of agrochemical parameters (pH, P (P_2O_5), K (K_2O), Hum (humus) in 1968 with modern data on the Wilcoxen criterion

Pair of Variables	No. of	Percent	Z	p-value
pH_{1968} & pH_{2020}	30	162,5	1,44	0,149
P_{1968} & P_{2020}	38	11,0	5,21	0,000
K_{1968} & K_{2020}	38	273,5	1,41	0,159
Hum_{1968} & Hum_{2020}	32	14,0	4,67	0,000

Table 8. Comparison of agrochemical indicators (pH, P (P_2O_5), K (K_2O), Hum (humus) in 1968 with modern data for Kingisepp (K), Gatchina (G) and Volosovsky (V) research areas

District	pH_{1968}	pH_{2020}	P_{1968}	P_{2020}	K_{1968}	K_{2020}	Hum_{1968}	Hum_{2020}
K	5,3±0,8	5,4±0,6	1069±607	207±93	144±52	135±64	5,4±1,7	5,0±2,5
G	5,7±0,6	5,9±0,5	988±653	319±88	127±34	140±29	5,8±1,1	5,5±1,2
V	5,8±0,4	5,8±0,4	1187±459	206±51	134±60	157±38	4,5±0,6	4,2±0,6
All Gr.	5,7±0,3	5,7±0,3	1081±324	254±55	133±27	146±21	5,2±0,7	4,9±0,7

According to the survey areas, pH has changed over 50 years only in the Volosovsky district (Fig. 12a and b). The phosphorus content decreased in all groups as a whole, but Gatchina district stands out with high values from Kingisepp and Volkhov districts with low values at $p < 0.05$. In terms of humus content, the differences are not statistically significant in terms of the difference in average values, but the nonparametric Wilcoxen criterion showed that the differences are significant.

A visual change in the area (ha) by gradation of indicators over the past 50 years in three districts is presented in Table 9. The size of the area and its percentage composition by pH has changed slightly, which corresponds to the estimates of Table 4. The structure of the area by gradation P_2O_5 has undergone significant changes. Soils with low and medium phosphorus content completely disappeared in 2020, but areas with high and very high phosphorus content increased, for example, in the 151-250 gradation tripled. There are no changes in the content of K_2O. The areas of land with a very high content of organic matter decreased, at the same time, the gradation of its content increased (2 – 4.9), forming soil aggregates of high strength (Zubkova and Karpachevsky, 2001).

Over 50 years of agricultural land exploitation, the structure of clusters allocated on the basis of agrochemical data has changed significantly (Fig.13). In the 70s of the last century, the C1 cluster with a high content of phosphorus and potassium and all other clusters in the C2-C6 group were allocated, which in turn is divided into two C2 subgroups-C3 and C4-C6 (Table 10). Cluster C2 stands out with a low phosphorus content against the background of a high potassium content, and cluster C5 has the opposite trend. C3 is close in phosphorus and potassium content to the C1 cluster, but differs from it by an increased pH value and a reduced humus value. C4 has values close to the average for the entire cluster group, among them C6 stands out for its low content of organic matter (Humus).

Figure 12. Change in agrochemical indicators (pH, P_2O_5, K_2O, Humus) from 1968 to 2020 for Kingisepp (K), Gatchina (G) and Volosovsky (V)

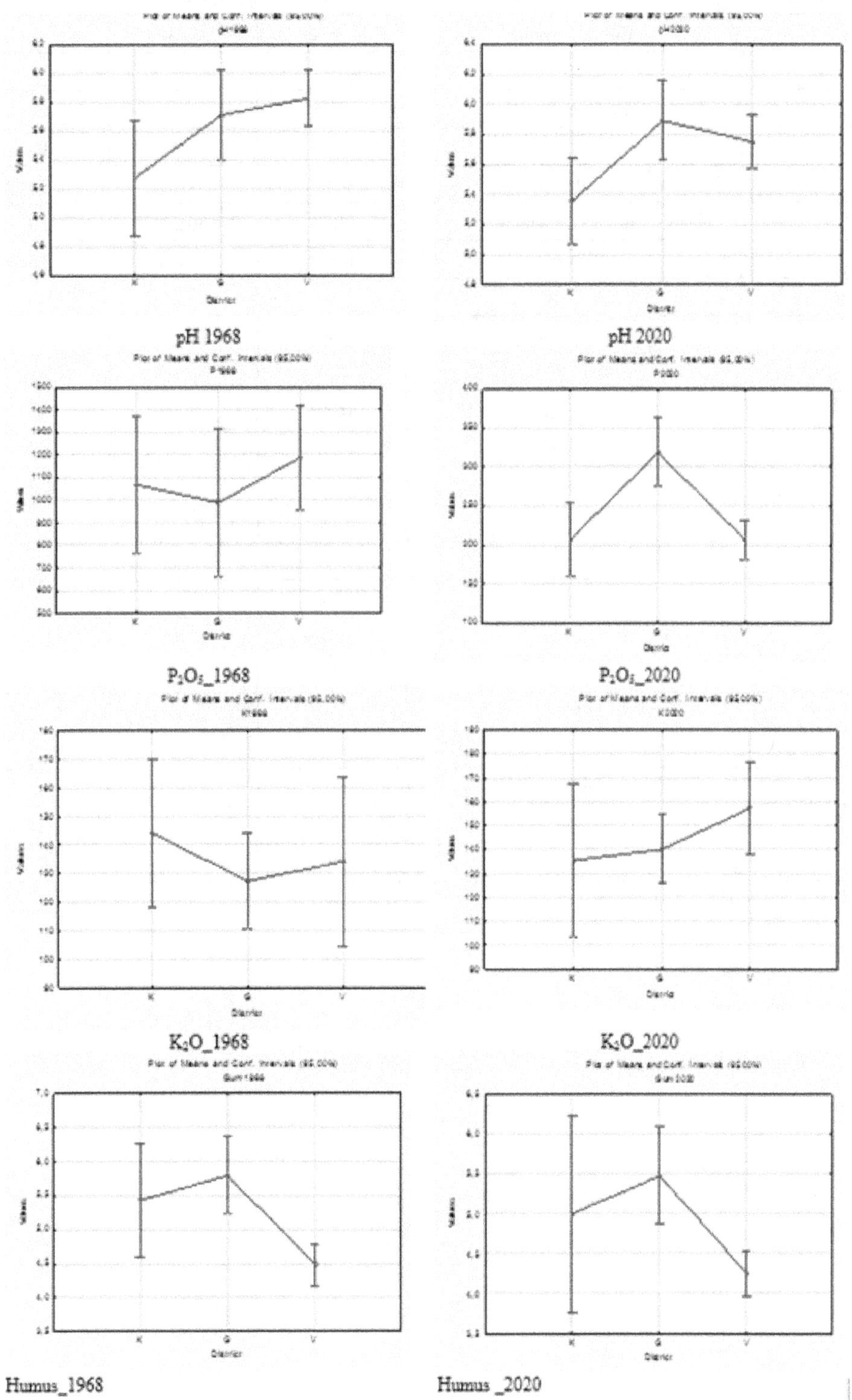

Table 9. Change in areas (ha) and shares (%) over 50 years from 1968 to 2020 in accordance with gradations of agrochemical indicators: pH, P_2O_5, mg/kg; K_2O, mg/kg; organic matter (Humus), %

Gradations of indicators		Sum of areas, ha		Area share, %	
Significant	**Characteristics**	**1968**	**2020**	**1968**	**2020**
pH					
< 4,5	Very strongly acidic	2645	0	0,7	0,0
4,6-5,0	Medium acid	10624	10521	11,2	9,0
5,1-5,5	Slightly acidic	38221	39137	28,2	33,3
5,6-6,0	Close to the neutral	51379	50827	46,7	43,3
> 6	Neutral	14591	16975	13,3	14,5
P_2O_5, mg/kg					
< 25	Very low	2142	0	1,8	0,0
26-50	Low	0	0	0,0	0,0
51-100	Average	44094	0	37,5	0,0
101-150	Elevated	45584	3925	38,8	3,3
151-250	High	23840	72900	20,3	62,1
> 250	Very high	1800	40635	1,5	34,6
K_2O, mg/kg					
41-80	Low	3306	5297	2,8	4,5
81-120	Average	21280	17705	18,1	15,1
121-170	Increased	68016	72797	57,9	62,0
171-250	High	24858	21661	21,2	18,4
Humus, %					
2–4,9	Среднее	5978	12373	5,1	10,5
4–5,9	High	82424	87307	70,2	74,3
6–10	Very high	29058	17780	24,7	15,1

Figure 13. The difference in the structure of clusters of agrochemical indicators between 1968 and 2020 on an area of 14 thousand hectares, for 49 observations, their average values for clusters are shown in Table 10

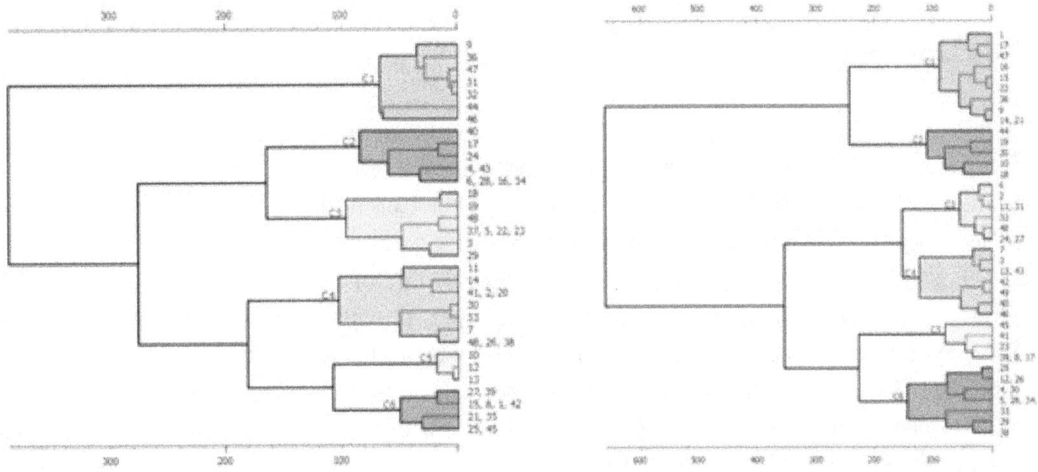

1968 2020

Table 10. Average values and confidence intervals ($p < 0.05$) of agrochemical pH, P2O5, mg/kg, K2O mg/kg, organic matter (Humus, %) for a number of clusters with a time interval of 50 years (from 1968 to 2020)

Clusters	Year	N	S, Ha	pH	P_2O_5, mg/kg	K_2O, mg/kg	Humus, %
C1	1968	10	1761	5,3±0,7	213±40	172±37	5,7±2,1
C2		11	2634	5,3±0,6	83±27	175±35	5,2±1,8
C3		9	2538	5,5±0,8	152±40	124±16	5,2±1,6
C4		9	3041	5,8±0,5	121±23	149±31	5,4±1,4
C5		7	1669	5,4±6,0	110±29	78±36	5,9±1,0
C6		3	2722	5,6±0,9	110±14	126±17	4,6±1,1
C1	2020	10	1908	5,8±0,6	324±24	141±34	5,5±1,8
C2		5	1146	6,2±1,1	418±79	144±103	5,8±2,9
C3		8	2845	5,6±0,5	257±26	138±15	5,2±1,3
C4		8	1797	5,2±0,5	248±57	85±26	4,4±1,7
C5		6	2127	5,4±0,3	145±43	103±48	4,3±2,7
C6		12	4153	5,7±0,6	190±34	170±39	4,5±0,9

In 2020, a more homogeneous cluster structure is observed. The sample is divided into two groups: C1–C2 and C3-C6. The second group is divided into two subgroups: C3-C4 and C5-C6. The first group includes all those with a high content of phosphorus and potassium, gum. They differ in high (alkaline) pH values of the aqueous extract of soils. The C2-C3 subgroup is united by a high phosphorus content, and shares a low potassium content in C4. The C5-C6 subgroup is united by a low phosphorus content, humus, and differ in potassium content. Cluster C6 contains the largest amount of it for the 2020 sample – 170 mg/kg.

CONCLUSION

Monitoring of the agroecological condition of agricultural lands is an essential component of global ecosystem sustainability management, which determines the relevance of the topic. At the regional level, these lands make up a fifth (20.3%) of all the lands of the Leningrad region, which significantly affects its economic and environmental sustainability. Regular examination in a five-year observation cycle makes it possible to identify trends in the agrochemical state of soils.

1. Monitoring of agrochemical indicators, such as pH, phosphorus, potassium, and organic matter, allows you to control changes in soil fertility in space and time, which affects yield.
2. It has been established that the pH of the aqueous extract of the soils of most farms belong to neutral or alkaline conditions (pH > 6). The soils of the farms of the Kingisepp and Pri-Ozersk districts are characterized by relatively low pH values (4.9-5.2). The phosphorus content in the soils of agricultural lands is highest in farms with favorable conditions. The remaining areas are characterized by an average content of mobile phosphorus. According to the content of K_2O in the soils of L1 and L2 Lomonosov districts. 123 elementary plots with a total area of 43 thousand hectares are characterized by a very high content of phosphorus and potassium, but low values of the sum of exchange bases and high hydrolytic acidity. These sites are characterized by an average or high content of organic carbon from 3.9% to 6.7%.
3. Reliable correlations were revealed between the used agrochemical indicators: the phosphorus content in soils positively correlates with the humus content (r = 0.57) and an increase in pH values (r = 0.54). The sum of the exchange bases increases with an increase in the humus content (r = 0.4). The regularity of a decrease in the amount of exchange bases with an increase in the hydrolytic acidity of soils (R^2= - 0.82) is noted. 4. The distribution of agrochemical indicators by five factors is considered: District, agricultural crop Vegetation, soil types Soil, granulometric composition, types of agrochemical lands Land. The soils of the farms of Volosovsky and Lomonosovsky districts differ in high pH values (alkaline extracts); the most acidic soils are found in the farms of Kirishsky and Priozeorsky districts. High potassium levels were detected in farms of Lomonosov and Priozersky districts with an increased content of P_2O_5; low values were attributed to acidic conditions of water extraction of soils. An abnormally high content of organic carbon is typical for the economy of the Lomonosov district (L4).
4. The arable land is characterized by neutral and alkaline conditions, low and medium pH values are characteristic of the soil of the drained hayfield. According to the content of mobile P2O5, the soil of drained arable land is allocated (at $p < 0.05$); according to the content of mobile potassium – pasture soils; humus (> 15) – soils of hayfields and pastures. The most alkaline (> 6) conditions were found in sod-carbonate leached and sod-podzolized, carbonate soils with a high content of carbonates. The most acidic (pH~5) soils are sod-podzolic, medium-sod soils. The maximum phosphorus content is observed in shallow podzolic and podzolic soils, and the minimum is in sod-carbonate leached soils. Shallow and shallow podzolic soils are distinguished by their high K_2O content. Very high values

of humus content (> 8%) are typical for soils: podzolic, peeled, shallow and shallow podzolic, and low – sod-carbonate. Agrochemical parameters depend on the granulometric composition.

5. Loamy soils have a neutral or slightly acidic aqueous extract from the soil; light loamy and sandy soils have an acidic environment. Sandy soils are distinguished by a high range of values for all agrochemical indicators. The highest content of P_2O_5 was found in loamy soils, while low levels are typical for heavy loamy, light loamy and sandy loam soils. According to the potassium content, loams are isolated. The highest humus content (> 10%) is present in heavy loamy soils (HL).
6. There have been changes in agrochemical properties over the past 50 years from 1968 to 2020. This concerns a decrease in phosphorus content, especially evident in Volosovsky and Kingisepp districts (by 3 times). According to the Wilcoxen criterion, a decrease in the content of organic matter in soils was revealed.
7. The changes in the surveyed areas by pH gradations are insignificant, the gradation of phosphorus content has undergone significant changes in the areas - soils with low and medium content have completely disappeared, but areas with high phosphorus content have increased. There are no changes in potassium. The areas with a very high humus content decreased, while at the same time the areas of medium and high humus content increased. 8. The hierarchical cluster identified 6 clusters, which were divided into two large groups: the first (2 clusters) – large in area with a slightly acidic aqueous content, low phosphorus and potassium contents, low amount of exchange bases and low organic matter content, and the second (4 clusters) - with a high content phosphorus and organic matter, an increased value of hydrolytic acidity.
8. The general trend leads to a decrease in the areas of acidic soils and an increase in chemically optimal slightly acidic and neutral ones. Areas of both low and very high phosphorus content are decreasing in the direction of economically and environmentally justified amounts close to the average content. The same thing should be noted for potassium - an increase in the area towards an increased content. The study revealed a tendency of change in agrochemical indicators, which is reflected in a decrease in areas with marginal properties and an increase in optimal ones from the point of view of agrochemical theory.

ACKNOWLEDGMENT

The research was carried within the framework of the state assignment of the Ministry of Science and Higher Education of the Russian Federation (theme No. FWRZ-2021-0006).

REFERENCES

Alterra, W. (2014). Mapping Soil Erosion Prevention Using an Ecosystem Services Modeling Framework for Integrated Land Management and Policy. *Ecosystems (New York, N.Y.)*, 17(5), 878–889. 10.1007/s10021-014-9766-4

BAFU (Ed.). (2017). Boden in der Schweiz, Zustand und Entwicklung, Stand 2017, Bundesamt für Umwelt, Bern. *BAFU, 2020, Bodenstrategie Schweiz, für einen nachhaltigen Umgang mit dem Boden.* BAFU. https://www.bafu.admin.ch/bafu/de/home/themen/boden/publikationenstudien/publikationen/bodenstrategie-schweiz.html

Belenkov, A.I., Tyumakov, A.Yu., Sabo, Umar, & Mokicheva D.S. (2013). Agrochemical and biological characteristics of soil fertility in the experimental site of the precision farming center. *News of TSHA*, 53-62.

Bispo, A. (2020). *Proposal of methodological development for the LUCAS programme in accordance with national monitoring programmes.* European Joint Programme Soil. https://ejpsoil.eu/fileadmin/projects/ejpsoil/WP6/EJP_SOIL_Deliverable_6.3_Dec_2021_final.p df) accessed 12 December 2022.

Borrelli, P., Poesen, J., Vanmaercke, M., Ballabio, C., Hervás, J., Maerker, M., Scarpa, S., & Panagos, P. (2022). Monitoring gully erosion in the European Union: A novel approach based on the Land Use/Cover Area frame survey (LUCAS)'. *International Soil and Water Conservation Research*, 10(1), 17–28. 10.1016/j.iswcr.2021.09.002

Bucci, G., Bentivoglio, D., & Finco, A. (2018). Precision agriculture as a driver for sustainable farming systems: State of art in literature and research. *Calitatea*, 19, 114–121.

Climate of St. Petersburg and Leningrad region. (n.d.). METEO. http://www.meteo.nw.ru/articles/index.php?id=2

De Vries, W., Schulte-Uebbing, L., Kros, H., Voogd, J. C., & Louwagie, G. (2021). Spatially explicit boundaries for agricultural nitrogen inputs in the European Union to meet air and water quality targets. *The Science of the Total Environment*, 786, 147283. 10.1016/j.scitotenv.2021.14728333958210

Decree of the Government of the Russian Federation dated 30.07.2010 No. 1292-r (ed. dated 30.05.2014) "On approval of the Concept for the development of state monitoring of agricultural lands and lands used or provided for agriculture as part of lands of other categories, and the formation of state information resources about these lands for the period up to 2020" Retrieved from https://lenoblinvest.ru

Decree of the Governor of the Leningrad Region dated October 22, 2021 N 100-pg "On approval of the scheme for the placement, use and protection of hunting grounds on the territory of the Leningrad Region. Retrieved from http://publication.pravo.gov.ru/document/4700202110260004

Digital Agriculture tecnnology, *Adaption & Attitudes,* November, 2019, 68. European Soil Data Center. Coordination of the LUCAS soil module. Retrieved from https://esdac.jrc.ec.europa.eu/euso/soil-monitoring

Federal Law. "On the transfer of lands or land plots from one category to another" dated December 21, 2004 N 172-FZ (latest edition). Retrieved from https://www.consultant.ru/document/cons_doc_LAW_50874

FOEN (Ed.). (2017). Soils in Switzerland, status and development, status 2017, Federal Office for the Environment, Bern. *FOEN, 2020, Swiss Soil Strategy for the Sustainable Use of Soils [Swiss Soil Strategy for the Sustainable Use of Soils].* Federal Office for the Environment. https://www.bafu.admin.ch/bafu/de/home/themen/boden /publicationsstudies/publications/bodenstrategie-schweiz.html

Gavrilova, O. I., & Shchukin, P. O. (2019). Characteristics of agricultural soils in the northwestern region of Russia, *Novaum.ru, 18*, 54-56. https://www.elibrary.ru/item.asp?id=38206285

Gerasimova, M.I. (2019). International classification of soils and the possibilities of its application in geographical research, *Bulletin of Moscow University, 5*, 49 – 56.

Grilli, E., Carvalho, S. C. P., Chiti, T., Coppola, E., D'Ascoli, R., La Mantia, T., Marzaioli, R., Mastrocicco, M., Pulido, F., Rutigliano, F. A., Quatrini, P., & Castaldi, S. (2021). Critical range of soil organic carbon in southern Europe lands under desertification risk. *Journal of Environmental Management*, 287, 112285. 10.1016/j.jenvman.2021.11228533725659

Guerra, C., Bardgett, R. D., Caon, L., Crowther, T. W., Delgado-Baquerizo, M., Montanarella, L., Navarro, L. M., Orgiazzi, A., Singh, B. K., Tedersoo, L., Vargas-Rojas, R., Briones, M. J. I., Buscot, F., Cameron, E. K., Cesarz, S., Chatzinotas, A., Cowan, D. A., Djukic, I., van den Hoogen, J., & Eisenhauer, N. (2021). Tracking, targeting, and conserving soil biodiversity. *Science*, 371(6526), 239–241. 10.1126/science.abd792633446546

Gulde, S., Chung, H., Amelung, W., Chang, C., & Six, J. (2008). Soil carbon saturation controls labile and stable carbon pool dynamics. *Soil Science Society of America Journal*, 72(3), 605–612. 10.2136/sssaj2007.0251

Hirte, J., Richner, W., Orth, B., Liebisch, F., & Flisch, R. (2021). Yield response to soil test phosphorus in Switzerland: Pedoclimatic drivers of critical concentrations for optimal crop yields using multilevel modelling. *The Science of the Total Environment*, 755, 143453. 10.1016/j.scitotenv.2020.14345333199000

Horn, R. (2022). Soil health and biodiversity: interactions with physical processes and functions. In Reyes Sanchez, L. (Eds.), *Sustainable soil management as a key to preserve soil biodiversity and stop its degradation*. International Union of Soil Sciences.

Joint Research Centre (JRC) of the European Commission, Zarco-Tejada, P.J., Hubbard, N., Loudjani, P. (2014). *Precision Agriculture: An Opportunity for EU Farmers—Potential Support with the CAP 2014–2020.* European Parliament; Brussels, Belgium:. Agriculture and Rural Development; Policy Department B: Structural and Cohesion Policies European Union.

Karpachevskii, L.O. (2005). *Ecological Soil Science.* GEOS, Moscow.

Kataev, M.Yu., Kartashov, E.Yu., Ryabukhin, V.V., Makarov, E.V., & Pasko, O.A. (2023b) Technique for segmenting images of unmanned aerial vehicles using neural networks. *Modern problems of remote sensing of the Earth from space, 20*(1), 55–66.

Kataev, M. Yu., Pasko, O. A., & Kartashov, E. Yu. (2023a). Analysis of the practical possibilities of using unmanned aerial vehicles in agriculture. *Bulletin of the Krasnoyarsk State Agrarian University*, 1(190), 54–62.

Kholodov, O. A. (2019). Integrated monitoring of agricultural land use in the modern period, *Bulletin of the Volgograd State University.Economy*, 3, 107–119.

Komarov, A. A. (2018), Assessment of the state of the herbage using the vegetation index NDVI. *Proceedings of the St. Petersburg State Agrarian University,2,* 124-129.

Kuhwald, M., Busche, F., Saggau, P., & Duttmann, R. (2022). Is soil loss due to crop harvesting the most disregarded soil erosion process? A review of harvest erosion'. *Soil & Tillage Research*, 215, 105213. 10.1016/j.still.2021.105213

Lavallee, J. M., Soong, J. L., & Cotrufo, M. F. (2020). Conceptualizing soil organic matter into particulate and mineralassociated forms to address global change in the 21st century'. *Global Change Biology*, 26(1), 261–273. 10.1111/gcb.1485931587451

Lehmann, J., Bossio, D. A., Kögel-Knabner, I., & Rillig, M. C. (2020). The concept and future prospects of soil health'. *Nature Reviews. Earth & Environment*, 1(10), 544–553. 10.1038/s43017-020-0080-833015639

Lessmann, M., Ros, G. H., Young, M. D., & de Vries, W. (2021). Global variation in soil carbon sequestration potential through improved cropland management. *Global Change Biology*, 28(3), 1162–1177. 10.1111/gcb.1595434726814

Martin, I., et al. (2022). *Derivation and use of soil screening values for assessing ecological risks (revised). Report – ShARE id26 (revised).* Environment Agency, Horizon House.

Minasny, B., McBratney, A. B., & Hartemink, A. E. (2010). Global pedodiversity, taxonomic distance, and the World Reference Base. *Geoderma*, 155(3-4), 132–139. 10.1016/j.geoderma.2009.04.024

Mudrykh, N. M. (2018). Spatial variation of properties in the agrosod-podzolic soil. *Anthropogenic Transformation of Nature*, 4, 82–84.

Panagos, P., Ballabio, C., Himics, M., Scarpa, S., Matthews, F., Bogonos, M., Poesen, J., & Borrelli, P. (2021). Projections of soil loss by water erosion in Europe by 2050. *Environmental Science & Policy*, 124, 380–392. 10.1016/j.envsci.2021.07.012

Panagos, P., Ballabio, C., Poesen, J., Lugato, E., Scarpa, S., Montanarella, L., & Borrelli, P. (2020). A soil erosion indicator for supporting agricultural, environmental and climate policies in the European Union'. *Remote Sensing (Basel)*, 12(9), 1365. 10.3390/rs12091365

Pasko, O., Staurskaya, N., Zakharchenko, A., Zharnikov, V., Larionov, Y. (2022). Bio-Farming as the Basis of Environmentally-Sustainable Arable Farming at the Time of Global Warming. *Research Anthology on Strategies for Achieving Agricultural Sustainability*, (pp. 600–627). Research Gate.

Pasko, O.A., Goryachkin, V.N., Ten, A.D., Nabieva, M.N. (2022a). Experience of agrarian reforms of the 20th century in Russia. *Economics, labor, management in agriculture,4*(86). 102-110.

Pasko, O. A., & Zakharchenko, A. V. (2020b). Cadastral and geoecological characteristics of especially valuable productive agricultural land. *Bulletin of Tomsk Polytechnic University.Georesources Engineering*, 331(11), 7–19.

Pasko, O.A., Zakharchenko, A.V., Pospelova, E.V. (2019b). Differentiation of agricultural lands by area on the example of the Tomsk region. *Bulletin of Tomsk Polytechnic University. Georesource engineering,330*(5), 100-112.

Paul, C., Kuhn, K., Steinhoff-Knopp, B., Weißhuhn, P., & Helming, K. (2020). Towards a standardisation of soil-related ecosystem service assessments. *European Journal of Soil Science*, 72(4), 1543–1558. 10.1111/ejss.13022

Pestryakov, V.K. (1973). *Soils of the Leningrad region*. Publishing house: Lenizdat.

Prasuhn, V. (2020). Twenty years of soil erosion on-farm measurement: Annual variation, spatial distribution and the impact of conservation programmes for soil loss rates in Switzerland'. *Earth Surface Processes and Landforms*, 45(7), 1539–1554. 10.1002/esp.4829

Prout, J. M., Shepherd, K. D., McGrath, S. P., Kirk, G. J. D., & Haefele, S. M. (2020). What is a good level of soil organic matter? An index based on organic carbon to clay ratio. *European Journal of Soil Science*, 72(6), 2493–2503. 10.1111/ejss.13012

Schimmel, H., & Amelung, W. (2022). *Organic soils. Encyclopedia of Soils in the Environment.* Elsevier. 10.1016/B978-0-12-822974-3.00073-2

Schjønning, P. (2021). Topsoil shear strength – measurements and predictions'. *Soil & Tillage Research*, 212, 105049. 10.1016/j.still.2021.105049

Soil Atlas of Europe. (2005). *European Soil Bureau Network.* European Commission. https://esdac.jrc.ec.europa.eu/euso/soil-monitoring

Sorokin, I. B., Kulizhsky, S. P., Pasko, O. A., & Merzlyakov, O. E. (2023). The influence of long-term use of straw and green manure on gray forest soil on the productivity of agrocenoses. *Агрохимия*, 9(9), 22–27. 10.31857/S0002188123090119

Steinhoff-Knopp, B., Kuhn, T. K., & Burkhard, B. (2020). The impact of soil erosion on soil-related ecosystem services: Development and testing a scenario-based assessment approach. *Environmental Monitoring and Assessment*, 193(S1), 274. 10.1007/s10661-020-08814-033988744

Timofeeva, Yu. R. (2020). *Assessment of the agro-resource potential of the soils of the Leningrad region.* Materials on the study of Russian soils: collection of scientific papers, St. Petersburg.

Tokareva, O. S., Pasko, O. A., Majid, S. M., & Cabral, P. (2020). Monitoring the state of vegetation cover in the territory of central Iraq using Landsat-8 satellite data. *Bulletin of Tomsk Polytechnic University. Georesources Engineering*, 331(6), 19–31.

Veerman, C. (2020). *Land Contamination & Reclamation*, 11, 31–36. 10.2462/09670513.617

Vogel, H. J. (2020). Quantitative evaluation of soil functions: Potential and state. *Frontiers in Environmental Science*, 22. 10.3389/fenvs.2019.00164

Vohland, M., Besold, J., Hill, J., & Fründ, H.-C. (2011). Comparing different multivariate calibration methods for the determination of soil organic carbon pools with visible to near infrared spectroscopy. *Geoderma*, 166(1), 198–205. 10.1016/j.geoderma.2011.08.001

Vonk, W. J., van Ittersum, M. K., Reidsma, P., Zavattaro, L., Bechini, L., Guzmán, G., Pronk, A., Spiegel, H., Steinmann, H. H., Ruysschaert, G., & Hijbeek, R. (2020). European survey shows poor association between soil organic matter and crop yields. *Nutrient Cycling in Agroecosystems*, 118(3), 325–334. 10.1007/s10705-020-10098-2

Zakharchenko, A., Pasko, O., Tigeev, A., & Moskovchenko, D. (2023). On the Background Values of the Geochemical Spectra of Microelements of the Solid Phase of Snow During their Comparative Analysis in the Cities of Tobolsk And Tyumen. *Ecology and Industry of Russia*, 27(4), 61–65. 10.18412/1816-0395-2023-4-61-65

Zakharchenko, A., Pasko, O., Tigeev, A., & Moskovchenko, D. (2023). On the Background Values of the Geochemical Spectra of Microelements of the Solid Phase of Snow During their Comparative Analysis in the Cities of Tobolsk And Tyumen. *Ecology and Industry of Russia*, 27(4), 61–65. 10.18412/1816-0395-2023-4-61-65

Zakharchenko A.V, Moskovchenko D.V., Tigeev A.A. (2022). Analysis of the heterogeneity of the chemical composition of snow cover using clustering methods (k-means) and GIS-technology. *Ice and snow, 62*, 2. 227-240. 10.31857/S2076673422020128

Zakharchenko, A. V., & Pasko, O. A. (2022). *Plants, soils and sustainable farming*.

Zakharchenko, A. V., Pasko, O. A., & Sorokin, I. B. (2021). The dynamics of the reduction of arable land areas according to their long-term monitoring in the Tomsk region. *Ecology and Industry of Russia*, 25(11), 30–35. 10.18412/1816-0395-2021-7-54-59

Zhang, Q. (2015). *Precision Agriculture Technology for Crop Farming*. CRC Press. 10.1201/b19336

Zharnikov, V. B., Larionov, Yu. S., & Pasko, O. A. (2019). On traditions and innovations in domestic land management and cadaster. *Interexpo Geo-Siberia*, 3(2), 73–79. 10.33764/2618-981X-2019-3-2-73-79

Zhu, Q., Liu, X., Hao, T., Zeng, M., Shen, J., Zhang, F., & de Vries, W. (2020). Cropland acidification increases risk of yield losses and food insecurity in China'. *Environmental Pollution*, 256, 113145. 10.1016/j.envpol.2019.11314531662249

Zhu, Q., Liu, X., Hao, T., Zeng, M., Shen, J., Zhang, F., & de Vries, W. (2020). Cropland acidification increases risk of yield losses and food insecurity in China'. *Environmental Pollution*, 256, 113145. 10.1016/j.envpol.2019.11314531662249

Zink, A., Fleige, H., & Horn, R. (2011). Verification of harmful subsoil compaction in loess soils'. *Soil & Tillage Research*, 114(2), 127–134. 10.1016/j.still.2011.04.004

Zubkova, T. A., & Karpachevsky, L. O. (2001). *Matrix organization of soils*. State University named after M.V. Lomonosov.

Chapter 8
Utilizing Soil Erosion Modeling and GIS Technologies to Develop Future Soil and Water Conservation Strategies

Haytham Mohamed Salem
http://orcid.org/0000-0002-8519-7601
University of Idaho, USA

Mohamed A. I. Abdalla
Desert Research Center, Egypt

ABSTRACT

Drought and soil erosion stand as predominant drivers of land degradation within arid regions. The infrequent occurrence of rainfall events and subsequent flash floods significantly diminish agri-food productivity and exacerbate soil erosion rates. Thus, employing a modeling approach becomes imperative for forecasting flash floods and soil losses. The aim of this book chapter is to provide an introduction to soil erosion modeling and to delve into the performance evaluation of the kinematic runoff and erosion model (KINEROS2) through a case study conducted in Egypt. The model utilizes satellite imagery of digital elevation maps, soil maps, and annual weather datasets. Runoff discharge data observed at the outlet of the watershed is integrated into the model. In conclusion, it is strongly advised to integrate satellite imagery with the KINEROS2 model, a step that promises significant benefits in detecting environmental changes and understanding their impact on the watershed.

INTRODUCTION

Problem Statement and Possible Solution

Approximately one-third of the Earth's land surface is categorized as arid or semiarid (Sen, 2008; Salem et al., 2020). Arid regions typically experience prolonged periods of aridity throughout the year, punctuated by infrequent but intense rainfall events that can lead to flash floods. These hydrological

DOI: 10.4018/979-8-3693-3374-7.ch008

conditions exhibit considerable variability, with irregular rainfall contributing to annual precipitation accumulations ranging from several millimeters to several centimeters, occasionally exceeding 100 mm. Egypt, being predominantly arid, relies heavily on rainfall to sustain its rapidly expanding population and support development initiatives (Salem et al., 2023). In Egypt's northwestern coastal zone, where water scarcity is prevalent, water harvesting systems play a crucial role in meeting domestic, industrial, and agricultural needs, while also facilitating the development of additional water resources to meet future demands (Salem, 2014; Salem et al., 2015; Meselhy et al., 2019). Wadi-catchment water harvesting has garnered significant attention in recent decades, particularly for enhancing crop production in arid regions.

Intermittently, intense rainfall events can trigger flash floods. These events are characterized by heavy rainfall over a short duration, steep terrain, limited infiltration capacity, and sudden release of impounded water, resulting in rapid runoff accumulation. This runoff often converges into Wadi-type channel networks, especially in lower areas with sandy alluvial-type sediments, where it can infiltrate and recharge groundwater reserves. During flash floods, surface water flows as a flood wave within irregular Wadi channels, with coarse alluvium in the channel beds facilitating high infiltration rates and reducing discharge volumes as floods progress downstream. In larger wadi systems, significant runoff volumes may be diverted directly from streams to crop fields for spate irrigation (Van Steenbergen et al., 2010). Crop productivity exhibits significant variability due to the wide range of Wadi runoff volumes experienced from year to year, season to season, and even day to day. The erratic nature of Wadi hydrology poses challenges in accurately estimating water storage in alluvial beds of terraces, thus complicating assessments of potential cropped areas.

Traditionally, when establishing terrace systems in Wadi catchments, farmers prioritize maximizing the area available for irrigation rather than considering crop water requirements. Decisions regarding terrace width and number are typically based on farmers' past experiences and local knowledge. However, reliance solely on local knowledge often leads to suboptimal outcomes, as it overlooks the actual water needs of crops. Consequently, in many instances, downstream terraces may not receive runoff water during rainfall events, as water is absorbed by upstream terraces. In such cases, seepage from the Wadi bed becomes the primary water source for plants downstream. The dominance of upstream terraces in diverting wadi flow can disrupt water infiltration downstream, potentially disadvantaging water users further down the catchment. Therefore, an integrated approach to Wadi irrigation planning is essential to ensure that development initiatives benefit the entire community within the catchment.

Understanding the entire hydrology of the Wadi system is crucial for successful agricultural development. The hydrological behavior of Wadi systems is primarily influenced by geological formations, geomorphology, soil types, and their infiltration properties. However, obtaining long-term data to understand Wadi hydrology is often challenging. Therefore, irrigation projects aimed at wadi improvement should incorporate hydrological (and possibly sediment) data collection to develop and calibrate rainfall-runoff models. These models should be tailored to the unique characteristics of Wadis, as using models developed for humid catchments may not accurately represent the hydrological processes in arid and semiarid regions due to data limitations.

The Kinematic Runoff and Erosion model (KINEROS2) is a distributed, physically based event model designed to simulate interception, dynamic infiltration, surface runoff, and erosion in watersheds primarily characterized by overland flow. It conceptualizes the watershed as a series of planes and channels, with flow routed using a finite difference solution of the one-dimensional kinematic wave equations. KINEROS2 can assess the impacts of various artificial features, such as urban developments or detention reservoirs, on flood hydrographs and sediment yield (Abdalla et al., 2022; Shmilovitz et al., 2023; Meles et al., 2024).

To operationalize KINEROS2 effectively, automated procedures are essential to leverage widely available spatial datasets and the computational capabilities of geographic information systems (GIS). The Automated Geospatial Watershed Assessment (AGWA) tool, developed in 2002, serves this purpose by utilizing spatial datasets to delineate watersheds, subdivide them into model elements, and derive parameter inputs for each element. AGWA also facilitates spatial visualization and comparison of model results, enabling the assessment of hydrologic impacts associated with landscape changes. Integration with GIS allows for linking model results with other spatial information, enhancing the understanding of watershed dynamics (Abdalla et al., 2022).

Rainfall-Runoff in Arid and Semi-Arid Regions

Rainfall variability presents a significant challenge in modeling hydrology in arid and semi-arid regions. Compared to humid areas, rainfall characteristics exhibit greater variability in space and time (Wheater et al., 1991; Al-Qurashi, 1995). The influence of rainfall spatial variability on runoff modeling at the catchment scale is contingent upon several factors, including rainfall patterns, catchment characteristics, and runoff generation processes (Emmanuel et al., 2015).

Mediterranean catchments experience pronounced hydrological fluctuations due to their unique climatic features. The Mediterranean climate is characterized by high variability, particularly in precipitation. Annual rainfall predominantly occurs in two distinct periods, autumn and winter, with the latter witnessing intense rainfall events leading to substantial runoff (Ramos and Martínez-Casasnovas, 2009; Salem et al., 2021; Salem and Meselhy, 2021). These catchments often undergo extended dry periods interspersed with intense rainfall events. Furthermore, intra-event variability can also be significant. The presence of high reliefs and local microclimates with rain shadows in these catchments results in an altitudinal gradient in rainfall distribution (Verdú et al., 2006; Frot et al., 2008). These characteristics significantly influence sediment production in headwaters and subsequent sediment transfer downstream through channel networks. Moreover, changes in land use in the headwaters of Mediterranean mountain catchments since the mid-twentieth century exert control over hydrological regimes and sediment delivery at various temporal and spatial scales (Garcia-Ruiz et al., 1997; Beguería et al., 2003; Buendia et al., 2015).

Additionally, many agricultural areas in the Mediterranean region experience erosion processes due to soil and rainfall characteristics, leading to significant soil and nutrient losses (Tropeano, 1983; Wicherek, 1991; Kosmas et al., 1997; Lorenzo et al., 2002; Ramos and Martinez-Casasnovas, 2006; Salem et al., 2024). Changes in land management practices, such as land leveling, exacerbate erosion processes and result in variations in water availability for plants (Bazzoffi et al., 2006; Capolongo et al., 2008; Ramos and Martínez-Casasnovas, 2010).

The investigation of rainfall distribution at the catchment scale is crucial for geomorphologists to comprehend the interplay among precipitation, runoff generation, and sediment transport. Traditionally, catchment-scale maps of rainfall variability have been derived from spot rainfall records using various interpolation methods (López-Tarazón et al., 2010; Ly et al., 2013; Szcześniak and Piniewski, 2015). Remote sensing data have significantly enhanced the spatial and temporal coverage of rainfall information. Radar-derived data offer opportunities not only for improved spatial resolution but also for enhanced temporal resolution of catchment rainfall data. Despite the need for calibration and validation, advancements in technology are enhancing the accuracy of these estimates (Fiener and Auerswald, 2009; Navratil et al., 2012; Borga et al., 2014). Rainfall in arid and semi-arid climate regions typically exhibits characteristics of scarcity, irregularity, and high variability (Goodrich et al., 1995; Ahrens, 2012).

Runoff generation in these regions is primarily governed by infiltration excess mechanisms, characterized by rapid rates of infiltration and a swift response owing to steep hill slopes, shallow soils, exposed rocks, and sparse vegetation (Wheater, 2002b; Greenbaum et al., 2006). In some arid and semi-arid catchments, flood water infiltrates through alluvial channel beds during flow and flood peaks, leading to volume reduction (Wheater, 2002a). However, this process is negligible in catchments with rocky channel beds. Precipitation can be intercepted and retained in the vegetation canopy or reach the soil surface. Surface water either infiltrates into the soil profile or runs off overland, swiftly moving towards stream channels, thereby contributing to short-term stream response (Salem et al., 2022). Infiltrated water may be retained in the soil, undergoing evapotranspiration, or slowly make its way to the surface-water system via subsurface pathways (Gobattoni, 2009).

Rainfall-Runoff Modeling

Hydrological models aim to simulate natural processes related to water movement, encompassing phenomena such as stream flow, evaporation, evapotranspiration, groundwater recharge, soil moisture, sediment transport, chemical transport, and microbial growth in water bodies. These processes occur across both time and space, with distributed models and lumped models representing two primary classifications based on their treatment of spatial derivatives (Mujumdar and Kumar, 2012). While lumped models assign parameters to individual sub-basins, semi-distributed models allocate parameters to grid cells, often grouping cells with similar characteristics. Lumped precipitation–runoff models treat catchments as single homogeneous units and are widely used due to their simplicity and good performance following calibration with historical data. However, distributed models offer a more realistic portrayal of spatial heterogeneity in hydrological processes, accounting for variability in precipitation and physical watershed properties (Beven, 2001).

Temporal differentiation of hydrological models is based on the importance of timescales, leading to two main approaches: event simulation and continuous simulation. Event-based models focus on single, isolated storm events, making them useful for designing storm water control facilities in small catchments. In contrast, continuous simulation models depict ongoing hydrological processes over extended periods, addressing long-term flow volumes and storage considerations. These models integrate flow and storage processes, accounting for factors such as solar radiation and air masses that influence energy input, evaporative processes, and vegetative growth (Gobattoni, 2009).

Spatial and temporal variability are inherent to hydrological processes, with distributed models often requiring daily meteorological data to simulate the hydrological cycle accurately. However, challenges arise due to the scarcity of long-term continuous data, particularly hourly or monthly data, in many regions worldwide. The availability and accuracy of data significantly impact model simulations, highlighting the importance of addressing data limitations in hydrological modelling (Shrestha et al., 2004; Schuol and Abbaspour, 2007).

Lumped Models

Lumped models integrate all parameters affecting watershed hydrology by spatially averaging them, creating uniformity across the basin. These models treat the watershed catchment as a single unit with a relatively small set of parameters and variables (Johnson and Miller, 1997; Refsgaard, 1997). Historically, hydrological modeling has predominantly employed lumped modeling techniques. However, this approach has inherent limitations. Lumped models assume uniform distribution of rainfall—mean areal precipitation—across the watershed basin both spatially and temporally, a condition rarely observed in reality (Reed et al., 2004; Smith et al., 2004).

Moreover, lumped models presuppose uniformity in soil types, vegetation types, and land-use practices across the basin, despite significant variations often present. Consequently, these parameters are averaged across the basin, resulting in uniform conditions and yielding a lumped model. Mean areal runoff for the drainage basin is computed based on abstractions from mean areal precipitation.

Conventionally, this runoff estimation is then applied to a unit hydrograph to ascertain the total stream flow at the basin outlet. The concept of the unit hydrograph was introduced by Sherman in 1932. It represents the runoff at the downstream outlet of a drainage basin resulting from a unit depth (e.g., 1 inch or 1 mm) of excess rainfall for a storm of uniform intensity and specified duration over the entire watershed drainage (Sherman, 1932).

Distributed Models

Hydrologic distributed models explicitly account for geographic spatial variations and processes across a watershed (Feldman, 2000). They aim to quantify the spatial variability of hydrologic parameters and analyze rainfall-runoff processes at specific locations within a watershed basin (Smith, 1993). These models consider the spatial variability of both input and output hydrologic variables within a watershed and the hydrologic response at ungauged sites (Smith et al., 2004). Parameters used in distributed models are directly linked to the physical characteristics of the watershed basin, such as topography, soil, vegetation, and geology. Additionally, they incorporate the spatial variability of meteorological conditions in the drainage basin (Refsgaard and Knudsen, 1996; Shultz and Corby, 2006).

In a distributed modeling approach, the watershed is represented by a grid that includes topography, precipitation, soils, and land use/cover, reflecting the variability of these parameters. Unlike lumped models, distributed models can capture changes in slope and drainage network that affect the hydrologic response of the basin (Vieux, 2001). Distributed modeling is an area of active research, facilitated by high-resolution datasets, advancements in GIS capabilities, and improved computational power. Previously, limitations such as computer processing constraints hindered the use of distributed models due to the complexity of solving physics-based equations and uncertainties in rainfall input, parameter errors, and model structure (Smith et al., 2004).

Several questions remain regarding how the variability of rainfall and drainage basin characteristics influence runoff and stream flow generation at the basin outlet (Smith et al., 1999; Carpenter and Georgakakos, 2004; Smith et al., 2004). The objective of distributed modeling is to improve the simulation of watershed hydrologic response by capturing the spatial and temporal characteristics governing the transformation of precipitation into runoff (Vieux and Moreda, 2003).

Rainfall – Runoff Processes

Rainfall

Rainfall distribution within a watershed exhibits both spatial and temporal variability. Spatial properties encompass the location and quantity of precipitation across the watershed, while temporal properties involve the timing of rainfall throughout the drainage basin. Precipitation gages and rainfall radar estimates are used to determine both spatial and temporal characteristics. Various climatic factors influence runoff, including rainfall intensity, duration, and distribution over the drainage basin, storm movement direction, type of storm, precipitation type, soil moisture content, antecedent precipitation, and other meteorological conditions (Salem et al., 2023).

Rainfall intensity directly affects peak flow rate and runoff volume. Higher intensity rainfall leads to increased peak discharge and runoff volume, provided it surpasses the soil infiltration rate. Variations in intensity impact the shape of hydrographs, especially in small basins, albeit minimally in large basins.

Duration of rainfall directly influences peak flow, surface runoff volume, and duration of surface runoff. Longer storm durations can generate significant runoff, particularly in small basin watersheds, where most precipitation may become runoff during prolonged storms. However, this equilibrium is seldom attained in very large basins.

The spatial distribution of rainfall across the drainage basin significantly influences hydrograph shape. Rainfall in upper reaches results in broader, lower hydrograph peaks at the basin outlet, whereas rainfall near the outlet leads to sharp, rapid peak rises followed by quick recession. The watershed size also affects hydrologic response, with large basins rarely experiencing uniform rainfall distribution, often resulting in high peak flows from multiple storms. Conversely, intense localized thunderstorms over small areas can cause significant peak flows in small drainage basins.

Soil moisture content directly affects surface runoff and groundwater recharge; high soil moisture leads to increased surface runoff, while soil at field capacity enhances infiltration and groundwater discharge. During high evapotranspiration periods, soil moisture depletion reduces surface runoff, although successive rainfall events over short periods can induce runoff (Shultz, 2007).

Runoff

Watershed runoff comprises three main components: surface runoff, interflow, and groundwater runoff (referred to as base flow). Surface runoff travels over the watershed surface and streams downstream to the basin outlet. Interflow represents runoff infiltrating the upper soil layers, laterally moving until it reaches the stream channel, albeit slower than surface runoff. Base flow percolates through soil to the water table, then laterally to the stream, exerting minimal influence on storm-induced flood peaks.

Surface runoff encompasses overland flow and channel flow. Overland flow occurs when storm precipitation surpasses canopy interception, soil infiltration capacity, and surface storage, flowing over the land surface to the stream channel (Salem et al., 2014). Channel flow denotes the translation of a flood wave downstream within the stream channel, with runoff undergoing modifications across both overland flow and channel flow paths.

Spatial characteristics of soil and vegetation can be discerned through GIS databases and field studies, including soil type and vegetation distribution (Salem et al., 2024). Surface runoff is influenced by climatic factors such as precipitation intensity, duration, type (e.g., convective, orographic), and form (e.g., rain, snow). Vegetative cover, evapotranspiration, and soil moisture content also respond to climate. Physiographic factors like watershed size, shape, slope, soil type, land use, geologic structure, and drainage network also affect surface runoff.

Excess rainfall denotes the portion of rainfall directly involved in runoff generation, without being retained in storage or lost through infiltration, interception, or evapotranspiration (Salem et al., 2015). Direct runoff ensues from excess rainfall once it traverses the surface as Hortonian overland flow. In flood analysis, deriving the direct runoff hydrograph (DRH) from an observed total runoff hydrograph is a critical step. Two commonly employed methods for estimating flood runoff and peak flows in hydrological designs, even with constrained data, are the SCS curve number method and the rational formula.

Hydrograph Shape

Estimation of Flood Peak Discharge

Estimating flood runoff stemming from a specific storm entails assessing the losses incurred during rainfall events. The Soil Conservation Service (SCS), now known as the Natural Resources Conservation Service (NRCS), curve number method is widely employed due to its simplicity and practical utility. This method revolves around quantifying infiltration losses based on the antecedent moisture content (AMC) and soil type. It assumes rainfall uniformly blankets the entire watershed during the storm. The core principle of the SCS curve number method is that runoff initiates after initial abstractions, encompassing interception by vegetation and built-up areas, surface storage in water bodies like lakes and depressions, and infiltration, are accounted for. A key assumption underlying the development of curve numbers is that the ratio of actual rainfall retention to potential retention (S) in the watershed equals the ratio of direct runoff to rainfall minus initial abstractions (Ia) before runoff initiation.

The parameter S hinges on the catchment attributes of soil, vegetation, and land constituting the soil-vegetation-land (SVL) complex, alongside the AMC. Employing a parameter, CN, to depict the relative water retention by a given SVL complex, the potential retention in a watershed with a specific SVL complex is computed using the equation:

S

$$= \frac{25400}{\mathrm{CN}} - 254\left(mm\right) \quad (1)$$

The CN parameter, referred to as the curve number, ranges from 0 to 100 and varies depending on the soil type and AMC in the watershed. CN lacks a physical interpretation. The runoff equation, also known as rainfall excess (Pe), is expressed as:

$$Pe = \frac{(p - 0.2S)2}{P + 0.8S} \quad (2)$$

Soils are categorized into four groups (A, B, C, and D) based on their runoff potential, with group A having the lowest and group D having the highest runoff potential. The AMC considers the soil moisture content preceding the storm for which runoff computation is performed. The watershed's AMC is classified into three categories (I, II, and III), based on rainfall in the preceding five days and whether it is a growing or dormant season.

Figure 1. Schematic diagram of KINEROS2

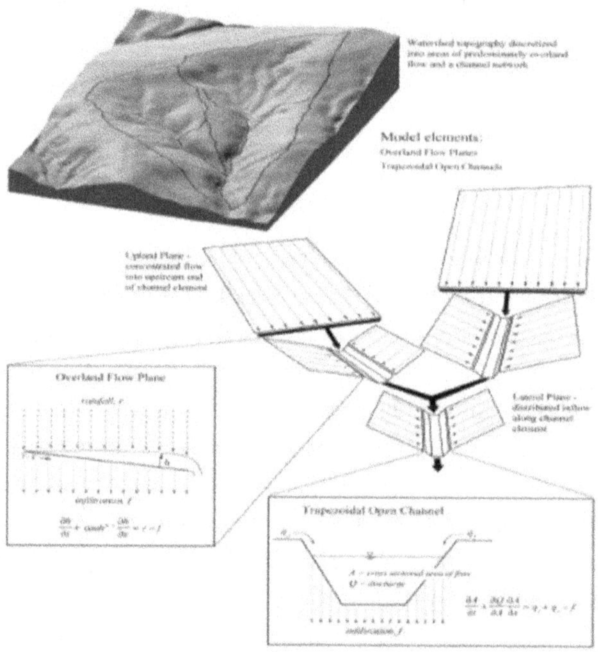

(source: USDA-ARS)

Figure 1 depicts a schematic representation of the process through which topographic data and channel network topology are simplified into the geometric elements of the KINEROS2 model. It's noteworthy that overland flow planes, or curvilinear surfaces, are dimensioned to maintain an average flow length. Consequently, planes contributing laterally to channels generally do not possess widths that correspond to the channel length (Goodrich et al., 2010).

In addition to computing infiltration capacity and surface runoff, the model also assesses sediment transport, which is determined by a mass-balance equation akin to that governing kinematic water flow. Splash erosion is considered utilizing the methodology outlined in Meyer and Wischmeier (1969). Hydraulic erosion is conceptualized as a kinetic transfer process, and sediment transport capacity is modeled based on the principles proposed by Engelund and Hansen (1967). Estimating these outcomes involves adjusting parameters to align the model's predictions with observed reality. These parameters and their definitions are presented in Table 1.

Table 1. Kineros2 (K2) model

Parameter	Definition
Width	Channel width (m)
Depth	Channel depth (m)
Ks	Saturated hydraulic conductivity (mm/hr)
Roughness n	Manning's roughness coefficient
G	Mean capillary drive (mm)
Interception	Interception depth (mm)
% Cover	Cover fraction of surface covered by interception
Pave	Fraction of surface covered by erosion pavement (0-1)
Splash	Rain splash coefficient
CV	Coefficient of variation of Ks
Si	Initial saturation index

Certain parameters, such as width and depth, are inferred from topographic data, while others like Ks (saturated hydraulic conductivity), G (Green-Ampt parameter), CV (coefficients of variation), and Splash are derived from soil characteristics. However, % Cover (percentage of ground covered by vegetation), Roughness, Interception, and Pave (percentage of impervious surfaces) are obtained from land cover data.

The performance of the KINEROS2 (K2) model has been evaluated in numerous studies. For instance, Woolhiser et al. (2006) utilized K2 in the Upper Split Wash watershed on Yucca Mountain, Nevada, to estimate plane and channel infiltration across 18 rainfall-runoff events spanning nine years. They found that soil depth significantly influenced surface runoff and infiltration rate, with average annual channel flow, runoff, and bedrock infiltration measured at 0.6, 3.6, and 8.7 mm/year, respectively. Al-Qurashi et al. (2008) applied K2 to 27 rainfall events in the Wadi Ahin watershed, Oman, discovering that soil hydraulic conductivity, Manning's roughness, and capillary length scale were crucial parameters affecting peak volume and timing, yielding runoff coefficients ranging from 0.04 to 0.79. In a study by Miller et al. (2002), K2 was employed in the Sierra Vista sub-watershed of the San Pedro Basin, southern Arizona, to estimate surface runoff and sediment discharge from six design storms and assess the impact of land cover changes. Calibration with flow data from the Walnut Gulch watershed indicated that runoff varied by 7.8 to 177.2% over a four-year period due to proposed alterations in land cover.

Kalin and Hantush (2006) conducted a comparative analysis between the KINEROS2 (K2) model and the Gridded Surface Subsurface Hydrologic Analysis (GSSHA) model within the W-2 watershed in Iowa. Their findings indicated that K2 performed well in simulating flow dynamics, yielding Nash Sutcliffe Efficiency (NSE) values ranging from 0.46 to 0.91 for calibration events and from 0.27 to 0.90 for validation events. They concluded that K2 exhibited superior performance in events characterized by initial dry soil conditions. Various sensitivity analyses were employed, including alterations in land cover, changes in contributing source areas, and assessment of parameter sensitivity.

Yatheendradas et al. (2008) applied the KINEROS2 model to eight events within a 6.4 km2 sub-catchment of Walnut Gulch to evaluate its efficacy in flash flood forecasting and to identify key sources of uncertainty. Radar rainfall data with approximately 1 km × 1 km grid resolution and measurements from eleven rain gauges were utilized, with input time-steps ranging from 4 to 6 minutes. A Monte Carlo-based sensitivity analysis experiment was conducted, sampling twenty-one parameters, two initial conditions, and rainfall numerous times (often exceeding 100,000 iterations). Their analysis revealed that the primary source of uncertainty stemmed from radar rainfall estimation, with cumulative rainfall varying significantly (ranging from approximately 15 mm to 35 mm) depending on the estimation method employed. Key parameters such as Ks (saturated hydraulic conductivity), r (Manning's roughness), n (exponent in Manning's equation), and v (capillary length scale) were identified as highly influential, along with the initial plane wetness parameter, s, in certain scenarios. Additionally, considerable uncertainty was observed when attempting to forecast events using parameter sets derived from other events.

CASE STUDY: IMPLEMENTATION OF THE KINEROS2 MODEL IN EGYPT

Description of the Study Area

The study area is situated within the North Western Coastal Region (NWCR) of Egypt, specifically at latitudes 31° 17' 11.5584" N and longitudes 27° 11' 38.7096" E, to the west of Marsa Matrouh city (Figure 2). It encompasses the sub-catchment of Wadi Kharouba, spanning an area of 6 km2, characterized by sandy clay loam and sandy clay soil types. Additionally, the Wadi El-Raml watershed, covering 20,388 ha, is predominantly classified as sandy loam. The region experiences an annual precipitation ranging from 100 to 190 mm, with a short rainy season occurring between November and March, followed by a dry period from May to September. According to the KINEROS2 requirements, the catchment was split into 18 planes, as illustrated in Figure 3, with seven associated channel reaches. The planes were based on topography, land cover and soil type. Channels represent the locations of identifiable channels. The slopes and geometries of planes and channels were estimated from the Digital Elevation Model (DEM).

Figure 2. North-west coastal zone of Egypt—the red star indicates the study area

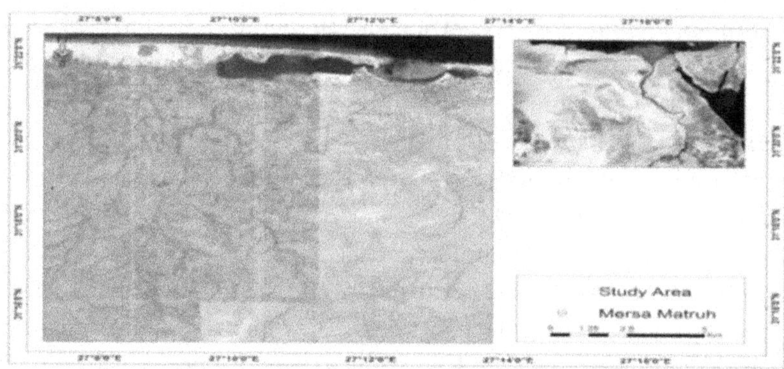

(source: NARSS – National Authority for Remote Sensing & Space Sciences)

Figure 3. General view of Wadi Kharouba layout

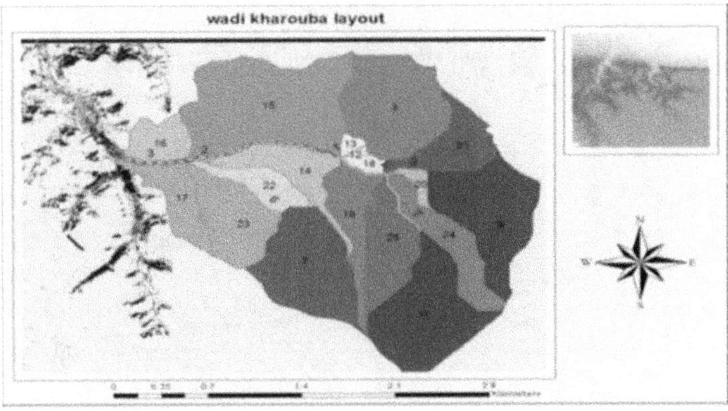

The Monitoring System and Sources of Data

The monitoring system implemented within the sub-catchment of Wadi Kharouba includes the establishment of a weir at the downstream outlet to monitor surface runoff not retained in the soil. An ultrasonic sensor (Rosemount 3108) was installed at the weir to automatically record runoff water levels for discharge calculations. The design of the weir was based on hydrological calculations to accommodate a peak discharge of 10 m3/s, featuring dimensions of 10 m wide, 0.40 m high at the bottom, and 0.90 m at the side walls.

Climate data were collected from three meteorological stations located within the watershed and surrounding areas. Daily precipitation measurements obtained from these stations were utilized as input data for the hydrological model. Each meteorological station is equipped with a rain gauge, temperature and hygrometer sensors, radiation sensors, and a sonic anemometer, all integrated into a data logger provided by DECAGON Company. The data logger is connected to a GPRS system for remote access, and climate data are collected at half-hour intervals three times a day.

The term “available water content” refers to the quantity of water that soil is capable of retaining and that is accessible for plant utilization. This parameter is defined as the volume of water held within the soil between its field capacity, the maximum amount of water retained after excess drainage, and the wilting point, which marks the moisture level at which plants are unable to extract water from the soil and begin to wilt. Refer to Figure 4 for visualization of this concept.

Figure 4. Available water holding capacity

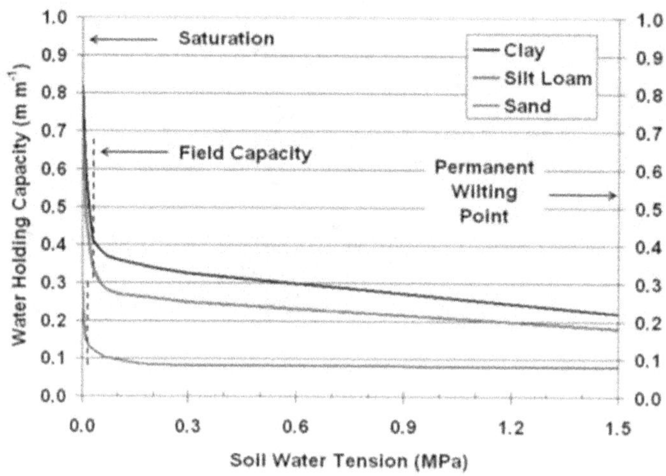

source: http://biomet.ucdavis.edu

The AGWA method for stream network extraction involves calculating the accumulated area upslope of each pixel using a network of cell-to-cell drainage paths. This results in a flow accumulation grid, which is then pruned by removing cells with accumulated flow areas smaller than a user-defined threshold drainage area, referred to as the Channel or Contributing Source Area (CSA). Subsequently, the watershed is divided into upland and channel elements based on the density of the stream network, allowing users to control the spatial complexity of the watershed discretization.

However, this approach often generates numerous spurious polygons and disconnected model elements. To address this, AGWA incorporates algorithms to refine watershed elements by eliminating spurious elements and ensuring downstream connectivity. Additionally, users have the option to customize their model during the discretization process. They can employ internal gauges to split the watershed at predetermined locations along the channel network, such as flow gauging stations, which is beneficial for model calibration if data are available. Users can also designate ponds or drainage-retention structures along the channel network. Stage discharge relationships are stored in database files, queried by AGWA when writing the parameter file. Furthermore, users can characterize the channel network using various hydraulic-geometry relations or define their own relationship between contributing area and channel geometry.

Parameter Estimation

AGWA delineates each overland and channel element, and these are translated into K2 with a specific set of parameter values, assumed to be consistent within each element. However, significant spatial variability exists in topographic, soil, and land-cover attributes across the watershed. To address this, AGWA employs an area-weighting approach to compute an average value for each parameter within an overland flow model element abstracted to an overland flow plane (Miller et al., 2007).

The Kinematic Runoff and Erosion Model (KINEROS2)

KINEROS2 (Kinematic Runoff and Erosion model, version 2) is a distributed hydrological model designed to simulate the processes of water infiltration, surface runoff, and soil erosion within a watershed, primarily in response to individual rainfall events (Semmens et al., 2008). These processes are influenced by various factors such as topography, soil properties, land use characteristics, and the duration and intensity of precipitation events. KINEROS2 is integrated into the Automated Geospatial Watershed Assessment (AGWA) tool, developed by the Agricultural Research Service of the United States Department of Agriculture (USDA), in collaboration with the Environmental Protection Agency (Miller et al., 2007). Utilizing geographic information system (GIS) technology, AGWA facilitates the parameterization of KINEROS2 by dividing the watershed into a series of overland flow planes and channel elements based on topographic data. This enables the model to analyze the response of the watershed to changes in land cover and management practices.

K2 first calculates the infiltration capacity when rainfall rate less than infiltration rate using the Smith-Parlange model as follow:

$$f_c(t) = K_s\left\{1 + \frac{\alpha}{e^{\alpha F(t)/[(G+h)(\Phi-\theta_i]}}\right\} \tag{3}$$

(Where: $f_c(t)$ refer to the infiltration capacity (L/T); K_s indicates soil hydraulic conductivity (L/T); α is the soil parameter, 0 for sand and 1 for well mixed loam; $F(t)$ is the calculated depth of infiltrated water (L); G refer the mean capillary drive (L); h refer to the depth of flow (L); Φ is porosity (L^3/L^3); θ_i indicates initial soil water content).

When the rainfall rate exceeds the infiltration capacity, the model uses the kinematic wave equation to calculate the over land flow as follow:

$$\frac{\partial h}{\partial t} + \alpha m h^{m-1}\frac{\partial h}{\partial x} = q(x,t) \tag{4}$$

(Where: t is the time (T); x is the distance along the slope direction (L); q is the lateral flow rate (L/T); α and m are parameters related to slope, surface roughness and flow regime). After the water reaches the channel, the model uses a similar approach to calculate the channel flow as follow:

$$\frac{\partial A}{\partial t} + \frac{\partial Q}{\partial x} = q_c(x,t) \tag{5}$$

(Where: A is the cross-sectional area (L^2); Q is the channel discharge (L^3/T); $q_c(x,t)$ is the net lateral inflow per unit length of channel (L/T)). The sediment movement is also calculated by the model, which is determined by the mass-balance equation.

The statistical criteria carried out to assess the performance of model, namely Nash-Sutcliffe efficiency (NSE), the coefficient of determination (R^2) and Relative Error (RE). The simulation of stream flow can be considered as satisfactory when NSE is higher than 75%, R^2 is close to one and RE values are lower than 20%, according to (Safari et al., 2012; Memarian et al., 2013; Ghimire et al., 2020).

$$NSE = 1 - \frac{\sum_{i-1}^{n}\left(Q_{Obs} - Q_{sim}\right)^2}{\sum_{i-1}^{n}\left(Q_{Obs} - Q_{Obs}^{-}\right)^2} \tag{6}$$

$$R^2 = \frac{\{\sum_{i-1}^{n}(Q_{sim} - Q_s^-)(Q_{abs} - Q_s^-)\}^2}{\sum_{i-1}^{n}(Q_{sim} - Q_s^-)^2 \sum_{i-1}^{n}(Q_{abs} - Q_s^-)^2} \tag{7}$$

$$RE = \frac{\sum_{i-1}^{n}(Q_{sim} - Q_{obs})}{\sum_{i-1}^{n} Q_{abs}} x\ 100 \tag{8}$$

(Where: Q_{sim} is simulated discharge water (m^3/s); Q_{obs} is Observed discharge water (m^3/s); Q^- is Average of observed discharge water (m^3/s); Q_s- is average of simulated discharge water (m^3/s); Q_{obs} is observed discharge water (m^3 /s); Q_o- is average of observed discharge water (m^3/s)).

Satellite Image and Input Data

The KINEROS2 model necessitates a range of input data and maps to effectively describe the watershed in terms of its geometric parameters. These inputs include a digital elevation model (DEM) map, a land use/cover classification map, a soil map, and precipitation data detailing duration and intensity. The hydrological model relies on a DEM with a resolution of 30 meters to derive key topographic characteristics of the watershed, including elevation, delineation of the watershed boundary, flow direction, slope, and contributing basins. Land use/cover classification was performed using Landsat imagery with a resolution of 30 meters, enabling the identification of five distinct classes: forest, agricultural land, grassland, desert scrub (bare land), and rocky mountain. Soil texture mapping within the Kharouba watershed resulted in the categorization of two major soil groupings based on the FAO Soil Map, denoted as Yk19-1/2ab and Yk22-1/2ab. These groupings encompass two soil type categories: Sandy Clay soils and Sandy Clay Loam soils.

Parameters Inputs

The parameters file compiles all the hydrological data essential for each overland flow plane and channel element. This data encompasses critical parameters such as saturated hydraulic conductivity (Ks), mean capillary drive (G), pore size distribution index (α), porosity (ф), and volumetric rock fraction. This file is structured as a text document, detailing the characteristics of each channel and plane element. AGWA automatically generates this file when a specific combination of watershed, land cover, topography, and soils is selected. Additionally, KINEROS2 relies on estimates of initial soil saturation (S) to determine runoff. The input parameters utilized for KINEROS2 were procured through the AGWA-GIS-based processor. The 30-meter Digital Elevation Model (DEM) sourced from the National Authority for Remote Sensing & Space Sciences (NARSS) in Egypt served as the foundational dataset for delineating watershed boundaries and determining other model-specific parameters, as illustrated in figures 5 to 7.

Figure 5. A DEM is used to subdivide the watershed into upland and channel elements

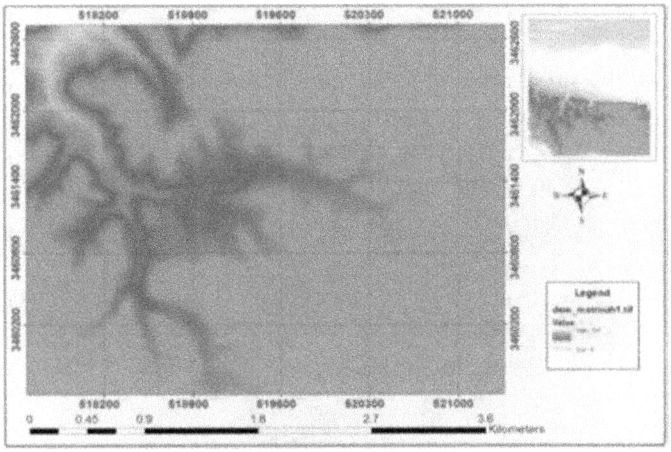

The Digital Elevation Model (DEM) serves as a fundamental tool for modeling and analyzing spatial-topographic information in various applications. DEM Applications: DEMs are utilized for modeling, analyzing, and interpreting terrain morphology, including tasks such as drainage basin delineation, hydrological runoff modeling, geomorphological simulation and classification, geological mapping, among others. Watershed delineation constitutes the initial step in the utilization of AGWA. This process entails generating a feature class that represents the entire area draining into a user-specified outlet, a crucial step for subsequent use in KINEROS2 modeling.

Output Location: This step necessitates the identification of the workspace name and the specification of a new name for the geodatabase.

Input Grids: In this phase, users are required to specify the grid surfaces to be utilized, including a DEM, a flow direction grid (FDG), and a flow accumulation grid (FACG), as depicted in figure 2.10. Optionally, a stream grid may be generated after selecting these inputs to aid in outlet location. The stream grid delineates all streams within a specific DEM, with a default threshold of 2500, designating cells that receive runoff from at least 2500 cells as streams. AGWA allows users to adjust this threshold to accommodate varying levels of network detail based on the resolution of the DEM.

Figure 6. The overland and channel elements delineated by AGWA

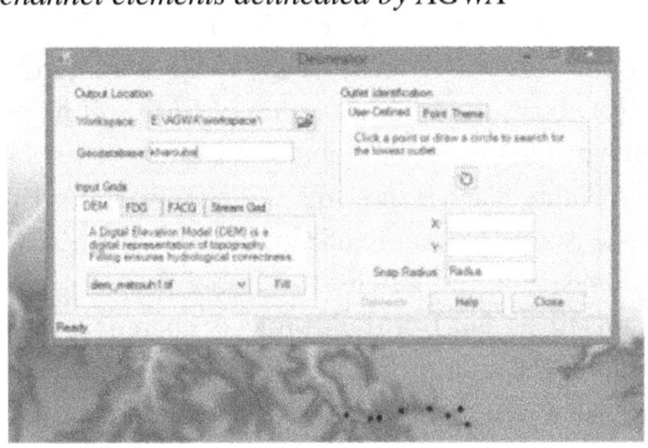

- Flow direction grid is an integer grid, specific to a given DEM, representing the direction of flow throughout a study area.

Figure 7. A flow direction grid (FDG) represents the direction of flow for a DEM

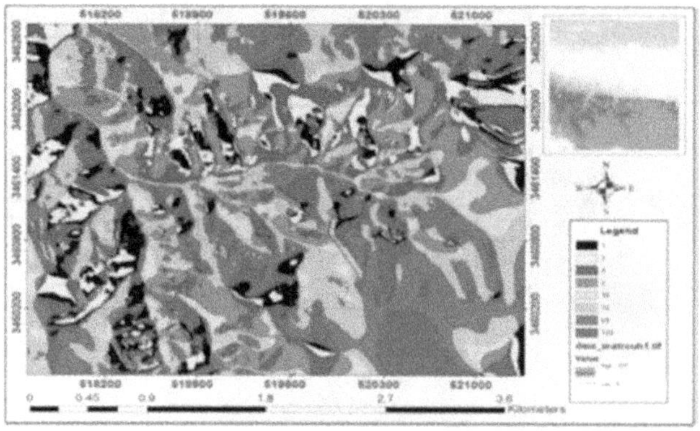

- Flow accumulation grid is an integer grid, specific to a given DEM, representing the accumulation of runoff throughout a study area.

The Discretizing function breaks up the delineation/watershed into model specific elements and creates a stream feature class that drains the elements as shown in fig.2.13. The CSA, or Contributing/ Channel Source Area, is a threshold value which defines first order channel initiation, or the upland area required for channelized flow to begin. Smaller CSA values result in a more complex watershed, and larger CSA values result in a less complex watershed.

Soil Parameters

Soil parameters essential for upland planes in KINEROS2 (K2), including characteristics like percent rock, suction head, porosity, and saturated hydraulic conductivity, are initially approximated based on soil texture data, following methodologies outlined by Rawls et al. (1982) and Woolhiser et al. (1990). The suitability of digital soil maps for various countries or regions can vary significantly in terms of their content and database organization. The automated utilization of soil maps for model parameterization is contingent upon this information structure, thus not all soil maps are compatible with AGWA. Consequently, specific procedures have been developed to effectively employ the United Nations Food and Agriculture Organization's (FAO) Digital Soil Map of the World, maximizing its geographic applicability. Despite the relatively coarse spatial resolution of FAO soil maps, K2 outcomes derived from them exhibit favorable comparisons with results obtained from higher resolution soil maps in the United States (Levick et al., 2004). Figure 8 and 9 show the soil map from FAO at a scale of 1:250,000 and the land cover from the Multi-sensor resolution Land Characteristics (MRLC) Consortium-National Land Cover Data (NLCD) at a 30m spatial resolution.

The FAO has amassed soil profile data from field projects, encompassing information found in volumes accompanying the Soil Map of the World (FAO-UNESCO, 1971-81), publications in Soil Taxonomy (Soil Survey Staff, 1972), and datasets released by ISRIC. A comprehensive analysis of 1700 soil profiles was conducted, categorized by FAO Soil Unit and Topsoil Texture group. Statistical methods, including weighted averages, were employed to derive mean values for both topsoil (0-30 cm) and subsoil (30-100 cm) layers across various chemical and physical parameters.

Parameters related to land cover, such as interception, canopy cover, Manning's roughness, and percentage of paved area, are estimated through expert judgment and reference to previously published lookup tables (Woolhiser et al., 1990). Slope characteristics for upland elements are determined based on the average plane slope, while geometric attributes like plane width and length are derived assuming a rectangular shape, with the longest flow length equating to the element's length. Geometric properties of stream channels are parameterized following the findings of Miller et al. (1996), who identified significant correlations between channel width and depth and watershed attributes. Channel parameters linked to soil characteristics assume a sandy bed, and uniformity is assumed across all channels. Channel slope is computed from a slope grid derived from the Digital Elevation Model (DEM) (Semmens, 2008).

Figure 8. The digital soil map of the world food and agriculture organization of the United Nations version 3.6

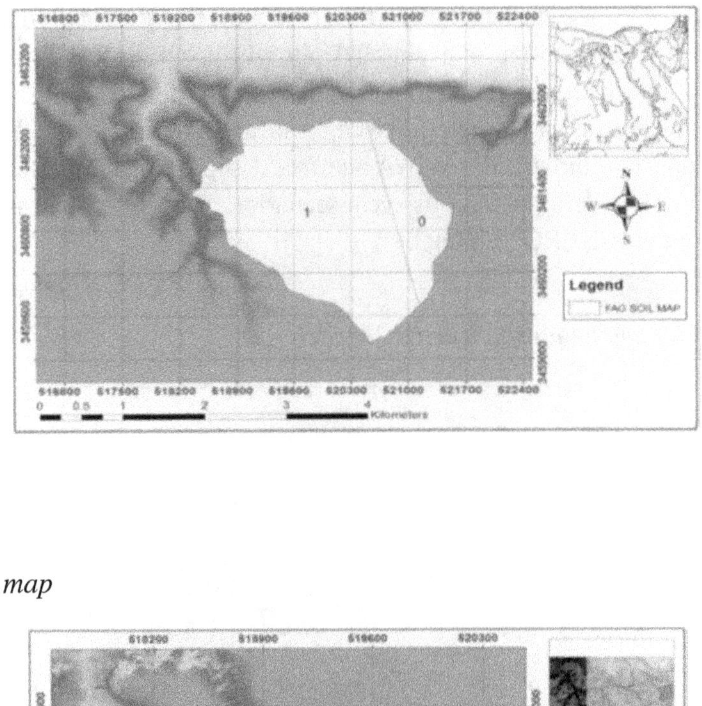

Figure 9. Land cover map

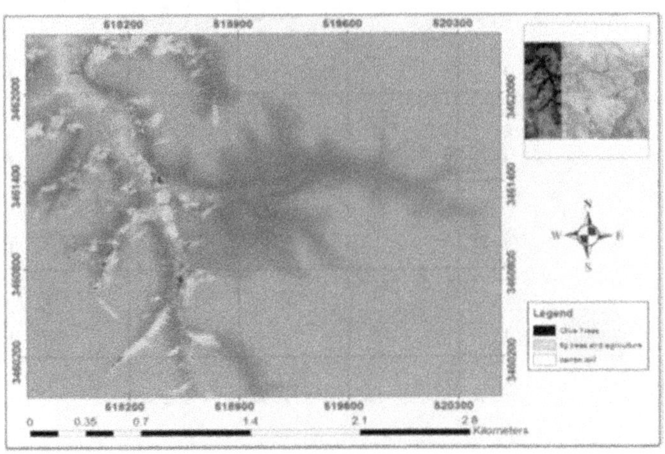

Rainfall Input

Uniform rainfall input files for KINEROS2 (K2) can be generated in AGWA through various methods. These include using gridded return-period rainfall maps, a database of geographically specific return-period rainfall depths provided within the tool, or inputting data manually by the user. While uniform rainfall is less suitable for quantitative modeling in arid regions, it serves well for assessing relative changes in land use and cover.

Return period rainfall depths are transformed into hyetographs using the Soil Conservation Service (SCS) methodology and a type II distribution. This distribution, although hypothetical, is commonly used to derive the time distribution of 24-hour rainfall for extreme events in many regions. However, it may lead to overestimations of peak flows, especially for shorter-duration events. When return-period rainfall grids are accessible, AGWA extracts the rainfall depth from the grid cell containing the centroid of the watershed and converts it into a hyetograph using the SCS methodology.

This process is automated to facilitate use with prevalent U.S. datasets and can be adjusted for other formats. If return-period rainfall maps are unavailable or specific depth and duration are preferred, the provided design-storm database can be readily edited to include new data. This database allows inputting location, recurrence interval, duration, and rainfall depth, with an option to incorporate an area-reduction factor, particularly useful in regions prone to convective thunderstorms.

Alternatively, if gauge observations of rainfall depth are accessible or a specific hyetograph is desired, users can manually input data through the AGWA interface. User-defined storms are entered as time-depth pairs, offering flexibility in defining various event scenarios. The figure 10 provides a synthesis of the inputs and outputs of the KINEROS2 model.

Figure 10. Main inputs and outputs in Kineros2 model

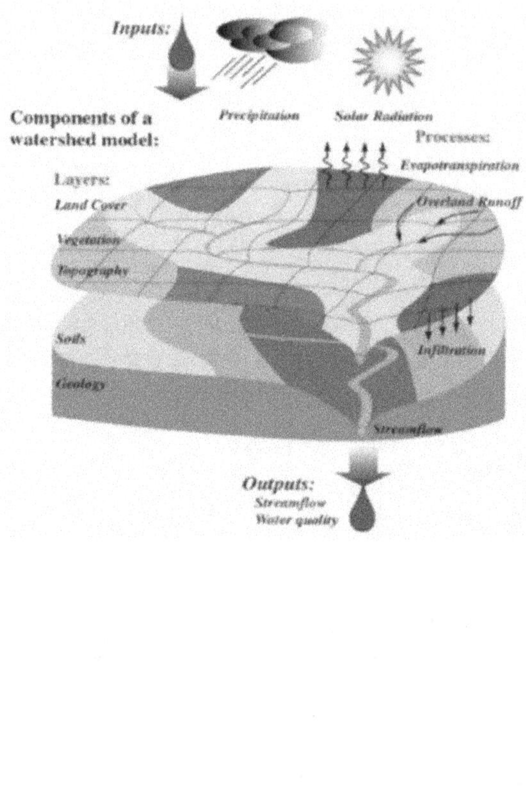

(Gobattoni, 2009)

Modeling

These datasets were employed to calibrate and validate the KINEROS 2 model, which is utilized for investigating hydrological surface processes and evaluating the availability of water resources under varying rainfall conditions. Consequently, it aids in the design of Wadi terraces and Wadi irrigation systems. For example, the graphs depicted in figures 11 and 12 illustrate a comparison between the observed and simulated values for the first and second events. Overall, the validation outcomes can be deemed satisfactory, considering the limited historical data available for calibration and validation. This initial assessment serves as a constructive starting point, as the integration of storage and discharge measurements is anticipated to enhance the model calibration substantially. Moreover, the inclusion of discharge measurements in the model tuning process is expected to alter the sensitivity of the model output, particularly contingent on the specific process being simulated, whether it pertains to infiltration or runoff dynamics.

Figure 11. Comparison between simulated and measured water storages (the partial and the total values) obtained for the best simulation at first event

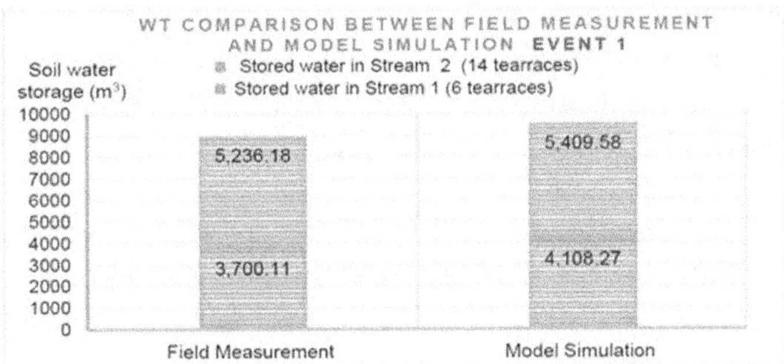

Figure 12. Comparison between simulated and measured water storages (the partial and the total values) obtained for the best simulation at second event

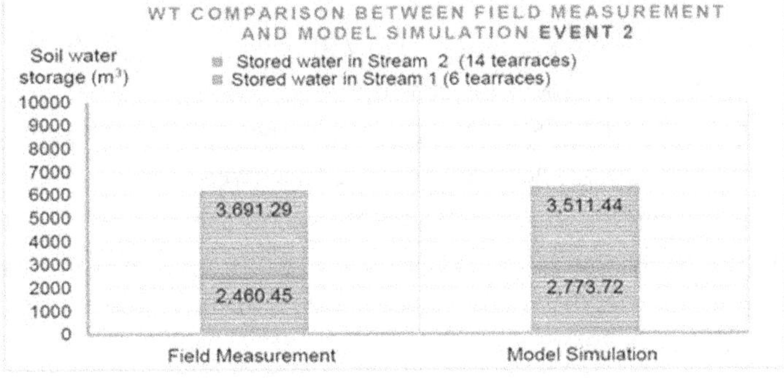

EXPECTED OUTCOME OF UTILIZING SOIL EROSION MODELING (KINEROS2 MODEL) AND GIS TECHNOLOGIES

- Improved Flood Forecasting: Enhanced understanding of hydrological processes and accurate flood forecasting capabilities enable early warning systems and proactive flood management strategies.
- Enhanced Water Management: The KINEROS2 model supports evidence-based decision-making in water allocation, irrigation scheduling, and water conservation initiatives, leading to more efficient water use and increased agricultural productivity.
- Sustainable Development: By integrating hydrological modeling into water resource management practices, Egypt can achieve more sustainable development outcomes, including enhanced water security, reduced environmental degradation, and improved resilience to climate change impacts.

CONCLUSION AND FUTURE RECOMMENDATION

The successful application of the KINEROS2 model in Egypt underscores its efficacy in addressing water-related challenges and advancing sustainable development goals. Leveraging advanced hydrological modeling tools can enable Egypt to mitigate water scarcity, manage flood risks, and bolster agricultural resilience, thereby contributing to socio-economic progress.

The implementation of the KINEROS2 model in Egypt necessitates several key steps, including:

1. Data Collection: Comprehensive data collection, encompassing topographic maps, soil characteristics, land use/land cover data, and rainfall data, forms the foundation for model parameterization and calibration. Continuous monitoring and updating of data are essential for model accuracy.
2. Model Calibration: Rigorous calibration of the KINEROS2 model involves fine-tuning model parameters to align simulated outputs with observed hydrological data, such as streamflow measurements, rainfall-runoff relationships, and sediment transport dynamics. Iterative calibration processes ensure model robustness and reliability.
3. Scenario Analysis: Conducting scenario analysis using the calibrated KINEROS2 model allows for the assessment of various land use changes, water management strategies, and climate scenarios on water availability, flood risk, and agricultural productivity. This facilitates informed decision-making and proactive planning for future challenges.
4. Stakeholder Engagement: Active engagement with diverse stakeholders, including government agencies, research institutions, local communities, and farmers, fosters collaboration, knowledge sharing, and consensus-building. Stakeholder involvement ensures the relevance and applicability of model outcomes to real-world decision-making processes.
5. Continuous Monitoring and Evaluation: Establishing a robust monitoring and evaluation framework is essential to continuously assess the performance of the KINEROS2 model. Regular validation exercises and sensitivity analyses help identify model limitations, refine parameterizations, and improve predictive accuracy over time.

6. Integration with Decision Support Systems: Integrating the KINEROS2 model with decision support systems (DSS) enhances its usability for water resource management and policy formulation. Developing user-friendly interfaces and dashboards facilitates access to model outputs, allowing decision-makers to make informed choices based on real-time data and scenario analyses.
7. Long-term Planning and Investment: Long-term planning and investment in KINEROS2 model implementation are essential for sustaining its benefits and impact. Securing funding for model maintenance, data collection, and capacity-building initiatives ensures the longevity and relevance of the model in addressing complex water management challenges in Egypt.

By adhering to these key steps and fostering a collaborative approach, Egypt can maximize the potential of the KINEROS2 model to achieve sustainable water resource management and contribute to the country's socio-economic development objectives.

REFERENCES

Abdalla, M. A. I., EL-WAHAB, M. A., Tawfik, M., & Khater, I. (2022). Hydrological Models Simulation in Agricultural Watershed of Wadi Kharouba in North-Western Coast Region–Egypt. *Fresenius Environmental Bulletin*, 31, 1063–1078.

Ahrens C. D. (2012). *Meteorology today: an introduction to weather, climate, and the environment: Cengage Learning*.

Al-Qurashi, A. (1995). *Rainfall–runoff relationships in arid areas*. [MSc thesis, Heriot-Watt University, Edinburgh, UK].

Bazzoffi P., Abbattista F., Vanino S. and Pellegrini S. (2006). Impact of land levelling for vineyard plantation on soil degradation in Italy. *Bollettino della Società Geologica Italiana. Volume speciale*(6), 191-199.

Beguería, S., López-Moreno, J. I., Lorente, A., Seeger, M., & García-Ruiz, J. M. (2003). Assessing the effect of climate oscillations and land-use changes on streamflow in the Central Spanish Pyrenees. *Ambio*, 32(4), 283–286. 10.1579/0044-7447-32.4.28312956594

Beven, K. (2001). How far can we go in distributed hydrological modelling? *Hydrology and Earth System Sciences*, 5(1), 1–12. 10.5194/hess-5-1-2001

Borga, M., Stoffel, M., Marchi, L., Marra, F., & Jakob, M. (2014). Hydrogeomorphic response to extreme rainfall in headwater systems: Flash floods and debris flows. *Journal of Hydrology (Amsterdam)*, 518, 194–205. 10.1016/j.jhydrol.2014.05.022

Buendia, C., Batalla, R. J., Sabater, S., Palau, A., & Marcé, R. (2015). Runoff trends driven by climate and afforestation in a Pyrenean basin. *Land Degradation & Development*.

Capolongo, D., Pennetta, L., Piccarreta, M., Fallacara, G., & Boenzi, F. (2008). Spatial and temporal variations in soil erosion and deposition due to land-levelling in a semi-arid area of Basilicata (Southern Italy). *Earth Surface Processes and Landforms*, 33(3), 364–379. 10.1002/esp.1560

Carpenter, T. M., & Georgakakos, K. P. (2004). Continuous streamflow simulation with the HRC-DHM distributed hydrologic model. *Journal of Hydrology (Amsterdam)*, 298(1), 61–79. 10.1016/j.jhydrol.2004.03.032

Emmanuel, I., Andrieu, H., Leblois, E., Janey, N., & Payrastre, O. (2015). Influence of rainfall spatial variability on rainfall–runoff modelling: Benefit of a simulation approach? *Journal of Hydrology (Amsterdam)*, 531, 337–348. 10.1016/j.jhydrol.2015.04.058

Engelund, F., & Hansen, E. (1967). *A monograph on sediment transport in alluvial streams. Technical University of Denmark 0stervoldgade 10*. Copenhagen K.

Feldman, A. D. (2000). Hydrologic modeling system HEC-HMS: technical reference manual. US Army Corps of Engineers, Hydrologic Engineering Center.

Fiener, P., & Auerswald, K. (2009). Spatial variability of rainfall on a sub-kilometre scale. *Earth Surface Processes and Landforms*, 34(6), 848–859. 10.1002/esp.1779

Frot, E., van Wesemael, B., Benet, A., & House, M. A. (2008). Water harvesting potential in function of hillslope characteristics: A case study from the Sierra de Gador (Almeria province, south-east Spain). *Journal of Arid Environments*, 72(7), 1213–1231. 10.1016/j.jaridenv.2007.12.009

Garcia-Ruiz, J. M., White, S. M., Lasanta, T., Marti, C., Gonzalez, C., Errea, M. P., Valero, B., & Ortigosa, L. (1997). Assessing the effects of land-use changes on sediment yield and channel dynamics in the central Spanish Pyrenees. *IAHS Publications-Series of Proceedings and Reports-Intern Assoc Hydrological Sciences*, 245, 151–158.

Ghimire, U., Agarwal, A., Shrestha, N. K., Daggupati, P., Srinivasan, G., & Than, H. H. (2020). Applicability of lumped hydrological models in a data-constrained river basin of Asia. *Journal of Hydrologic Engineering*, 25(8), 05020018. 10.1061/(ASCE)HE.1943-5584.0001950

Gobattoni, F. (2009). *Hydrological models for watershed planning and water resources management: a Decision System Support for Marta River Basin*. Central Italy.

Goodrich, D., Unkrich, C., Smith, R., Guertin, D., Hernandez, M., Burns, I., Massart, J., Levick, L., Miller, S., & Semmens, D. (2010). The AGWAKINEROS2 suite of modeling tools.In: (eds). *Proc. 2010 Watershed Management Conf.* 10.1061/41143(394)116

Goodrich, D. C., Faurès, J.-M., Woolhiser, D. A., Lane, L. J., & Sorooshian, S. (1995). Measurement and analysis of small-scale convective storm rainfall variability. *Journal of Hydrology (Amsterdam)*, 173(1), 283–308. 10.1016/0022-1694(95)02703-R

Greenbaum, N., Ben-Zvi, A., Haviv, I., & Enzel, Y. (2006). *The hydrology and paleohydrology of the Dead Sea tributaries* (Vol. 401). Geological Society of America Special Papers. 10.1130/2006.2401(05)

Johnson, D. L., & Miller, A. C. (1997). A spatially distributed hydrologic model utilizing raster data structures. *Computers & Geosciences*, 23(3), 267–272. 10.1016/S0098-3004(96)00084-2

Kalin, L., & Hantush, M. H. (2006). Comparative assessment of two distributed watershed models with application to a small watershed. *Hydrological Processes*, 20(11), 2285–2307. 10.1002/hyp.6063

Kosmas, C., Danalatos, N., Cammeraat, L. H., Chabart, M., Diamantopoulos, J., Farand, R., Gutierrez, L., Jacob, A., Marques, H., Martinez-Fernandez, J., Mizara, A., Moustakas, N., Nicolau, J. M., Oliveros, C., Pinna, G., Puddu, R., Puigdefabregas, J., Roxo, M., Simao, A., & Vacca, A. (1997). The effect of land use on runoff and soil erosion rates under Mediterranean conditions. *Catena*, 29(1), 45–59. 10.1016/S0341-8162(96)00062-8

Levick, L., Semmens, D., Guertin, D., Burns, I., Scott, S., Unkrich, C., & Goodrich, D. (2004). Adding global soils data to the automated geospatial watershed assessment tool (AGWA). *Proc. 2nd SAHRA,(-Sustainability of Semi-Arid Hydrol. and Riparian Areas), U. of Ariz., Internat'l. Sym. on Transboundary Water Manage.* University of Arizona.

López-Tarazón, J., Batalla, R. J., Vericat, D., & Balasch, J. (2010). Rainfall, runoff and sediment transport relations in a mesoscale mountainous catchment: The River Isábena (Ebro basin). *Catena*, 82(1), 23–34. 10.1016/j.catena.2010.04.005

Lorenzo, L., Casalí, J., López, J., & Del Valle de Lersundi, J. (2002). *Long term assessment of soil erosion rates in vineyards, and its application for USLE model evaluation*. European Society of Agronomy Conference, Córdoba.

Ly, S., Charles, C., & Degré, A. (2013). Different methods for spatial interpolation of rainfall data for operational hydrology and hydrological modeling at watershed scale. A review. *Biotechnologie, Agronomie, Société et Environnement*, 17(2), 392.

Meles, M. B., Goodrich, D. C., Unkrich, C. L., Gupta, H. V., Burns, I. S., Hirpa, F. A., Razavi, S., & Guertin, D. P. (2024). Rainfall distributional properties control hydrologic model parameter importance. *Journal of Hydrology. Regional Studies*, 51, 101662. 10.1016/j.ejrh.2024.101662

Memarian, H., Balasundram, S. K., Talib, J. B., Teh Boon Sung, C., Mohd Sood, A., & Abbaspour, K. C. (2013). KINEROS2 application for land use/cover change impact analysis at the H ulu L angat B asin, M alaysia. *Water and Environment Journal : the Journal / the Chartered Institution of Water and Environmental Management*, 27(4), 549–560. 10.1111/wej.12002

Meselhy, A. A., Salem, H. M., & Elhagarey, M. E. (2019). Manufacturing of machine for crop cultivating under rainfed conditions. *Bioscience Research*, 16(4), 3903–3926.

Meyer, L. D., & Wischmeier, W. H. (1969). Mathematical simulation of the process of soil erosion by water. *Transactions of the ASAE. American Society of Agricultural Engineers*, 12(6), 754–0758. 10.13031/2013.38945

Miller, S., Semmens, D., Miller, R., Hernandez, M., Goodrich, D., Miller, W., Kepner, W., & Ebert, D. (2002). GIS-based hydrologic modeling: the automated geospatial watershed assessment tool. *Proceedings of the Second Federal Interagency Hydrologic Modeling Conference.*

Miller, S. N., Semmens, D. J., Goodrich, D. C., Hernandez, M., Miller, R. C., Kepner, W. G., & Guertin, D. P. (2007). The automated geospatial watershed assessment tool. *Environmental Modelling & Software*, 22(3), 365–377. 10.1016/j.envsoft.2005.12.004

Mujumdar, P., & Kumar, D. N. (2012). *Floods in a changing climate: hydrologic modeling*. Cambridge University Press. 10.1017/CBO9781139088428

Navratil, O., Evrard, O., Esteves, M., Legout, C., Ayrault, S., Némery, J., Mate-Marin, A., Ahmadi, M., Lefèvre, I., Poirel, A., & Bonté, P. (2012). Temporal variability of suspended sediment sources in an alpine catchment combining river/rainfall monitoring and sediment fingerprinting. *Earth Surface Processes and Landforms*, 37(8), 828–846. 10.1002/esp.3201

Ramos, M., & Martinez-Casasnovas, J. (2006). Nutrient losses by runoff in vineyards of the Mediterranean Alt Penedès region (NE Spain). *Agriculture, Ecosystems & Environment*, 113(1), 356–363. 10.1016/j.agee.2005.10.009

Ramos, M., & Martínez-Casasnovas, J. (2009). Impacts of annual precipitation extremes on soil and nutrient losses in vineyards of NE Spain. *Hydrological Processes*, 23(2), 224–235. 10.1002/hyp.7130

Ramos, M., & Martínez-Casasnovas, J. (2010). Effects of field reorganisation on the spatial variability of runoff and erosion rates in vineyards of Northeastern Spain. *Land Degradation & Development*, 21(1), 1–12. 10.1002/ldr.958

Rawls, W., Brakensiek, D., & Saxtonn, K. (1982). Estimation of soil water properties. *Transactions of the ASAE. American Society of Agricultural Engineers*, 25(5), 1316–1320. 10.13031/2013.33720

Reed, S., Koren, V., Smith, M., Zhang, Z., Moreda, F., Seo, D.-J., & Participants, D. (2004). Overall distributed model intercomparison project results. *Journal of Hydrology (Amsterdam)*, 298(1), 27–60. 10.1016/j.jhydrol.2004.03.031

Refsgaard, J. C. (1997). Parameterisation, calibration and validation of distributed hydrological models. *Journal of Hydrology (Amsterdam)*, 198(1), 69–97. 10.1016/S0022-1694(96)03329-X

Refsgaard, J. C., & Knudsen, J. (1996). Operational validation and intercomparison of different types of hydrological models. *Water Resources Research*, 32(7), 2189–2202. 10.1029/96WR00896

Safari, A., De Smedt, F., & Moreda, F. (2012). WetSpa model application in the distributed model intercomparison project (DMIP2). *Journal of Hydrology (Amsterdam)*, 418, 78–89. 10.1016/j.jhydrol.2009.04.001

Salem, H. M. (2014). *Effect of conservation tillage on soil physical properties, in-situ rainwater harvesting, and erosion control in arid and semi-arid regions* [Doctoral dissertation, Agronomos].

Salem, H. M., Ali, A. M., Wu, W., & Tu, Q. (2021). Initial effect of shifting from traditional to no-tillage on runoff retention and sediment reduction under rainfall simulation. *Soil Research (Collingwood, Vic.)*, 60(6), 547–560. 10.1071/SR21082

Salem, H. M., Meselhy, A., Elhagarey, M., Ali, A. M., & Wu, W. (2022). Soil erosion control and wheat productivity are improved by a developed ridge-furrow and reservoir tillage systems. *Archives of Agronomy and Soil Science*, 68(2), 273–282. 10.1080/03650340.2020.1832655

Salem, H. M., & Meselhy, A. A. (2021). A portable rainfall simulator to evaluate the factors affecting soil erosion in the northwestern coastal zone of Egypt. *Natural Hazards*, 105(3), 2937–2955. 10.1007/s11069-020-04432-8

Salem, H. M., Meselhy, A. A., Elhagarey, M. E., Abdalla, M. A., El Salem, A., Ali, A. M., & Schott, L. R. (2023). In-Situ Rainwater Harvesting Using Innovative Engineering Technologies for Rehabilitation of Degraded Lands and Climate Change Adaptation. In *Intelligent Engineering Applications and Applied Sciences for Sustainability* (pp. 92–108). IGI Global. 10.4018/979-8-3693-0044-2.ch006

Salem, H. M., Muñoz-García, M. Á., & Rodríguez, M. G. (2020). Soil physical properties and soil water tension monitoring by wireless sensor network after reservoir and minimum tillage practices. *Soil Research (Collingwood, Vic.)*, 59(3), 309–317. 10.1071/SR20129

Salem, H. M., Schott, L. R., Piaskowski, J., Chapagain, A., Yost, J. L., Brooks, E., Kahl, K., & Johnson-Maynard, J. (2024). Evaluating Intra-Field Spatial Variability for Nutrient Management Zone Delineation through Geospatial Techniques and Multivariate Analysis. *Sustainability (Basel)*, 16(2), 645. 10.3390/su16020645

Salem, H. M., Valero, C., Muñoz, M. Á., & Gil-Rodríguez, M. (2015). Effect of integrated reservoir tillage for in-situ rainwater harvesting and other tillage practices on soil physical properties. *Soil & Tillage Research*, 151, 50–60. 10.1016/j.still.2015.02.009

Salem, H. M., Valero, C., Muñoz, M. Á., Gil-Rodríguez, M., & Barreiro, P. (2014). Effect of reservoir tillage on rainwater harvesting and soil erosion control under a developed rainfall simulator. *Catena*, 113, 353–362. 10.1016/j.catena.2013.08.018

Schuol, J., & Abbaspour, K. (2007). Using monthly weather statistics to generate daily data in a SWAT model application to West Africa. *Ecological Modelling*, 201(3), 301–311. 10.1016/j.ecolmodel.2006.09.028

Sen, Z. (2008). *Wadi hydrology*. CRC Press. 10.1201/9781420061550

Sherman, L. K. (1932). The relation of hydrographs of runoff to size and character of drainage-basins. *Eos (Washington, D.C.)*, 13(1), 332–339.

Shmilovitz, Y., Marra, F., Wei, H., Argaman, E., Goodrich, D., Assouline, S., & Morin, E. (2023). Assessing the controlling factors on watershed soil erosion during intense rainstorm events using radar rainfall and process-based modeling. *Catena*, 231, 107282. 10.1016/j.catena.2023.107282

Shrestha, D., Zinck, J., & Van Ranst, E. (2004). Modelling land degradation in the Nepalese Himalaya. *Catena*, 57(2), 135–156. 10.1016/j.catena.2003.11.003

Shultz M. J. (2007). *Comparison of distributed versus lumped hydrologic simulation models using stationary and moving storm events applied to small synthetic rectangular basins and an actual watershed basin*. ProQuest.

Shultz, M. J., & Corby, R. (2006). Development, calibration, and implementation of a distributed hydrologic model for use in real-time river forecasting.In: (eds). *Proceedings of 3rd Federal Interagency Hydrologic Modeling Conference*. Research Gate.

Singh, V. P. (1992). *Elementary hydrology*. Pearson College Division.

Smith, M. B. (1993). A gis-based distributed parameter hydrologic model for urban areas. *Hydrological Processes*, 7(1), 45–61. 10.1002/hyp.3360070106

Smith, M. B., Koren, V., Finnerty, B., & Johnson, D. L. (1999). *Distributed modeling: Phase 1 results*. US Department of Commerce, National Oceanic and Atmospheric Administration, National Weather Service.

Smith, M. B., Seo, D.-J., Koren, V. I., Reed, S. M., Zhang, Z., Duan, Q., Moreda, F., & Cong, S. (2004). The distributed model intercomparison project (DMIP): Motivation and experiment design. *Journal of Hydrology (Amsterdam)*, 298(1), 4–26. 10.1016/j.jhydrol.2004.03.040

Szcześniak, M., & Piniewski, M. (2015). Improvement of Hydrological Simulations by Applying Daily Precipitation Interpolation Schemes in Meso-Scale Catchments. *Water (Basel)*, 7(2), 747–779. 10.3390/w7020747

Tropeano, D. (1983). Soil erosion on vineyards in the Tertiary Piedmontese basin (northwestern Italy): Studies on experimental areas. *Catena Supplement*, 4, 115–127.

Van Steenbergen, F., Lawrence, P., Mehari, A., Salman, M., & Faurés, J. (2010). *Guidelines on spate irrigation. FAO Irrigation and Drainage Paper 65: ISBN 978-92-5-106608-09*. FAO.

Verdú, J., Batalla, R., & Martínez-Casasnovas, J. (2006). Estudio hidrológico de la cuenca del río Isábena (Cuenca del Ebro). II: Respuesta hidrológica. *Ing Agua*, 13(4), 331–343. 10.4995/ia.2006.10498

Vieux, B. E. (2001). *Distributed hydrologic modeling using GIS*. Springer. 10.1007/978-94-015-9710-4

Vieux B. E. and Moreda F. G. (2003). Ordered Physics-Based Parameter Adjustment of a Distributed Model. *Calibration of Watershed Models*.

Wheater H. (2002a). Progress in and prospects for fluvial flood modelling. *Philosophical Transactions of the Royal Society of London A: Mathematical, Physical and Engineering Sciences*, *360*(1796), 1409-1431.

Wheater, H., Butler, A., Stewart, E., & Hamilton, G. (1991). A multivariate spatial-temporal model of rainfall in southwest Saudi Arabia. I. Spatial rainfall characteristics and model formulation. *Journal of Hydrology (Amsterdam)*, 125(3), 175–199. 10.1016/0022-1694(91)90028-G

Wheater H. S. (2002b). Hydrological processes in arid and semiarid areas. *Hydrology of wadi systems*, 5-22.

Wicherek, S. (1991). Viticulture and soil erosion in the north of Parisian basin. Example: The mid Aisne region. *Zeitschrift für Geomorphologie*, 83(Supplementband), 115–126.

Woolhiser, D. A., Fedors, R. W., Smith, R. E., & Stothoff, S. A. (2006). Estimating infiltration in the upper split wash watershed, Yucca Mountain, Nevada. *Journal of Hydrologic Engineering*, 11(2), 123–133. 10.1061/(ASCE)1084-0699(2006)11:2(123)

Woolhiser, D. A., Smith, R., & Goodrich, D. C. (1990). *KINEROS: a kinematic runoff and erosion model: documentation and user manual*. US Department of Agriculture, Agricultural Research Service.

Yatheendradas, S., Wagener, T., Gupta, H., Unkrich, C., Goodrich, D., Schaffner, M., & Stewart, A. (2008). Understanding uncertainty in distributed flash flood forecasting for semiarid regions. *Water Resources Research*, 44(5), 2007WR005940. 10.1029/2007WR005940

Chapter 9
Determination of Soil Color Characteristics Using RGB Images

Mikhail Yurievich Kataev
http://orcid.org/0000-0002-7710-5463
Department of Control Systems, Tomsk State University of Control Systems and Radioelectronics, Russia

Evgeny Yurievich Kartashov
http://orcid.org/0000-0003-3056-4555
Tomsk State University of Control Systems and Radioelectronics, Russia

Igor Borisovich Sorokin
Agrochemical Service Station, Russia

ABSTRACT

The purpose of the research is to identify color patterns of different types of soils to assess the conditions for identifying types using RGB images. To achieve this goal, the following tasks were solved: 1) Identify problems in the agricultural plan where soil plays an important aspect; 2) identify the characteristics of the soil composition and the relationship between the composition and color characteristics; 3) determine the features of processing RGB images during preliminary and thematic processing; 4) evaluate the color characteristics of several types of soils; and 5) determine criteria for distinguishing the color characteristics of different types of soils. The object of research is the soil type and its RGB images. Research methods included the collection of statistical information about brightness variations in RGB spectral channels for different types of soils.

INTRODUCTION

Currently, for many countries there is a problem of food independence and security due to the negative economic background, the emergence of low quality natural and synthetic food products, which is a consequence of deteriorating public health. Soil is an important asset of any country, given its use to ensure

DOI: 10.4018/979-8-3693-3374-7.ch009

food independence and obtain high-quality food products, which determines the relevance of soil research (Zhukova, 2021, Godin, 2020, Androkhanov, 2004, FAO, 2006, McBratney, 2003, Dobrovolsky, 2015).

The emergence of a technological base for work in agricultural areas makes it possible to intensify agricultural production, and this, on the other hand, requires the development and implementation of highly effective soil research methods. When studying soils, it is necessary to achieve high efficiency and accurate, objective information about the state and properties of the territory under study and the soils located on it. Since agricultural areas are usually large in size, it is preferable to solve the problem of soil research using remote sensing methods. Remote sensing research approaches include satellite approaches (Lagacherie, 2007), aircraft (Andronikov, 1979) and unmanned approaches (Colomina, 2014).

One of the most common methods for studying the surface of the earth and soil in particular is the optical spectrometric method, which allows one to obtain quantitative characteristics of individual soil properties. The studies of the spectral reflectivity of soils, which is the basis of the spectrometric method, are the subject of the works of many authors (Orlov, 2002, Kostenko 2009, Bowers, 1965, Karavanova, 2003, Viscarra, 2016). These studies show that the formation of soil reflectivity is influenced by several factors, which can be divided into two groups - physical and chemical. The physical group includes the ratio of soil areas with different colors, structural state (texture, lumpiness), humidity, etc., and the chemical group of factors includes the content of carbonates, iron compounds, salts, mineral and granulometric composition, humus content, etc. (Aparin, 2016, Shishov, 2004). The basis of the spectrometric method is that the optical indicators of soils and their components are related to the material composition of the organic and mineral parts of soils and are indicative parameters for identifying the type of soil.

Conducting a comprehensive study of the optical properties of soils is most relevant for regions where agriculture is an important industry and, therefore, the load on the soil is increased. It is known that the use of soils in agricultural production is accompanied by intensified processes of degradation of soil quality (Dobrovolsky, 2002). Therefore, the use of simple approaches that will allow agricultural workers to control soil characteristics is important and highly relevant. In this regard, the goal of the work is related to identifying the features of optical parameters expressed in color for different types of soils and the possibility of constructing a bank of soil color indicators to determine trends in indicators.

The high level of agricultural land use for various crops causes intensive soil cultivation, which causes the gradual destruction of the soil structure. Changes in the structure of the soil lead to a deterioration in its water, air, thermal and nutrient regimes, as well as increased erosion. Therefore, regular and systematic study of soil resistance to technogenic (agricultural) impacts is the basis for assessing the degree of soil degradation throughout the entire territory of the agricultural field being studied. Such studies make it possible to establish an acceptable limit for the impact load on the soil, allowing for compliance with the time frame for intensive land use and guaranteeing the preservation of soil functions.

Research of the entire surface of agricultural fields, as a whole, will allow us to implement an approach leading to the following effects: 1) economic (reducing the cost of growing crops for all processes existing for this); 2) environmental (increasing the content and improving the balance of organic matter, preserving moisture, reducing CO2 losses from the soil, maintaining soil structure, reducing the threat of erosion); 3) organizational (reducing the number of technological operations when growing crops).

Currently, there are approaches that make it possible to obtain promptly objective data on the agrochemical characteristics of agricultural lands. Such approaches include satellite remote sensing (ERS) (Kataev, 2017) and the use of unmanned aerial vehicles (UAVs) (Kataev, 2021). Among satellite remote sensing devices, one can distinguish optical and radar ones, which allow obtaining images with varying regularity (from a day to two weeks). However, optical satellite instruments are significantly hindered by clouds and there is a lack of available processing software. Moreover, the information obtained in this way is indirect and data verification using ground-based measurements is required. The indirectness of the results obtained suggests that the results obtained are related to soil fertility indicators by certain indices that change their values depending on the measurement conditions (solar illumination, meteorology). UAVs make it possible to carry out measurements under clouds, but, like satellites, there is no software available to agricultural workers.

The color of the soil carries important information about the chemical and physical properties of the soil (Altynov, 2010). The generally accepted practice of measuring soil color involves taking samples in the field and processing them in the laboratory using spectrometers (Orlov, 1997). This approach is accurate, but not effective in analyzing large-area CX fields in terms of time, high cost and description of the results obtained. This paper proposes a method for obtaining color characteristics of soil samples obtained directly in the field using a digital camera. It is shown that RGB values from digital images allow the construction of indices that correlate with soil characteristics. The article we bring to your attention is based on the authors' experience in image processing and conducting agrochemical soil analysis (Kataev, 2019).

An important argument in favor of introducing a system for assessing the color state of the soil in agricultural tasks is its significant soil-protective effect, which allows you to quickly detect changes in the state of the soil cover. Soils and soil cover of agricultural lands are the objects of this work in the context of studying color characteristics. The relevance of the work on studying soil characteristics lies in the fact that when monitoring the condition of the soil and especially soil degradation and a decrease in its fertility, monitoring the condition becomes important, and this requires appropriate signs, which include color signs. In addition, the problem of rational environmental management and land protection is one of the most important problems of countries around the world.

SOIL AND ITS CHARACTERISTICS

Soil is the surface fertile layer of the Earth's lithosphere, which arose from the formations of decay and weathering of rocks, as well as as a result of the vital activity of organisms. Soil is the most valuable resource that the Earth has, being the main source of food for humans, besides water. Soils across the planet vary greatly and, in addition to serving as food, play an important role in maintaining ecological balance. As a result of human activity, the state of the environment, including the soil, changes due to the occurrence of geochemical processes that are not typical for nature.

Soil is an independent natural formation of the surface layer of the earth, which is the source of 90% of the food resources of our planet, and also provides industry with raw materials. The soil consists of solid (mineral and organic particles), liquid (soil solution, water with dissolved organic and mineral compounds) and gaseous (soil air, fills pores) parts. The solid part makes up 80-98% of the mass and consists of sand, clay, and silt particles remaining from the parent rock. The content of these particles determines the mechanical composition. The liquid part contains up to 40-60%. It supplies plants with water and dissolved chemical elements. The gaseous part contains more carbon dioxide than atmospheric air and less oxygen, methane, volatile organic compounds, etc.

What is important for soils is fertility, the sum of properties that ensure the productivity of agricultural plants and the biological productivity of phytocenoses. Fertility depends on various factors: the content of nutrients (nitrogen, phosphorus, potassium, iron, manganese, boron and other micro and macroelements), moisture and acidity, agricultural technology, erosion. Soil fertility is determined by the presence of humus in it - organic substances formed in the processes of biochemical decomposition of organic residues. The distribution of soil humus is related to climate, natural areas, and the nature of the vegetation cover.

The chemical composition of the soil consists of two fractions, one of which is associated with macroelements, which are found in abundance in the soil, these are: nitrogen, phosphorus, potassium, calcium, magnesium, iron, sulfur, and the second fraction, which is contained in smaller quantities in the soil (microelements), these are: boron, molybdenum, manganese, copper, etc. These fractions provide plants with chemical elements, since nitrogen is included in the composition of proteins, nucleic acids, chlorophyll and other organic substances of plants. If there is a lack of its compounds, plants grow poorly, the leaves change color and become light green. For example, phosphorus enters the roots of plants from the soil, and if it is deficient, the leaves of the plants turn reddish-purple.

The synthesis of organic substances in plants enhances potassium, participating in the reactions of the synthesis of complex carbohydrates, and potassium deficiency is detected by changes in the edges of leaves, which become brown and curl. Calcium is also necessary for the roots of plants; if it is deficient, brown spots form on the leaves and, as a result, after some time the leaves turn yellow and die. Another component of the soil, magnesium, provides redox reactions and is included in chlorophyll and when it is deficient, yellow-whitish spots appear on the leaves of plants. With a lack of sulfur, yellowing of the upper and lower leaves occurs. Copper delays the aging of leaves and when it is deficient, the tips of the leaves turn white. Thus, the color of plants is an indicator of the condition of the soil as a source of water, micro- and macroelements.

Soils have acidity, pH value (hydrogen value). Neutral soil reaction pH = 7, pH above 7 - alkaline soils, below - acidic, which are more common. Acidic soils are: very acidic - pH 3.8 - 4.0, strongly acidic - pH 4.1 - 4.5, moderately acidic - pH 4.6 - 5.0, slightly acidic - pH 5.1 - 5.5, close to neutral - pH 5.6 - 6.9. Decreasing the pH value by one increases the acidity of the soil by 10 times! Many plants grow well in slightly acidic or neutral soil conditions. Optimal soils are from slightly acidic pH = 5.5 to neutral pH = 7. In acidic soils, chemicals are poorly absorbed, the root system develops poorly, substances harmful to plants accumulate, beneficial microorganisms that increase soil fertility do not multiply, elements turn into forms inaccessible to plants . Soil acidity is also reflected in the color and structure of the soil.

Thus, the color of the soil is related to its mineral composition and soil formation processes. One of the options for understanding soil color is the Zakharov triangle (Figure 1). Color (color) is an important morphological feature of the soil and its changes are influenced by the composition of soil-forming rocks and the type of soil formation. Soils have all colors and shades, from black to white, there are no bright greens and blues. Humic substances give the soil black, dark gray and gray color. The reddish-rusty color is the presence of various forms of iron (III) oxide, another form of iron (II) oxide leads to the appearance of bluish tones. Silica (SiO2), calcium carbonate (CaCO3), gypsum (CaSO4) and easily soluble salts color the soil gray and whitish. Black spots and layers on a reddish-brown background indicate the presence of manganese hydroxide in the soil.

Figure 1. Basic soil colors

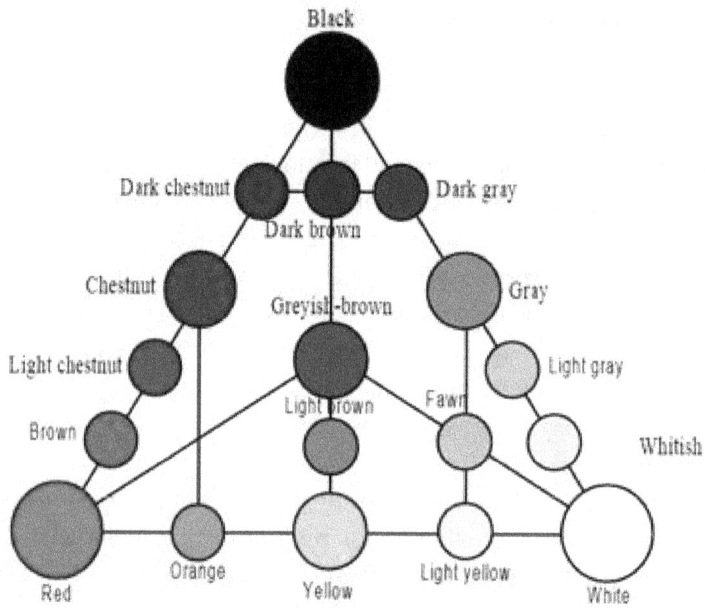

(Zacharov, 1931)

The whitish color is given by fine-grained quartz grains, and the white color is created by the accumulation of carbonates and sulfates. The color of the soil strongly depends on the degree of moisture and the light source, so the true colors are determined from samples in an air-dry state, in diffuse daylight. However, collecting samples and preparing them in a special way takes a long time and therefore there is a desire to find conditions under which understanding the color of the soil in the field will be the basis for understanding the type of soil and its structure.

The earth's surface consists of different types of soils, which differ significantly in color, structure, mechanical, chemical and physical properties. To study and practically apply information about soils, appropriate classification methods are required. Agrochemical laboratories around the world carry out a complete survey of the territory and the identification of different soil types, creating soil atlases. However, given the diversity of soil properties, they are classified (divided) into several categories: 1) according to the conditions of their formation and appearance; 2) on the mechanical and chemical properties of soils; 3) by color.

The first two methods are costly in terms of time and analysis resources, but are accurate in terms of the determined characteristic values. The last category is most convenient for conducting primary, field research, but this approach is fast, but has certain limitations. One of the limitations is that the perception of color differs from person to person and therefore the transition to a quantitative approach to color assessment using digital cameras is important for study and practical application.

Color in the field is influenced by soil moisture, determining a darker shade than dry soils of the same type; there is also a dependence of color on the time of day (the height of the Sun above the horizon), the presence of shadows from a person or nearby plants, and various other factors. To objectify data on soil color, it is necessary to measure not only images, but also knowledge of the time of day, meteorological parameters and geographic coordinates. The color of the soil is directly related to the presence of various chemical components, which are part of various oxides, hydroxides, soluble and insoluble salts.

Soils do not have a uniform structure in depth and there is a soil profile consisting of several layers that differ from each other in properties, including color. The uppermost part - the litter - has only certain soil components that are associated with organic plant residues, but is not soil. The soil itself is located in the surface horizon, which is saturated with processed organic matter with mineral deposits. This layer is followed by a subsurface horizon and then several more layers that lie quite deep. In this work, we will be interested only in the surface soil horizon, as it is the most practically significant.

The presence of different proportions of chemical elements determines the color tone of the soil, which was taken into account when compiling another color palette of soils, the Munsell atlas (Turk, 2020) (Figure 2), which has received international distribution. In this atlas, each soil color is characterized by three indicators: 1) tone (shades based on the following color components: red, yellow, green, blue and violet), 2) color intensity and 3) color purity. Colors of the plates were developed, which are compared visually with the soil sample and, by comparison, select a specific type of soil. This approach allows you to obtain a fairly accurate soil type, however, its results are subjective, since when comparing the color of a plate and a soil sample, the assessment of similarity depends on the researcher's perception of the color. In addition, the selection of paints for painting plates in a mass (printed) version leads to a distortion of the color of the plates, which affects the accuracy.

The color system and atlas of A. Munsell was introduced to the scientific community in 1905 and listed its advantages, including the following: 1) color names based on natural objects whose color often changes are replaced by a specific designation; 2) each color has its own designation and it can be written down and transmitted, which makes it possible to determine the color quite accurately; 3) the system can be expanded to accommodate other colors.

Figure 2. Basic soil colors of Mancell atlas

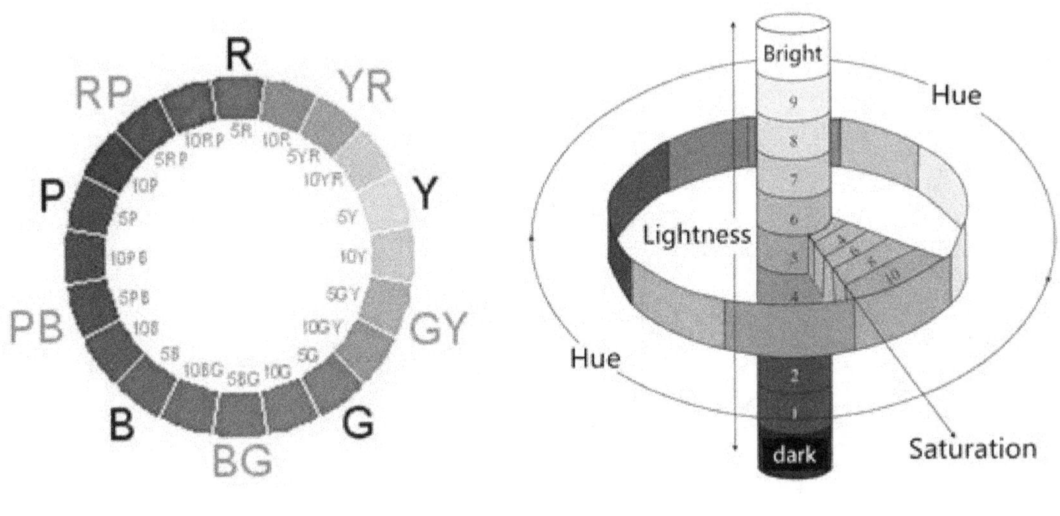

A committee of the Optical Society of America has developed the Munsell atlas into a standard system with which researchers compare many new systems and check the uniformity of color spaces. In the Munsell system, the colors of reflective objects are determined by three characteristics: 1) hue, 2) saturation and 3) lightness. In his system there are no primary colors, but there are primary color tones. He chose five primary color tones: red, yellow, green, blue and purple. Then he introduced five additional, intermediate between the main color tones and brought them into a circle, which he divided into 10 intervals of color tones, which he designated: R (red), YR (yellow-red), Y (yellow), GY (yellow-green).), G (green), BG (blue-green), B (blue), PB (magenta-blue), and RP (red-magenta). He divided the sector of each color tone into 10 parts, thus obtaining a circle of 100 color tones. Primary color tones are designated 5R, 5Y, etc. This is the middle of the interval of a given color tone. The Munsell color wheel is evenly divided and the differences between adjacent colors remain equal along the entire color wheel. For practical use, the Munsell optical system is inconvenient, since it requires a certain lighting to evaluate the color palette, which introduces some subjectivity in assessing the color of soils.

It should be noted that according to the physics of soil color formation, the most accurate source of color when using photographic technology is the reflection spectrum (figure 3). Considering that one of the most important sources of information, which is recorded by a digital camera, is the recording of the spectrum of solar radiation reflected from the surface (for the measurement method in the daytime). The structure of the soil determines the scattering pattern of radiation in different directions, which for soil is close to diffuse reflection into a hemisphere. This law is predominant for dry or wet soil due to the fact that there are gaps between small soil lumps, which form reflections in different directions. Technically accurate measurement of reflectance requires a specific technique that allows the radiation to be collected from a hemisphere. Reflectance is defined as the ratio of the intensity of reflected light to the intensity of radiation incident on the soil surface. Measuring the reflectance at different wavelengths allows one to construct the spectral reflectance. The reflection coefficient is associated with the characteristics of the soil sample being studied and weakly depends on external factors, however, such a dependence exists and varies depending on the wavelength.

Figure 3. Spectral reflectance of some surface types

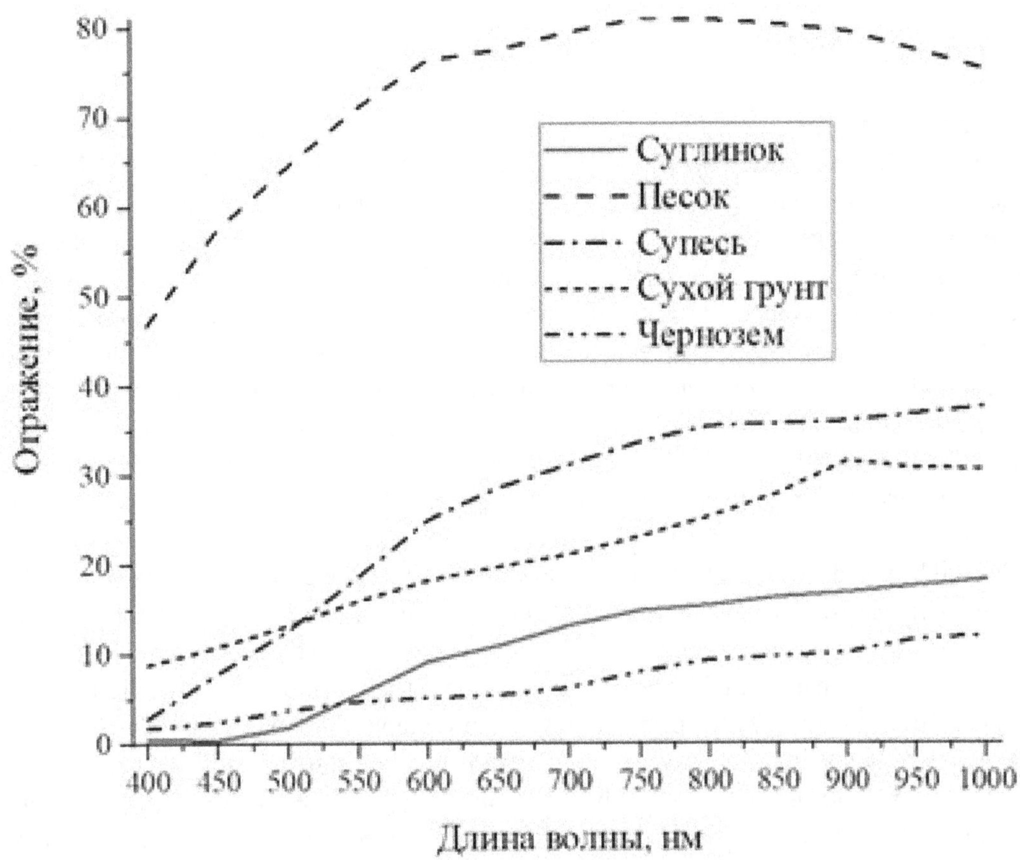

In addition, the reflection coefficient depends on the angle of the field of view of the measuring equipment, as well as the direction of the optical observation axis, for the remote measurement method. Photographic technology is associated with obtaining information in the visible range of the spectrum, which ranges from 350 to 650 nm. The relationship between the spectral reflectance and the content of various elements in the soil is determined by typically nonlinear relationships. The values of the parameters of nonlinear dependencies differ for different types of soil.

One of the first who, in the 40s, classified surface types according to the properties of spectral reflection was E.L. Krinov. He carried out mass measurements of the spectral brightness coefficients of reflected radiation in the spectral wavelength range from 0.4 μm to 2.4 μm for objects such as soil cover and soils, woody and herbaceous vegetation, water surfaces and snow (ice) cover.

It should be noted that measurements on a spectrometer differ significantly from measurements using a digital camera due to the significantly spectrally wide RGB channels and their overlap with each other. The chapter proposes to consider the features of obtaining soil color characteristics using digital RGB cameras, as well as the methodology for determining estimates of spectral reflectance coefficients from these values. The theoretical foundations of the proposed approach and the results of soil image processing for one of the regions of the Tomsk region, Russia are presented.

SPECTRAL CHARACTERISTICS OF SOIL

The spectral characteristics of soil are determined by the chemical and physical properties associated with the soil. Many of these properties can be detected not only in the visible (0.4–0.7 mm) region of the spectrum, but also in the near-infrared (0.7–2.5 mm) region of the spectrum. The color of the soil in the human visible range is associated with the presence of elements that absorb and scatter radiation of different wavelengths. Organic substances, water molecules, iron oxides, carbonates, clay, and minerals are the main chemical components that affect the color of the soil (Mikhailova, 1986).

Digital cameras have several advantages over other types of sensors as they allow the transfer of technologies from a variety of remote sensing applications. However, they do not require the expensive need for calibration and precise knowledge of the physical characteristics of the scene being measured (terrain, vegetation types, etc.). Digital cameras are used to measure the color of food surfaces (Lakiza, 2015), determine the ripening or diseases of plants (Shumilov, 2018), search for weed habitats (Lunev, 2009), etc.

It should be noted, however, that when using spectrometers operating in the range of 0.3-2.5 μm, it is possible to evaluate reflection features unique to each soil type at a quantitative level (Lesaignoux, 2010). However, using a digital camera, based on only three wide RGB bands, spectral indices can be obtained that allow one to evaluate individual soil characteristics (Kirillova, 2017). Thus, digital cameras offer a new and simple opportunity to characterize soil from samples directly in the field (Ben-Dor, 2009).

As with satellite or aircraft remote sensing instruments, care should be taken when analyzing resulting images from digital cameras. Typically, the feature of each image is associated with an uncontrolled light source, which is the Sun, changing the brightness of images taken at different times of the day and year. Accurate assessment of soil characteristics requires calibration, which requires accurate knowledge of the spectral transmittance curves of each RGB channel, however, these are often not supplied by digital camera manufacturers. Another obstacle for digital cameras is scattered radiation, the brightness of which falls on the lens at different angles and, in addition to the main one, is recorded in the image. These aspects require the development of a clear methodology for obtaining soil images on the ground or using UAVs.

PRECISION AGRICULTURE

Precision agriculture began its development in the early 1980s, when a variety of devices began to be used in the practice of various types of work. Currently, for precision farming tasks, satellite data (images and signals), aircraft, unmanned aerial vehicles, and Internet of Things components are used, where sensors exchange data with each other. Understanding plant growth and other agricultural tasks contributed to the development of business ideas and, after the 1990s, agriculture began to develop on a commercial basis.

Precision agriculture is associated with several stages in development, which are associated with the emergence of a large volume of data and methods for obtaining, storing, processing, analyzing and visualizing it. These stages are still developing at the technical and information level in modern times. One of the currently poorly solved problems is the problem of monitoring the surface of agricultural fields, which occupy a large space and traditional inspection methods do not allow covering them entirely. Only satellite and, more recently, unmanned aerial vehicles can successfully solve such problems. Computing systems and their development make it possible to set tasks for collecting and processing monitoring data in near real time. One of the most important and most common sources of data for agriculture is images. The development of photographic technology has led to an increase in the accuracy of images, the number of pixels in the image has become larger, and additional spectral channels (multispectral cameras) have appeared, which makes it possible to evaluate the properties of soil and vegetation in agricultural fields at high spatial resolution.

The large volume of data and related information received from agricultural fields (for example, meteorology) makes it possible to obtain more accurate data processing results and, therefore, successful application in the tasks being solved. Successful implementation of tasks allows you to obtain a better harvest in terms of volume and quality, increase labor productivity and preserve the environment. Monitoring of agricultural fields allows for spatio-temporal processing and analysis of the results obtained.

Existing modern tasks of agricultural development require that production be more economical, high-quality and environmentally friendly products be produced, which, in turn, leads to the need to optimize technological systems and processes. A very important stage in agricultural production is understanding the possible results of harvesting, which requires achieving high precision in field and logistics operations. Therefore, fulfilling all these requirements requires increasing automation and the involvement of scientific research. One of the areas of scientific application of knowledge in certain tasks is computer vision technology, despite the fact that this technology remains expensive and difficult to use in agricultural practice. Another important component of the implementation of computer vision requires highly qualified agricultural workers.

Since field conditions in agriculture are complex and variable, computer vision equipment (digital cameras) must meet the appropriate standards. At the same time, the software must solve complex problems of data processing and analysis in such a way as to minimally affect the knowledge of agricultural workers in physics, chemistry and mathematics. A computer vision system should have minimal technical configuration and an intuitive user interface for agricultural workers.

Modern agricultural control usually includes a visual assessment of the condition of the land and crops in order to obtain sufficient information for making management decisions (for example, the time of sowing, harvesting, etc.). During the control, some quantitative measurements are also carried out, which do not evenly and simultaneously cover the required territory. Vision systems allow access to a large portion of the field of interest, potentially allowing automated condition assessment at high frequencies (up to several times per day). This ability to cover large areas in a relatively short time, coupled with the introduction of appropriate equipment (airborne carrier (e.g., UAV), technical knowledge, storage and processing) can potentially allow agricultural producers to carry out operations more efficiently.

Vision, being a potentially low-cost technology with a low risk of failure, is suitable for the agricultural producer. The resulting video information, supported by information received from a large number of sensors located in the fields, allows you to obtain sufficient information for timely decision-making. Let us note that the scientific contribution of various directions in the processing and analysis of incoming information plays a huge role. As a result, the application of the technology in question allows us to obtain a guaranteed yield of agricultural products, depending on natural factors.

In its simplest form, a conceptual computer vision system may consist of a single monochrome or color digital camera with image analysis algorithms designed to identify and monitor the crop of interest under environmental conditions (e.g., at a minimum, taking into account lighting and meteorological parameters). In a more complex form, it is possible to use stereo vision technologies, multispectral imaging, and laser sensing, which could potentially contribute to a more accurate determination of the parameters of agricultural objects. The information obtained using sensors is of interest in obtaining the data necessary to reduce the dependence on the influence of environmental factors on image quality. The advantage of this approach is simpler and more reliable data processing for the entire life period of an agricultural facility (from field preparation at the planting stage to harvesting and autumn field preparation). The design of a computer vision system for measuring plant characteristics depends on many factors, such as the scale of the plant and the measurement space (e.g., laboratory, greenhouse, or field).

Imaging systems designed to measure plants in agricultural fields or greenhouses require high-resolution color or multispectral images. The objectives of such systems are related to monitoring, analysis and forecasting of crops and evaluation of decision-making during the agricultural season. One of the important challenges of measuring during a growing season is obtaining a representative sample of virtually every plant in the field, which involves obtaining a significant amount of data. The latter fact is not a problem at present, since there are software and hardware capabilities for this. The resulting data set makes it possible to use modern artificial intelligence approaches that do not create computational difficulties and complexity, while ensuring automation, speed and quality of data processing. However, the external environment presents difficulties (changes in meteorological parameters, solar illumination, cloudiness, precipitation, drought, insects, etc.) that make such automation only partially possible. Although, solving automation problems, for some conditions, will already be a great achievement (the main elements of precision agriculture). The goals of precision agriculture are to maximize profits, rationalize agricultural resources, and reduce environmental damage by adjusting agricultural practices.

The development of computer vision technologies is desirable to occur while monitoring the plant's environment (e.g., laboratory and greenhouse), which eliminates many of the random variations that complicate systems for actual plant monitoring on the ground. In laboratory conditions, the regulation of lighting and meteorological parameters, the positioning of free-standing individual plants can be reliably controlled, so the disadvantages of variable natural sunlight, irregular spacing between plants, and varying soil quality are minimized. As a result of this, mathematical models can be built that make it possible to detect patterns of plant growth under various environmental conditions and, therefore, can be successfully transferred to the outdoor, real environment. Software-algorithmic technical vision technologies can emphasize features of interest (select the plant itself against a complex background and determine its features (color, shape, spectral)) in the resulting image, and significantly increase the reliability of systems in a real environment.

The article discusses literature data on methods for studying plant growth based on computer vision technologies. Five measurement methods can be distinguished, namely: - monochrome or color image (RGB); - stereo vision (a pair of RGB images); - depth camera (RGBD); - multispectral image; - laser probing.

Nowadays, images are actively used everywhere. New films are constantly being shot, sports matches are being recorded, and information from CCTV cameras is being stored. Thanks to modern technology, every phone has a photo and video camera, which allows anyone to create a large number of photos and videos almost every day. When creating images, a person pursues a specific goal: to save them as a memory, put them on public display, or use them to solve some problems specific to a particular area. In order for images to be of any use, they must be processed (enlarged, cropped, selected objects, etc.).

Image processing refers to any processing of information whose input data is photographs or video frames. The results of data processing can be both images (for example, preparation for television broadcasting, for printing replication, etc.) and any information of interest about the data (for example, counting the number and type of cells in the microscope field, text recognition, etc.). Processing is required not only for static two-dimensional images, but also for images that change over time (video files).

In the modern world, images and computer vision are used in many areas of life (for example, medicine, security, etc.). The most important thing today is object recognition in a multimedia video stream. Thanks to easily accessible technologies, people often look at the world not with their own eyes, but through the prism of video cameras, which allows them to monitor several scenes at the same time. So, for example, a security guard working in a large enterprise, who needs to monitor the entire territory and monitor any changes using CCTV cameras, can sit at his workplace and watch all the scenes on one screen. But the images contain a huge amount of unnecessary detail that distracts from the true purpose of observation or interferes with the analysis carried out on these images. There are very few analogues of a software product that allows you to select the necessary objects from the unnecessary background. Due to the fact that this area is very young and rapidly developing, such analogues have a large number of shortcomings and shortcomings. That is why image processing, and in particular, isolating an object from a video stream, is a relevant area of activity.

An image is information. It conveys, carries or communicates information about an object, image or phenomenon. This information is obtained using physical laws that make it possible to obtain an image (optical image) or using the results of the work of the image creator (stage image, sculpture, drawing, painting). Basically, an image means a work in graphics or painting. The source of the optical image is a video camera. Video cameras are generally divided into three categories: 1) Professional (cameras for professional use have significant weight, they can be either portable or stationary); 2) Household (cameras intended for people who do not have professional skills, due to their ease of operation and compactness); 3) Special (chambers specialized in a certain area: medicine, security, etc.

A webcam is a digital video or photo camera that allows you to capture images in real time for later use on the Internet. To communicate between the webcam and the web server, you must connect the camera to your computer or use a camera that allows access to the web server. Delivering images over the Internet, webcams upload them to a web server at regular intervals, continuously or on demand. Webcams are used for various purposes. For example, cameras used in security companies may include various functions and devices (connection of external sensors, motion detectors, etc.). Using special programs for organizing video communications, you can conduct video conferences, which allows you to be present in real time at a conference, meeting, or see a person while on the other side of the world. In addition to video conferencing, webcams are also used as a means by which some Internet users are able to see the world through the cameras of other Internet users.

A digital camera is a device that is a type of camera. It contains a photosensitive material (matrix or several matrices), consisting of individual cells with three photocells with a color filter (red, green and blue). The signal from the photocells is stored, processed and presented in the device itself in digital form.

In the modern world, the border between photographic equipment and video equipment can be said to be arbitrary. A large number of video recording equipment have the function of obtaining static images, and digital cameras, on the contrary, have the function of recording sound and video and outputting the video signal in television format. As a result, we can conclude that the choice of digital equipment depends not on the functions of the camera, but on the task assigned by the operator.

Digital cameras can be divided into several classes: 1) Cameras with built-in optics: 2) Ultra-compact or miniature. They are distinguished not only by their small size, but often by the absence of a screen and viewfinder; 3) compact. They are characterized by low weight and size. Due to the small size of the matrix, the camera has low sensitivity or high noise levels. These cameras also have no or insufficient flexibility for manual exposure settings; 4) built into other devices. These cameras do not have their own controls; 5) digital SLR is a camera that uses a viewfinder to select the subject. The optical design of the viewfinder consists of a mirror that redirects the light flux from the lens to the eyepiece or onto ground glass.

The photosensitive element is a matrix, which consists of photocells. In the first cameras of this type, the LCD screen was intended to access the camera menu and to view captured frames. When shooting in crowds above heads or at ground level, such cameras caused significant inconvenience for photographers due to the fact that framing was only possible through an optical viewfinder. Therefore, DSLR camera owners had to spend a lot of money on inconvenient viewfinder attachments or shoot at random, while many compact camera owners could use a rotating LCD screen in such situations. In the modern world, almost all major manufacturers have in their product line SLR digital cameras with the ability to view on an LCD screen.

A color space is a color representation model that is based on the use of color coordinates. It is constructed in such a way that any color is represented by a point that has certain coordinates. Usually, one set of coordinates is matched to one color, but there are exceptions. For example, in the CMYK model or when a color tone is encoded (the tone data is "loop" and the minimum and maximum values are the same). Color spaces are described using rules for constructing colors and sets of color coordinates. In turn, the number of coordinates determines the dimension of space. There are a huge number of color spaces of various dimensions - from one-dimensional (they describe only a monochrome image), to six and ten-dimensional (for example, CMYK space).

RGB (Red, Green, Blue - red, green, blue) is an additive color model that describes the method of color synthesis. When choosing primary colors, we relied on the physiology of color perception by the retina of the human eye. The RGB color model is widely used in modern technology.

The model is called additive because colors are created by adding them to black. That is, if the color of the screen that is illuminated by a color spotlight is denoted in RGB as (r1, g1, b1), and the color of the same screen that is illuminated by another spotlight is (r2, g2, b2), then when illuminated by two spotlights, the color screen will be denoted as (r1+r2, g1+g2, b1+b2). Images in the RGB color model consist of three channels. To obtain the required color, it is enough to combine the primary colors (red, green, blue). So, for example, mixing blue and red will produce magenta, mixing green and red will produce yellow, and mixing green and blue will produce cyan. Mixing all three primary colors will result in white.

The HSL color space (Hue, Saturation, Lightness) is based on an intuitive way to describe color (that is, it operates with the concept of basic hue, color saturation and color brightness). Options: 1) Hue is a color. Determined using degrees from 0 to 360 based on the colors of the rainbow ring; 2) Saturation is saturation. Varies from 0 to 1; 3) Lightness is lightness. Varies from 0 to 1. The parameters hue, saturation and lightness are quite convenient to use when constructing a spectrum, since from them it is easy to obtain the main parameters of the function that describes the spectrum.

MONOCHROME OR COLOR IMAGE CAPTURED BY RGB CAMERA

In its simplest form, a plant vision system consists of a single camera capturing a natural scene so that the resulting image displays the same information that a person observes. It follows that there is scope for algorithms to analyze images and extract objects (plants) in an image that can be identified by humans. This entails the implementation of algorithms that allow color and shape to be determined in various environmental situations.

Common applications of plant vision systems using color imaging include counting plant numbers, estimating shape and biomass, determining plant health, and identifying species. To solve these problems, it is necessary that the parts of the image (pixels) containing plants are reliably separated from the background pixels. Background pixels also require processing and extraction of soil characteristics (texture, color, etc.), which may be associated with agrochemical monitoring data. Segmentation can be achieved by various methods: based on color indicators, texture, due to additional mathematical transformations, etc. Shape and size algorithms can be used to describe individual leaves or the plant itself as a whole.

When segmenting plants, a problematic situation arises when several small plants are usually extracted as a single object, but for large pieces, the leaves of neighboring plants may interfere. Individual plant leaves are potentially isolated from a whole plant using methods based on shape (e.g., using a watershed algorithm), color, or texture. The segmentation task, at different stages, includes methods such as automatic thresholding and edge detection. However, these approaches are highly dependent on specific meteorological conditions and efficient techniques are required to reduce the complexity of image analysis in these cases.

Modifying the processing technique of a standard digital camera provides another alternative to segmenting vegetation in an image. To do this, you need to imagine the camera as a spectral device operating in the visible region of the spectrum. Given the location of the RGB spectral channels, it is possible to highlight vegetation in an image using NDVI (Normalized Difference Vegetation Index). Because vegetation has higher reflectivity than soil at near-infrared wavelengths, it is possible to distinguish the plant from the soil and other background objects in the image. The use of an additional spectral channel located in the near-IR region of the spectrum allows the segmentation procedure to be performed more confidently. Calibration of such a system can be performed in laboratory conditions. Identification of various plants and their condition is also possible in the case of determining the vegetation index, since it is known that when the condition of plants changes, NDVI decreases.

IMAGE PROCESSING AND ANALYSIS SOFTWARE

Digital image processing is performed using a computer to manage flows of information (images, documents, digital data), process and analyze the resulting information, or visualize it. Image processing can consist of four stages: 1) image improvement, 2) extracting information from the image, 3) segmentation, classification and identification, 4) visualization.

Image enhancement is typically applied to digital images to correct problems such as brightness correction, poor contrast or noise, or geometric distortion. To improve the image, procedures such as morphological operations, filters and pixel-by-pixel operations are used. Statistical characteristics of the image are considered, such as mean, standard deviation and dispersion, or more complex ones, for example, wavelet, Gabor, etc. Image segmentation is one of the first steps towards understanding the content of the image by identifying different objects. Digital image processing refers to finding, extracting and analyzing information about objects in an image. Image analysis mainly focuses on using various image processing techniques such as noise filters to enhance the contrast or quality of an image. Obtaining the visual part is associated with the use of 6 degrees of freedom, in space (x, y, z) and orientation (θ, ψ, φ) of the plant in the agricultural field.

The use of technology for agricultural development is growing rapidly and is widely practiced in many parts of the world where agriculture serves as a major source of income and livelihood. Technology is referred to as the application of scientific knowledge to practical purposes or the use of machinery to better facilitate a process and reduce the intensive manual labor required for agricultural production.

Technological applications in the agricultural sector eliminate the stress and tedious manual labor involved in farming. It also increases profitability and helps in proper management of farm entry into production. The benefits of using modern technology to increase agricultural productivity: more people will eat better, while eradicating hunger and reducing malnutrition from increased production; improved nutrients in food; reduction of physical pressure on the environment; improving quality of life and living standards as food costs decrease; increased savings since most people spend most of what they earn on food.

Space technology plays a key role in its use to improve agricultural productivity through precision farming and aerial imagery of farms. Precision agriculture deals with obtaining accurate readings of certain specific events, such as: weather forecasts to help prepare lands for farming; and Global Positioning System (GPS) monitoring, as well as the use of unmanned aerial vehicle (UAV) systems that can provide GPS guidance to farmers to place applicators (variable speed technology) by applying specific amounts of chemicals in specific locations. Precision agriculture is a viable solution for agricultural production because: 1) the continuing reduction in the amount of arable land; 2) projected increase in world population; 3) reducing the cost of agriculture by avoiding losses when using chemicals or water. In this regard, the information obtained from images about the state of the soil will be very useful in all remote monitoring and ground-based survey approaches.

PROPOSED APPROACH

The purpose of optical sensors operating in the visible range of 0.35-0.7 microns, which includes a digital camera, is to register solar radiation reflected from the earth's surface (Kobernichenko, 2016). However, the DN values recorded by the sensor do not coincide with the reflectivity of the Earth's surface. This is due to sensor parameters such as color processing algorithms, camera settings, vignetting, sensitivity (depending on flight altitude), etc. This is also due to changes in such factors as the azimuth and altitude of the sun, atmospheric conditions (fog or aerosol filling of the atmosphere (for example, smoke from fires, etc.)), and the reflectivity of the types of surface being studied.

For this reason, it is necessary to convert DN values to surface reflectance. This process is known as radiometric calibration (Mendoza, 2006) and is performed during the pre-processing of the measured images. The measured value depends on the illumination of the surface by solar radiation passing through the atmosphere (Rice, 2006):

$$E(\lambda,\varphi) = Eo(\lambda)\bullet\cos(\varphi)\bullet\rho(\lambda,\varphi)Surf(h,\alpha,\beta)T1(\lambda,\varphi)\bullet T2(\lambda), \quad (1)$$

$$DNi=c1+c2A\bullet\omega\bullet\tau \int(E(\lambda)\bullet Ai(\lambda)\bullet d\lambda)/\int Ai(\lambda)\bullet d\lambda, \quad (2)$$

where DN is the measured value of reflected solar radiation from the surface in spectral channels (i=R,G,B) for each image pixel, $E(\lambda)= E1(\lambda)+E2(\lambda)$ is the surface illumination at wavelengths λ in the visible spectral range due to the effects of reflection from the surface and scattering in the air and $Ai(\lambda)$ – hardware spectral function for each channel (i=R,G,B), A,ω,τ – detector area, field of view angle, exposure time, φ – zenith angle of illumination by the Sun, h – altitude above sea level, (α, β) – angles characterizing the inclination and rotation of the relief area, T1, T2 – transparency of the atmosphere on the optical path throughout the entire thickness of the atmosphere at the illumination angle φ and the optical path of the reflected from the radiation surface to the UAV, c1, c2 are the linear signal conversion constants. For each digital camera, the hardware spectral function is unique and depends on the technical and optical characteristics (Lin, 2004).

Figure 4 shows the spectral reflectance of some types of surface (Altynov, 2010).

Figure 4. Spectral functions of channels for two sensors (digital cameras)

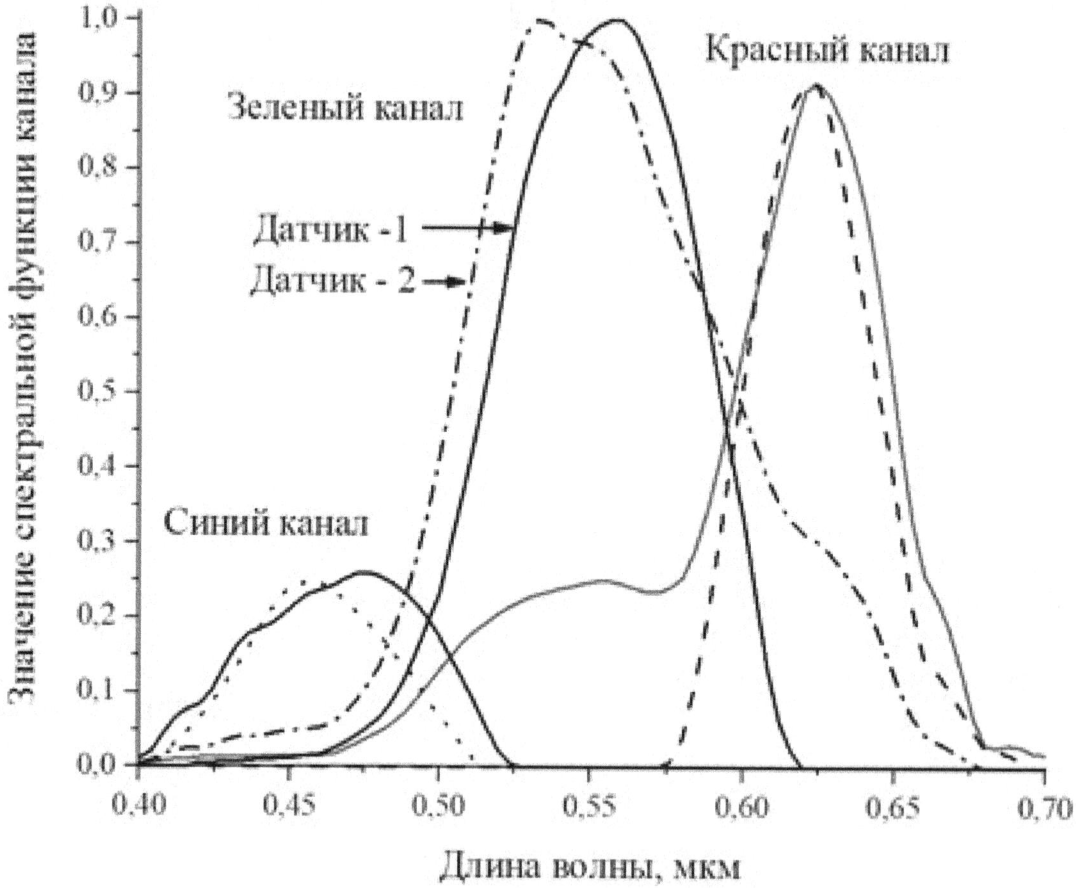

The surfaces of agricultural fields (vegetation and soil) reflect sunlight anisotropically (the reflected radiation changes depending on the viewing angle and the geometry of the illumination). In addition, the following factors play a significant role in reflection: the type of soil and vegetation (density of soil and foliage, water and nitrogen content in leaves, soil structure (humidity and roughness). This information is recorded in the RGB images of the UAV and can be used to highlight and use in practice of agricultural work. Extracting reflectivity information from RGB images is a complex process that is influenced by lighting conditions, scattering processes in the atmosphere (haze and cloudiness) and objects adjacent to the field (forest or buildings), as well as the topography of the field.

In addition, shading areas may form on the field in which the components E1 and E2 have different values from the place where direct illumination is observed. Therefore, obtaining important quantitative information from an RGB image requires conducting comprehensive studies on the impact and significance of these factors under various conditions. This article allows us to only partially cover the numerous questions that arise due to the complexity and scope of the topic under consideration.

This article presents a simple method for using a digital camera to measure soil color from samples taken directly from the field. The methodology was developed taking into account the following criteria: 1) Images must be taken with a standard digital camera in natural daylight; 2) The tilt of the camera to the surface must be perpendicular; 3) The distance from the camera to the soil is approximately 1 meter; 4) The sample is obtained by removing the top layer of turf to expose only the soil; 5) For statistics, at least three photographs of the same sample are taken.

Soil samples (selected options are shown in Figure 4), which met the above criteria, were obtained for different areas of an agricultural field in the Tomsk region. A total of 70 soil samples were obtained. A 12-megapixel Canon digital camera was used to acquire images under diffuse daylight conditions where the Sun was not directly illuminating the soil samples. Figure 5 shows two images of the same type of soil, taken in the central and edge areas of the field. It can be seen that the soil in image 3a is drier than in Fig. 3b. In addition, a different soil structure (clumpiness) is noticeable, which is most likely determined by soil moisture.

Figure 5. Obtained soil samples for two neighboring fields in the Tomsk region (soil type - dark gray thick forest (a, b - two different places of the field, center and edge) 07/30/2019)

RGB imaging is concerned with the measurement of reflected incident solar radiation on a surface (ground, plants and man-made objects). Based on the image channels, it is possible to calculate indices, by the value of which it is possible to identify individual types of surfaces and study their color characteristics independently. The quality of the image in the visible range of the spectrum strongly depends on the amount of illumination (solar zenith and azimuthal angle), the state of the atmosphere and the topography. Thus, in order to express the true color of the soil, it is necessary to correct the images, taking into account expressions (1) and (2).

In this study, a simple and effective color correction method based on the "gray world" technique was used.:

$$Qi = \sum_x \sum_y E(x,y)/(m \bullet n), \qquad (3)$$

where Qi = (i=R, G, B) are the average values for the R, G and B image channels, respectively, m, n are the size of the image in question (pixels) and (x,y) are the pixel coordinates.

Under standard lighting conditions, the average R, G and B value from the gray card should be 128 respectively; therefore, the registration coefficients R, G and B were R'=128/R, G'=128/G', B'=128/B', respectively. Thus, the R, G and B values of each pixel after registration can be expressed as follows:

$$Q0i(x,y) = Q'i(x,y) \bullet Qi(x,y), \qquad (4)$$

where Q0={R0, G0, B0} are the adjusted values of R, G and B, respectively.

Thus, information about the color of each pixel for the entire color imagc obtained in natural light conditions can be recorded using equations (3) - (4), which were subsequently used to calculate color indicators.

Figure 6. Histograms of brightness distribution in each channel for soil images shown in Figures 5a and 5b

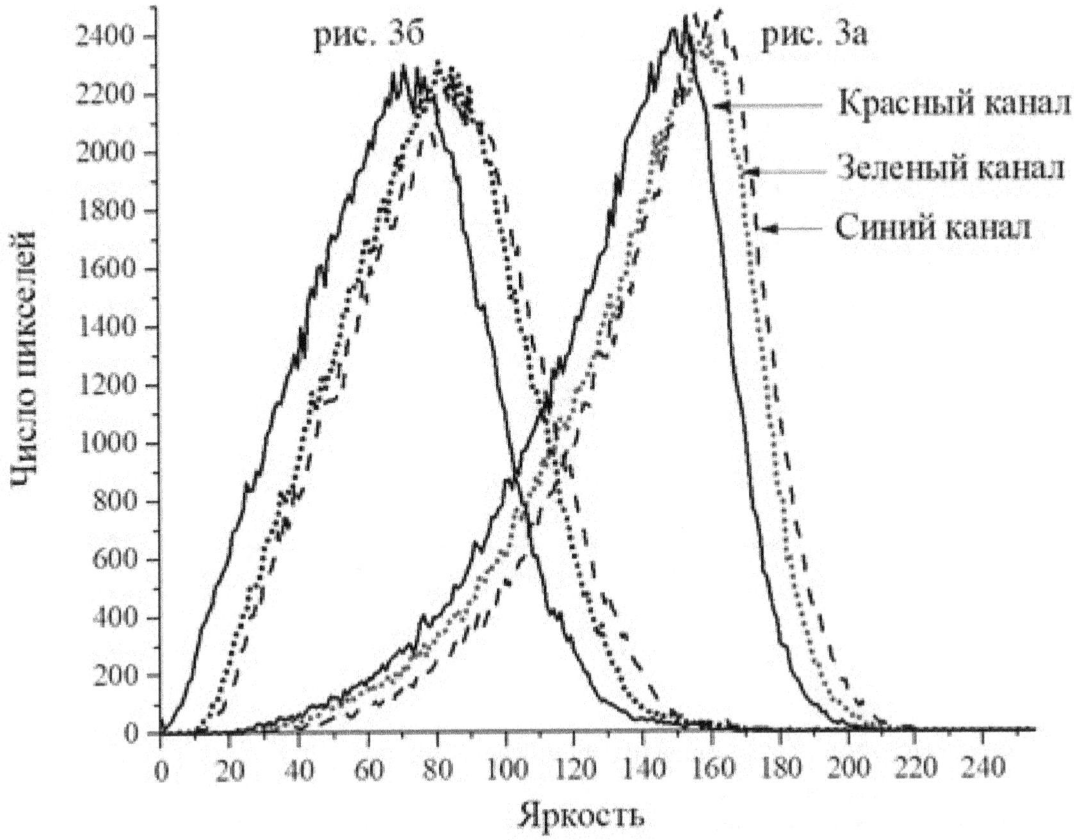

Figure 6 shows that the histograms of brightness distribution in each channel of images 5a and 5b differ significantly from each other. The shape of the histogram for soil image 5a is asymmetrical, as for image 5b. These differences allow us to speak about the diagnostic capabilities of soil using RGB images. One of the options for comparing histograms is to find their parameters.

Figure 7. Fitting an asymmetrical contour to the green channel brightness histogram (Figure 6)

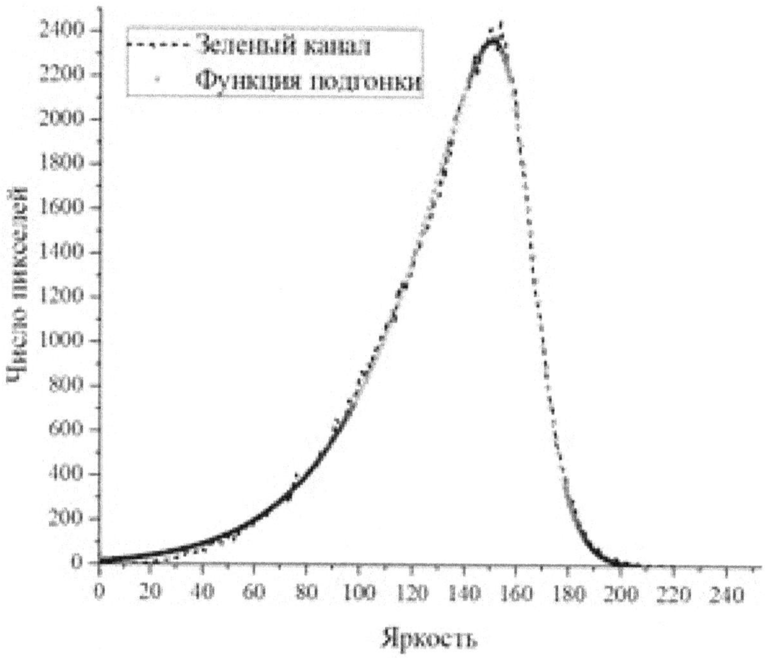

An asymmetrical contour is written in the following form:

$$Y(x)=y0+A*(1/(1+\exp(-(x-xc+w1/2)/w2)))\bullet(1-1/(1+\exp(-(x-xc-w1/2)/w3))), \quad (5)$$

where y0 is the background substrate parallel to the x-axis, A is the amplitude of the curve at the maximum point, xc is the maximum position, w1 is the displacement, w2 is the contour width on the right, w3 is the contour width on the right, at half the amplitude.

From Figure 7 it can be seen that fitting the asymmetric contour to the brightness histogram of the green channel of the image is successful and the accuracy is R2=0.99. The results of fitting an asymmetrical contour to all brightness histograms shown in Figure 5 are shown in Tables 1 and 2.

Table 1. Results of fitting the asymmetric function into the contour of the brightness histogram of each channel for Fig. 6a (soil type - deep dark gray forest)

Channel	**Red**	**Green**	**Blue**
Parameter	Рис. 3а		
Xc	162,52 ± 1,1	155,07 ± 3,7	170,94 ± 1,9
W1	19,98 ± 1,6	17,46 ± 1,8	3,85 ± 0,5
W2	28,32 ± 0,5	27,62 ± 0,5	27,47 ± 0,6
W3	6,59 ± 0,1	6,67 ± 0,1	7,41 ± 0,1

Table 2. Results of fitting the asymmetric function into the contour of the brightness histogram of each channel for Figure 3b (soil type - dark gray forest powerful)

Channel	Red	Green	Blue
Parameter			
Xc	68,51 ± 0,2	82,41 ± 0,2	79,03 ± 0,2
W1	15,94 ± 1,2	17,79 ± 1,4	17,34 ± 1,2
W2	15,46 ± 0,4	15,11 ± 0,4	14,93 ± 0,4
W3	10,13 ± 0,2	11,30 ± 0,3	10,24 ± 0,3

From tables 1 and 2 it is clear that for both images 3(a,b), the brightness histograms of the RGB channels are asymmetrical and the accuracy of determining the parameters is high, which indicates the possibility of using these parameters to solve the problem of recognizing soil types.

Figure 8. Obtained soil samples for two neighboring fields in the Tomsk region (soil type - light gray thick forest, a, b - two different places of the field, center and edge) 07/30/2019

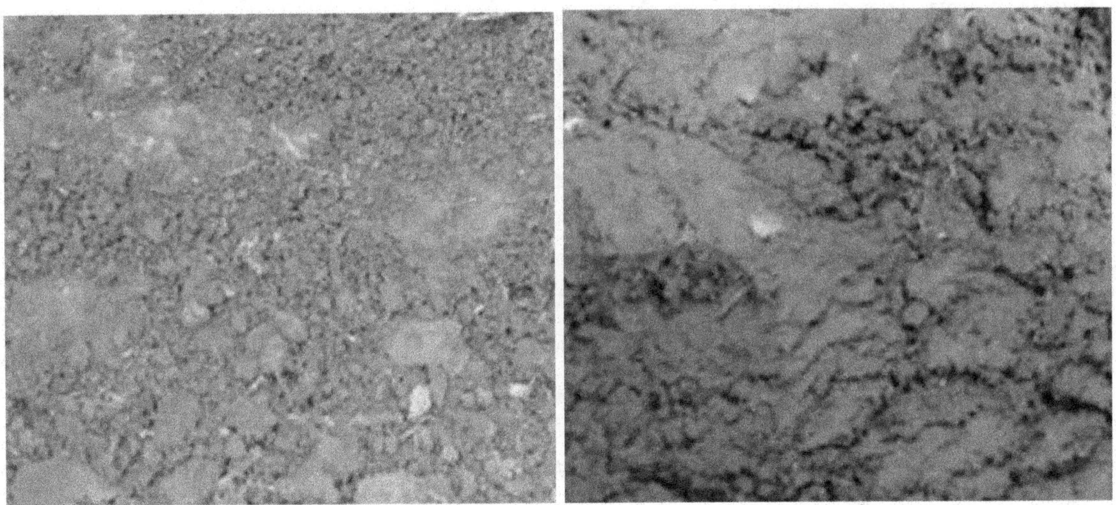

a b

Table 3. Results of fitting the asymmetric function into the contour of the brightness histogram of each channel for Fig. 8a (soil type - light gray forest rich)

Channel	Red	Green	Blue
Parameter			
Xc	145,69 ± 0,2	137,45 ± 0,3	152,39 ± 0,2
W1	0,0	0,0	0,0
W2	18,42 ± 0,2	18,10 ± 0,2	18,26 ± 0,2
W3	5,57 ± 0,1	5,73 ± 0,1	5,78 ± 0,1

Table 4. Results of fitting the asymmetric function into the contour of the brightness histogram of each channel for Figure 8b (soil type - light gray thick forest)

Channel	Red	Green	Blue
Parameter			
Xc	120,59 ± 0,2	116,17 ± 0,3	126,59 ± 1,4
W1	5,41 ± 0,8	6,19 ± 0,9	5,3 ± 0,8
W2	25,56 ± 0,3	26,18 ± 0,3	27,02 ± 0,3
W3	6,25 ± 0,1	5,36 ± 0,1	6,25 ± 0,1

From tables 3 and 4 it is clear that for both images 8(a,b), the brightness histograms of the RGB channels are also asymmetrical and the accuracy of determining the parameters is high. It is important to note that the values of the soil parameters of the dark gray thick forest soil and the light gray thick forest soil differ from each other, which makes it possible to perform the procedure for recognizing soil types using the usual threshold method.

Calculation of soil color indicators

Saturation index SI=(R-B)/(R+B)

Hue index HI=(2 R-G-B)/(G-B)

Color index CI=(R-G)/(R+G)

Table 5. Values of soil color indicators

	SI	HI	CI
Таб.1	-0,025	-0,061	0,024
Таб.2	-0,071	-7,225	-0,092
Таб.3	-0,023	-0,103	0,029
Таб.4	-0,024	-0,151	0,018

Figure 9. Mutual luminance curves between R-G and B-G channels for the soil image shown in Figure 8a

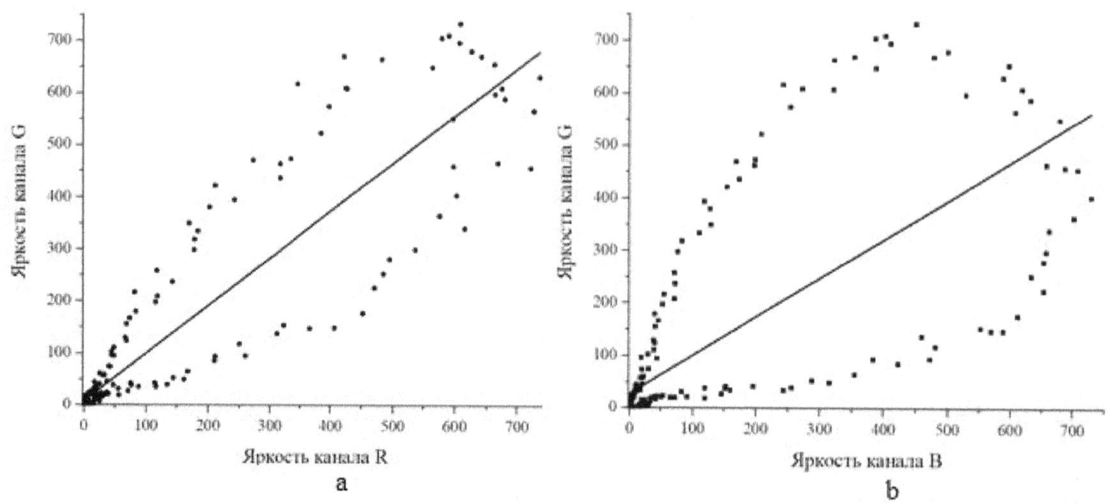

Intercept 27,31481 ± 9,30634 Intercept 8,96912 ± 5,69026
Slope 0,73635 ± 0,04259 Slope 0,91343 ± 0,02608

Figure 10. Mutual luminance curves between R-G and B-G channels for the soil image shown in Figure 8b

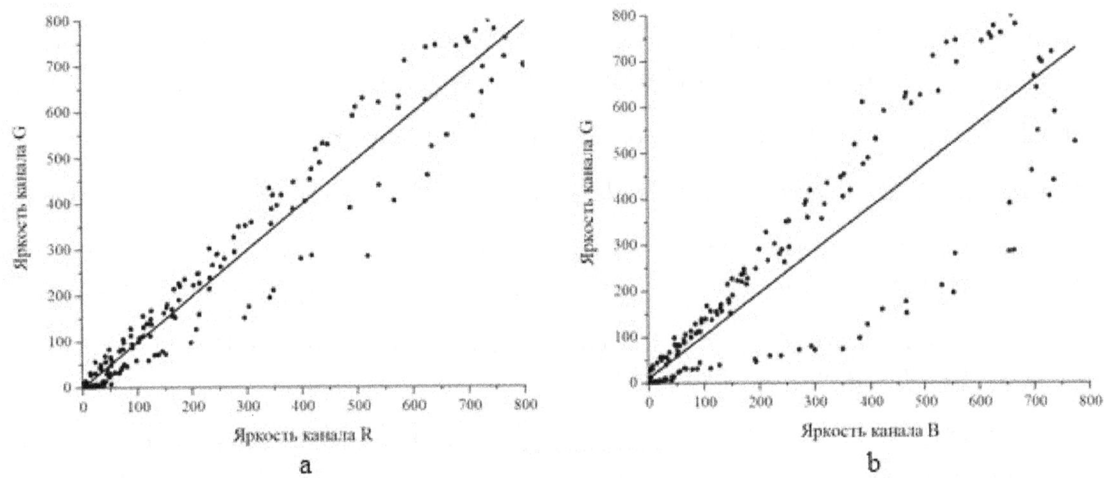

Intercept 0,64735 ± 3,68286 Intercept 11,3443 ± 7,53712
Slope 0,99586 ± 0,01362 Slope 0,9274 ± 0,02806

It is clearly seen from Figures 8(a,b) and Figures 9,10 that the shown mutual brightness curves differ significantly from each other not only in appearance, but also in digital indicators, which makes it possible to construct threshold algorithms for distinguishing these types of soil.

CONCLUSION

It is noted that soil is the most important element of agriculture, which contributes to the production of the bulk of natural food. The work provides information about the characteristics of the soil and the relationship of these characteristics with color. Among all the methods for analyzing soil characteristics, spectral and color characteristics have been studied for a long time, about 80 years, and with the advent of the first digital cameras, this method is only beginning to develop extensively. Knowledge of the color differences between soil types makes it possible to develop methods for identifying soil in an image (Lefevre C., 2017, Lal R., 2004, Turk J., 2020). The work proposes different methods for analyzing color differences between soil types and shows the features of the approaches and the results obtained, which clearly demonstrate the capabilities of the approaches.

REFERENCES

Altynov, A. E. (2010) *Spectrometry of landscapes*. Moscow: Publishing house MIIGAiK.

Altynov, A.E., Malinnikov, V.A., Popov, S.M., & Stetsenko A.F. (2010). Spectrometry of landscapes, M.: Publishing house. *MIIGAiK. UPP Reprography*, 120.

Androkhanov, V. A., Kulyapina, E. D., & Kurachev, V. M. (2004). *Soils of technogenic landscapes: genesis and evolution of Ros. acad.* Institute of Soil Science and Agrochemistry; Novosibirsk: Publishing house.

Andronikov, B.JI. (1979). *Aerospace methods for studying soils*. M. Kolos.

Aparin, B. F. (2008). *Evolution of soils and soil cover of reclaimed lands*. St. Petersburg Publishing House.

Ben-Dor, E., Chabrillat, S., Dematte, J.A.M., Taylor, G.R., Hill, J., Whiting, M.L., Sommer, S. (2009) Using imaging spectroscopy to study soil properties. *Remote Sensing of Environment*. Research Gate.

Bowers, S. A., & Hanks, R. J. (1965). Reflection of radiant energy from soils. *Soil Science*, 100(2), 130–138. 10.1097/00010694-196508000-00009

Colomina, I., & Molina, P. (2014). Unmanned aerial systems for photogrammetry and remote sensing: A review. *ISPRS Journal of Photogrammetry and Remote Sensing*, 92, 79–97. 10.1016/j.isprsjprs.2014.02.013

Dobrovolsky, G. V. (2002). *Soil degradation and protection*. Moscow: Moscow State University Publishing House.

Dobrovolsky, G. V., & Urusevskaya, I. S. (2015). *Geography of soils*. Moscow State University.

FAO. (2006) *World reference base for soil resources 2006: a framework for international classification, correlation and communication*. FAO, 145.

Godin, V.V., Belousova, M.N., Belousov, V.A., & Terekhova, A.E. (2020) *Agriculture in the digital era: challenges and solutions Electronic management in industries*, no. 1, 4–15

Karavanova, E. I. (2003). *Optical properties of soils and their nature*. M.: Moscow State University Publishing House, 187.

Kataev, M.Yu., Kartashov, E.Yu., & Smirnov, D.S. (2021). Methodology and program for atmospheric correction of images of unmanned aerial vehicles in the problem of safe vegetation location. *TUSUR Reports*, T.24, No. 4, 73–78. TUSUR.

Kataev, M.Yu., Skugarev, A.A., Sorokin, I.B. (2017) Possibilities of space monitoring for agricultural purposes in the Tomsk region. *TUSUR Reports,* T.20, No. 3, 186–190. TURSUR.

Kirillova, N. P., & Sileva, T. M. (2017). Colorimetric Analysis of Soils Using Digital Cameras. *Moscow University Soil Science Bulletin*, 72(1), 13–20. 10.3103/S0147687417010045

Kobernichenko, V. G. (2016). *Radio-electronic systems for remote sensing of the Earth*. Ekaterinburg: Ural Publishing House.

Kostenko, I.V. (2009) Study of the optical properties of samples of soddy-steppe sandy soils in the south of Ukraine using a scanner. *Soil Science*, (9), 090–1098.

Lagacherie, P., McBratney, A. B., & Voltz, M. (2007). *Digital Soil Mapping: An introductory perspective*. Elsevier.

Lakiza, N. V., & Neudachina, L. K. (2015) *Analysis of food products*. Ekaterinburg: Ural Publishing House.

Lal, R. (2004). Soil carbon sequestration impacts on global climate change and food security. *Science*, 304(5677), 1623–1627. 10.1126/science.109739615192216

Lefevre, C., Rekik, F., Viridiana, A., & Liesl, W. (2017). *Soil Organic Carbon. The Hidden Potential, Food and Agriculture Organization of the United Nations*. FAO.

Lesaignoux, A., Fabre, S., Briottet, X., & Olioso, A. (2010) Estimation of Soil Moisture Content of bare soils from their spectral optical properties in the 0.4 – 12 µm spectral domain, *IEEE International Geoscience and Remote Sensing Symposium*, (pp. 3861-3864). IEEE. 10.1109/IGARSS.2010.5649907

Lin, S., Gu, J., Yamazaki, S., & Shum, H.-Y. (2004). Radiometric calibration from a single image. In *Proc. IEEE Comp. Vision and Pattern Recognition*, (pp. 938–945). IEEE.

Luneva, N. N. (2009). *Technological methods for accounting and monitoring of weeds in agroecosystems, High-performance and high-precision technologies and methods of phytosanitary monitoring*. VIZR.

McBratney, A., Mendonça, S. M., Minasny. B. (2003) On digital soil mapping. *Geoderma, 1–2* (117), 3–52.

Mendoza, F., Dejmek, P., & Aguilera, J. M. (2006). Calibrated color measurements of agricultural foods using image analysis. *Postharvest Biology and Technology*, 41(3), 285–295. 10.1016/j.postharvbio.2006.04.004

Orlov, D. S. (1997). Color and diagnostics of soils, Soros Educational Journal. No. 4, 45-51. 23. Kataev, M.Yu., Dadonova, M.M. (2019) Vegetation recognition technique based on color and texture analysis of RGB images. *Light & Engineering*, (2), 34–39.

Orlov, D. S., Sukhanova, N. I., & Rozanova, M. S. (2002). *Spectral reflectance of soils and their components*. M.: Moscow State University Publishing House.

Rees, W.G. (2006). *Fundamentals of remote sensing*. Moscow, Tekhnosphere, 336.

Shishov, L. L., Tonkonogov, V. D., & Lebedeva, I. I. (2004). *Classification and diagnostics of soils in Russia*. Smolensk: Oikumena.

Shumilov, Yu.V., Luneva, N.N., Ermolenko, S.A., Savva, A.P., Zakota, T.Yu., Mysnik, E.N., & Danilov, R.Yu. (2018). Use of unmanned aerial vehicles for phytosanitary monitoring of weeds. *Plant Protection Bulletin, 4* (98), 22-27.

Turk, J., & Rebecca, Y. (2020). Field Conditions and the Accuracy of Visually Determined Munsell Soil Color. *Soil Science Society of America Journal*, 84(1), 163–169. 10.1002/saj2.20023

Viscarra, R. R. A., Behrens, T., & Ben-Dor, E. (2016). A global spectral library to characterize the world's soil. *Earth-Science Reviews*, 155, 198–230. 10.1016/j.earscirev.2016.01.012

Zakharov, S. A. (1931) *Soil science course*. Moscow, Moscow State University.

Zhukova, M. A., & Ulezko, A. V. (2021). *Prospects for the digital transformation of agriculture*. Voronezh: Voronezh State Agrarian University.

Chapter 10
Automated Detection of Plant Diseases and Chlorophyl With Image Processing Using Mobile Application

Dharshan Y.
http://orcid.org/0000-0002-4992-6344
Sri Ramakrishna Engineering College, India

Devasena D.
Sri Ramakrishna Engineering College, India

Sharmila B.
Sri Ramakrishna Engineering College, India

Nandhanaa K. S.
Sri Ramakrishna Engineering College, India

ABSTRACT

Green plants are capable of producing their own sustenance. They accomplish this through a process known as photosynthesis, which makes use of a chlorophyll-based green pigment. A pigment is a chemical that has a specific color and, depending on that color, may absorb light at various wavelengths. The chlorophyll of the leaf may be predicted from leaf color, which has the potential to be an excellent indicator of plant health. Chlorophyll a (Chl-a) and Chlorophyll b are the two main types of chlorophyll found in the photosystems of green plants (Chl-b). Each kind has a unique function as well as advantages and a unique chemical composition. So, it is crucial to anticipate the chlorophyll of the takes off as it provides valuable information regarding plant health and edit management. In the proposed work, a non-destructive image processing technique is employed to estimate the chlorophyll content of tomato plants using images of the leaves. The chlorophyll content of the acquired image is predicted using a deep learning (DL) algorithm. Using the camera of an Android smartphone, samples of tomato plant leaves were collected from the field, and the color vegetation indices for all three RGB channels were then abstracted. The Chlorophyll detector, which is aimed at both small- and large-scale farms, has

DOI: 10.4018/979-8-3693-3374-7.ch010

finally been built. For easy deployment, the mobile app combines an image from a smartphone camera with cloud-hosted software.

INTRODUCTION

Green plants are capable of producing their own sustenance which is accomplished through a process known as photosynthesis, using a chlorophyll-based green pigment. Chlorophyll stands out among the many different kinds of pigments found in nature because it gives plants the ability to absorb the energy they require to develop tissues. In a plant, chlorophyll's function is to absorb light, often sunlight to turn water and carbon dioxide from the air into glucose, a form of sugar, through photosynthesis, for the purpose of producing new leaves and other plant parts. Destructive approaches based on laboratory procedures are available which are time and money consuming. Non-destructive methods, on the other hand, are less precise but neither time nor money consuming. In order to estimate foliar chlorophyll or N for numerous species, non-destructive approaches have been employed often. Plant Ch and N have been identified using digital colour picture analysis based on the RGB colour model. Many research employs RGB color models to discover relationships between plant Ch and N status. There aren't many research that combine Ch and N status with other color models. The study shows that HSI and Lab color models be used to identify nutrient deficiencies in tomato and lady's finger plants.

LITERATURE SURVEY

Various researches has been carried out on the typical plant diseases identification, as researchers has proposed different techniques and methods which could help us identify the parameters that are utilized in the process.

E Yongli Zhu, Hanping Mao, Pingping Li & Yanyou Wu (2011) presented Vegetable nitrogen nutrition may be quickly and accurately diagnosed, which is helpful for managing fertilization and raising product quality. Spectral data from a visible and near-infrared (Vis-NIR) spectrophotometer and a chlorophyll meter were collected in order to develop a method for quickly detecting the leaf nitrogen level in lettuce (SPAD 502). The optimal chlorophyll content prediction model was chosen using a variety of spectra pre-processing approaches, including moving average filter (MAF), first derivative (FD), normal normalization (NM), second derivative (SD) and their combination. The characteristic wavelength was confirmed to be at 368 nm, 694 nm, 744 nm, and 825 nm, respectively. As a result, it was determined that the Vis-NIR methodology would be a viable option for quickly determining the chlorophyll content of lettuce leaves.

Y. Sheng, Z. He and H. Zhang (2010) proposed it's crucial to learn how to accurately and quickly identify the chlorophyll content of crops in agricultural research for crop yield estimation and surveillance. We set up an absolute quantitative analytical model for determining the chlorophyll concentration using stoichiometry. Mathematical language MATLAB was utilized, and it has a user-friendly graphical interface for ease of usage. According to the experiment's findings, there is a strong positive association between the anticipated value and the metrical value determined by chemical means. This technique investigates the viability of non-destructive testing for in vivo leaf chlorophyll content that can be utilized for quick detection of the chlorophyll content in field crops.

A. Y. Khaled, S. A. Aziz, S. K. Bejo, N. M. Nawi, I. A. Seman and M. A. Izzuddin (2018) A typical plant disease called basal stem rot (BSR) is substantially to blame for the significant financial losses in the oil palm industry. For the purpose of detecting BSR illness in oil palm fields, several innovative approaches have recently been developed and published in the literature. There are no research on the use of electrical characteristics to identify BSR disease in oil palm, though. This research seeks to add to the body of knowledge. Several leaf samples, including healthy, mildly, moderately, and seriously diseased samples, were collected for the investigation. Principal component analysis (PCA), Duncan's multiple range test (DMRT), and ANOVA were utilized for statistical analysis. The study's findings demonstrated a statistically significant correlation between DC and various BSR illness severity levels (p ; 0.0001). The degree of BSR disease in each sample that was taken was specifically distinguished based on DC. Contrarily, only two groups—one for each stage of BSR disease—could be distinguished based on the chlorophyll content (healthy and BSR-infected).As a result, the findings showed that DC and chlorophyll concentration might be used as sensing parameters for Ganoderma disease detection to a certain extent.

D. Nath, P. K. Dutta and A. K. Bhattacharya (2021) The location of crop diseases and their classification are of tremendous importance in countries like India, whose economies are primarily dependent on agriculture. Programmed disease identification and classification of plant/crop infection has developed into a crucial viewpoint nowadays to support horticulture. Manual infection detection involves an enormous amount of labor and takes too long to process. In order to identify crop/plant diseases from images of sick plants, image processing is crucial. The suggested approach is based on the automated classification and division of various infected leaf locations from plant/crop images. It involves three steps: The collected leaf image is subjected to middle channel and differentiation stretching as part of the initial step of pre-processing in order to remove digitization noise and improve each image's distinctness. For the leaf image's L*a*b* color show, which divides pixels into uniform sets, K-means clustering analysis is used in the current stage. These pieces are then converted to an HSV color display, and the infected fragment that has the highest tone value among the other pieces is extracted. Next, we will use machine learning, namely the Multi SVM classifier, to classify the edited picture into various disease categories.

H. Andrianto, Suhardi and A. Faizal (2017) Data innovation is used in "Smart Farming" to assist farmers in carrying out their daily tasks. A chlorophyll meter is one of the tools that may be used to promote careful cultivation. This device appears to be used to monitor plant nitrogen shortage. The major element needed for the production of chlorophyll is nitrogen. By employing a chlorophyll meter, we can maximize the use of nitrogen fertilizer to lengthen the growing season, reduce production losses, and prevent unsuitable environmental pollution. While being available within the exhibit, the chlorophyll metre is nonetheless rather expensive. Making a low-cost, reliable chlorophyll metre is therefore the task at hand. This study offers a comprehensive explanation of how chlorophyll content is calculated using the morphological traits of the plant. 28 articles were selected as key factors using an effective writing survey approach. On the basis of these important factors, it is understood that the amount of chlorophyll may be calculated by detecting the light that is transmitted or reflected from the leaf using a phantom sensor in an open and near infrared area.

Y. Sari, M. Maulida, R. Maulana, J. Wahyudi and A. Shalludin (2021) like other common plants, corns too need nutrients to survive. Except for corn, all plants always need phosphorus, potassium, and at least three other basic nutrients. These three maize supplements may be examined using a variety of techniques, including the Leaf Color Chart (LCC), Chlorophyll Meters Soil Plant Examination Advancement (SPAD), and Soil Test Pack. As it fetched less than the other two, the Leaf Color Chart (LCC) is a common method that agriculturists use to examine the supplements material in corn clears out. A fantastic solution to this problem may be digital image preparation, which agriculturists may use to examine their plants' nutritional demands in a less time-consuming and more affordable manner. For an advanced image preparation framework for maize takeoff photographs, the Tone, Immersion, Esteem (HSV) RGB extraction approach is suggested in this study. Back Vector Machine (SVM) is used as a classification approach for this consideration in order to categorize its image results.

Le Jing, Li Yun, Hua Dengxin, Tan Linqiu and Cao Ning (2011) A method of LED-induced chlorophyll fluorescence examination entangled picture processing is shown in order to highlight the absence of data substance and determination using chlorophyll fluorescence spectrometry. The theory of active-inducing chlorophyll fluorescence is used to successfully activate plant leaf fluorescence using small, long-life, and high-intensity blue (460nm) Driving light for the close field site. By using a spectrometer and picture sensor settled channel to get spectra and picture data independently, the method of extracting chlorophyll fluorescence properties is studied. By examining data from the fluorescence spectra and images of chlorophyll, researchers can show that there is a strong correlation between the two. This research establishes the groundwork for in-place, real-time, online estimate of a living body.

K. Bhosle and V. Musande (2017) Ruddy Edge Point (R E P) and chlorophyll foliar concentration and content are quite closely connected. Chlorophylls a and b are deeply assimilated, which affects the abrupt shift in location of the green vegetation reflectance range from 680 nm to 800 nm. Ruddy Edge Point can identify the perception's greenness zone. Ruddy Edge Point is included in the vegetation that is provided by further detecting techniques. REP can also be observed through laboratory or field studies. An A S D Field Spec Master spectro radiometer was used to measure the canopy's ghostly reflectance. This instrument provides estimates for the ghostly range between 350 and 2500 nm with otherworldly resolutions of 3 nm and a testing step of 1 nm It is possible to locate undesirable crops using unavailable sensing information. Refraction and reflection information from a spectrum radiometer is similar to that from remote sensing. This is possible if Ruddy Edge Point is prepared to be established. This method is used to identify plant dryness. Finding stretches of mulberry, cotton, and sugarcane plants is currently being done for this article, and results are being evaluated using the crest subsidiary, direct introduction, and direct extrapolation strategies. The final outcome is compared using all of the techniques.

Shoaib M, Shah B, EI-Sappagh S, Ali A, Ullah A, Alenezi F, Gechev T, Hussain T and Ali F (2023), has reviewed on the various techniques implemented using Machine Learning and Deep Learning in the identification of plant diseases. The author has also inferred the various limitations when using the machine learning and deep learning in this process. Various challenges were identified in the process of implementation, where some are the quantity and the quality of the data which is available, the quality of the image utilized in the training and testing process, and the classification of the plants based on their diseases. The author has concluded by stating the advancement and the Machine and Deep Learning has both advantages and certain considerable challenges that could be resolved by developing a proper model, with good understanding of the problem and data for the system development.

W. Kong, W. Huang and X. Zhou (2016) has reviewed a vision system based on the colorimetric sensor to identify the quality and defects of the food products. The authors has presented their perspective as the proposed machine vision based calorimetric sensor has been applied into various sectors of the food industry, which has proven to have implemented for estimating the freshness of the food, adulterations or safety conditions of the food and also the quality of the food. Rongrong Li, Shuchang Zhao & Bokai Yang (2023) has reviewed the different based on applications such as detection, sorting operations, material handling and management. The authors have utilized the data from publications for the year

2011 to 2022. The authors have proposed that usage of machine vision system has increased the rate of production and also the creating a safe and intelligent manufacturing process.

Ramazan Hadipour-Rokni, Ezzatollah Askari Asli-Ardeh, Ahmad Jahanbakhshi, Iman Esmaili paeen-Afrakoti, & Sajad Sabzi (2023) presented a convolution neural network based detection for identification of pests in citrus fruits. Data set for the identification has been carried out on 3 stages and in acquired images 70% of images for the training CNN and rest 30% is for testing and validation. They have used four model of CNN models in which AlexNet model with the help of SGD has got better results.

DETECTION OF PLANT DISEASES

India is a country which consists of a large area in the field of agriculture, where different types of crops are grown at different climatic condition, and different land surface. To identify the crops and the types of diseases, various methods are currently being utilized. The general way of identify the healthiness of a crop is by the green color of the leaf and the stem. The color of the leaf determines the chlorophyll content of the plant that is being produced. As chlorophyll content is good the plant health is in good condition and it has enough nutrient for its growth. When the plant could not develop enough chlorophyll in them, them the plant is identified to be not in health condition, which in turn defines that it is either affected by any diseases or its mineral content has reduced.

The primary green pigment in plants which defines chlorophyll continually provides reliable information about the well-being, nourishment, and photosynthesis of plants. The amount of chlorophyll in a leaf is roughly proportional to how close nitrogen is to the soil. One of the most important nutrients necessary for healthy plant growth, development, and reproduction is nitrogen. Deficiency of nitrogen results in the unfortunate green color diminishment of the leaf zone as well as the escalating difficulty of photosynthesis in the leaf. It has been discovered that the total amount of chlorophyll appears to have a relationship to the total amount of nitrogen used as fertilizer. The methods used in research facilities to quantify plant nitrogen and chlorophyll are time and money consuming.

Certain methods of determining the chlorophyll content in a leaf is by using a Spectrophotometer and Soil Plant examination Advancement meter (SPAD).

Extraction of Chlorophyll by Spectrophotometer

The spectrophotometer is a destructive type test of chlorophyll estimation and it takes more time to appraise chlorophyll of a leaf. The chlorophyll of the leaf may be predicted from leaf color, which has the potential to be an excellent indicator of plant health. Chlorophyll a (Chl-a) and Chlorophyll b are the two main types of chlorophyll found in the photosystems of green plants (Chl-b). Each kind has a unique function as well as advantages and a unique chemical composition. So, it is crucial to anticipate the chlorophyll of the takes off as it provides valuable information regarding plant health and edit management. The chlorophyll can be anticipated by analyzing the dissemination of color components of the leaf.

The method of measurement using spectrophotometer is that it intends to make adjustments for turbidity provided by the channels, foundation assimilation, and/or any particle matter. The commitment of the channel to the clear with solvent layer channels must be properly determined. Centrifugation is essential when using glass fibre channels to keep strands from transferring into the optical cuvette. For stability during the period needed to measure the test, the clear (750 nm) should be examined. The presence of particles is indicated by shrinking According to Moed and Hallegraeff's 1978 study, fermentation techniques should aim to keep the corrosive concentration under tight control (pH 2.6-2.8 within extricate). With [H] = 0.003 mol l-1 (Holm-Hansen and Riemann, 1978) to 0.006 mol l-1 (Nusch, 1980) inside the of 2-3 min, ideal conditions are achieved. Figure 1 displays a graphical depiction of the chlorophyll concentration of a leaf measured using spectrophotometer.

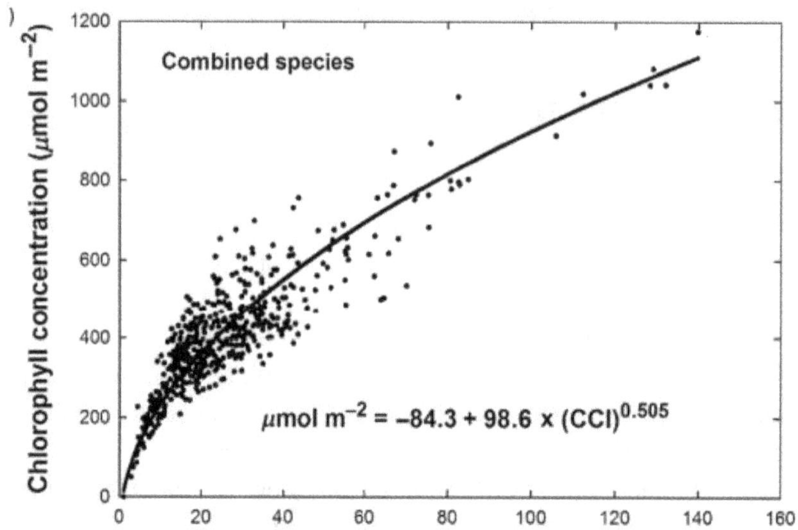

Extraction of Chlorophyll Using SPAD

The SPAD meter give a basic, fast, and non - damaging strategy for assessing leaf chlorophyll substance. The SPAD meter has been broadly utilized in measurement of the chlorophyll and also in the measurement of color pigmentation in a leaf. The calculation of SPAD readings is based on two transmission values: the transmission of reddish light at 650 nm, which chlorophyll absorbs, and the transmission of infrared light at 940 nm, where there is no assimilation of chlorophyll.

The use of SPAD metres to measure leaf chlorophyll concentration has become widespread, however converting SPAD readings into coordinate units of chlorophyll concentration is still difficult, and thus it is crucial to comprehend how these two parameters relate to one another. The correlation between SPAD measurements and chlorophyll content per leaf across entirely distinct species has been evaluated by a number of factors.

Figure 2.

The chloroplasts receive roughly 80% of the nitrogen in leaves, and the photosynthetic proteins in leaves receive about 50% of the nitrogen. Yet, depending on the species and the environment in which the plant develops, 0.5-1.5% of the nitrogen in leaves is devoted to chlorophyll. In several species, the increased amount of leaf N supplied to chlorophyll-protein complexes with declining irradiance has been observed. This layout is consistent with the best nitrogen splitting design that maximizes daily human waste removal of CO2 carbon.

Moreover, the circumstances of N supplementation have an impact on the assignment fraction of leaf N to chlorophyll. When the SPAD-502 is used to guide N administration in horticulture systems, understanding the effects of leaf features and natural factors on SPAD readings as well as the link between chlorophyll substance and leaf N substance per leaf zone will be crucial. Figure 2.3 displays the graphical depiction of SPAD units and chlorophyll concentration.

Figure 3.

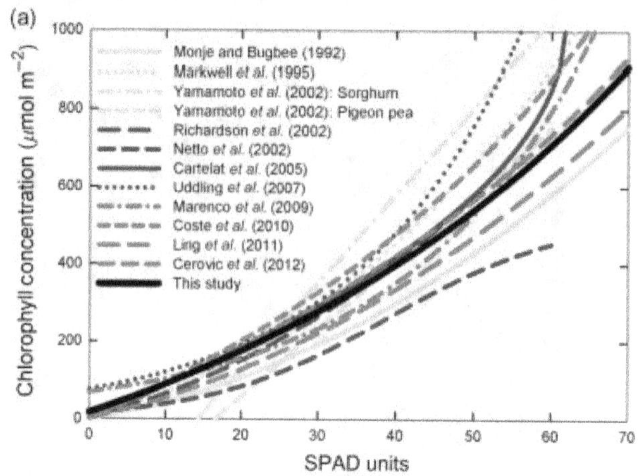

PROPOSED METHODLOGY

Leaf Sample Collection

Based on how far the leaf is from the branch terminal, the leaves are gathered and sorted into three categories: tender, immature, and mature. The first three leaves from the branch terminal are regarded as delicate, while the next four to ten leaves are regarded as immature. Mature leaves are those that are located closer to the branch's base. As a part of the study, tomato plants are identified for the classification and the identification process. Unlike most of the existing image processing methods that use digital cameras to capture image, to capture leaf images of tomato plant using an android smart phone and are shown in figure 4.

The common point of Picture Procurement is to convert an optical picture (Genuine World Information) into a cluster of numerical information which may be afterward controlled on a computer, sometime recently any video or picture handling can commence an picture must be captured by camera and changed over into a sensible substance.

The Picture Securing prepare comprises of three steps:-

1. Optical framework which centres the vitality
2. Vitality reflected from the protest of intrigued
3. A sensor which degree the sum of vitality.

Based on how far the leaf is from the branch terminal, the leaves are gathered and sorted into three categories: tender, immature, and mature. The first three leaves from the branch terminal are regarded as delicate, while the next four to ten leaves are regarded as immature. Mature leaves are those that are located closer to the branch's base. Picture Procurement is the primary step in any picture handling framework and the method stream is appeared within the figure 5.

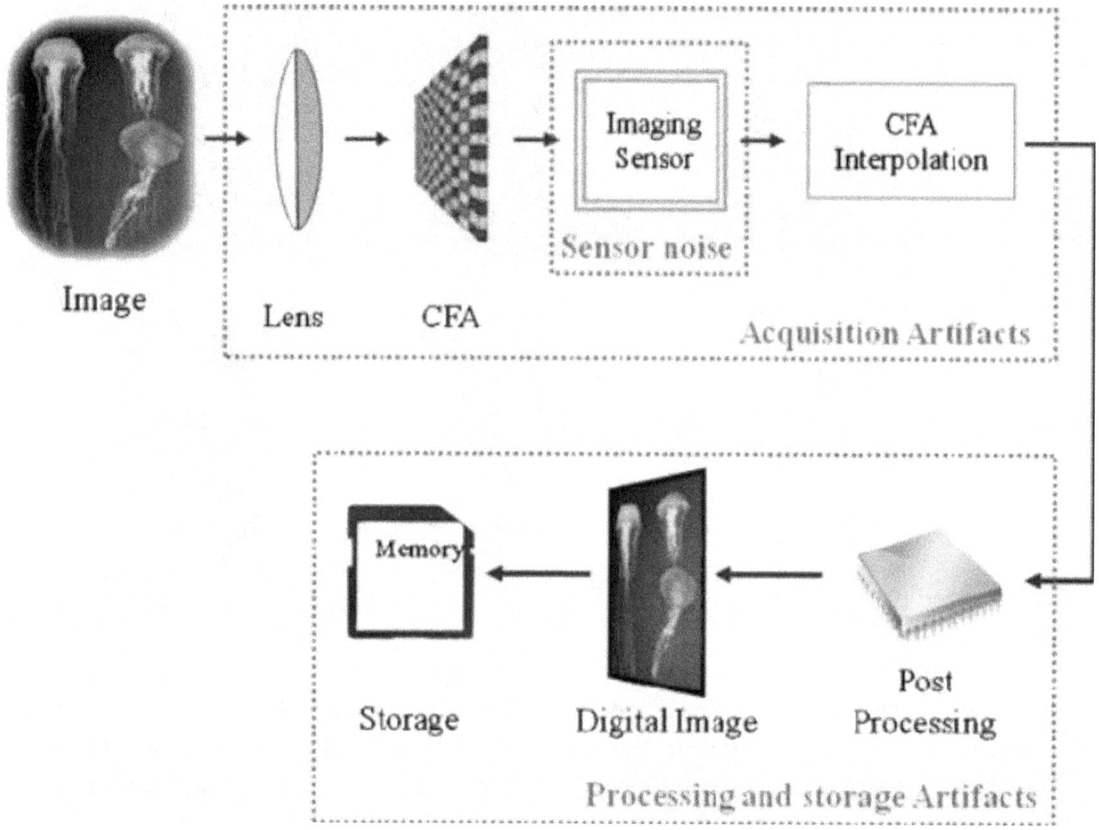

Image Pre - Processing and Segmentation

Images of tomato leaves were taken using an Android smartphone in 8-bit RGB mode. JPEG is the format used to save images. All of the recorded photographs have a high resolution of 3120 x 4160 and 72 dpi.

The phrase "image pre-processing" refers to actions on pictures at the most basic abstraction level. If entropy is a measure of information, then these actions diminish rather than enhance the information content of the picture. By minimizing undesirable distortions or boosting particular visual features that are crucial for later processing and analysis activities, pre-processing seeks to enhance the picture data.

There are 4 different types of Image Pre-Processing techniques and they are listed below.

1. Brightness adjustments and pixel brightness changes
2. Transformations of Geometry
3. Image Segmentation and Filtering
4. Image restoration and Fourier transform

In the proposed method, the picture pre-processing method is connected in tomato leaf pictures to move forward the quality of the picture. The resizing method is utilized to resize the picture into a 300x300 measurement and white adjusting approach is utilized to alter the brightness of the color picture. After pre-processing picture division is done by utilizing foundation subtraction calculation for recognizing a clear leaf picture. A boundary-based approach is utilized for isolating tomato leaf from white background. The strategy explore for pixels that lie on a boundary locale, these pixels are called edges. The nearness of an edge is characterized by a critical nearby alter in picture force and by analyzing neighboring pixels. The boundaries of the pictures are characterized by utilizing the distinction between the neighborhood pixels. Foundation dark cover is connected to redress foundation and frontal area protest. The result of pre - processed images are shown in figure 6.

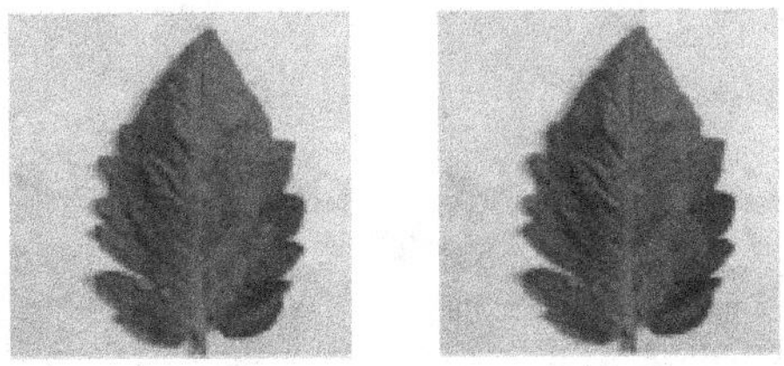

Image segmentation is a popular technique in digital image processing and analysis to separate an image into multiple portions or zones, sometimes depending on the characteristics of the picture's pixels. Segmentation allows for the separation of foreground and background in a picture, and it also allows for the grouping of pixels based on shared characteristics. Two categories of picture segmentation techniques exist such as "Thresholding" is the most straightforward non-contextual segmentation technique. It converts a grayscale or color image into a binary image that can be thought of as a binary region map with just one threshold. "Texture segmentation" In many image analysis or computer vision applications, texture is the most crucial element. Four categories can be used to separate the texture problem procedures. The result of segmented images are shown in figure 7.

After tomato leaf separation, several tomato leaf parameters are extracted using a separated RGB model. The length of the leaf determines the leaf area; longer leaves produce more values, while shorter leaves produce less. Image pixel-based mean median values vary, with the highest representation for color starting at 255 for a correct green leaf and falling off for other leaves. The collected dataset after image processing and segmentation is shown in figure 8.

The block diagram for detection of chlorophyll content in leaves using image processing is shown in figure 9.

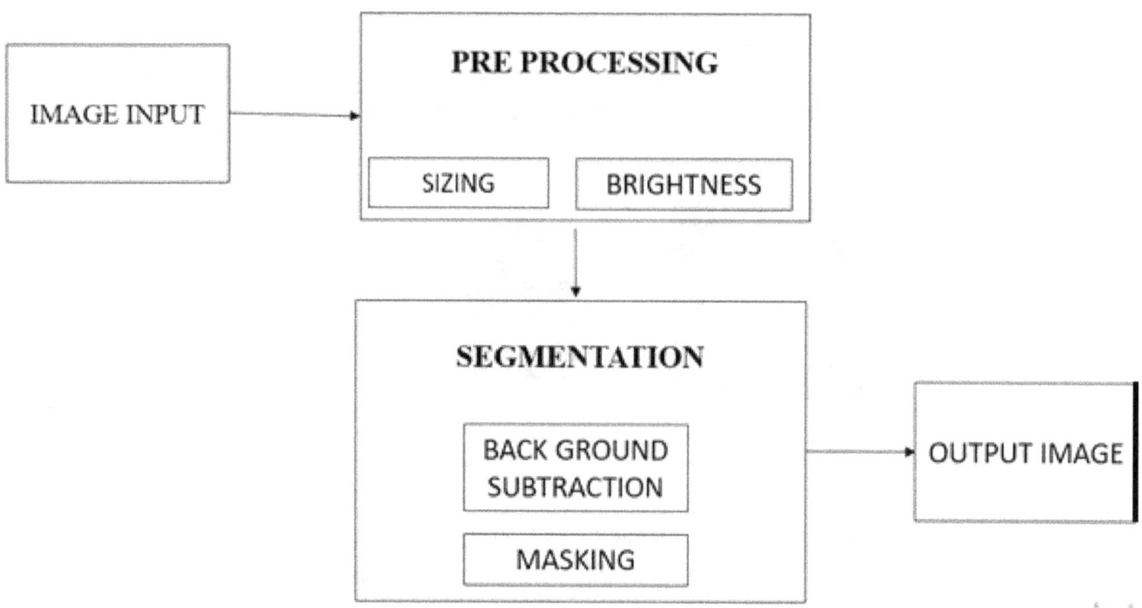

In image pre-processing sizing and brightness adjustment of the leaf images are done. After pre-processing the image is segmented. In segmentation the image under goes background subtractions and masking. Moving elements in videos are compared to background and foreground images to perform background removal. The pre-processed image is acquired after segmentation.

Figure 10.

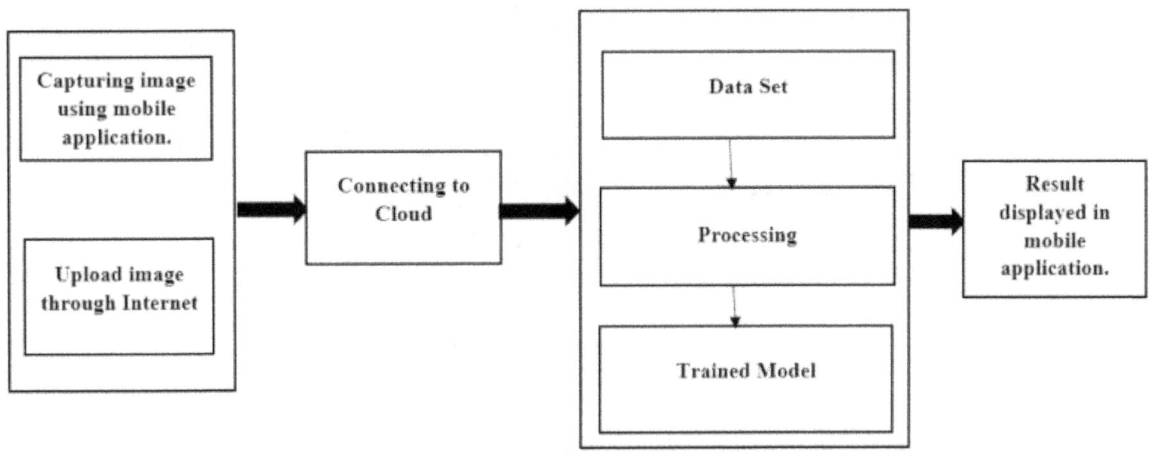

A mobile app was used to take the picture. The features are extracted from the taken image before it is sent to a cloud through the internet. The trained neural network is then fed the retrieved features as input. The chlorophyll of the provided image was predicted by the neural model. The estimated chlorophyll value was transmitted to a mobile device through the internet. The anticipated chlorophyll values are shown in the mobile application. An app for smartphones with a photography program was developed to guarantee that every photo is the same size and resolution. I randomly chose a plant leaf to include in my photo. With the white paper close to the plant, the leaf was placed in the sample glass. Hence, the background can be easily distinguished from the leaf samples. The image was saved in the JPG format and occupied roughly 60 KB of memory. This image size was clear enough for color resolution and needed little memory or processing time.

The virtual server was built using Microsoft Azure. The virtual server uses the learned neural model to recognize chlorophyll. A virtual server retrieves the values of the color index for a neural network when it receives an image. The recovered features are subsequently sent into the trained neural network as input. The neural model predicted the chlorophyll in the presented image. A mobile device received the estimated chlorophyll value via the internet. Through a mobile application, the predicted chlorophyll is displayed on the smart phone. Because processing occurs on the virtual server, this mobile application can be utilized on a smart phone with even the most basic capabilities. The flow process of the proposed system is shown in the figure 11.

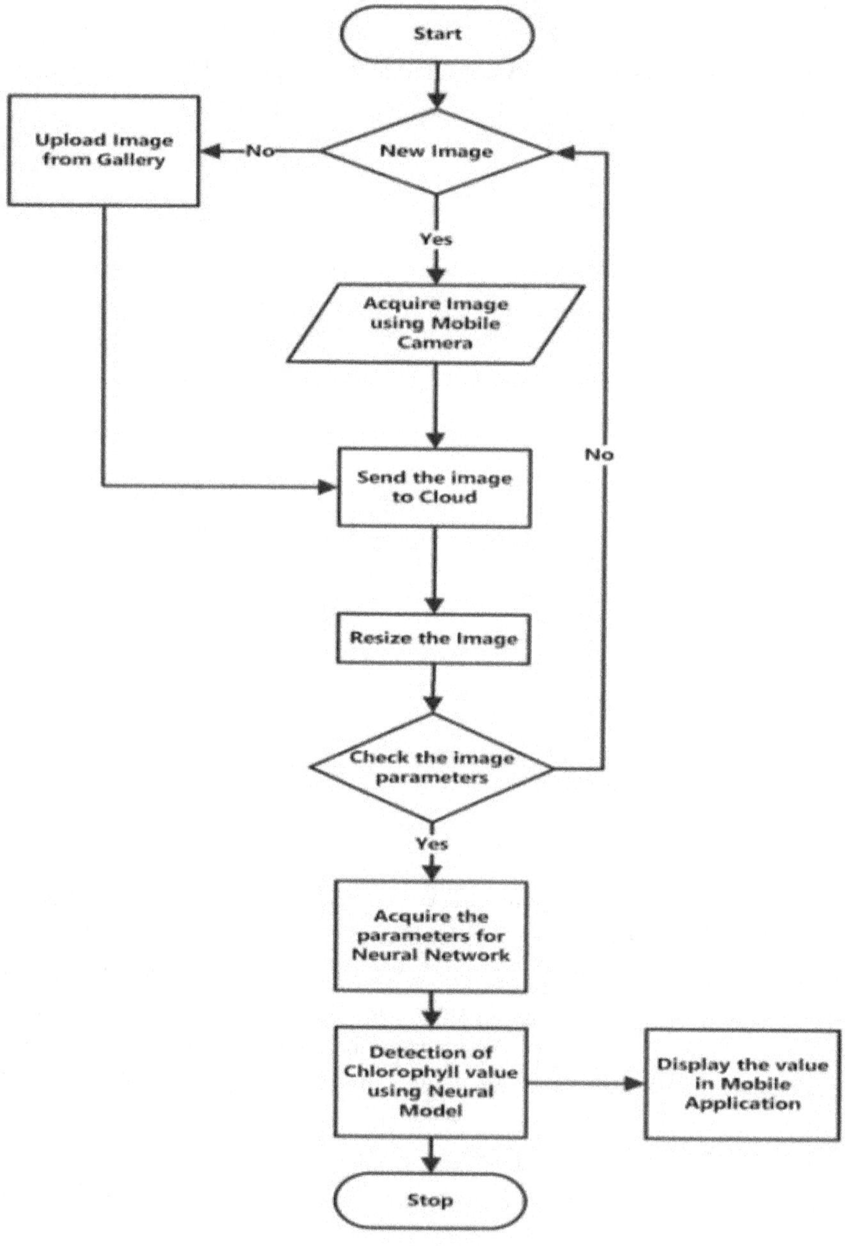

RESULTS AND DISCUSSIONS

To analyse and create the systems and technologies that will alter our planet, engineers and scientists can use the programming environment MATLAB. The core of MATLAB is the matrix-based MATLAB language, which permits the most organic representation of computer mathematics. For a variety of applications in business and academia, including deep learning and machine learning, signal processing and communications, image and video processing, control systems, test and measurement, computational finance, and computational biology, millions of engineers and scientists use MATLAB.

Figure 12.

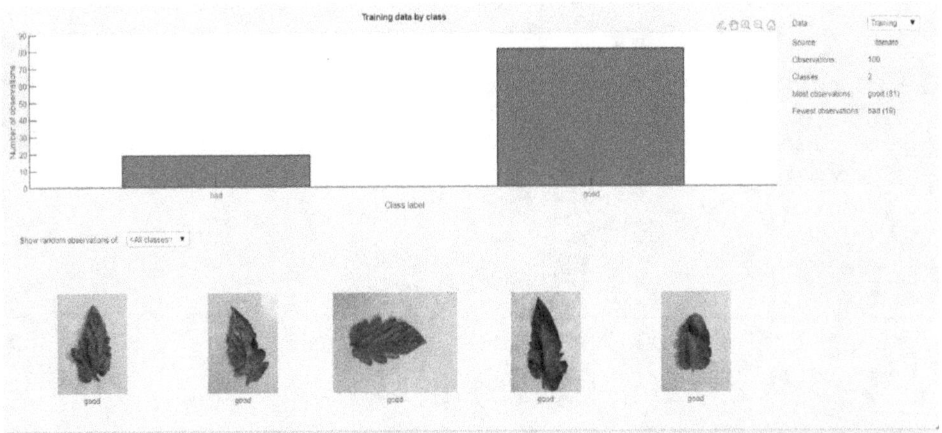

Extracting relevant information from photos, such as discovering forms, counting items, recognizing colors, or measuring object attributes, is the process of image analysis. For image analysis tasks including statistical analysis and property measurement, the toolbox offers a complete array of reference-standard algorithms and visualization features. For image processing, analysis, visualization, and algorithm development, Image Processing Toolbox offers a complete range of reference-standard algorithms and workflow tools. Deep learning and conventional image processing methods can be used to accomplish image segmentation, enhancement, noise reduction, geometric changes, and registration. Processing of 2D, 3D, and arbitrary big images is supported by the toolbox.

An extensive range of reference-standard workflow tools and toolbox features are available for algorithm creation, image processing, analysis, and visualization. They also allow C/C++ code generation for desktop prototype and embedded vision system deployment. Deep learning and conventional image processing methods can be used to accomplish image segmentation, enhancement, noise reduction, geometric changes, and registration. Processing of 2D, 3D, and arbitrary big images is supported by the toolbox. This project proposes a new method for computing leaf chlorophyll content via image processing. Tomato plants' chlorophyll content is evaluated using a non-destructive image processing technique. The estimation of chlorophyll in plants is accurate, efficient, and takes less time. The algorithm is used to forecast the amount of chlorophyll in the acquired image. ML model estimation of the chlorophyll levels and image processing techniques are used in the non-destructive method of chlorophyll estimation.

Figure 13.

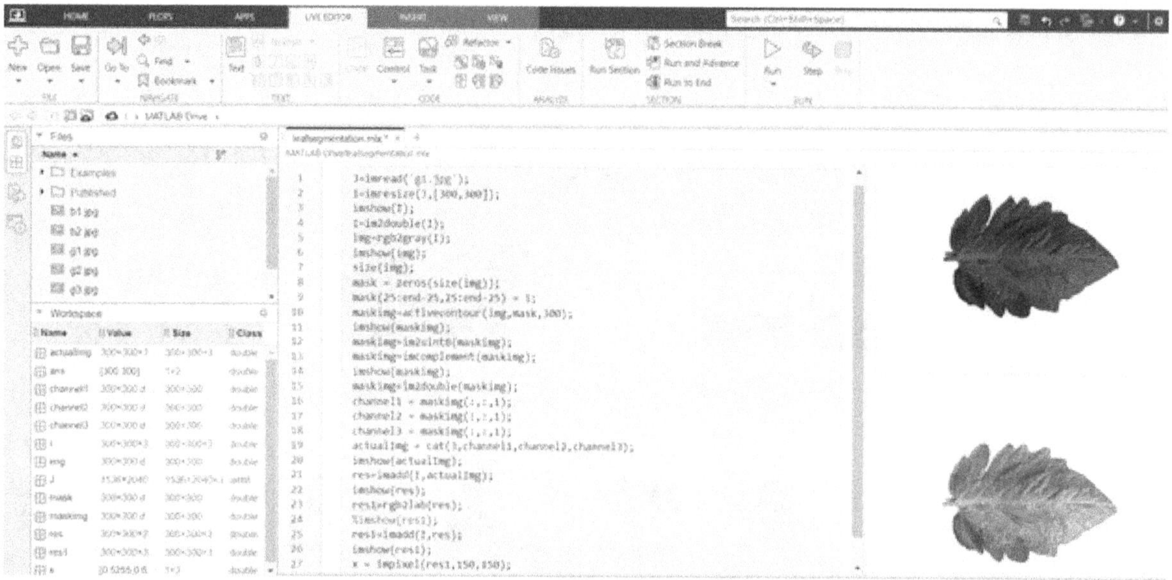

The R G B values for the chlorophyll-rich and chlorophyll-unrich leaf pictures are displayed in the Table 1. Photographs are taken of both chlorophyll-rich and chlorophyll-poor leaves. The photos' R G B values are extracted. The R G B values retrieved from the photos taken using the mobile app are compared to and placed above the extracted samples. The values and the outcome of comparisons are displayed in the mobile application.

Figure 14.

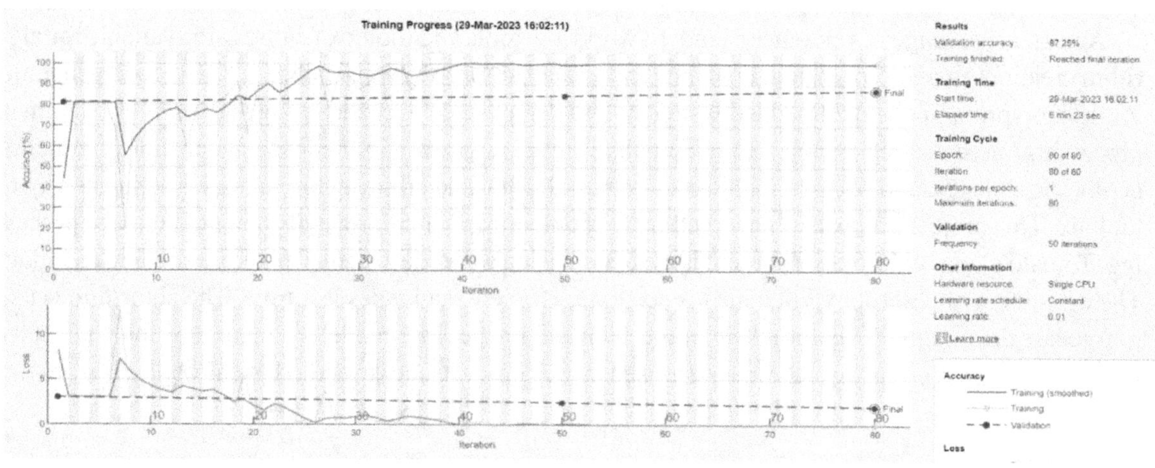

Figure 15.

The mobile app displays the chlorophyll values of the leaf images. Mobile app is opened in mobile phone the image is captured using mobile camera. The captured image is uploaded to the cloud. In cloud neural network predicts the chlorophyll values of the uploaded images. The values are displayed in the mobile app.

CONCLUSION AND FUTURE SCOPE

In the proposed work, a non-destructive image processing technique is employed to estimate the chlorophyll content of tomato plants using images of the leaves. The chlorophyll content of the acquired image is predicted using a deep learning (DL) algorithm. Using the camera of an Android smartphone, samples of tomato plant leaves were collected from the field, and the colour vegetation indices for all three RGB channels were then abstracted. The images of the tomato leaves are pre-processed. In image pre-processing sizing and brightness adjustment of the leaf images are done. After pre-processing the image is segmented. In segmentation the image under goes background subtractions and masking. Moving elements in videos are compared to background and foreground images to perform background removal. The pre-processed image is acquired after segmentation.

The Chlorophyll detector, which is aimed at both small- and large-scale farms, has finally been built. For easy deployment, the mobile app combines an image from a smartphone camera with cloud-hosted software. Using the MATLAB App Designer programme, the DL model for chlorophyll detection is constructed and evaluated. MATLAB C Coder is used to implement in real time Cloud platform. MATLAB utilizes the cloud service for image storage and deployment.

AUTHOR'S NOTE

This research was supported and carried out at Department of Electronics and Instrumentation Engineering, Sri Ramakrishna Engineering College, Coimbatore. We would like to thank our Management, Principal for supporting us with the infrastructure to carry out the research work.

REFERENCES

Andrianto, H. (2017). *Measurement of chlorophyll content to determine nutrition deficiency in plants: A systematic literature review*. International Conference on Information Technology Systems and Innovation (ICITSI), Bandung, Indonesia.

Bhosle, K., & Musande, V. (2017). Red edge point detection for mulberry leaf. *1st International Conference on Intelligent Systems and Information Management (ICISIM),* Aurangabad, India. 10.1109/ICISIM.2017.8122160

Hadipour-Rokni, R., Askari Asli-Ardeh, E., Jahanbakhshi, A., Esmaili paeen-Afrakoti, I., & Sabzi, S.Ramazan Hadipour-Rokni. (2023). Intelligent detection of citrus fruit pests using machine vision system and convolutional neural network through transfer learning technique. *Computers in Biology and Medicine*, 155, 106611. 10.1016/j.compbiomed.2023.10661136774891

Khaled, A. Y., Aziz, S. A., Bejo, S. K., Nawi, N. M., Seman, I. A., & Izzuddin, M. A. (2018). *Dielectric constant and chlorophyll content measurements for basal stem rot (BSR) disease detection.* August 2018 International Conference on Signals and Systems (ICSigSys), Bali, Indonesia. 10.1109/ICSIGSYS.2018.8373570

Kong, W., Huang, W., & Zhou, X. (2016). *Remote estimation of carotenoid/chlorophyll ratio in vertical layers using canopy multi-angle spectral data.* Fifth International Conference on Agro-Geoinformatics (Agro-Geoinformatics), Tianjin, China. 10.1109/Agro-Geoinformatics.2016.7577684

Le Jing, L. Y., Hua, D., Tan, L., & Ning, C. (2011). Research on method of LED-induced chlorophyll fluorescence spectrum and image information acquisition. *IEEE 2011 10th International Conference on Electronic Measurement & Instruments*. IEEE. 10.1109/ICEMI.2011.6037913

Li, R., Zhao, S., & Yang, B. (2023). Research on the Application Status of Machine Vision Technology in Furniture Manufacturing Process. *Applied Sciences (Basel, Switzerland)*, 13(4), 1–14. 10.3390/app13042434

Nath, D., Dutta, P. K., & Bhattacharya, A. K. (2021). *Detection of plant diseases and nutritional deficiencies from unhealthy plant leaves using machine learning techniques.* 4th Smart Cities Symposium (SCS 2021), Online Conference, Bahrain. 10.1049/icp.2022.0368

Sari, Y., Maulida, M., Maulana, R., Wahyudi, J., & Shalludin, A. (2021). *Detection of Corn Leaves Nutrient Deficiency Using Support Vector Machine (SVM).* 2021 4th International Conference of Computer and Informatics Engineering (IC2IE), Depok, Indonesia. 10.1109/IC2IE53219.2021.9649375

Sheng, Y., He, Z., & Zhang, H. (2010). *The research on chlorophyll content in vivo-leaf based on non-destructive testing by fiber spectrometer*. April 2010 3rd International Congress on Image and Signal Processing, Yantai, China. 10.1109/CISP.2010.5646171

Shoaib, M., & Shah, B. (2023). An advanced deep learning models-based plant disease detection: A review of recent research. *Frontiers in Plant Science, 14*, 1-22.

Zhu, Y., Mao, H., Li, P., & Wu, Y. (2011). *Measurement of lettuce leaf chlorophyll content by means of VIS-NIR spectroscopy.* International Conference on Remote Sensing, Environment and Transportation Engineering, Nanjing.

APPENDIX

Table 1. Extracted sample RGB value from tomato leaves

CLASSIFICATION	IMAGES	R Value	G Value	B Value
Chlorophyll rich	Image 1	0.7137	0.9020	0.6118
	Image 2	0.4078	0.5569	0.3216
	Image 3	0.4863	0.6431	0.3529
	Image 4	0.3294	0.5098	0.1882
	Image 5	0.2980	0.4706	0.1647
Chlorophyll unrich	Image 6	0.5333	0.6980	0.3843
	Image 7	0.3608	0.4706	0.2039
	Image 8	0.5098	0.6275	0.2510
	Image 9	0.2980	0.4235	0.2196
	Image 10	0.5098	0.6275	0.3216

Chapter 11
Algorithm for Using Artificial Intelligence to Monitor Agrogenic and Anthropogenically Disturbed Soils

Olga Anatolyevna Pasko
National Open Institute, St. Petersburg, Russia

Alexander Viktorovich Zakharchenko
http://orcid.org/0000-0002-1559-7925
National Research Tomsk State University, Russia

Vladimir Nikolaevich Goryachkin
Agrochemical Service Center, Russia

Nadezhda Anatolievna Lebedeva
http://orcid.org/0000-0003-4095-2631
International Personnel Academy, Germany

ABSTRACT

An analysis of the possibilities of using intelligent systems to study the condition of soils was carried out. The use of intelligent systems scheme for studying the condition of soils using the example of agrogenic and anthropogenically disturbed soils is proposed in this chapter. It includes a number of successive stages, containing field research, analysis of results and selection of deciphering features and conditions. This work focuses on the first stages, since they are specific to soil research. Other stages of statistical data processing are unified for a wide variety of objects in different fields of knowledge. They include creating a training sample, calculating statistical indicators and analyzing the decipherability of object properties from training samples, adjustment, segmentation and the use of artificial intelligence to solve average problems in the chosen direction.

DOI: 10.4018/979-8-3693-3374-7.ch011

INTRODUCTION

In 2019, by Presidential Decree No. 490, the National Strategy for the Development of Artificial Intelligence for the period until 2030 was approved in the Russian Federation." (Decree of the President..., 2019). It is aimed at bringing the country to a leading position in the world. Interest in AI is driven by the emergence of new technologies and graphics processors, the growth in the computing power of modern computers, cloud computing, and the explosive growth of big data. The high speed and accuracy of models obtained through automated machine learning (Kataev et al., 2023a) has opened up new opportunities for process automation and digital transformation in different areas of human life.

The emergence of unmanned aerial vehicles, digital cameras and display sensors made it possible to continuously collect multi-temporal and multispectral data with high resolution (Kataev et al., 2023b; Tokareva et al., 2018, 2020, 2021; Pasko et al., 2019a,b, 2020a, b). Their analysis using GIS and AI represents a huge potential for solving modern problems facing the natural sciences, the agro-industrial complex, etc. The use of AI in the field of agriculture, soil science and agrochemistry is a relatively new, but promising task, since it allows to improve management efficiency, implement optimal preventive measures based on forecasts and prompts of AI systems, automate routine operations and identify hidden factors affecting the effectiveness of measures and offer information available for decision-making by ordinary users.

The synergistic use of artificial intelligence, remote sensing of the Earth and geoinformation systems opens up great opportunities for solving scientific and applied problems in the field of soil ecology monitoring and agricultural production management (Agroecological state, 2018). It allows you to quickly obtain and analyze objective and reliable data on the state of soil and vegetation according to unified methods, at a time – from sites significantly distant from each other, or information about their condition in previous years. The accumulated experience will form a database for machine learning of neural networks and the selection of significant criteria. The compiled algorithm will become the basis for modeling the state of the agroecosystem and forecasting its development in regulated and/or natural conditions.

As part of the implementation of the project on the use of artificial intelligence for monitoring agrogenic and disturbed soils, the following are provided:

1. Preliminary processing and analysis of spatial data obtained in the field and from remote sensing data;
2. Automated classification without training;
3. Interpretation and verification of results;
4. Field survey of the territory;
5. Selection and description of key areas for creating training samples. Creating training standards;
6. Evaluation of the quality of standards;
7. Choosing the decisive classification rule;
8. Classification;
9. Post-classification processing using the smoothing method and generalization.

The scientific literature available to us does not contain information about the use of intelligent systems for studying soils. **The purpose of the chapter** is to apply the capabilities of intelligent systems to solve urgent problems of agrochemistry and soil science, as well as to model crop yields. The issues of anthropogenic soil degradation and the use of agrogenic soils of agricultural lands are selected as urgent problems.

Anthropogenically transformed soils receive a lot of attention in modern science (Agroecological state..., 2018).

The demand for land degradation assessment under different natural conditions is increasing (Šamonil et al, 2015; Vågen et al., 2016). Anthropogenically changed soils are of particular interest in Russian soil classification (Shishov et al., 2004; Kaskevich et al.,2003). Anthropogenically produced soil has lost its natural structure; however, natural soils that have been anthropogenically changed preserve some of those characteristics. These types of soils have not been sufficiently studied to date (Volungevičius and Skorupskas, 2011).

Transformed soils are changed purposefully. For example, under a power line, such soils are found only under the supports. In addition to them, there are anthropogenic soils that changed their status accidentally. Anthropogenically created (men-made) soils that have lost their natural status and anthropogenically modified soils that have partially preserved it are poorly studied objects today (Shishov et al., 2004). Soil-forming factors in the territories of the sanitary protection zone are under the control of production conditions and such thing determines the direction of soil formation.

Specific environmental conditions are formed under overhead power lines (Derbentseva et al., 2015). Unlike quarry soils, anthropogenic soils of sanitary protection zones of linear objects have only partially lost their parts, which is one of the reasons for soil biodiversity. Power transmission lines, including substations, are objects that slightly pollute the environment with chemicals, therefore, they are classified as a "mildly" affecting technical object, although multifaceted negative impacts on the environment are noted (Volungevičius et al., 2011; Lvov, 1986). On forest lands of the small-leaved subzone, facts of soil degradation have been established that are not taken into account by designers of energy facilities in the boreal forest zone (Kartashev, 2000). This is due to additional operations involving heavy equipment: cutting, skidding, transporting wood, and uprooting stumps.

Studies of the soil surface and topsoil thickness make it possible to establish the shrinkage of the humus horizon of the soils in the NWZ relative to the forest soil, caused by the decay of tree roots (Zakharchenko et al., 2022) and the remains of the soil of the forest parcel were found (Zakharchenko, 2005a, b). The impact of trees on the soil is described in many works (Phillips et al., 2005; Vågen, 2015). Traces of the growth of trees on the soils of the Northwest Territories after 20 years of its operation of the VL SVN were revealed by the size of the network of vertical cracks of the eluvial horizon (Zakharchenko, 2014). It turns out that the cross-sectional area of the cell of the crack network is significantly smaller in the zone of impact, where the tree grew earlier. It is necessary to develop new technologies for the diagnosis of soil heterogeneity at different scale distances in 3-D format, the hierarchical structure of forest soils (Zinck et al., 2016).

The study of agrogenic soils of agricultural lands has become especially relevant in recent years (Pasko et al., 2020c; Zakharchenko et al., 2021). Due to objective reasons of an organizational, legal, economic, social and ecological nature in Russia and other countries of the former Soviet Union, a number of enterprises of the agro-industrial complex were closed (Pasko et al., 2022a). Their lands fell into the category of unused (Karavaeva, 2008; Platonycheva et al., 2013) and began to lose their economic properties (Rusanov et al., 2011): the soil structure is destroyed, a plow sole is formed, water-physical properties deteriorate, etc. The transition of arable land to fallow occurs over a long period of time, the larger the area of the land plot, the slower it becomes overgrown with forest (Platonycheva et al., 2013). According to the data of agroecological monitoring (Karavaeva, 2008), the dependent lands contain 0.4% less humus in the former arable layer than the original soils.

On the other hand, they have an increased content of organic carbon. There is more humus in fallows on chernozems (6%) than in arable land (3.4%) (Rusanov et al., 2011; Sorokin et al., 2023). The rows of decreasing humus content are shown: a forest under a young forest - a pine forest (25 years) - arable land developed from under a forest. The content of organic substances changes insignificantly; the C/N ratio of the active organic matter of the soil reaches a stable state at the age of the forest of 55 years; the restoration process has been going on for 170 years at a very slow pace. During the removal of soil from arable land to fallow, soil development proceeds in the direction of either degradation or growth of humus content (Zakharchenko et al., 2021). In order to reveal the patterns and direction of these processes, complex monitoring of the soil condition and an objective assessment of the ongoing processes according to many indicators are necessary. In this regard, intellectual systems are an extremely promising tool.

OBJECTS AND METHODS

The work uses a set of methods, including monographic, descriptive, cartographic, statistical methods, as well as field research data. Specific methods are applied to each of the two research options, which will be described in the corresponding section.

In general, the research stages included: preliminary processing and analysis of spatial data; automated classification without training; interpretation and verification of results; field studies and description of key areas, training standards for creating training samples.

In further studies, it is planned to develop this base and conduct classification with training; quality assessment of standards; the choice of the decisive classification rule; automated classification with training; post-classification processing, including smoothing and generalization; development of models.

The object of research of degraded soils is man-made disturbed soil cover of the sanitary protection zone of the power transmission line. A field survey of man-made disturbed soil composition was carried out, covering a distance of 760 m between three supports at a width of 65 m of the SVN overhead line. The research was carried out near the village. Lomachevka (latitude 56.136758, longitude 86.835315). The HV-500 kV transmission line of the "Itatskaya"-PS "Novo-Anzherskaya" No. 524 (cadastral number of the sanitary protection zone - 42.00.2.41) was put into operation in 1967 (Fig. 1).

Figure 1. Localization in the locality of the studied part of the SZZ VL SVN (2016 photo published by Google®). The landscape situation is indicated on the tab. The lines in the center of the NWZ show the distances between the supports (from left to right - 340 m, 368 m), the black line shows the position of the trench where the thickness of the soil horizons was measured, the white spots of the NWZ are degraded soils

The territory under the VL SVN is located in a sanitary exclusion zone with a width of 60-65 m and is used to a limited extent by farmers for driving, grazing livestock and mowing hay. Geographically, the research site is located on the macroslope of the Archekassky ridge of the western exposure. Absolute height marks of the watershed of the ridge increase from west to east. In the study area, they vary in the interval from 180 m to 220 m above sea level. The relative height of the ridge reaches 100 m. The upper part of the soil slope is developed on Quaternary deposits, and the lower part on deposits of Neogene age (presumably, Miocene).

In the NWZ, texturally differentiated turf-podzolic soils (Ultisol) have an eluvial horizon (El) morphological feature in the form of gray spots in the lower part of the eluvial horizon, second humus horizon (Elhh), illuvial Bt (argillic). The morphological structure, chemical and physical properties of these soils have been studied in sufficient detail (Zakharchenko et al., 2006). The sod-podzolic soils studied are azonal and are located in the form of a strip on the gentle middle part of the slope of the Archekas range.

The vegetation cover of the SZZ has a mosaic character, which is determined by a complex of both anthropogenic and natural microcenoses. In undisturbed or slightly disturbed sections of the clearing, the abundance of forest herbaceous species decreases and the participation of forest-meadow and meadow species increases. The abundance of meadow species increases significantly, which can be seen both in the increase in their projective coverage and in the height of vegetative shoots.

During the construction and operation of the SVN overhead line, the soil of the clearing was exposed to the impact of aggregates and running systems of heavy machinery. Tracks, spots with the absence of a humus horizon or its small thickness, and bulk soil layers are observed on the surface of the soil of the clearing. With the age of the territory, the pits are laid down, covered with turf and overgrown with herbaceous vegetation so that it becomes difficult to distinguish the borders of the disturbed areas. Embryozems are formed. The research is conducted in the spring period. The absence of vegetation at this time allows you to visually detect soil disturbances by the characteristic shape of the relief.

The choice of this part as a key one is due to the characteristic features of the territory for the southern taiga of Western Siberia - a flat, gently sloping relief, a mixed birch-aspen-pine forest, meadow herbaceous vegetation, favorable climatic conditions, sod-podzolic soils, characteristic of the subzone of small-leaved forests.

On the basis of calculations of the hierarchical structure of soil heterogeneity of the NWZ (Zakharchenko et al., 2022), a grid of observed anthropogenically altered soils with a distance of 40 m between nodes is recommended. The boundaries of soil deposits were identified on a scale of 1:1000. In order to map the contour of a disturbance with a complex shape or the presence of many small pits, the square was divided into 4 parts. A compass and a 40-meter metal measuring tape were used to mark the territory. The morphological structure of disturbed soil composition is described inside the square. The composition of anthropogenic soils and the areal percentage of the presence of a separate type of disturbed soils in the general contour are visually estimated. Assessment of the degree of disturbance within the contour is determined by the prevailing soil type. If possible, the contours of the plots were drawn on a graph paper with an indication of the genesis - construction or operational.

Positions of the contours of the plots using AUTOCAD are plotted on a high-resolution aerial photograph. Each circuit is indexed and listed in a table. The characteristics of the plot are used, such as the area, perimeter and coordinates of the starting point of the contour. The volume of the sample is 384 plots.

To characterize the shape of selected polygons, a calculated variable is used - the shape factor. The form factor is calculated according to the formula:

$$F= 4\pi S/P^2, \qquad (1)$$

where S is the area of the object, P is the perimeter of the object.

The shape factor makes it possible to estimate how close the borders of the contours are to the shape of a circle. Ideally, at $F = 1$, the shape of the contour is close to a circle, for a triangle $F = 0.64$, the more elongated the shape, the cut border, the smaller the value of F.

Random values of contour area characteristics and perimeters have a sharp non-normal distribution, or rather a power-law distribution with heavy tails. Therefore, structural characteristics and non-parametric statistical analyzes such as the Mann-Whitney U-test and the Kruskel-Wallis test are used for statistical analysis.

When preparing a map of the anthropogenically disturbed lands of the NWZ, the publicly available Google aerial image of 2016 was used as a topographic basis. Construction of maps and statistical processing of the obtained data were carried out using AUTOCAD and STATISTICA.

The study of cultured soils was carried out on the example of agricultural land in the Kingisepp district of the Leningrad region, one of the main suppliers of food production for the city of St. Petersburg. The general physical and chemical properties of soils were determined by standard methods adopted in agrochemistry and soil science (Arynushkina, 1970; Vadyunina et al., 1986).

Kingiseppsky municipal district is located in the western part of the Leningrad Region. The area is 290.7 thousand ha (3.9% of the territory of the Leningrad Region). It ncludes 11 municipal entities, 2 cities (Kingisepp, Ivangorod), 9 rural settlements, 188 rural settlements. The population is 73,714 people.

Figure 2. Location of the Kingisepp district on the map of the Leningrad region

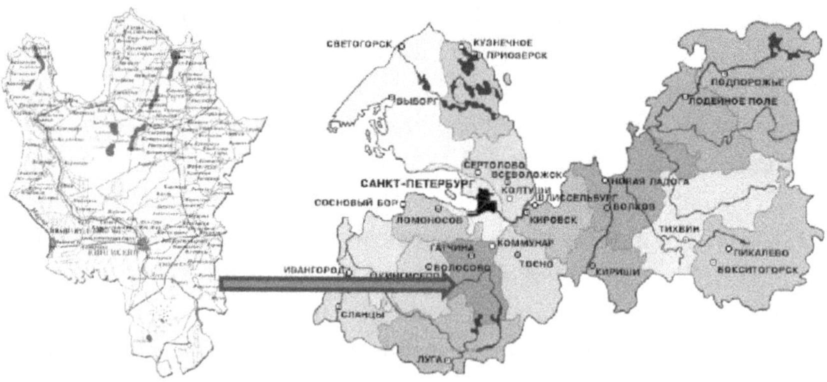

The Kingisepp municipal district has access to the Gulf of Finland, to the state border of the Russian Federation with Estonia, and borders with the Lomonosov, Volosovsky and Slantsev municipal districts of the Leningrad region (Fig. 2).

The territory of the district has a flat relief, absolute heights of 0 - 150 meters above sea level. To the east of Luzhskaya guba is the Soykinskaya upland, and in the valleys of large rivers - lowlands (Luzhskaya and Plyusskaya). The Baltic-Ladozhsky escarpment passes through the territory of the district. There are deposits of phosphorites, construction sand, glass sand, and diatomite on the territory of the district.

The climate is Atlantic-continental. Winter is moderately mild; summer is moderately warm. Average temperatures: January is -4.9o, July - +18.2o. Absolute minimum - 40 o, absolute maximum +35 o. The average annual amount of precipitation is 557-609 mm. Maximum precipitation falls during the growing season. The number of days with precipitation is 18 per month. The northwest wind prevails. The height of the snow cover is 16-63 cm (on average - 41 cm). The duration of the stay is 127-152 days. The depth of soil freezing is 0.5 m. The moisture reserve is 63 m3/ha. The duration of the frost-free period is 126-155 days. Characteristic: inconstancy of weather associated with frequent changes of air masses and dangerous weather phenomena: strong winds, fog, short-term intense showers and prolonged rains, thunderstorms, etc. In the area there is a developed hydrographic network belonging to the Baltic Sea basin. Main rivers: Narva, Luga, Sista, Rosson. There are 11 lakes on the territory. The largest are Kopanskoe, Lipovskoe, Beloe, Glubokoe, Babinskoe, Khabolovskoe. There are reserves of sapropel (2180 and 460 thousand cubic meters) in lakes Khabolovskoe and Muraveyskoye. Sapropel is a centuries-old bottom sediment used in agriculture as a fertilizer. It is especially effective on acidic soils.

Mineral resources of the Kingisepp district. Phosphorite, limestone, dolomite, peat, clay and sand deposits. After modernization, industrial group "Phosforit" increased the production capacity of nitrogen-phosphorus fertilizers (ammophos and diammophos) by 30% to 1 million tons per hour. The EuroChem company intends to supply about 95% of mineral fertilizers to Latin America and the Baltic countries. Soils in the western part of the district are mostly sod-podzolic, marshy and peaty, alluvial in the Luga river valley, sod-carbonate in the east, weakly podzolic and medium-podzolic in the north. The main soil-forming rocks are sands and loams, peat, clays and loams.

Features of the relief, soil-forming rocks, moistening, atmospheric water runoff and the level of standing groundwater contribute to a great variety of the soil cover of this territory. Podzolic, sod-podzolic, and swampy soils that require liming and the introduction of sufficient doses of organic and mineral fertilizers predominate. In almost the entire territory of the region, the soil is stony. After the accident at the Chernobyl nuclear power plant, there are local areas with cesium-137 soil contamination in the area of the Kingisepp district in the range of 1-5 Ki/ /km^2. The content of heavy metals, sulfur, and pesticides generally does not exceed the MPC.

Secondary aspen-birch forests prevail on the territory of the Kingisepp district. Primary spruce and pine forests cover a relatively small area. The level of forest cover in the southern part of the district is high, in the northern part it is medium. Significant territories are occupied by agricultural land. Among the mammals living in the area are moose, wild boar, hare, squirrel, muskrat, common beaver, wolf, raccoon dog, and fox. From birds - capercaillie and gray partridge. The fauna of the Gulf of Finland is represented by the local duck, the migratory duck, the seal, and other.

According to Rosstat data for 2023, the total area of agricultural land in the Kingisepp district is 62,200 hectares, incl. The area of agricultural land is 35,313 hectares, of which 22,167 hectares are arable land, 6,926 hectares are pastures, 4,958 hectares are hayfields, and 1,262 hectares are perennial crops.

Agroclimatic zoning. The basis of the agroclimatic zoning of the Leningrad Region is the provision of heat to the territory during the growing season and the characteristics of the winter period, which determines the overwintering of crops. The following are taken into account: the sum of active temperatures above +10 °C, the frost-free period, the average of the absolute minimums, the sum of soil temperatures at a depth of 10 cm above +10 °C for June - September, the Selyaninov hydrothermal coefficient, the mechanical composition of the soil and physical and geographical conditions (Fig. 3).

Figure 3. Agroclimatic zones of the Leningrad Region

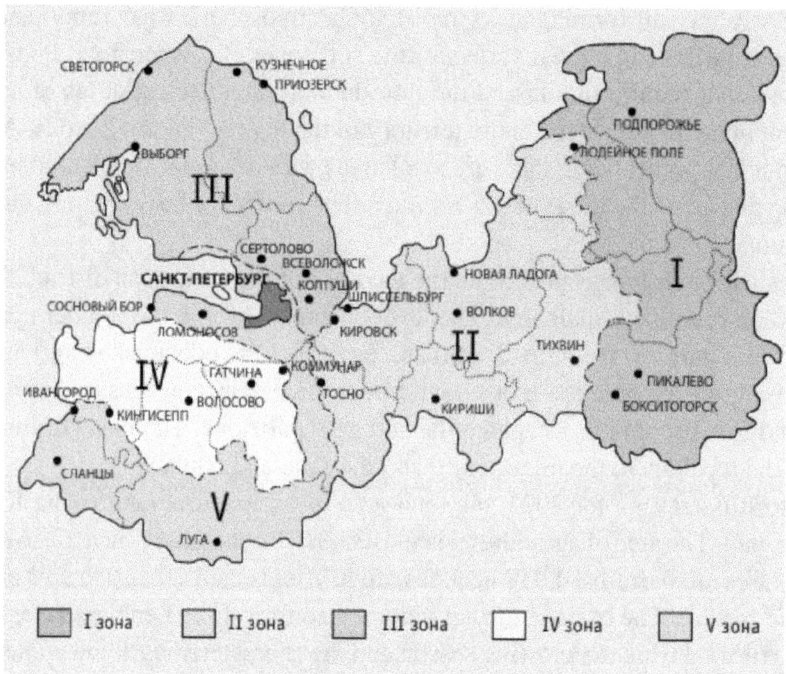

Lands of agricultural use in the Kingisepp district are assigned to the Fourth and Fifth agro-climatic districts.

The IV district is characterized by a greater duration of the frost-free period and a greater supply of moisture than in the other districts.

The V district is the warmest, with the highest sums of active temperatures (1700 - 1900 °C) and the longest period with average daily temperatures above +10 °C (120 - 125 days). It is characterized by the best wintering conditions for fruit and winter crops in the region (tables 1, 2).

Table 1. Characteristics of the agroclimatic potential of the Leningrad region

Agroclimatic district	T10	MF	CC	APAK
Fourth (partially Kingisepp district)	1600	1,12	138,8	7,50
Fifth (partially Kingisepp district)	1800	1,05	138,8	7,91
Leningrad region	1600	1,05	144,0	7,08

Notes: T10 is the sum of positive temperatures for the period with an average daily temperature higher than + 10 degrees Celsius; KU – moisture factor; CC – coefficient of continentality; APAK - agroclimatic potential according to I.I. Karmanova.

Table 2. Distribution of lands in agricultural areas of administrative districts by agrogenetic soil groups

Regions	Area of agricultural land, thousand Ha	Areas of agrogenetic groups of soils in % of the area of agricultural lands							
		1	2	3	4	5	6	7	8
Kingiseppsky	35,4	24,4	18,5	16,4	8,2	18,9	0,5	2,7	10,4
Luzhsky*	70,5	37,5	15,5	8,6	0,7	22,5	0,3	9,6	5,3
Tykhvinsky**	22,8	37,8	20,7	12,3	3,4	0,4	5,1	8,5	11,8
Leningrad region	617,5	27,5	19,1	17,3	4,7	17,7	1,7	6,6	5,4

Note – * maximum and ** minimum values for the Leningrad region.

Soils are found in the Kingisepp district:

1. Sod-podzolic
2. Sod-podzolic weakly gleyish
3. Turf-podzolic gleyish
4. Turf-podzolic gleys
5. Turf-carbonate
6. Sod carbonate
7. Peat-podzolic
8. Peat lowlands
9. Alluvial

Figure 4. Examples of soils in the Kingisepp district

Figure 5. Land fund. In the Kingysepp district, forest fund, and agricultural land predominate

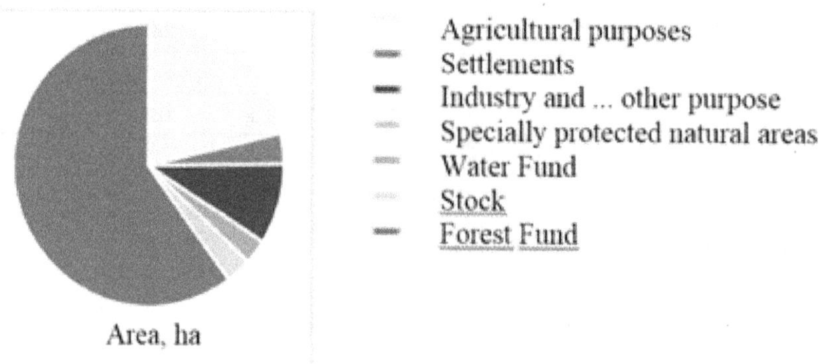

The area of agricultural land is 61,424 hectares, including the area of agricultural land (arable land, hayfields, pastures) – 35,643 hectares. About 82 percent of agricultural lands (50,555 ha) are owned, permanently (perpetually) used, or leased. Among the plots, lands under forests and tree-shrub vegetation and under surface water are most often found (Fig. 6).

Figure 6. Distribution of the land fund of the Kingisepp district by land

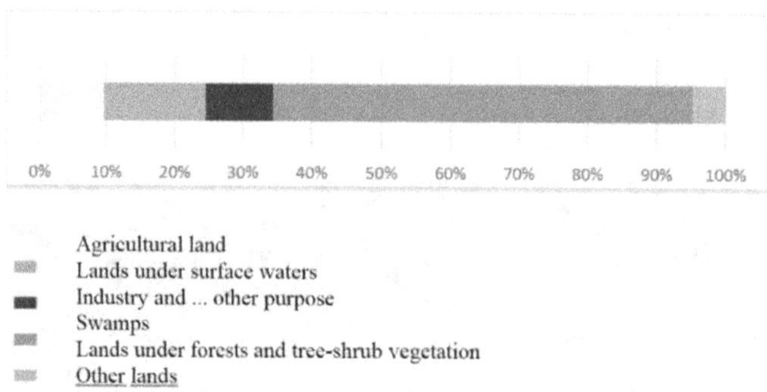

The area of agricultural land in the Kingisepp district is 1.7 times larger than the average for the region (Fig. 7).

Figure 7. Classification (%) of agricultural land by land quality classes in Kingisepp district and Leningrad region (arable land is shown separately). Land quality: I – best (credit rating > 75 points), II – good (61–75 points); III – medium (41–60 p.); IV – lower middle (26–40 p.); V - the worst (< 25 b)

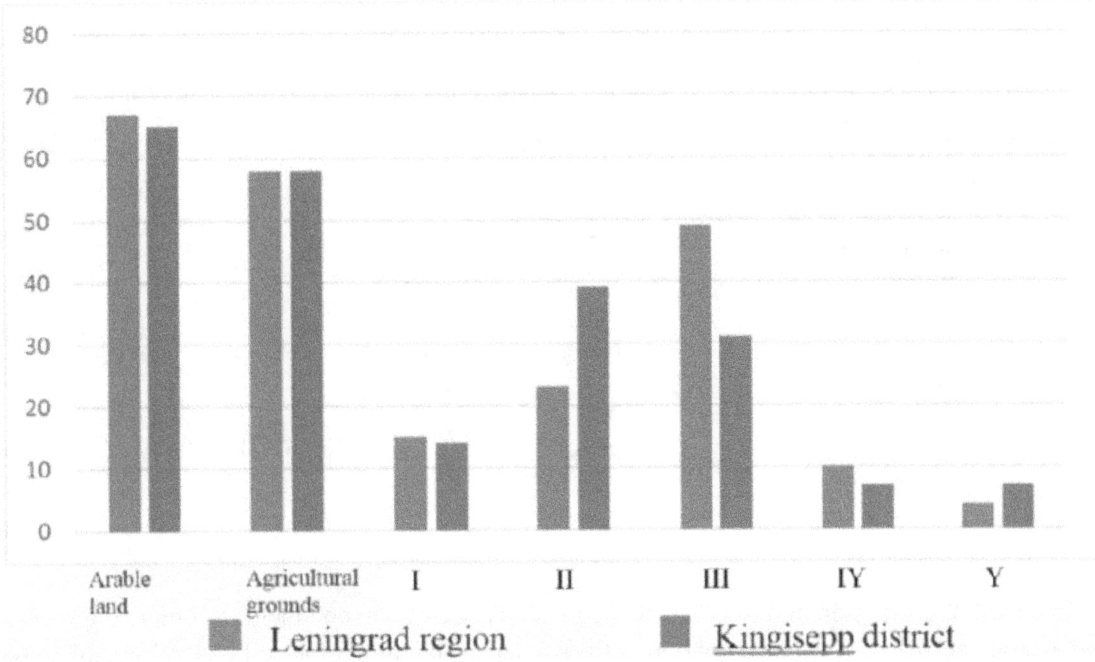

This indicates a high intensity of their use. 47% of the agricultural land is meliorated, however, due to the lack of reconstruction of the drainage network, the reclamation network is overgrown. In general, the ratio of lands of agricultural agreements in terms of land quality in the Kingisepp district and the Leningrad region is comparable. In Kingiseppsk, the land inventory was carried out more than 35 years ago, and the soil science is 21% (fig. 8), which does not allow making effective decisions on their management, controlling measures to improve, protect the land and protect the soil from wind and water erosion and the prevention of other processes and other negative effects on the environment, deteriorating the quality of land.

Figure 8. Soil study of lands in the Leningrad region

In 2020-2022, the "Leningradsky" Agrochemical Service Center conducted a complete agrochemical and ecological-toxicological examination of the soils of agricultural plots with a total area of 26,646.79 hectares. In the borders of the former agricultural enterprises of the late eighties. Their effective targeted use at the present time can become the most important factor in the stable development of agricultural production, therefore the current results of a complete agrochemical survey of the soils of agricultural plots represent a high scientific and practical value. The area of agricultural land in the Kingisepp district is 1.7 times larger than the average in the region (Fig. 9).

Figure 9. The territory of the land survey of agricultural enterprises in the Kingisepp district of the Leningrad region

RESULTS AND DISCUSSIONS

We have proposed a scheme for using intelligent systems for soil monitoring, in particular, studying the state of agrogenic and anthropogenically disturbed soils (Fig. 10). It includes a number of successive stages. In our work, emphasis is placed on the first stages, since they are specific specifically for soil research (field research, analysis of results and selection of decoding features. Further stages of statistical data processing are unified for a variety of objects in different fields of knowledge. They include the creation of a training sample, calculation statistical indicators and analysis of the decipherability of object properties based on training samples, adjustment, segmentation and the use of artificial intelligence to solve average problems in the chosen direction.

Figure 10. Algorithm of using intelligent systems in soil monitoring

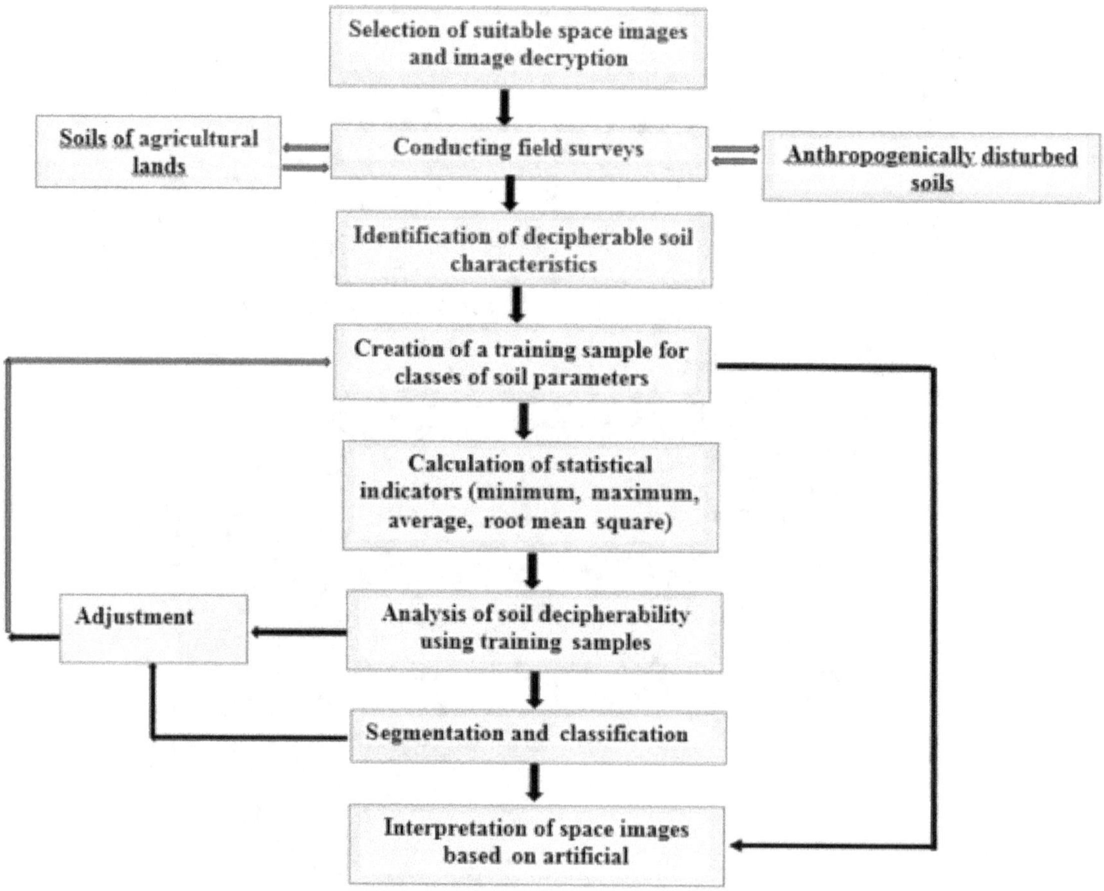

As a result of a survey of anthropogenically disturbed soils, it was established that the heterogeneity of the soil cover of the sanitary protection zone formed on the soils of the forest zone consists of two components: inherited from the parcel structure of the forest biocenosis and superimposed technogenic ones. In turn, technogenic heterogeneity is determined by a planned and random component, creating a mosaic of areas of disturbed land. The soils of the sanitary protection zone of the overhead line SVN have a characteristic set of altered parts of the profile. Accordingly, the classification of anthropogenic soils is adapted to the prevailing type of natural soils and the specifics of mechanical disturbances.

Resectozems are soils that have lost part or all of the upper horizon, abrazems are soils that have lost part or all of the soil profile. Stratolites are outcrops of underlying rocks, mounds of them. Soils under power line supports develop on fill soils. Such soils are classified as stratozems if the thickness of the bulk layer exceeds 40 cm. In plan they have smooth, rounded boundaries.

The most active process of humus accumulation occurs on bulk soils, where the volumetric mass is comparable to what is observed in the humus horizon of undisturbed soils and amounts to 0.8-1.0 g/cm3.

Soils in which the integrity of the diagnostic horizon is damaged or is completely absent were considered abrazem. In abrazem, the density of topsoil is 1.0-1.5 g/cm3. The high density of surface horizons prevents the rapid colonization of vegetation, making it difficult for these soils to recover naturally.

Resectozems of soddy-podzolic soils are divided into two subgroups: slightly disturbed (1) and moderately disturbed (2). Slightly disturbed soils in plan form these soils form sporadically scattered spots throughout the sanitary protection zone, but sometimes form complexes along the edge of abrazems. Soil density fluctuates around 1.0 g/cm3, which contributes to their rapid recovery and reduction of their areas by 20 years of age. The moderately disturbed ones have completely lost their humus horizon, and the eluvial horizon is exposed to the surface, which gives them a whitish color. The volumetric mass of the top soil (topsoil) of the surface horizon ranges from 0.8-1.2 g/cm3

Under the power line there are soils that have been subjected to repeated passage of equipment undercarriages in one place, which has led to the formation of ruts. Tracks can be formed by tractor running units during suspension operations and tension of overhead line wires. Tracks and dirt roads have high density and low porosity, and they occupy insignificant areas, so their properties were not studied, but were recorded in morphological descriptions of soils (tracksol).

Anthropogenic soils are azonal, but their analogs that have undergone natural restoration acquire zonal features.

The identified disturbances were conditionally divided into four groups: bulk (stratozem department), moderately disturbed (resectozem of soddy-podzolic soil), severely disturbed (abrazem department, partial and complete types). Slightly disturbed soils include texturally differentiated resectozems of soddy-podzolic soil, in which the upper horizon has been partially or completely removed. The resectozem subtype is absent in the soil classification (Shishov et al., 2004), but can be formed by analogy with agrozems: in the former it is cut off, and in the latter it is mixed (Table 3).

Table 3. Taxonomic scheme for dividing anthropogenic soils of the sanitary protection zone

Hierarchy	**Taxonomia**		
Order	**Postlitogenic**	**Synlithogenic**	**Notsoil**
Family	Structural differentiated	Stratozemia	Stratolitia
Type	1) Abrazem, 2) Dernovo-podzolic (Ultisol)	Stratozem	Stratolit
SubType	1) Abrazem full, 2) Dernovo-podzolic resectozem (Ultisol resectozem)	Stratozem light-gray	Stratolit light-gray

Embryozems are young soils that form on the surface of technozems. At the time of surveying, the disturbed areas were 20 years old. The rate of restoration of the upper horizon depends on the substrate on which development occurs. Resectozems (1) are almost completely restored, newly formed soddy soils, formed on resectozems of soddy-podzolic soil (2) - 1-3 cm, abrazems - 2-5 cm, and on stratoliths under supports up to 5-7 cm.

Anthropogenic soils are shown on the map (Fig. 11). Construction and operational disturbances of the soil cover are shown separately. The map also shows tracks left by running systems of technical devices and dirt roads.

Figure 11. Spatial distribution in the territory of the sanitary protection zone of the main divisions of the taxonomy of anthropogenic soils: blue - stratozems and stratolites, green - resectozems 1, yellow - resectozems 2 and red - abrazems, dark blue are tracks and dirt roads, the contours are completely shaded - operational soil disturbances, the black line indicates the boundaries of the sanitary protection zone

The number of stratozems turns out to be small, both in area (4%) and in the number of areas per unit area (5.7 specimens/ha) (Table 4). Soils are stored near supports, but there are not so many soils with a bulk layer thickness of more than 40 cm.

Table 4. Biodiversity of anthropogenic soils of the SPZ VL SVN

Departments, soil type	S disturbances, m²	% of total S	N	Occurrence of specimens/ha
Startozems, stratozems on tertiary, quaternary deposits	2847.82	4%	41	5.7
Texture-differentiated resectozem (1)	1555.3	2%	52	7,2
Texture-differentiated resectozem (2)	7989.86	11%	125	17.3
Abrazems, Abrazems complete	7498.581	10%	165	22.8
Area of the studied SPZ	72314.84	100	383	53.0

Texturally differentiated resectozems of soddy-podzolic soils (1) have the smallest amount by area. This is explained by the fact that the survey of anthropogenic soils was carried out 20 years after the construction of the overhead power line. During this time, humus horizons in most cases have already restored their thickness close to natural. The occurrence of this type of soil is higher than that of stratozems due to the larger number of small-sized preserved areas sporadically distributed over the area of the sanitary protection zone. Sometimes they surround contours subject to moderate to strong mechanical stress.

Texture-differentiated resectozem (2) occupies the maximum area (11%), the occurrence is 17.3 specimens/ha. These soils are the least likely to be populated by vegetation and are the slowest to recover.

Abrazem occupies 10% of the SPZ area, but the occurrence is 22.8 specimens/ha, which is the maximum for this property. This is explained by the maximum number of abrazem areas.

The degree of degradation of the SPZ territory can be assessed using the criterion "area of exposed rock" (Methodology for determining..., 1994). With a degradation degree of 0, it is 0-2% of the total area, 1 – 3-5%, 2 – 5-10%, 3 – 11-25, 4 – >25. If we approach it objectively, then the abrazem, although it has the remains of an illuvial horizon (Bt), but it essentially represents a completely disturbed soil profile. If the cutting of the soil layer occurs accidentally, then material from the overlying horizons always remains on the cut surface. If we consider abrases to be cutting off the entire soil, then the degradation of the soil cover of the SPZ VL SVN is between degrees 2 and 3, we take degree = 3.

Random variables of sample values of area and perimeters have a power-law distribution, therefore nonparametric analysis methods were used.

The area of land disturbance during construction is 1.43 hectares, and operational - 0.38 hectares, which as a percentage of the studied area is 19.73% and 5.26%, respectively. Comparison of samples by size of areas of disturbed areas using the Mann-Whitney U-test. showed that their difference is significant at $p>0.95$. The size distribution of areas during construction has a greater variety in size and shape than areas of operational disturbances.

Comparison of disturbed areas by type of anthropogenic soil allows us to identify significant differences between partial abrazem and complete abrazem ($p>0.95$). During construction, disturbances are more extensive and more varied in shape (at $p > 0.95$) than those formed during operation. The shape factor of disturbances during construction has a modal value of 0.8-0.9, during operation - 0.7-0.8, which indicates that in the first case rounded shapes predominate, in the second - elongated and angular. According to the shape factor, the contours of the startozem areas are significantly more rounded ($p>0.95$) than the abrazem areas.

It is necessary to create standards for the areas of formation of men-made soils during the construction and operation of UHV overhead lines; Also, measures should be provided for the reclamation of lands of the sanitary protection zone of the overhead power line in the boreal forest zone.

Thus, the anthropogenic impact on the lands of the SPZ VL SVN leads to the formation of biodiversity of soils of different scale, age and origin of soil cover heterogeneities, differing in morphology, general physical and chemical properties.

It has been established that the biodiversity of the soils of the SPZ VL SVN consists of stratozems in area - 4%, resectozems of soddy-podzolic soils (1) - 2%, texturally differentiated resectozems (2) - 11% and abrazems 10%. The remaining 73% of the SPZ area is occupied by anthropogenically modified soils of undisturbed composition. The gradation of the degree of degradation of the soil cover is 3. It is shown that the distribution of random variables of the area and perimeter of disturbed areas has a power-law form of distribution. During construction, disturbances are more extensive and more varied in shape (at $p > 0.95$) than those formed during operation. A comparison of the distribution of plot areas by type of anthropogenic soil shows that there are significant differences between complete and partial abrazems in terms of the area of disturbance ($p>0.95$). According to the shape factor, bulk soils are significantly more rounded ($p>0.95$) than abrazem areas.

WRB soil classifications need to be refined so that they reflect the full biodiversity of man-made soils (Volungevičius et al., 2003).

Selection of standards for agrogenic soils. Soil acidity. According to a survey in the Kingisepp district, the area of agricultural land with a very acidic soil reaction (pH less than 4.0) is 74.84 hectares, which is 0.3% of the area of agricultural land in the entire region. Soils with a strongly acidic reaction occupy 1004.91 hectares (3.8%) of farmland. Moderately acidic soils (pH 4.6-5.0) are typical for 5627.56 hectares, which corresponds to 21.1% of the total land area; soils on an area of 11,235.36 hectares (42.2%) have a slightly acidic reaction (pH 5.1-5.5); Soils on an area of 6141.42 hectares (23.0%) have a close to neutral environment (pH 5.6-6). Soils with a neutral environmental reaction (pH> 6.0) occupy 2562.7 hectares (9.6%) of the surveyed territories (Table 5, Fig. 12).

Table 5. Share of agricultural land areas by acidity level

Degree of acidity	pH of salt extract	Area	
		ha	%
Very strongly acidic	< 4,0	74,84	0,3
Strong acid	4,1-4,5	1004,91	3,8
Medium acid	4,6-5,0	5627,56	21,1
Slightly acidic	5,1-5,5	11235,36	42,2
Close to neutral	5,6 -6,0	6141,42	23,0
Neutral	> 6,0	2562,7	9,6
TOTAL		26646,79	100,0

Figure 12. Area of agricultural land in the Kingisepp district by degree of acidity Haymaking, pastures, arable land

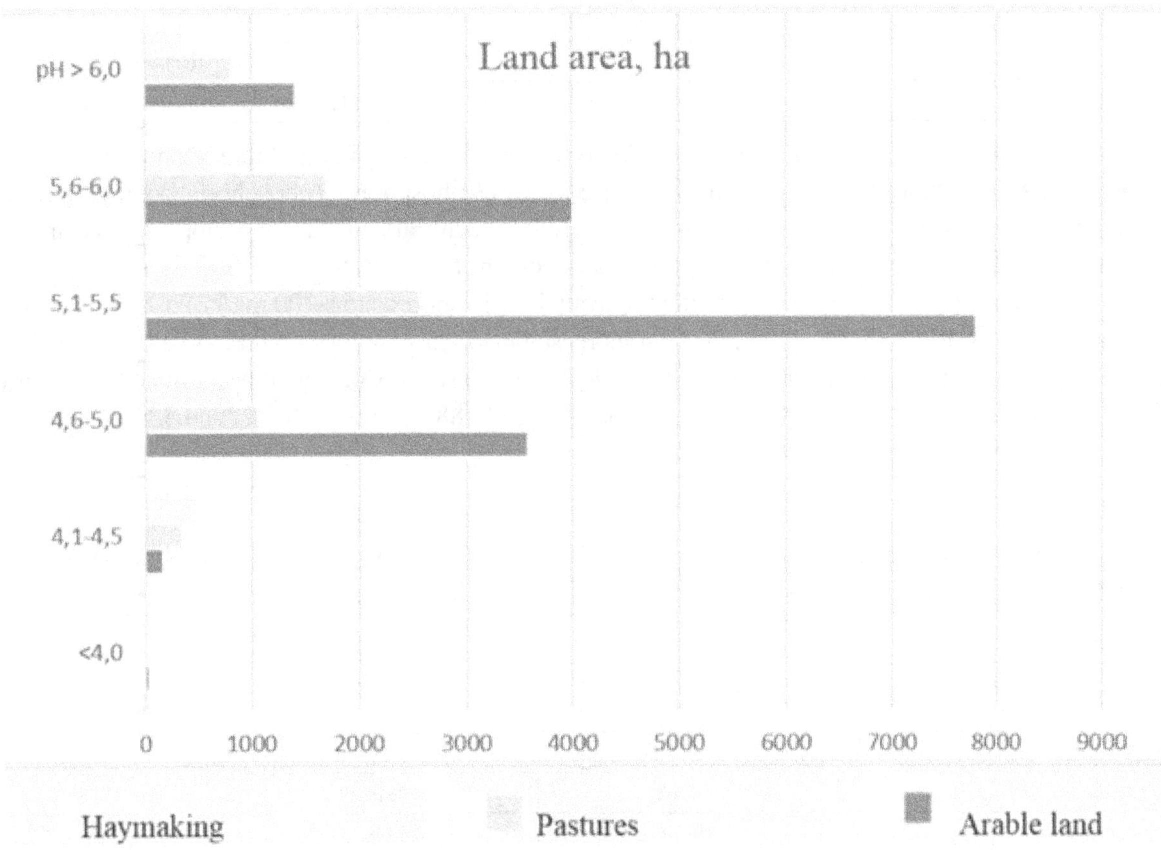

In table 6 shows the grouping of soils on lands of agricultural enterprises according to the degree of acidity

Table 6. Grouping of soils of former farms in the Kingisepp district by degree of acidity, ha

Former farms, conventional names	Very much sour	Strong sour	Medium sour	Weak sour	Close to neutral	Neutral
	pH < 4,0	*pH 4,1-4,5*	*pH 4,6-5,0*	*pH 5,1-5,5*	*pH 5,6-6,0*	*pH > 6,0*
A	-	-	378,57	773,58	1217,80	622,47
B	-	-	347,02	2187,54	828,20	275,73
C	-	101,99	1378,11	2011,66	1751,53	458,75
D	40,63	147,27	244,05	382,46	239,95	-
E	34,21	666,98	1954,72	1200,05	318,34	32,84
F	-	88,67	593,06	2534,06	688,30	30,14

continued on following page

Table 6. Continued

Former farms, conventional names	Very much sour	Strong sour	Medium sour	Weak sour	Close to neutral	Neutral
	pH < 4,0	*pH 4,1-4,5*	*pH 4,6-5,0*	*pH 5,1-5,5*	*pH 5,6-6,0*	*pH > 6,0*
G	-	-	732,03	2146,01	1097,30	1142,77
TOTAL	74,84	1004,91	5627,56	11235,36	6141,42	2562,70

Content of mobile forms of phosphorus in soil. From the data given in table. 7 and fig. 14, it follows that soils with a very low content of mobile forms of phosphorus (phosphorus content less than 25 mg/kg) occupy 0.1% of the territory of all surveyed agricultural lands in the region and this amounts to only 25.42 hectares. Soils with low phosphorus content (from 26 to 50 mg/kg) occupy an area of 411.13 hectares (1.6%). The average content of available phosphorus (51-100 mg/kg) was noted in soils on an area of 3819.29 hectares (14.3%), and increased (101-150 mg/kg) – 5488.35 hectares (20.6%) . The largest area in the surveyed territory of the Kingisepp district is occupied by farmland with a high content and very high content of mobile phosphorus: 7540.88 hectares (28.3%) and 9361.62 hectares (35.1%), respectively.

Table 7. Share of agricultural land areas by content of mobile forms of phosphorus P2O5 content, mg/kg Area

P_2O_5 content, mg/kg		Area	
		ha	%
Very low	< 25	25,42	0,1
Low	26-50	411,13	1,6
Average	51-100	3819,39	14,3
Increased	101-150	5488,35	20,6
High	151-250	7540,88	28,3
Very high	> 250	9361,62	35,1
TOTAL		**26646,79**	**100,0**

Figure 13 shows a diagram of the percentage of agricultural land areas by the content of mobile forms of phosphorus in general, and Figure 15 shows the percentage of agricultural l

Figure 13. Ratio of agricultural land areas by content of mobile forms of phosphorus

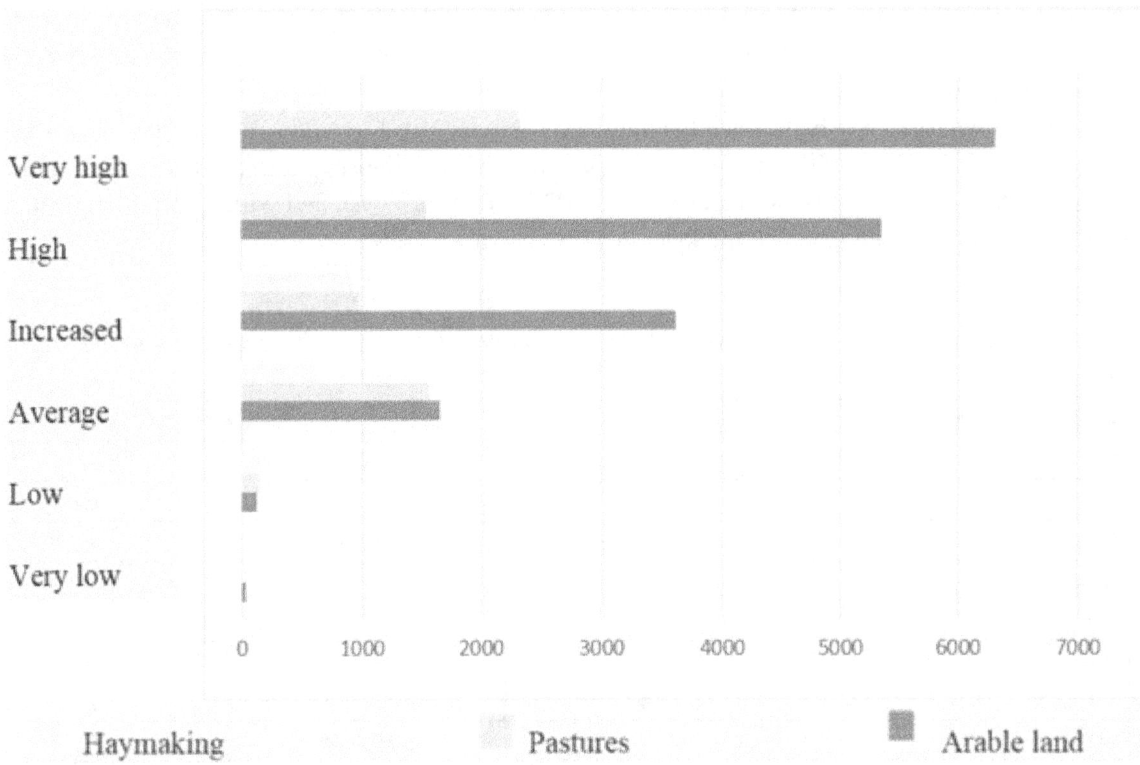

Table 8 shows the grouping of soils according to the content of mobile forms of phosphorus in the context of agricultural enterprises, in Fig. 16 – corresponding grouping of areas in percentage terms, in Fig. 17 – in absolute terms (ha).

Table 8. Grouping of soils according to the content of available phosphorus in the Kingisepp region, (ha)

Former farms, conventional names	P_2O_5 content, mg/ha					
	Very low	Low	Average	Increased	High	Very high
	< 25	*26-50*	*51-100*	*101-150*	*151-250*	*> 250*
A	-	13,95	123,04	173,04	741,65	1940,74
B	25,42	187,15	1270,51	1264,24	692,51	198,66
C	-	16,01	326,26	843,42	1857,94	2658,41
D	-	21,17	407,01	104,85	356,77	164,56
E	-	60,59	306,36	937,58	1283,23	1619,38
F	-	35,62	706,61	1272,09	1441,16	478,75
G	-	76,64	679,60	893,13	1167,62	2301,12
TOTAL	25,42	411,13	3819,39	5488,35	7540,88	9361,62

Table 9 presents data on the content of exchangeable potassium in the soils of the surveyed areas. Soils with very low potassium content (less than 40 mg/kg) occupy less than one percent and amount to about 119.71 hectares. Soils with low potassium content (41-80 mg/kg) occupy an area of 4497.94 hectares (16.9%). The average potassium content (81-120 mg/kg) was found in soils on an area of 8038.26 hectares (30.2%).

Soils characterized by a high content of mobile potassium (121-170 mg/kg) occupy an area of 7537.11 hectares, which is 28.3% of the entire surveyed territory of the Kingisepp district. The area of land with a high content of K2O (171-250 mg/kg) is 4436.03 hectares (16.6%), and with a very high content of mobile potassium (more than 250 mg/kg) - 2017.74 hectares (7.6%).

Table 9. Percentage ratio of areas by the degree of content of exchangeable potassium to the total area of the survey

K_2O,content, mg/ha		**Area**	
		ha	**%**
Very low	< 40	119,71	0,4
Low	41-80	4497,94	16,9
Average	81-120	8038,26	30,2
Increased	121-170	7537,11	28,3
High	171-250	4436,03	16,6
Very high	> 250	2017,74	7,6
TOTAL		**26646,79**	**100,0**

The percentage of soils in terms of the content of mobile forms of potassium in the Kingisepp district can be seen in Figure 14.

Figure 14. Areas of agricultural land with different contents of mobile potassium, ha

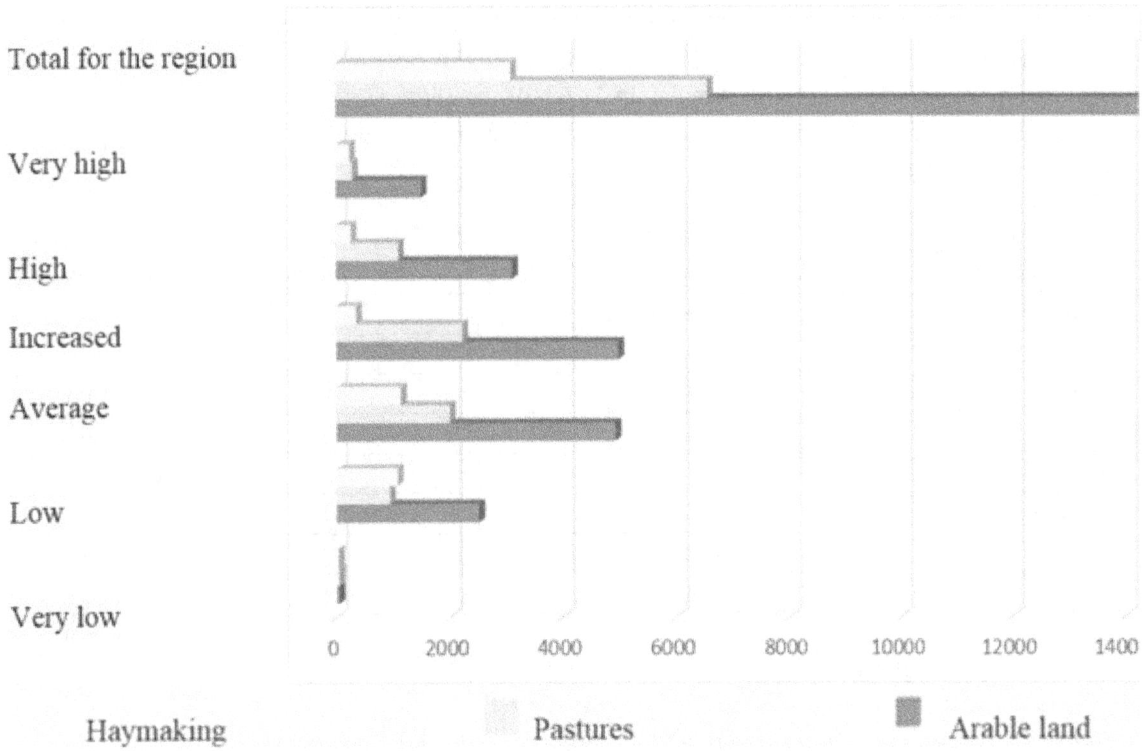

In table Table 10 shows the grouping of soils according to the content of exchangeable potassium in the context of agricultural enterprises. An increased content (6.1-8.0%) of humus is observed in soils on an area of 4549.89 hectares (17.4%), high (8.1-10.0%) – 1400.25 hectares (5.4%), and a very high humus content (more than 10%) on an area of 988.49 hectares (3.8%).

Table 10. Grouping of soils according to the content of exchangeable potassium in the Kingisepp region

Former farms, conventional names	K_2O,content, mg/ha					
	Very low	Low	Average	Increased	High	Very high
	< 40	*41-80*	*81-120*	*212-170*	*171-250*	*> 250*
A	-	746,19	898,08	380,99	633,93	333,23
B	9,14	1169,16	1288,79	671,62	410,29	89,49
C	23,47	866,00	1736,76	1584,63	864,73	626,45
D	-	33,35	169,89	296,89	314,21	240,02
E	87,10	1401,45	1436,14	842,65	263,80	176,00
F		188,06	834,62	1675,78	846,23	389,54
G		93,73	1673,98	2084,55	1102,84	163,01
ВСЕГО TOTAL	119,71	4497,94	8038,26	7537,11	4436,03	2017,74

Content of organic matter in the soil. Table 11 presents the percentage of the areas of surveyed lands according to the humus content of soil organic matter. Analyzing the presented data, we can conclude that most of the soils in the Kingisepp region are characterized by an average level of humus supply. The percentage of soils with a very low (less than 2%) supply of organic matter is 0.4% (106.28 ha). Soils with low humus content (2.1-4.0%) make up 21.6% of the entire surveyed territory (5635.84 hectares). More than half of the farmland is occupied by soils with an average supply (4.1-6.0%) - 13,491.70 hectares (51.5%) (Fig. 15).

Table 11. Ratio of areas according to the degree of organic matter content to the total survey area

Security	Humus content, %	Area	
		ha	%
Very low	< 2	106,28	0,4
Low	2,1-4,0	5635,84	21,6
Average	4,1-6,0	13491,70	51,5
Increased	6,1-8,0	4549,89	17,4
High	8,1-10,0	1400,25	5,4
Very high	> 10	988,49	3,8
TOTAL		**262172,45**	**100**

Figure 15. Areas of agricultural land with different organic matter content

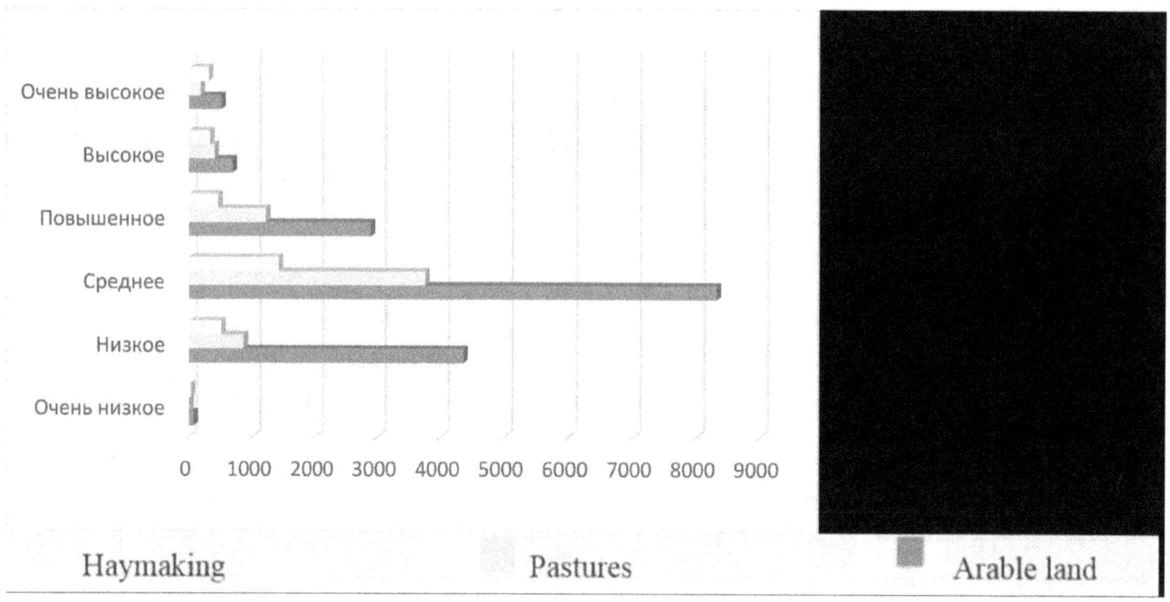

In table Figure 12 shows the grouping of soils according to the provision of organic matter in the context of agricultural enterprises.

Table 12. Grouping of soils of former farms in the Kingisepp district by organic matter content, (ha)

Former farms, conventional names	Humus content, %					
	Very low	Low	Average	Increased	High	Very high
	< 2	*2,1-4,0*	*4,1-6,0*	*6,1-8,0*	*8,1-10,0*	*>10*
A	-	1940,41	776,12	223,46	13,35	9,08
B	-	348,08	2915,07	234,86	71,79	12,94
C	6,72	673,87	1643,96	1930,60	763,82	482,03
D	26,70	60,11	636,69	269,10	37,22	24,54
E	62,71	1092,95	1702,26	626,41	323,91	262,73
F	-	1382,69	2162,91	362,73	-	7,98
G	10,15	137,73	3654,69	902,73	190,16	189,19
TOTAL	106,28	5635,84	13491,70	4549,89	1400,25	988,49

Thus, a comprehensive agrochemical survey conducted by the Leningradsky CAC showed that most soils on agricultural land are characterized by:

- the reaction of the soil solution from very strongly acidic to neutral. The most common soils are those with a slightly acidic reaction. They occupy an area of 11,235.36 hectares (42.2% of the surveyed areas);
- content of available phosphorus from very low to very high. Most of the surveyed soils are characterized by high and very high phosphorus content (7540.88 ha and 9361.62 ha, respectively)
- the content of mobile potassium ranges from low to high, mainly medium and high values (8038.26 hectares and 7537.11 hectares, respectively);

-the content of organic matter (humus) varies from very low to very high. More than half of the soils in the surveyed area are characterized by an average supply of humus (13,491.70 ha).

The soils of the surveyed areas of the Kingisepp district are characterized by the absence of contamination with heavy metals. They are suitable for growing crops.

In order to preserve and increase organic matter in the soil, crop residues should be used as much as possible, green manure fallows should be used, and organic fertilizers should be applied. The systematic use of organic (including straw, green manure) and the application of mineral fertilizers for the planned yield, taking into account the agrochemical characteristics of the soil, the correct scientifically based rotation of crops for a given zone will allow obtaining high and stable yields (Sorokin et al., 2023), the same is the main measure to preserve and increase soil fertility.

RESULTS

To study anthropogenically disturbed soils using the example of power lines, we recommend the following: bulk (stratozem section), moderately disturbed (resectozem of soddy-podzolic soil), heavily disturbed (abrazem section, partial and complete types). Slightly disturbed soils include texturally differentiated resectozems of soddy-podzolic soil. The resectozem subtype, which is absent in the classical soil classification, can be formed by analogy with agrozems: in the former it is cut off, and in the latter it is mixed.

For the study of agrogenic soils of agricultural lands, areas with very low, low, medium, high, high and very high content of organic matter, mobile phosphorus, mobile potassium, with a soil solution reaction from very strongly acidic to neutral, as well as areas covered water, weeds, areas free of vegetation. The condition and productivity of vegetation should be assessed using training standards for further use in processing using artificial intelligence.

A promising assessment of the potential economic productivity of land plots using artificial intelligence and the proposed set of influencing factors (agroclimatic, biological, organizational, technical, etc. conditions) and geoinformation mapping using Earth remote sensing data. The data obtained can be used to develop recommendations for optimizing the use of soil resources in economic, environmental, scientific and economic activities. The created web service will give users wide access to information about the soils of a particular region, as well as other spatial, quantitative and qualitative information on the Internet. It will allow you to effectively carry out economic, inventory, monitoring, research, educational, coordination and management functions.

In general, to monitor thc growth and development of agricultural crops and quickly predict their yield, we recommend the use of the following indicators.

Natural factors:
agroclimatic characteristics by plant development phases:
sums of air temperatures, including active, effective, etc. – above 5°C; above 10°C; above 15oC;
amount of precipitation, mm;
temperature of the top layer of soil;
topsoil moisture;
stroke (temperature difference), incl. frost;
length of daylight hours (Pasko, 2022b);
orographic characteristics of the area height above sea level,
latitude,
steepness (angle of inclination of the terrain relative to the horizon), exposure (position of the terrain relative to the cardinal points);
soil (Zakharchenko, 2007; Zakharchenko et al., 2016)
type;
fertility;
agrochemical characteristics:
degree of soil acidity,
content:
mobile forms of phosphorus,
exchangeable potassium,
organic matter;
pollution
Biological factors:
properties of the variety and/or hybrid:
precocity;
productivity;

resistance to diseases and pests;
Organizational and technological factors (Zharnikov et al., 2019)
fertilizers
organic
mineral
ameliorants
predecessors;
timing of sowing and planting;
tillage
agricultural technology
mineral fertilizers
plant protection products
deadlines
doses.

The main topics for the use of intelligent systems in studying the condition and quality of soils in the future, in our opinion, are:

- Classification of suitability of land plots for economic purposes. Inventory and monitoring of land plots.
- Mapping of soil differences in agricultural lands and forests based on long-term monitoring data.
- Protection of soil resources, assessment of the rate of their degradation and restoration.
- Forecast of the development of the situation with natural disasters (drought, soaking of crops, frost, etc.).
- Agricultural production and monitoring of agroecosystem development.
- Soil fertility management (calculation of doses and timing of fertilizer application).
- Digital remote control of agricultural land.
- Data collection and processing using spatial digital technologies for precision and intelligent farming.

This will require new technologies based on AI, such as:

- Development of deep and machine learning applications of remote sensing data for crop monitoring and management.

Mathematical modeling and algorithms for simulating plant growth in conditions close to real time based on non-contact images and taking into account the complex of factors described above.
Development and improvement of machine learning methods for remote sensing of soil varieties.
Development of remote sensing technologies, artificial intelligence applications and robotic systems to enable digital agricultural production and restoration of degraded soils.
Synthesis of AI and remote sensing imagery for agriculture and soil ecology.
Development of 3D modeling and mapping technologies for agriculture and soil ecology.

The development of these technologies will accelerate the implementation of intelligent systems in the science and practice of agriculture; simplify the management of its progress by unifying and implementing the most common approaches; increase management efficiency by delegating information processing to neural networks; ensure greater sustainability of agricultural production, especially in areas of risky farming; maximize crop yields and improve product quality by taking into account all possible impact factors; increase soil fertility and monitor it; optimize costs and reduce production costs.

REFERENCES

Arinushkina, E. F. (1970). *Guide to chemical analysis of soils. M.* From Moscow State University.

Decree of the President of the Russian Federation dated October 10, 2019 No. 490. On the development of artificial intelligence in the Russian Federation. https://base.garant.ru/72838946

Derbentseva, A. M., Chernovalova, A. V., Surzhik, M. M., Nesterova, O. V., Semal, V. A., Ribachuk, N. A., & Mayorova, L. P. (2015). Podgorodenka natural-technogenic soil catena: Morphological, physicomechanical, and chemical properties. *Contemporary Problems of Ecology*, 8(1), 99–111. 10.1134/S1995425515010023

Karavaeva, N. A. (2008). *Agrogenic memory of soils. Soil as a memory of biosphere-theosphere-anthropogenic interactions.* M., LKI.

Kartashev, A. G. (2000). *Electromagnetic ecology*. TSU.

Kaskevich, E. P., & Plekhanov, G. F. (2003). *Environmental characteristics of super- and ultra-high voltage power transmission.* International scientific and practical conference Transmission of energy by alternating current over long and ultra-long distances, Novosibirsk, Russia.

Kataev, M.Yu., Kartashov, E.Yu., Ryabukhin, V.V., Makarov, E.V., Pasko, O.A. (2023b) Technique for segmenting images of unmanned aerial vehicles using neural networks. *Modern problems of remote sensing of the Earth from space*, *20*(1), 55–66.

Kataev, M. Yu., Pasko, O. A., & Kartashov, E. Yu. (2023a). Analysis of the practical possibilities of using unmanned aerial vehicles in agriculture. *Bulletin of the Krasnoyarsk State Agrarian University*, 1(190), 54–62.

Lvov, Yu.A. (1986).Environmental changes in the construction and operation of overhead power transmission lines of high and ultra-high voltage. *Ultra-high voltage power transmission and ecology.* ENIN im. Krkikanovsky.

Methodology for determining the extent of damage from soil and land degradation. Committee of the Russian Federation on Land Resources and Land Management.

Pasko, O.A., Zakharchenko, A.V. (2021). The Degradation of the Territories of Snow Polygons. *Ecology and industry of Russia*, *25*(11), 30-35.

Pasko, O. A. (2022b). Dependence of potato yield on weather conditions. *Bulletin of KrasGAU.*, 1(178), 56–61. 10.36718/1819-4036-2022-1-56-61

Pasko, O.A., Goryachkin, V.N., Ten, A.D., & Nabieva, M.N. (2022a). Experience of agrarian reforms of the 20th century in Russia. *Economics, labor, management in agriculture,4*(86). 102-110.

Pasko, O.A., Tokareva, O.S., Alshaibi, A.D.A., Chernikova, T.Yu., Kabral, P. (2019a). Assessment of the condition of near-village cedar forests of the Tomsk region using Earth remote sensing data. *Izvestiya Tomsk Polytechnic University. Georesource engineering*, *330*(1), 98-109.

Pasko, O.A., Tokareva, O.S., Ibragimov, E.A. (2020a). Analysis of the ecological hazards of snow bars on the example of the city of Tomsk. *Modern problems of remote sensing of the Earth from space,17*(3),135-144.

Pasko, O. A., & Zakharchenko, A. V. (2020b). Cadastral and geoecological characteristics of especially valuable productive agricultural land. *Bulletin of Tomsk Polytechnic University.Georesources Engineering*, 331(11), 7–19.

Pasko, O.A., Zakharchenko, A.V., Pospelova, E.V. (2019b). Differentiation of agricultural lands by area on the example of the Tomsk region. *Bulletin of Tomsk Polytechnic University. Georesource engineering,330* (5), 100-112.

Platonycheva, Yu.N., Polyakova, N.V., Berchuk, A.V., Bogomolova, Yu.A. (2013). The influence of fallow plowing methods on the mobility of organic matter in light gray forest soil. *Agricultural science of the Euro-North East, 4* (35), 39-43.

Rusanov, A. M., Teslya, A. V., & Sayagfarova, A. M. (2011). Restoration of the humus state of steppe chernozems under long-term fallow land. *Bulletin of Orenburg State University*, 12(131), 132–138.

Shishov, L.L., Tonkonogov, V.D., Lebedeva, I.I., Gerasimova, M.I. (2004). Classification and diagnostics of soils in Russia. *Oikumena Smolensk.*

Sorokin, I. B., Kulizhsky, S. P., Pasko, O. A., & Merzlyakov, O. E. (2023). The influence of long-term use of straw and green manure on gray forest soil on the productivity of agrocenoses. *Агрохимия*, 9(9), 22–27. 10.31857/S0002188123090119

Tokareva, O. S., Alshaibi, A. D. A., & Pasko, O. A. (2021). Assessment of restoration dynamics of burnt forest area vegetation using landsat satellite data. *Bulletin of the Tomsk Polytechnic University. Geo Assets Engineering*, 332(7), 191–199.

Tokareva, O. S., Pasko, O. A., Majid, S. M., & Cabral, P. (2020). Monitoring the state of vegetation cover in the territory of central Iraq using Landsat-8 satellite data. *Bulletin of Tomsk Polytechnic University. Georesources Engineering*, 331(6), 19–31.

Tokareva, O.S., Pasko, O.A., Ushakova, N.S., Makartsova, E.S., Fedorova, L.A. (2018). Comprehensive assessment of the development of vegetation degradation of snow dumps. *Modern problems of remote sensing of the Earth from space, 15*(2), 75-83. .10.21046/2070-7401-2018-15-2-75-83

Vadyunina, A. F., & Korchagina, Z. A. (1986). *Methods for studying the physical properties of soils* (Vol. M). Agropromizdat.

Vågen, Š. P., Daněk, P., Schaetzl, R. J., Vašíčková, I., & Valtera, M. (2016). Soil mixing and genesis as affected by tree uprooting in three temperate forests. *European Journal of Soil Science*, 66(3), 589–603. 10.1111/ejss.12245

Volungevičius, S. J., & Skorupskas, R. (2011). Classification of anthropogenic soil transformation. *Geologija. Vilnius,4*(76), 165–177. 10.6001/geologija.v53i4.1904

Zakharchenko, A.V. (2005a). Morphological features of the surface nanorelief of soddy-podzolic soils of forest biogenocenoses. *Bulletin of Tomsk State Pedagogical University, Series: Natural and exact sciences (information technologies)*, 7(51), 100-108.

Zakharchenko, A.V. (2005b). Natural restoration of anthropogenically modified soils. *Bulletin of Tomsk State Pedagogical University, Series: Natural and exact sciences (information technologies)*, 136-144.

Zakharchenko, A.V. (2007). Spatial conjugacy of morphological soil surfaces. *Bulletin of Tomsk State Pedagogical University, Series: Natural and exact sciences (information technologies)*, 146-152.

Zakharchenko, A. V., Alekseev, V. I., & Ipatova, D. V. (2016). Hierarchical concept of soil heterogeneity and planning of study scale. *Bulletin of Tomsk Polytechnic University.Georesources Engineering*, 327(4), 149–163.

Zakharchenko, A. V., & Pasko, O. A. (2022). *Plants, soils and sustainable farming.*

Zakharchenko, A. V., Pasko, O. A., & Sorokin, I. B. (2021). The dynamics of the reduction of arable land areas according to their long-term monitoring in the Tomsk region. *Ecology and Industry of Russia*, 25(11), 30–35. 10.18412/1816-0395-2021-7-54-59

Zakharchenko, A. V., & Zakharchenko, N. V. (2006). Three-Dimensional Surface Morphometry of Soil Horizons in Field Studies. *Eurasian Soil Science*, 39(2), 134–140. 10.1134/S1064229306020037

Zharnikov, V. B., Larionov, Yu. S., & Pasko, O. A. (2019). On traditions and innovations in domestic land management and cadaster. *Interexpo Geo-Siberia*, 3(2), 73–79. 10.33764/2618-981X-2019-3-2-73-79

Zinck, J. A., Metternicht, G., & Bocco, G. (2016). *Geopedology. An Integration of Geomorphology and Pedology for Soil and Landscape Studies.* Springer International publishing.s 10.1007/978-3-319-19159-1_1

Chapter 12
Optimizing the Solid Waste Landfill Resource Management Model:
Land Usage Efficiency Increase and Negative Environmental Impact Mitigation

Victoria Merkusheva
http://orcid.org/0000-0003-0987-417X
Russian New University, Russia

Maria Liubarskaia
http://orcid.org/0000-0001-9843-1949
Russian New University, Russia

Natalia Klunko
Russian New University, Russia

Tatiana Regent
Russian New University, Russia

ABSTRACT

The chapter presents the results of the study related to the solid waste landfill resource management models used in various regions of the world. Studied parameters include usage of land, collection and utilization of landfill gas and control of landfill's negative impact on the environment. The relevance of the study is due to the fact that all over the world, solid waste landfills are among the largest environmental polluters; they have a negative impact not only on the condition of the ground water and surface water and soils of adjacent territories, but also on the efficiency of use of land resources in the region. The authors consider technological and organizational innovations that contribute to this optimization of solid waste landfill management. The drivers of these innovations are both changes in legislation and technologies that are proven, effective from an economic, environmental and social point of view, and available to the landfill staff. Another important factor is the availability of up-to-date and reliable information about all types of resources involved in the operation process. The suggestions developed by

DOI: 10.4018/979-8-3693-3374-7.ch012

the authors make it possible to take into account the dynamics of investments, as well as the inequality of costs and results carried out and received in different periods of time. The results are of interest both to the solid waste landfill staff and to regional and local authorities, as well as to investors and specialists involved in the development of regional waste management infrastructure.

INTRODUCTION

Even with the use of innovative recycling technologies, a significant portion of waste in various regions of the world still ends up at landfills, which poses a long-term environmental problem. Al-Hazmi H.E., Hassan G.K., Kurniawan T.A., Sniatala B., Josph T.M., Majtacz J., Pietchota G., Li X., El-Gohary F.A., Saeb M.R., and Makinia J. (2024) have summarized the researches executed during the period from 1998 to 2023 by an international society of scientists. They found out that 191 researches had been published witnessing the global negative environmental impact of inadequate solid waste management models implemented locally (Al-Hazmi et al., 2024). This negative impact implies, but are not limited to the release of harmful gases, contamination of soil, ground and storm waters, and depletion of resources. Landfills in different regions of the world occupy huge areas and exclude them from useful economic application for a long time, and often forever.

Despite the fact that humanity is aware of the environmental dangers of landfills, landfilling is expected to remain a major waste disposal option in the near future (Awasthi et al., 2022; Niu et al., 2023).

The aim of the research is to determine the world trends and regional specific in the implementation of solid waste landfill resource management models in order to justify ways to increase land usage efficiency and mitigate negative environmental impact of existing and future landfills.

In the process of reaching the above-mentioned aim of the research, the authors accomplished the following objectives:

- ✓ Survey current world practices based on the academic publications examining the cases of Asian, African, and European countries. Such an analysis allows us to combine the geographical diversity of the practical objects under consideration with the scientific analysis of experts from the relevant locations;
- ✓ Analyse the factors of the negative impact of existing landfills on various environmental components, including the negative impact of an ineffective solid waste landfill management model on soil characteristics;
- ✓ Explore the solid waste landfill resource management model of Krasnoyarsk Territory of the Russian Federation, taking into consideration the distribution of existing waste storage and disposal facilities, as well as incoming waste streams and the impact of their characteristics on the functioning of landfills;
- ✓ Develop and justify proposals for increasing land usage efficiency and mitigating negative environmental impacts of existing and future landfills based on the economic and mathematical optimization model with the criteria of capital productivity of equipping the landfill zones with a biogas collection system. This criterion is directly related to the volume of planned work and reflects the intermediate commissioning and equipping of individual zones of the landfill during the process of creating a biogas collection system.

Research objects represent regional specific of the implementation of solid waste landfill resource management models with emphasis on one of the remote regions of the Russian Federation - Krasnoyarsk Territory. Research methodology imply using mostly secondary data, including official websites and academic journals, as well as statistics. Various programs were used to process the data: the Microsoft Office package, the graphic editor Photoshop, that helped to add translation into English to the map with the locations of waste storage and disposal facilities retrieved from the open sources. The research approach of the article employs inductive reasoning, where conclusions are rooted in the specific cases analysis and brought to the level of general conclusions on the current state of land usage efficiency and negative environmental impact of existing landfills. These conclusions provided authors with the base for the proposed changes to widely implemented solid waste landfill resource management models in the part of implementation of landfill gas collection.

BACKGROUND

World Practices of Implementation of Solid Waste Landfill Resource Management Models

Landfill disposal of solid municipal, industrial and construction waste still remains the most common method of handling them in the world. In modern conditions of urban growth and an increase in the capacity of industrial enterprises, the issues of creating new landfills and waste processing complexes are increasingly being discussed (Ortskhanov, 2010). According to Mvula R.L.S., Mundike J., and Nguvulu A. (2023), one of the big problems is that many of the sites for landfills were selected based on land availability and not on land suitability.

At the same time, the majority of landfills are approaching the limits of their production capacity for filling, moving to the final stage of their life cycle (Gorbovskaya & Gonina, 2015). At this stage, landfills are closing and land is reclaiming. The reclamation process is complicated by the fact that these waste management facilities contain a huge number of components that pose an environmental hazard both to the population living near these landfills and to the environment (Maslikov et al., 2012). In particular, the liquid and gaseous fractions formed in the body of the landfill are chemical and fire hazard factors (Baburina, 2017; Ityshev & Potapova, 2018).

Experience of countries with advanced environmental policies shows that landfill sites can be rationally used to obtain energy resources (Busch et al., 2009; Deublein & Steinhauser, 2011). During the recent years, due to the implementation of national and international policies on sustainable development and growth of public awareness of climate change the demand for renewable energy resources is growing globally (Aleshina, 2010; Consonni et al, 2005; Banar et al., 2009; Awasthi et al., 2022). As Barraqan-Escandon A., Terrados-Cepeda J., Zalamea-Leon E., and Arias-Reyes P. (2018) note, creating conditions for sustainable urban development is impossible without taking into account the regenerative capacity of materials and energy. The tensions between economic growth and environmental protection are becoming increasingly apparent due to rapid industrialization and urbanization. On the one hand, the goal is sustainable development, but on the other hand, energy consumption is increasing. This contradiction can be resolved by involving renewable energy resources in economic circulation. The growth rates of production and consumption waste achieved to date increase the relevance of taking into account their potential in reproduction processes.

According to numerous contemporary researches (Ghinea et al., 2010; Mvula et al., 2023; Mor & Ravindra, 2023) in lower- and middle-income countries solid waste landfill resource management models often include some unorganized and disordered processes. Improper waste disposal in landfills in these countries have severe environmental consequences in terms of groundwater, soil and air contamination, resulting in numerous health hazards. For example in India, landfills require 1240 ha of land per year and only 21% of solid waste is adequately managed (Mor & Ravindra, 2023). The remaining solid waste is not reprocessed or recovered through treatment technologies and is generally disposed of at the landfills. Urban or rural solid waste is usually dumped in low-lying areas, which adds to the greenhouse gas emissions, odor production and causes fire hazards that adversely affect the local ecosystem (Kannankai & Devepiya, 2024). Landfills contribute 29% to CH_4 production and more than 15% to all greenhouse gas global emissions (Mor & Ravindra, 2023).

Based on the analysis of the secondary resources, the following trends in existing solid waste landfill resource management models could be outlined:

- They show low land usage efficiency due to their affection to occupy huge amount of territories in different regions and extract them from the economic circulation;
- Not all of them contribute to the renewable energy resource market despite the increased demand for these resources globally.

Negative Impact of Landfills on the Land Usage and Resource Management

The negative impact of solid waste on the environment is related to the composition of waste streams entering landfills, as well as chemical reactions that arise within the landfill. This fact increases the importance of studying both waste streams entering landfills and the negative consequences of placing various materials in landfills.

The decomposition of solid waste occurs under the influence of microorganisms. During the decomposition process, large molecules of organic waste break down into smaller ones, and inorganic compounds of non-biological waste are oxidized, which releases new hazardous substances. If the waste is exposed to oxygen, they become a source of non-decomposable material and a kind of chemical and biological reactor in which landfill gas is released, decomposition products accumulate and pathogenic microorganisms multiply. Ideally, the residues of solid waste should eventually become a nutrient medium for plants, but often this mass is toxic and unsuitable for living organisms. Due to precipitation and the release of moisture from waste, horizontal and vertical migration of harmful substances occurs, which end up first in the soil and then in surface and ground waters, ultimately the value of maximum permissible concentrations of various pollutants in ecosystems can increase to enormous values. For example, the phenol content can be 900 times higher (Kostyleva & Racheva, 2016).

The occurrence of waste fires in unscientifically managed landfill sites has become a pressing environmental issue in the urban centers of developing economies (Kannankai & Devepiya, 2024). The toxic compound benzopyrene is formed in the temperature range from 800° C to 1000° C during the combustion of any organic material; therefore, it is formed in abundance during open combustion of solid waste. From soil and water bodies, benzopyrene migrates first to plants, and then through the food chain into the bodies of animals and humans. The toxin enters the human body not only through the digestive system, but also through breathing and skin, the placenta, and easily accumulates in fatty tissues. This substance has a carcinogenic, mutagenic and teratogenic effect on organisms due to the interaction of its derivatives with proteins and nucleic acids, and is highly stable in the environment and living organisms (Kostyleva & Racheva, 2016).

Some of solid waste components contain heavy metals, which, together with dust, spread over long distances. Heavy metals from the soil and environment enter plants, inhibit their growth and development, disrupt the functioning of the photosynthetic apparatus and respiration, negatively affect the transport of substances, and sometimes lead to the complete death of plant communities (Elizarova & Ivanushkina, 2023). The toxic effect of heavy metals has a great impact on the productivity of aquatic communities and the quality of water in general, which reduces the amount of fresh water suitable for drinking and poses a threat to the future of humanity. They have a teratogenic effect, accumulate in tissues, bones and blood, contribute to the development of gastrointestinal cancer, and reduce immunity and cause mental retardation in children.

Ochs C., Garrison K., Saxena P., Romme K., and Sarkar A. (2024) emphasize the significance of landfills as sources of persistent organic pollutants (POPs). These organic chemical substances threaten human health and the planet's ecosystems due to their toxicity and ability to remain intact for a long time, wide distribution throughout the environment, and accumulation and magnification in living organisms through the food chain.

Solid waste landfills create a favorable environment for the formation and development of pathogenic flora and potential carriers of various diseases - flies, rodents of various species. Waste from research and medical organizations (dentistry, surgery) is especially dangerous. Each material brings extra concerns about the future of landfill sites. For example, the processes of chemical corrosion of glass occurring on the outside of glass containers remain without due attention, because practically have no effect the consumer properties of products packaged in glass. These processes begin to play a decisive role after the glass containers are placed to the landfills as a part of solid waste (Gurevich et al., 2013). From this point on, in addition to the simple accumulation in the environment of non-decomposing and, often traumatic, cullet waste, the problem of chemical pollution of soils, surface and groundwater with filtrate runoff from solid waste storage sites arises due to their alkalization after the chemical interaction of glass and atmospheric precipitation.

In their research, Shi X., Chen Z., Wu L., Wei W., Ni B.J. (2023) noted that while many artificial and natural biotops have documented the presence of micro plastics, there is still a lack of knowledge about these tiny particles in municipal solid waste landfills. Micro plastics serve as important vectors inevitably interact with other pollutants in landfills. These vectors carrying hazardous contaminants aggravate the adverse impact on the environment.

Solid waste may also include petroleum industry products: used oils and old oil filters, dirty work clothes and various cleaning materials. Oils from solid waste penetrate into the soil, after which they migrate to the lower layers, envelop plant roots and enter groundwater. Under the influence of oils, soil particles stick together and form large hydrophobic blocky particles larger than 1 cm in size, because of this the number of small soil particles valuable for agriculture sharply decreases (Kostyleva & Racheva, 2016). Soil pores become sealed and lose the ability to absorb water, access to oxygen becomes difficult, the water-air balance and root nutrition of plants are disrupted, the nitrogen regime of the soil worsens, the number of mobile forms of phosphorus decreases, and the carbon content increases. Due to oxygen deficiency, the activity of anaerobic microorganisms that absorb nitrogen and phosphorus from the soil increases. Residues of petroleum products penetrate into plants and, passing through cell membranes, inhibit growth and development, reduce the activity of photosynthesis and lead to the death of crops.

In the process of all above-mentioned interactions between different components of solid waste landfill gas generation occurs. The main components of landfill gas are methane (40-75%) and carbon dioxide (25-45%), as well as nitrogen dioxide (Busch et al., 2009).

Russian Practices of Implementation of Solid Waste Landfill Resource Management Models

The strategic goal of state environmental policy of the Russian Federation is to ensure environmentally oriented economic growth, preserve a favorable environment, biological diversity and natural resources to satisfy the needs of current and future generations (Fundamentals of State Policy, 2012; Strategy for the socio-economic development, 2021).

In early 2000s, the Russian Federation was actively developing solid waste management legislation and adopted many state standards related to resource conservation (GOST R 1.0-2004; GOST R 53691-2009; GOST R 54098-2010). However, the historical practice of designing and constructing solid waste landfills in the Russian Federation and neighboring countries did not involve equipping them with biogas collection systems at the initial stage of the life cycle (Aleshina, 2010; Kirsanova & Filippova, 2008). As a result, with the growing popularity of the ideas of sustainable low-carbon development and the transition to the principles of a circular economy, the need to create such systems at subsequent stages began to arise in various regions (Oganesyan et al., 2005).

Lyubinskaya T.V. (2010) notes that in Russia is just at the initial stage of industrial development of the production of biogas and energy from solid waste. There is experience in operating solid waste landfills with biogas utilization in the cities of Serpukhov and Mytishchi in the Moscow region of Russia. Since the beginning of the 2000s, several projects have been implemented in the Leningrad region of Russia to organize the collection of biogas at landfills.

In 2013–2015, the Prodex group of companies implemented a pilot investment project at the Novy Svet-ECO landfill in the Gatchina region of Russia to create a generating facility operating on landfill gas as a renewable energy source. The volume of electricity generation is 4.8 MW. The generated energy is transferred to the electrical networks of PJSC Lenenergo and purchased at the approved tariff (Kysyydak & Salchak, 2013). The implementation of such projects contributes to the achievement of the strategic goals of the Government of the Leningrad Region of Russia to ensure environmental safety, including by preventing the harmful effects of production and consumption waste on human health and the environment. At the same time, companies investing in the creation of biogas collection systems at solid waste landfills, with a rational organization of the process, receive not only an environmental and social effect that is important for authorities and local communities, but also an economic effect that is significant for the owners.

In the last few years, the issue of rational use of solid waste landfill resources at the final stages of its life cycle has become relevant to the Northern Samarka landfill located in the Leningrad region of Russia. The landfill was built almost 50 years ago and was originally intended exclusively for construction waste. Due to the increased demand for waste disposal facilities in the Leningrad region of Russia, during operation the landfill began to accept waste of all hazard classes. The location of the landfill on swampy soils and the heterogeneity of the waste buried on it increases its potential chemical and fire hazard. At the same time, around the landfill there are several large settlements with a demand for electricity. Landfill gas generated at that landfill could be successfully sold to the local greed, but such projects are still under consideration. As analysis shows, even in the central regions of Russia, such as

Moscow and Leningrad regions, solid waste landfill resource management models doesn't employ landfill gas collection systems to increase the efficiency of land usage and mitigate the negative impact on the environment. Krasnoyarsk Territory have a huge amount of solid waste and a great potential to become pioneers in landfill gas collection.

Specifics of Regional Solid Waste Management Models: The Example of the Krasnoyarsk Territory

The policy of the Krasnoyarsk Territory in the field of environmental protection is focused on achieving the general goal of the State Program of the Russian Federation *Environmental Protection* (Decree of the Government of the Russian Federation#326, 2014). The goal of the Program is to increase the level of environmental safety and preserve natural systems.

A number of historical circumstances complicate the functioning of the waste management system in the Krasnoyarsk Territory. Among them, the following characteristics should be noted.

1. Uneven distribution of population across the region. According to data for 2023, population density ranges from 1.20 people/sq. km in certain areas to 2.8 people/sq. km in the city of Krasnoyarsk. This variation in population density makes it difficult to provide low-population areas with waste collection and recycling systems. Filling containers with solid municipal waste in such areas is slow, and organizing the processing of small volumes is not economically feasible.
2. Low provision of the region with road infrastructure. Krasnoyarsk Territory occupies one of the last places both in Russia and in the Siberian Federal District in terms of the density of public roads and railways. The low value of these indicators is explained by the fact that most of the area of the Krasnoyarsk Territory is made up of Taimyr Dolgano -Nenets and Evenki municipal districts, where transport is very poorly developed. This fact has an impact on the solid waste removal system of the Krasnoyarsk Territory.
3. The presence of the largest rivers in the world on the territory of the region, including the Yenisei and its tributaries. This circumstance increases attention to the selection of locations for organizing new solid waste landfills, since any failure in the operation of the landfill in terms of collection and treatment of leachate can lead to contamination of ground and storm waters, spreading over significant distances. The population of the Krasnoyarsk Territory as of January 1, 2024 was 2,847,826 people and is a fairly stable indicator.

In 2016, the Institute of Design, Ecology and Hygiene, LLC (Saint-Petersburg, Russia), developed the Territorial Waste Management Plan for the Krasnoyarsk Territory. The total amount of municipal solid waste generated in the Krasnoyarsk Territory was estimated at 970,415 tons, including 46,580 tons of bulky waste; the average density was estimated at 251.75 kg per cubic meter.

In accordance with this document, the accumulation of municipal solid waste can be carried out in several different ways:

- in containers located on container sites;
- using a garbage chute;
- in containers for the accumulation of bulky waste;
- in packages placed in designated places;
- by accepting waste upon request;
- by driving around the territory and accepting waste according to a schedule;

- in containers separately for different types of waste.

At present, in the Krasnoyarsk Territory, the most common system is one-stage removal of municipal solid waste (MSW) with preliminary collection in the metal containers of various volumes. Current legislation requires that the accumulation and storage of waste of production and consumption generated as a result of the activities of business entities is carried out by business entities independently in places specially equipped for these purposes on their own territories. Waste removal is carried out by specialized enterprises in accordance with concluded agreements. While developing the Territorial Waste Management Plan of the Krasnoyarsk Territory, 14 technological zones have been identified. One Regional Operator was selected for each technological zone (Table 1).

Table 1. Regional operators of the Krasnoyarsk Territory by technological zones

Name of Technological Zone	Regional Operator
Achinskaya	Eco-Transport LLC
Zheleznogorskaya	RostTech LLC
Zelenogorskaya	Ecoresurs-PromTech LLC
Kansko-Abakanskaya	Recycling Company LLC
Krasnoyarsk left bank	Krasnoyarsk Recycling Company LLC
Krasnoyarsk right bank	RostTech LLC
Lesosibirskaya	Recycling Company LLC
Minusinskaya	MUP “Minusinsk Municipal Economy” LLC
Nazarovskaya	Eco-Transport LLC
Norilskaya	RostTech LLC
Rybinskaya	Agropromkomplekt LLC
North	JSC “Avtospetsbaza”
North Yenisei	MUP “UKRR”
Taimyrskaya	RostTech LLC

On the land of the Krasnoyarsk Territory, waste disposal facilities are mostly represented by solid waste landfills. There are also some solid waste sorting facilities in the region (Table 2).

Table 2. Solid waste sorting facilities at the Krasnoyarsk Territory

Name of Sorting Facility	Technology	Design Capacity, tons per year
Prosoft LLC	Manual sorting of MSW	No data
Eco-Transport LLC	Sorting of MSW	100 000
MSK South-West LLC	Manual selection of useful fractions from MSW moving along a conveyor line	35 000
MSK Vostok LLC	Manual selection of useful fractions from MSW moving along a conveyor line	25 000

continued on following page

Table 2. Continued

Name of Sorting Facility	Technology	Design Capacity, tons per year
Recycling Company LLC	Manual selection of useful fractions from MSW moving along a conveyor line	35 000
RostTech LLC	Waste sorting complex	400 000
Ecoresurs-PromTech LLC	Technological complex of MSW sorting line using a conveyor of the KPS1M feeding and sorting series	40 000
LLC "Improvement and landscaping"	Mobile sorting complex "Ural-Sot-3"	25 000

To identify ways to improve existing waste management methods, including the use of landfill resources, it is necessary to conduct a qualitative and quantitative assessment of incoming waste streams. In order to minimize accumulated environmental damage, waste disposal, processing, neutralization and disposal facilities must comply with the requirements of environmental legislation, taking into account established restrictions and prohibitions, including those provided for in of the Federal Law On Production and Consumption Waste # 89-FZ (1998), mandatory requirements for solid waste disposal facilities include the following:

- recording incoming waste characteristics;
- measuring weight of incoming waste;
- maintaining access roads, fencing, accumulation and removal of filtrate, biogas, disinfection baths;
- implementing an environmental control program.

The main problems of solid waste collection in the Krasnoyarsk Territory led to the formation of unauthorized dumps include:

- the high cost of an individual container and, as a consequence, the lack of supply of containers in most rural settlements;
- the lack of incentives for owners of individual housing to pay for solid waste collection containers during their projects (e.g., construction),
- the high cost of individual hauling of solid waste from remote rural settlements, which is aggravated due to the one-stage removal scheme and the large transportation distance.

The sources of solid waste generation at the Krasnoyarsk Territory include both: the population and organizations (municipal solid waste - MSW), and enterprises (industrial solid waste). Table 3 presents the results of forecasting of MSW generation in the areas of 14 technological zones accomplished during the development of the Territorial Waste Management Plan for the Krasnoyarsk Territory (2016).

Table 3. Forecast of municipal solid waste generation at the Krasnoyarsk Territory by technological zones

Name of Technological Zone	Weight, thousand tons		Volume, thousand cubic meter	
	2025	2030	2025	2030
Achinskaya	68,83	69,81	273,40	277,29
Zheleznogorskaya	47,42	48,10	188,36	191,04
Zelenogorskaya	22,20	22,51	88,17	89,42
Kansko-Abakanskaya	75,87	76,95	301,38	305,66
Krasnoyarsk left bank	276,31	280,24	1 097,56	1 113,17
Krasnoyarsk right bank	179,64	182,20	713,56	723,71
Lesosibirskaya	29,13	29,55	115,72	117,37
Minusinskaya	97,63	99,02	387,82	393,34
Nazarovskaya	73,40	74,44	291,54	295,69
Norilskaya	47,43	48,10	188,40	191,08
Rybinskaya	36,28	36,80	144,12	146,17
North	28,99	29,40	115,15	116,79
North Yenisei	2,90	2,95	11,53	11,70
Taimyrskaya	8,36	8,48	33,20	33,67

The morphological composition and density are two most significant characteristics of municipal solid waste. Detailed data on the morphological composition of MSW and its dynamics constitute the initial information for assessing the feasibility of extracting components from the waste stream. It allowing determination of the economic benefits of using valuable components of solid waste, which make it possible to obtain products from recycled materials that are in demand on the market. The latest waste characterization executed during the development of the Territorial Waste Management Plan for the Krasnoyarsk Territory (2016) is presented in the Table 4.

Table 4. Morphological composition of MSW in the Krasnoyarsk Territory

Component Name	Share in the composition of MSW, %
Organics (food waste)	21,4
Waste paper and cardboard	12,4
Waste metals	3,1
Waste glass	12,1
Waste polymers	10,7
Textile	3,4
Bones (biological waste)	0,2
Waste rubber, leather, shoes	2,2
Diapers	5,6
Wood waste	2,2
Construction and demolition waste	12,1
Waste of hazard classes 1-3	2,6

continued on following page

Table 4. Continued

Component Name	Share in the composition of MSW, %
Landscape waste	6,1
Residues from street sweeping	5,9
Total	**100**

With the development of technology and industrial growth, environmental pollution began to increase, reaching the limits of the environment's adaptive ability to cope with huge volumes of industrial waste, which are disposed of in landfills without proper treatment. The industrial sector consumes an increasing amount of resources, which leads to the generation of a corresponding amount of solid waste. However, with increasing environmental awareness, recycling and reuse of industrial waste has undergone significant changes (De Feo & Malvano, 2009; Isakova, 2020; Bukrinskaya, 2020).

The main types of economic activity at the Krasnoyarsk Territory are the following: mining, manufacturing, generating electric power, gas and steam, water supply and wastewater treatment, waste collection and disposal, and agriculture.

In accordance with federal legislation, legal entities and individual entrepreneurs have responsibility for the environmentally safe management of industrial waste they generated. According to reporting in Form 2-TP (waste), the volume of waste generated at enterprises in the Krasnoyarsk Territory is about 8,488,422 tons.

Industrial waste is characterized by:

- heterogeneity of composition;
- variety of types;
- inconsistency in the amount of generation.

The Territorial Waste Management Plan for the Krasnoyarsk Territory (2016) proposes:

- strengthening control by municipalities over legal entities in the field of waste storage and removal;
- maximum use of the resource potential of waste at the enterprises where it was generated, focusing on the use of waste in their own technological processes and/or its processing into secondary raw materials and secondary products;
- processing of industrial waste within the framework of the municipal waste management system (when concluding contracts with licensed organizations on market conditions).

The accumulation of industrial waste on the production site is carried out on a workshop basis or centrally. The conditions for collection and accumulation are determined by the hazard class of the waste, the packaging method, taking into account the physical state and reliability of the container.

Accumulation of solid waste is permitted:

- for class I - exclusively in sealed returnable containers;

- for class II - in securely closed containers (plastic bags, plastic bags);
- for class III - in special containers, in paper or textile bags;
- for class IV - in containers, in bulk.

The maximum accumulation of waste at each enterprise at a time is determined by that enterprise based on the balance of materials, and the waste inventory, taking into account their macro- and micro-composition, physical and chemical properties, including their state of aggregation, toxicity and levels of migration of waste components into the air. The storage conditions are specific to various types of industrial waste (Table 5).

Table 5. Requirements for storing the main types of industrial waste

Type of Waste	Storage Conditions
Used batteries	Temporary storage should be carried out in a room inaccessible to unauthorized persons, in a stack or on racks
Used oils	Waste should be stored in metal or plastic barrels placed on metal pallets or on a reinforced concrete cover, and, as accumulation occurs, transported to a specially designated storage area
Waste paints and varnishes	Waste should be stored in metal or plastic barrels placed on metal pallets or on a reinforced concrete covering, and, as it accumulates, transported to a specially designated storage area
Non-ferrous and ferrous metal waste	Waste should be stored in special metal containers or on a hard surface
Waste contaminated with petroleum products	Waste should be stored in metal boxes away from sources of possible ignition
Tires, rubber products	Waste should be stored on an equipped area with a hard surface in stacks, or in a special room on racks
Broken glass	Waste should be stored in separate containers
Paper and cardboard waste	Waste should be collected in separate containers or on a hard surface area in bales
Polymer waste	Waste should be accumulated in separate containers or on a hard surface area
Wood waste	Waste should be collected in separate containers or on a hard surface area

Agricultural waste includes:

- organic waste from livestock, field farming and greenhouses;
- waste from agricultural processing industries;
- fertilizers and insecticides used in field farming.

Every year in the Krasnoyarsk Territory, according to 2-TP reports, 589,421 tons of agricultural waste of III – V hazard classes are generated. There are a number of foreign studies showing the feasibility of using certain types of agricultural waste together with municipal solid waste in the generation of renewable energy. This requires organizing the collection and transportation of agricultural waste to MSW treatment facilities, while keeping agricultural producers interested in such cooperation. The cooperation imply concluding agreements with agricultural producers and carrying out educational work on the most rational methods for selecting agricultural waste for subsequent energy utilization.

Current methods of agricultural waste utilization at the Krasnoyarsk Territory include:

- composting;
- fermentation of manure together with crop waste;
- vermicomposting of manure using earthworm colonies;
- thermal or vacuum drying of manure to produce dry concentrated fertilizer;
- anaerobic digestion in reactors to produce biogas.

In accordance with the Territorial Solid Waste Management Plan for the Krasnoyarsk Territory (2016), significant volumes of industrial waste are still going to landfills. Table 6 presents the volume of waste of production and consumption, including above-mentioned municipal solid waste and industrial waste, going to storage and disposal at the landfills at the Krasnoyarsk Territory.

Table 6. Volume of waste of production and consumption sent for storage or disposal at the Krasnoyarsk Territory

Waste Classification	Local Public Facilities, thousand tons		Other Special Facilities, thousand tons	
	2025	2030	2025	2030
Waste from agriculture, forestry, fish farming and fishing unit 1 of FCWC)				
Class III	0	0	2,8	1 340,4
Class IV	0	57,5	28 200,3	1 030 254,5
Class V	0	63 350,3	378 103,0	29 059 671,4
Waste from mining (unit 2 of FCWC)				
Class III	0	0	0	8 986,2
Class IV	0	50,3	18 596,0	37 155,9
Class V	0	25 000 206,9	5 403,0	562 620 779,2
Waste from processing industries unit 3 of FCWC)				
Class III	0	105 408,7	45,5	523 715,5
Class IV	0	25 581,4	1 674,0	25 307 368,0
Class V	39 163,2	1 794 193,9	20 935,7	166 811 434,0
Waste from industrial and non-industrial consumption (unit 4 of FCWC)				
Class III	0	1 471,4	119,4	302,0
Class IV	106 667,3	0,5	3 089,9	254 179,5
Class V	228 199,5	49 584,4		110 919,0
Waste from supplying electricity, gas and steam (unit 6 of FCWC)				
Class III	0	0	2,8	1 340,4
Class IV	0	57,5	28 200,3	1 030 254,5
Class V	0	63 350,3	378 103,0	29 059 671,4
Waste from water supply and waste water treatment (unit 7 of FCWC)				
Class III	0	0	52,8	3 175,2
Class IV	0	0	1 970,4	15 787,2

continued on following page

Table 6. Continued

Waste Classification	Local Public Facilities, thousand tons		Other Special Facilities, thousand tons	
	2025	2030	2025	2030
Class V	0	464,7	18 860,4	167 764,8
Construction and demolition waste (unit 8 of FCWC)				
Class III	0	0	1 923,7	154,0
Class IV	0	2 173,3	64 652,6	5 528,0
Class V	0	14 915 573,8	49 677,8	22 673 127,6
Waste from other activities (unit 9 of FCWC)				
Class III	0	2,4	665,9	1 781,1
Class IV	28,2	2 123,6	18 979,7	2 806,9
Class V	0	24,4	2 205,3	135,0

The data in Table 6 shows that significant volumes of industrial waste have not found the involvement in the secondary utilization processes and were sent for storage or disposal at the Krasnoyarsk Territory.

Along with this, it is worth noting the fact that the resource potential of waste going to landfills is not used, since there are no biogas collection systems at landfills in the Krasnoyarsk Territory (Razumnykh et al., 2020). Such solid waste landfill resource management models are characterized by the high level of environmental risk and low economic and social efficiency (Stepkin & Gaidukova, 2018). The cost of their operation is high, which creates significant problems in the regional budget, which is forced to allocate additional resources for the maintenance of landfills. Considering the large territory of the region and, accordingly, the large number of solid waste landfills, the total amount of funds required is significant.

To plan the implementation of changes to existing solid waste landfill resource management models, it is necessary to consider the location of landfills and solid waste storage facilities across the regions of the Krasnoyarsk Territory.

Figure 1. Distribution of solid waste storage and disposal facilities in the different regions within Krasnoyarsk Territory

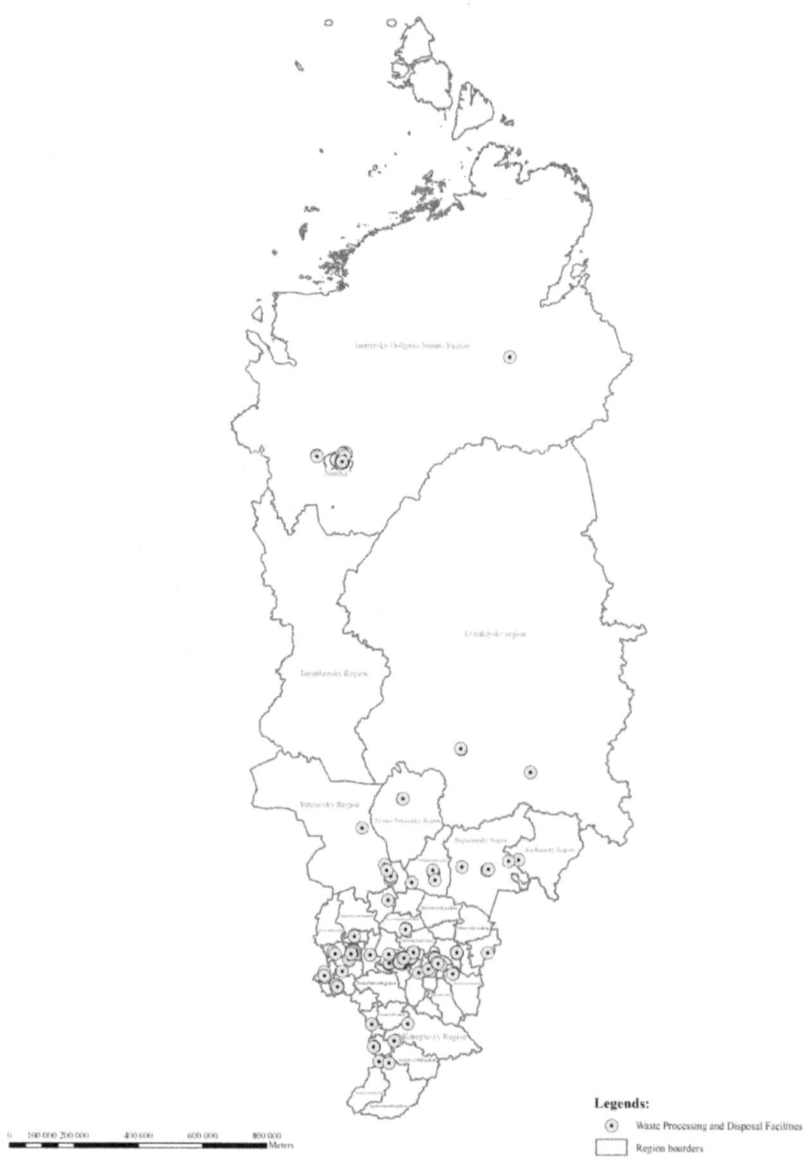

As shown at Figure 1, the distribution of solid waste storage and disposal facilities across the Krasnoyarsk Territory is uneven. The main concentration of facilities can be observed near larger cities and highly populated areas. Land resources in such parts of the Krasnoyarsk Territory are more expensive than in rural areas. Among the facilities, landfills predominate and occupy large and relatively expensive plots of land. Placing landfills in uninhabited areas remotes them from the main places of waste generation, increases the hauling distances and the cost of solid waste management in general. To solve this problem, it is necessary to pay attention to a more rational use of landfill resources, one of the directions of which is the creation of biogas collection systems and use of generated from landfills renewable energy.

SOLUTIONS AND RECOMMENDATIONS

Several options for using the land resources of a solid waste landfill at the final stage of its life cycle can be considered. In the opinion of the authors, it seems appropriate to compare options such as the construction of a waste sorting and recycling facility, or the organization of a biogas collection and use system at an existing facility.

According to expert estimates given in the study by T.V. Lyubinskaya (2010), if you collect and use half of the generated biogas, it will be equivalent to recycling 10% of the waste delivered to the landfill. Her research shows that the possible level of using waste potential as a resource at the expensive waste sorting complexes is equal to the level of using its resource potential while collecting biogas at the landfills. Moreover, the amount of energy consumed during sorting (25-30 kWh per ton of solid waste) versus renewable energy (biogas) generated through the collection systems at the landfill (50 -60 kWh per ton of solid waste). These arguments make it possible to focus further research on the option of equipping the existing landfill with a biogas collection system, but it is necessary to apply a differentiated approach to different zones of the landfill. Experts Shaimova A.M., Nasyrova L.A. and Faskhutdinova R.R. (2011) suggest to imagine the landfill as a biological reactor of periodic action, in which the destruction of organic substances contained in waste occurs and their transformation into gaseous (biogas), liquid (filtrate) and solid (decomposed mass) fractions.

The results of the work carried out by V.I. Maslikov, A.N. Chusov, and D.V. Molodtsov. and Ryzhakova M.G. (2012) analysis of quantitative and qualitative information showed significant difference by zone of biogas emissions at the solid waste landfill. The authors note that in exploited areas, biodegradation processes are in an active phase, and within 15-20 years after their closure, biogas containing 50-60% methane will be intensively released. It is in these areas that it is advisable to create a system for collecting and using biogas in the first place.

A number of experts (Busch, 2009; Deublein & Steinhauser, 2011) have shown the feasibility of monitoring the quantitative and qualitative composition of incoming solid waste at the stage of operation of a landfill for the purpose of its zoning (dividing, mapping into zones of rapidly and slowly decomposing waste). This information about the location of the zones will make it possible in the future to identify zones of the landfill site for placing a biogas collection system (Shilovskikh & Kostarev, 2007).

In addition to identifying the location of areas with potential stable methane formation, it is advisable to determine the priority of implementing projects to equip them with a biogas collection system. To optimize the use of land resources of solid waste landfill sites as a source of energy resources, it is necessary to build an economic and mathematical model that determines the stages of project implementation.

As a criterion for the optimality of the model, the capital productivity indicator Z of equipping the landfill zone with a biogas collection system was chosen. This criterion is directly related to the volume of planned work and reflects the intermediate commissioning (equipping) of individual zones of the landfill during the process of creating a biogas collection system. It takes into account the dynamics of investments, as well as the inequality of costs and results carried out and received in different periods of time.

In the proposed economic and mathematical model for optimizing the use of landfill land for organizing a biogas collection system, it is necessary to introduce the following notations:

S_i – cost of equipping zone i of the landfill land plot with a biogas collection system;

t_i – duration of equipping the zone i of the landfill plot, in months;

$t_i^{н}$, $t_i^{о}$ – calendar dates for the start and end of equipping the zone i of the landfill site with a biogas collection system;

$y_i^t \in \{0;1\}$ – an integer variable that means if $y_i^t = 1$, then it is planned to equip the zone i of the landfill site with a biogas collection system in calendar month t, if $y_i^t = 0$, then equipment is not planned in month t;

X_i^τ – volume of investments planned for the zone i equipped with a biogas collection system, in month τ, ordinal from the beginning of the project (τ=1,2,…, ν_i);

k_t – investment limit allocated for work to equip the zone i of the landfill site with a biogas collection system for a calendar month t;

ν_i – the number of months during which the zone i of the landfill site will be equipped with a biogas collection system (equipment of each land plot can begin and end at any time within the calendar month t);

T – duration of the planned period for equipping the site with a biogas collection system in months (t=1, 2,..,T);

$T_{дир}$ – target term for creating a biogas collection system at the landfill;

$P = \{P_i, i = 1, 2, \ldots, m\}$ – many zones allocated as part of the landfill site;

$T_q^{н}$, $T_q^{о}$ – start and end dates for work at a step q, where $q = 1, 2, \ldots, Q$;

E - discount rate, which takes into account the different times of costs and commissioned zones equipped with a biogas collection system.

In accordance with the accepted notations and formulated tasks, the economic and mathematical model for optimizing the use of landfill land for organizing a biogas collection system can be formulated as follows.

We need to find such non-negative numbers y_i^τ, X_i^τ so that the objective function is maximized while taking into account a certain number of restrictions.

$$Z = \frac{\sum_{t=0}^{T} \sum_{i=0}^{m} y_i^t \cdot S_i \cdot \alpha \cdot (T^o - t_i^o)}{\sum_{t=0}^{T} \sum_{i=0}^{m} \sum_{\tau=1}^{\nu_i} y_i^t \cdot X_i^\tau \cdot \alpha \cdot (T^o - t_i^{н} - \tau)} \rightarrow max\ (1),$$

where $\alpha_t = \frac{1}{(1+E)^{(t_i^{н}+t_i)}}$ - discount coefficient.

Under the following restrictions:

$y_i^t \in \{0;1\}$

,

$i = 1, 2, \ldots, m$

,

$\mathrm{t} = 1, 2, \ldots, T$ (2)

$$\sum_{t=1}^{T} y_i^t \leq 1 \quad (3)$$

$$t_i = t_i^{\text{O}} + t_i^{\text{H}}$$

,

$$i = \overline{1, m} \quad (4)$$

$$X_i^\tau \geq 0$$

,

$$i = \overline{1, m}$$

,

$$\tau = \overline{1, \nu} \quad (5)$$

$$X_i^\tau \leq K_t \quad (6)$$

$$T_\lambda^{\text{H}} = T_q^{\text{O}}$$

,

$$q = \overline{1, Q}$$

,

$$\lambda = q + 1, q + 2 \ldots R \quad (7)$$

$$T_q^{\text{o}} \leq T_{\text{дир}} \tag{8}$$

The formulated economics-mathematical model (1) - (8) is characterized by the following:

- the objective function is integer and nonlinear, non-additive with respect to time, since in the numerator and denominator each next term depends on the previous ones;
- acceptable control at each solution step represents a set of zones of the landfill site with a biogas collection system that are not included in the plan of the corresponding solution step;
- model restrictions are a system of linear equations and inequalities.

The objective function can be solved by a stepwise optimization method, which consists of obtaining an optimal plan for creating a landfill with a biogas collection system using a multi-stage decision-making process. The capital return criterion shows the share of capital investments spent during the implementation of work on creating a biogas collection system, which provide intermediate inputs to zones of the landfill and, thus, characterizes their return as a result of earlier receipt of the product (biogas) sold to consumers. Therefore, the application of the resulting economic and mathematical model will help to increase the economic efficiency of investments in equipping solid waste landfills with biogas collection systems.

FUTURE RESEARCH DIRECTIONS

The results of the analysis of the specific of regional solid waste management models on the example of the Krasnoyarsk Territory covers just the part of the problems. Further research and calculations are possible after the accumulation of statistical information on the creation and operation of biogas collection systems in the Krasnoyarsk Territory or neighboring regions. Future research results will also be influenced by changes in the Russian and world renewable energy market. In future studies, more significant attention should be paid to reducing the mitigation of the negative impact of the leachate discharges from landfills. The reactions occurring in the body of the landfill cause the formation of complex hazardous compounds, including carcinogens and PFAS. Future research should focus on sustainable and cost-effective treatment technologies for landfill leachate. Advanced control of incoming waste flow and landfill resource management can mitigate the adverse environmental and health effects of inadequate waste disposal.

CONCLUSION

The study shows that all over the world, solid waste landfills are among the largest environmental polluters; they have a negative impact not only on the condition of the ground water and surface water and soils of adjacent territories, but also on the efficiency of use of land resources in the region. The drivers of these innovations are both changes in legislation and technologies that are proven, effective from an economic, environmental and social point of view, and available to the landfill staff.

Based on the analysis of the secondary resources, current trends in existing solid waste landfill resource management models in the world characterized by some disadvantages. The first disadvantage is the low land usage efficiency due to the fact that landfills occupy huge amount of territories in different regions and extract them from the economic circulation. The second disadvantage is formed under the missed opportunities by number of landfills to contribute to the renewable energy resource pool despite the increase demand for these resources globally due to the implementation of national and international policies on sustainable development and growth of public awareness of climate change.

A number of experts have shown the feasibility of monitoring the quantitative and qualitative composition of incoming solid waste at the stage of operation of a landfill for the purpose of its zoning (mapping, dividing into zones of quickly and slowly decomposing waste). This information about the location of the zones allowing identification of landfill zones for placing a biogas collection system. In addition to identifying the location of areas with potential stable methane formation, it is advisable to determine the priority of implementing projects to equip them with a biogas collection system. To optimize the use of land resources of solid waste landfill sites as a source of energy resources, it is necessary to build a mathematical model that determines the stages of project implementation.

As a criterion for the optimality of the model, the indicator of capital productivity of equipping the landfill zone with a biogas collection system was chosen. This criterion is directly related to the volume of planned work and reflects the intermediate commissioning and equipping of individual zones of the landfill during the process of creating a biogas collection system. It takes into account the dynamics of investments, as well as the inequality of costs and results carried out and received in different periods of time.

The suggestions developed by the authors make it possible to take into account the dynamics of investments, as well as the inequality of costs and results carried out and received in different periods of time. The results are of interest both to the solid waste landfill staff and to regional and local authorities, as well as to investors and specialists involved in the development.

The introduction of such a model makes it easier to find investors for projects, since the gradual investment of funds in the development of biogas collection systems allowing expanding the potential range of investment sources. In addition, connecting one of the sections of the landfill generating biogas to the local grid making possible the partial use of incoming cash flows to finance work in subsequent sections.

REFERENCES

Al-Hazmi, H. E., Hassan, G. K., Kurniawan, T. A., Sniatala, B., Josph, T. M., Majtacz, J., Pietchota, G., Li, X., El-Gohary, F. A., Saeb, M. R., & Makinia, J. (2024). Technological solutions to landfill management: Towards recovery of bio methane and carbon neutrality. *Journal of Environmental Management*, 354, 120414. 10.1016/j.jenvman.2024.12041438412730

Aleshina, T. A. (2010). Extraction and utilization of landfill gas to increase the efficiency of the domestic electric power industry. *Bulletin of MGSU*, 2, 140–143.

Awasthi, S. K., Sarsaiya, S., Kumar, V., Chaturvedi, P., Sindhu, R., Binod, P., Zhang, Z., Pandey, A., & Awasthi, M. K. (2022). Processing of municipal solid waste resources for a circular economy in China: An overview. *Fuel*, 317, 1–13. 10.1016/j.fuel.2022.123478

Baburina, T.A. (2017). Environmental assessment of the negative impact of wastewater from a solid waste landfill (filtrate) on the environment. *Questions of science and education*, *6*(7), 177-178.

Banar, M., Cokaygil, Z., & Ozkan, A. (2009). Life cycle assessment of solid waste management options for Eskisehir, Turkey. *Waste Management (New York, N.Y.)*, 29(1), 54–62. 10.1016/j.wasman.2007.12.00618280731

Barraqan-Escandon A., Terrados-Cepeda J., Zalamea-Leon E., & Arias-Reyes P. (2018) Electricity production using renewable resources in urban centers. *Energy*, *171*(1), 12-25 ().10.1680/jener.17.00003

Bukrinskaya, E. M. (2020) Application of a logistics approach to the organization of extended producer responsibility. *News of the St. Petersburg State University of Economics*, *2*, 141-145.

Busch, G., Grobmann, J., Sieber, M., & Burkhardt, M. A. (2009). New and sound technology for biogas from solid waste and biomass. *Water Air and Soil Pollution Focus*, 9(1-2), 89–97. 10.1007/s11267-008-9195-5

Consonni, S., Giugliano, M., & Grosso, M. (2005). Alternative strategies for energy recovery from municipal solid waste Part B: Emission and cost estimates. *Waste Management (New York, N.Y.)*, 25(2), 137–148. 10.1016/j.wasman.2004.09.00615737711

De Feo, G., & Malvano, C. (2009). The use of LCA in selecting the best MSW management system. *Waste Management (New York, N.Y.)*, 29(6), 1901–1915. 10.1016/j.wasman.2008.12.02119168344

Deublein, D., & Steinhauser, A. (2011). Biogas from Waste and Renewable Resources. (2nd ed.) Deggendorf: Wiley VCH.

Ghinea, C., & Gavrilescu, M. (2010). Decision support models for solid waste management – An overview. *Environmental Engineering and Management Journal*, 9(6), 869–880. 10.30638/eemj.2010.116

Gorbovskaya, A. D., & Gonina, A. V. (2015). The need to use waste for energy production in the Udmurt Republic. *Eurasian Union of Scientists*, 11(20), 6–8.

GOST R 1.0-2004. *Standardization in the Russian Federation. Basic provisions.*https://docs.cntd.ru/document/1200038794.

GOST R 53691-2009. *Resource conservation. Waste management. Waste certificate for hazard class 1-4. Basic requirements.*

GOST R 54098-2010. Resource conservation. Secondary material resources. Terms and definitions. https://docs.cntd.ru/document/1200084875#7D20K3.

Gurevich P.A., Shavaleeva S.M., Glebov A.N., Bayanova L.N. (2013) Chemical corrosion of colleate glass and glass container waste as a factor of negative influence on the environment. *Bulletin of Kazan Technological University*, *4*.

Isakova A.V. (2020) Legal support for industrial environmental control from the investment project to operation. *Current issues of Russian law*, *15*(116), 207-217.

Ityshev I.K., Potapova S.O. (2018) On the problems of fire safety of solid household waste and their storage places. *Fire safety: problems and prospects*, *9*(1), 292-300.

Kirsanova, A. Yu., & Filippova, T. M. (2008). Energy potential of landfill gas at the solid waste landfill in Angarsk. *AGTA Bulletin*, 1(2), 82–87.

Kostyleva, N. V., & Racheva, N. L. (2016). *Characteristics of pollutants: reference book*. Federal State Budgetary Institution Ural Research Ecology Institute.

Kysyydak, A. S., & Salchak, A. D. (2013). Energy potential of solid waste landfills. *Bulletin of the Tuva State Technical University*, 3, 34–39.

Lyubinskaya T.V. (2010) Reducing biogas emissions from solid waste as the most important element in reducing the "greenhouse" effect. *Bulletin of RUDN University. Series: Ecology and life safety*, *1*, 76-81.

Maslikov, V. I., Chusov, A. N., Cheremisin, A. V., & Ryzhakova, M. G. (2012). Assessment of the geoecological risk of air pollution from solid waste landfill emissions for the selection of reclamation measures. *Scientific and Technical Journal of St. Petersburg State Polytechnic University.Science and Education*, 2(1), 239–243.

Maslikov, V. I., Chusov, A. N., Molodtsov, D. V., & Ryzhakova, M. G. (2012). Zonal determination of biogas emissions at a solid waste landfill to assess the geo-ecological state and substantiate the management of waste decomposition processes during reclamation. *Scientific and Technical Journal of St. Petersburg State Polytechnic University.Science and education*, 2(1), 260–265.

Mor, S., & Ravindra, K. (2023). Municipal solid waste landfills in lower- and middle-income countries: Environmental impacts, challenges and sustainable management practices. *Process Safety and Environmental Protection*, 174, 510–530. 10.1016/j.psep.2023.04.014

Mvula, R. L. S., Mundike, J., & Nguvulu, A. (2023). Spatial suitability analysis for site selection of municipal solid waste landfill using hybrid GIS and MCDA approach: The case of Kitwe, Zambia. *Scientific African*, 21, e01885. 10.1016/j.sciaf.2023.e01885

Niu X.X., Wang S.Z., Niu Y.C., Wei L.F., Yu L.Y. Improvement and optimization for the first order decay model parameters at typical municipal solid waste landfills in China. *Advances in Climate Change Research*, *14*(4), 605-614 ().10.1016/j.accre.2023.07.002

Oganesyan, A., Kurita, H., Maslov, D., & Take, A. (2005). Project for collecting landfill biogas and generating energy in Yerevan. *Alternative Energy and Ecology*, 12(32), 88–93.

On approval of the State Program of the Russian Federation “Environmental Protection” (2014). Decree of the Government of the Russian Federation #326 https://docs.cntd.ru/document/499091755

On Production and Consumption Waste Federal Law # 89-FZ (1998). https://legalacts.ru/doc/FZ-ob-othodah-proizvodstva-i-potreblenija/

Ortskhanov T.A. (2010) Urban planning aspects aimed at ensuring the environmental safety of the territories where solid waste landfills are located. *Academia. Architecture and construction*, 3, 150-153.

Razumnykh, A. V., Kalinina, K. V., & Vlasov, V. A. (2020). Implementation of the reform of solid waste management in the Krasnoyarsk Territory: Selected issues of theory and practice. *Bulletin of Science and Education*, 23(101), 39–42.

Shaimova, A. M., Nasyrova, L. A., & Faskhutdinov, R. R. (2011). Study of methane generation factors in the conditions of a solid waste landfill. *Bashkir Chemical Journal*, 18(2), 172–176.

Shi, X., Chen, Z., Wu, L., Wei, W., & Ni, B. J. (2023). Microplastics in municipal solid waste landfills: Detection, formation and potential environmental risks. *Current Opinion in Environmental Science & Health*, 31, 100433. 10.1016/j.coesh.2022.100433

Shilovskikh, P. A., & Kostarev, S. N. (2007). Conceptual approaches to modeling process management at waste disposal sites. *Bulletin of the Perm National Research Polytechnic University*, 2, 120–125.

Stepkin, Yu. I., & Gaidukova, E. P. (2018). Risk assessment and management in waste management. *Hygiene and Sanitation*, 8(97), 693–698. 10.18821/0016-9900-2018-97-8-693-698

Chapter 13
Problems of Formation of Specially Protected Natural Areas in the Arctic Zone of Russia

Vasily Kovyazin
St. Petersburg Mining University of Empress Catherine II, Russia

Elizaveta Bogdanova
http://orcid.org/0009-0006-4430-7567
St. Petersburg Mining University of Empress Catherine II, Russia

ABSTRACT

The chapter discusses the problems of forming specially protected natural areas in the Arctic zone of the Russian Federation. Research shows the need to preserve and restore the natural environment, awareness of the importance of environmental protection of nature, especially in the context of economic globalization. As part of the study, data on specially protected natural areas of the Arctic zone of the Russian Federation were collected from open sources of information. Data processing and analysis was carried out, which made it possible to determine the current state of the system of specially protected natural areas. The authors identified a number of problems in the system of specially protected natural areas of the Arctic zone of the Russian Federation and proposed practical ways to solve them.

INTRODUCTION

In its centuries-old history, humanity has not always realized the importance of maintaining the natural balance and has put its own interests above nature, not caring about the consequences of its actions and unlimitedly using the planet's natural resources. At the present stage of development, society has realized the need to preserve and restore the natural environment and reduce the anthropogenic impact on it.

The era of economic globalization obliges humanity to ensure environmental protection of the natural environment from economic activity and its consequences. The Russian Federation, as the largest country with diverse natural ecosystems, plays a significant role in solving global environmental problems. In

DOI: 10.4018/979-8-3693-3374-7.ch013

this regard, the Environmental Doctrine of the Russian Federation was developed and approved, which is aimed at improving the quality of life, creating sustainable development of the country and ensuring national security (The Government of the Russian Federation, 2002). According to the Environmental Doctrine, the conservation and restoration of the natural environment is possible through the creation and development of a system of specially protected natural areas (SPNA).

Currently, as a result of the active activities of mankind, the anthropogenic impact on the natural systems of the Arctic Zone (AZ) of the Russian Federation (RF) has increased. Today, the Arctic faces serious challenges caused by climate change and human activities. In conditions of increasing anthropogenic impact on Arctic ecosystems, the role of protected areas is increasing.

Protected areas are an important aspect in the protection and conservation of unique landscapes, biodiversity of species and natural complexes in general. These territories ensure the preservation of natural ecosystems and their main components and contribute to the sustainable development of the Russian Arctic. Protected areas in the Arctic provide favorable conditions for the life and reproduction of rare and endangered species of plants and animals, and help maintain the integrity of natural ecological systems, which helps ensure the environmental safety of the region. These territories serve as a kind of buffer that protects natural complexes from the destructive effects of natural and anthropogenic factors. The relevance of the problem under consideration is determined by the role of protected areas of the AZ of the RF in preserving the natural landscape and reducing the negative impact of technogenic and anthropogenic loads on existing natural complexes.

Purpose of research is to identify problems of the formation of specially protected natural areas in the Arctic zone of the Russian Federation.

To achieve this goal, the following tasks were solved:

- Conduct research, collect and analyze data on the number, location and boundaries of protected areas of the AZ of the RF, using the example of the Arkhangelsk region;
- Assess the current state of the protected areas system in the Russian Arctic;
- Identify current problems associated with the formation of protected areas in the AZ of the Russian Federation.

The object of research is the specially protected natural areas of the Russian Federation in the Arctic.

Research methods – collection of information about protected areas of the Russian Arctic; analysis of the obtained data; review of literary sources and media materials; study of legal acts of the Russian Federation; generalization of identified problems in the system of protected areas of the Russian Federation.

BACKGROUND

Arctic Zone of Russia: Nature and Production

Preservation of the natural environment and its biodiversity is a leading idea in the 21st century: as society develops, in the era of scientific and technological progress, the need to protect and restore the environment and reduce the anthropogenic impact on it is becoming more and more clearly realized. In the era of economic globalization, humanity is faced with the task of ensuring environmental protection of the natural environment from the consequences of economic activity. The Russian Federation, as a country with diverse, rich ecological systems and extensive biodiversity, plays a key role in solving global environmental problems. Thus, in Russia, the Environmental Doctrine (The Government of the Russian Federation, 2002) was developed and approved, it is a strategic document aimed at improving the quality of life, achieving sustainable development and ensuring national security. The environmental doctrine highlights the creation and development of a system of protected areas as the main way to preserve and restore the natural environment. The system of protected areas in Russia is represented by a wide range of categories, including nature reserves, national parks, sanctuaries, natural parks, natural monuments, etc. Each of these categories has a set of specific features in the field of protection and use of natural resources. Within the framework of the Environmental Doctrine, the task has been set to further strengthen and develop the system of protected areas, including the formation of new protection objects, such as nature reserves, national parks in particularly valuable natural areas.

Land relations in the Russian Federation are regulated by comprehensive land legislation: the Land Code, federal laws and laws of the constituent entities of the Russian Federation. From the point of view of the legislation of the Russian Federation, the territory of the country is considered as a natural object, resource (as a means of production) and real estate. However, when considering land as an object, one should adhere to the principle of protecting land as a basic and integral component of the environment and means of production. This in turn means that the use of land as real estate must be carried out in strict compliance and taking into account environmental requirements.

Specially protected areas of the Russian Federation play a key role in preserving the natural wealth and cultural heritage of the country. The Land Code of Russia has a separate chapter for the regulatory regulation of specially protected areas and sites (SPAs), which emphasizes their importance and significance for the country. Today, protected areas include lands of protected areas, environmental, recreational, historical and cultural significance and other valuable areas of land. These territories are not only unique natural resources, but also important sites for tourism, science and education.

The Arctic zone of Russia is a unique territory that attracts attention not only for its natural beauty, but also for its significant mineral reserves. Nine regions of the Russian Federation are included in the AZ of the RF, of which four regions belong to it in full, and five - partially, by including a number of municipalities (Fig. 1).

Figure 1. Map of AZ RF

Source: Il Tumen's press service, 2016

Subsequently, the active activities of mankind led to increased anthropogenic impact on the natural ecosystems of the AZ of the RF. The Arctic is an investment-attractive region with a unique, fragile ecological system. Researchers have identified changes in the structure of industrial production in the Arctic regions in the period from 2010 to 2016 as a result of an analysis of structural changes and innovative development of the economy of the Arctic regions of Russia (Berezikov, 2019). Researchers noted a new approach to economic activity in AZ through the development of innovative infrastructure and support for industry in the region. (Zuckerman, 2022).

Today, the Arctic faces a number of significant problems that require immediate intervention and are caused by climate change and human activities. The Arctic territory of the Russian Federation is of great importance for the strategic development of the country, since it forms from 12 to 15% of the gross domestic product and provides approximately a quarter of the export of its natural resources (Bogachev et al., 2019). Modern researchers have identified the increased exploitation of natural resources in the region, namely the development of mineral deposits, which has led to serious consequences - air, water and soil pollution - and has a negative impact on the environment, posing a threat to living organisms and biodiversity of the Arctic (Aleynikova et al, 2020). Research on the impact of production on carbon dioxide emissions in the regions of the RF is important. In the course of ongoing research, it was found that the greatest influence on the level of carbon dioxide emissions is exerted by the population size, by the level of electricity consumption and the development of the production sector, which made it possible to draw appropriate conclusions and propose a number of recommendations for reducing emissions (Tsvetkov et al, 2024). In particular, it should be noted that modern researchers are carrying out various sets of activities that study production processes, the influence of industry on the state of the natural environment and the possibility of its restoration. For example, one can cite studies on the prospects of using the algolization method in purifying quarry waters of northern mining agglomerations from nitrogen pollution (Matveeva et al, 2023).

Expansion of production, development of infrastructure, solution of social problems and creation of economic mechanisms – all this plays a significant role in the development of the Arctic zone. The development of the Arctic zone is a complex multicomponent and multifunctional mechanism in which it is necessary to take into account a huge range of interests: it is necessary to extract minerals without destroying the integrity of the sustainable ecosystem of the region, to exploit natural resources to meet the needs of humanity, preserving the environment for present and future generations, to take into account and solve the problems of the indigenous peoples of the Far North, the difficulties of the urban population of the Arctic (Vasilev et al, 2020).

In modern conditions of progress, infrastructure is becoming a fundamental factor for humanity as a system of connections for the transfer of products, energy and information under conditions of division of labor, which allows us to reduce the problem of ensuring the living conditions of the country's population to providing territories with infrastructure (Bykova, 2021). Spatial planning should rely on research on where people go, how passengers move, and which areas need improved services (Baltyzhakova & Romanchikov, 2021). In the case of the Russian Arctic, infrastructure takes on special importance. This vast region is characterized by its remoteness from developed areas and the lack of a developed transport network (Vasilev et al, 2020). Difficult natural and climatic conditions aggravate the transport and geographical situation. Providing the Russian Arctic with the necessary infrastructure is the key to its sustainable development: the creation of high-quality infrastructure will allow the region to be integrated into the country's economic system and open it to investment. However, it is necessary to understand that the development of infrastructure in this region is not only a matter of economic growth, but also of ensuring safety, environmental protection and preserving the Arctic heritage for future generations.

The construction of new cities and comfortable settlements, as well as the construction of roads and pipelines, destroy natural geosystems and threaten unique ecosystems and wildlife (Aleynikova et al, 2020). Despite the decline in industrial activity in the Arctic in the 1990s, many consequences remain relevant and cause irreparable damage to Arctic geosystems (Aleynikova et al, 2020). That is why the system of protected areas plays a key role in the conservation of natural biodiversity, landscapes and unique natural complexes of the Arctic.

The definition and establishment of a special regime for environmental protection actually leads to the cessation of industrial and other activities in such a territory. Protected areas make it possible to ensure the preservation of the natural environment of the region and lead to its restoration to its original state. Researchers in the region believe that protected areas, in addition to preserving and restoring the natural environment, perform research and monitoring functions, studying its condition under anthropogenic influence (Dolgikh & Kusticova, 2019). The current pace of industrial development in the Arctic zone obliges people to carry out serious environmental measures aimed at protecting and monitoring the state of the environment.

Any activity in the AZ that differs from traditional environmental management causes conflicts, contradictions in the way of life and use of land resources among indigenous peoples (Lipsky, 2018). According to S.A. Lipsky, a feature of the AZ land fund in Russia is the need for a balance between the interests of the industrial sector, the conservation of natural systems and the indigenous peoples of the region (Fig. 2) (Lipsky, 2018). To ensure sustainable development of the Arctic territory and preserve it as a unique object for future generations, it is necessary to take into account all existing factors and take into account their interaction.

Figure 2. Balance of interests in relation to the land fund of the Russian Arctic

The land fund of the Arctic zone of Russia differs from the rest of the country in the need to balance interests between:	industrial development of these lands and the inevitable negative impact on the environment and on the process of traditional environmental management;
	the need to preserve unique ecosystems, which are very vulnerable due to extremely low temperatures and low rates of biological processes;
	traditional environmental management practiced by the indigenous peoples and their communities living here.

Source: Lipsky, 2018.

In recent years, there has been a significant increase in economic activity in the AZ of the Russian Federation, which is accompanied by serious negative consequences for the environment: a significant expansion of the area of disturbed lands and fragmentation of natural ecosystems and habitats (Tishkov, 2012). In addition, there are other indicators that are significant for determining the state of ecosystems in the Arctic regions of Russia (Fig. 3).

To achieve a balance of interests, it is necessary to take an integrated approach: it is important to conduct geological exploration and mining, taking into account the environment and the principles of environmental sustainability. Technologies and methods used in these industries must be as safe as possible and minimize negative impacts on ecosystems.

Figure 3. Factors influencing the state of ecosystems in the Arctic regions of Russia

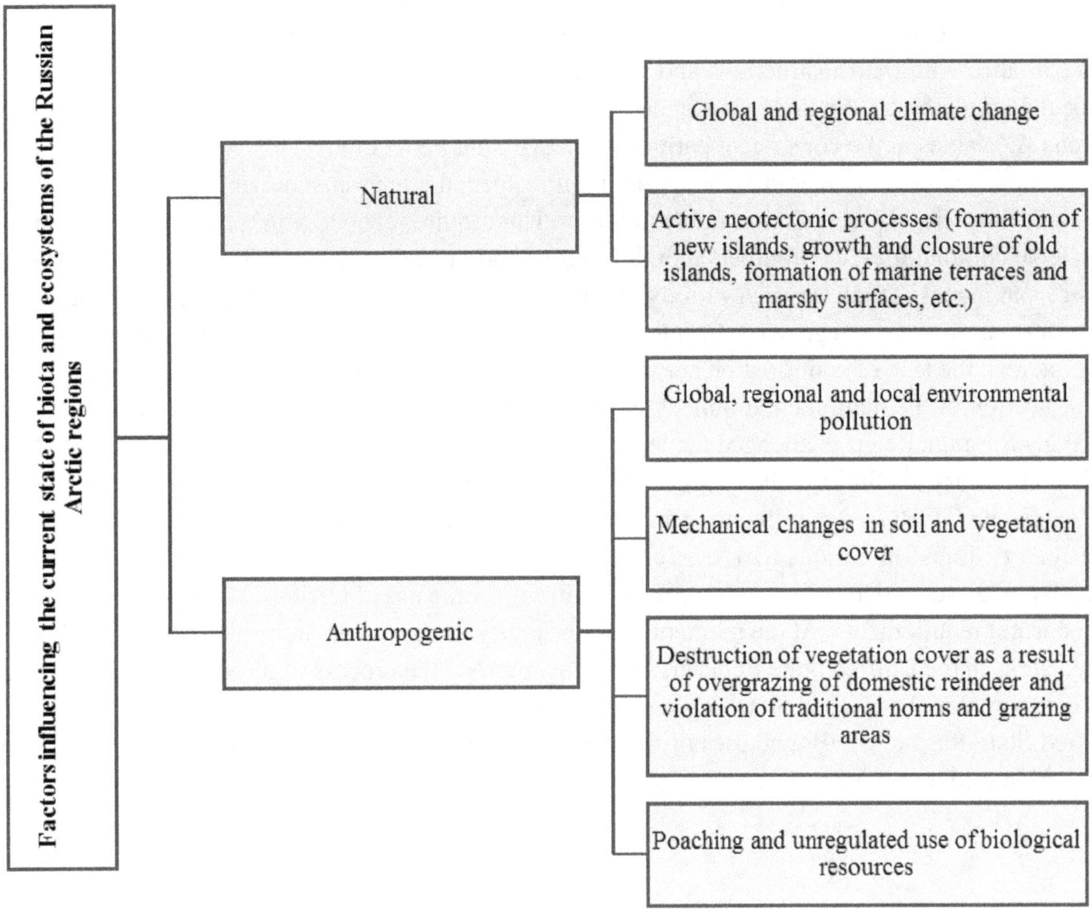

Source: Tishkov, 2012.

Research in the Arctic zone is becoming more active at the present time: over the past decade, mining organizations have achieved new results in geological mapping of the Arctic shelf and adjacent deep-sea areas of the Arctic Ocean (Gusev, 2022), which contributes to the study and discovery of unique landscape objects that require protection.

Formation of Specially Protected Natural Areas of the Region

Nature conservation has become one of the key priorities of modern society. Improving the system of protected areas plays a key role in the environmental process. In Russia, protected areas can be classified as federal, regional or local in accordance with their affiliation. Determining the status of a territory depends on the significance and role of such a territory. Federal authorities, such as the Government of the Russian Federation, play a key role in defining federal protected natural areas: national parks, reserves, natural monuments and other particularly valuable natural sites. The executive power of the constituent entities of the Russian Federation has the authority to determine territories to be protected at the regional level. They can make decisions on the creation of regional parks, reserves and other protected natural sites, taking into account the characteristics of their territories and priorities in nature conservation. Local governments also play an important role in identifying protected areas. They operate within the framework of federal laws, laws of constituent entities of the Russian Federation and regulations, acting in accordance with their instructions and requirements to determine the status and level of protection of local natural objects. Figure 4 shows the percentage distribution of the number of protected areas of various AZ values in the constituent entities of the Russian Federation.

It is important to note that protected areas require special attention and management. They must be protected from illegal use, pollution and destruction. This requires cooperation between government agencies, local communities, environmental organizations and the public at large. The fragmentation of data storage and inconsistency between various government bodies and their departments lead to difficulties in the formation and management of protected areas. One of the precedents in the Numto Natural Park is of interest: the lack of coordination between Rosnedra and Rosprirodnadzor led to a conflict between the protection of the natural environment and the extraction of natural resources, and therefore the oil company Surgutneftegaz challenged the legal status of the Numto Natural Park (Pristupa et al, 2017).

It is necessary to eliminate the fragmentation of data on protected areas in the Unified State Register of Real Estate (USRN). According to open information from the public cadastral map (PCM), protected areas can be found in various registers: protected areas as land plots of the specially designated land category, as zones and as zones with special conditions for the use of territories, which does not comply with current requirements. At the moment, it is necessary to correct the information entered about protected areas in the Unified State Register, as well as improve the process of recording the boundaries of protected areas in the Unified State Register, since they must be entered in the appropriate section of the Unified State Register – «Boundaries of Specially Protected Natural Areas», namely, as the boundary of the protected area zone.

Figure 4. Diagrams of the percentage of the number of specially protected natural areas of (a) federal significance, (b) regional significance and (c) local significance of the Arctic zone by constituent entity of the Russian Federation

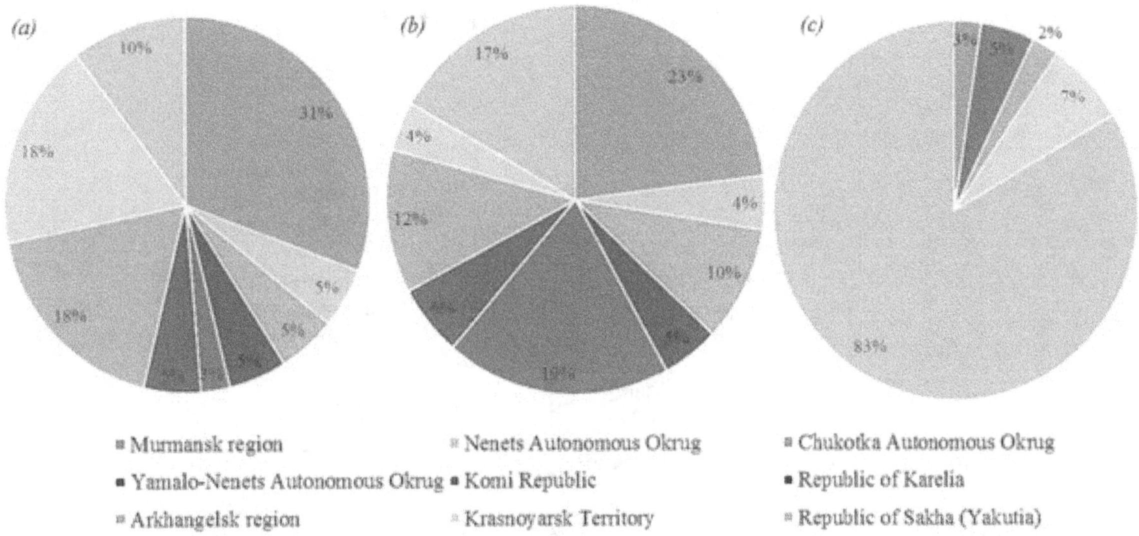

Source: Author's development.

Protected areas at different levels and regimes of protection play a key role in the conservation of unique landscapes, animal species and natural complexes in general. The year 2019 received special attention to protected areas in the RF:

- The National Project «Ecology» was launched, aimed at combining the main environmental directions of state policy and ensuring an integrated approach to the management of the country's natural resources in 2019 – 2024 (National projects, 2022);
- Within the framework of the National Project «Ecology», the Federal Project «Preservation of Biological Diversity and Development of Eco-Tourism» was launched, which involves the creation of new protected areas of federal significance in the amount of 24 units, due to which the total area of protected areas will increase by 5 million hectares (Fig. 5) (National projects, 2023);
- The President of the Russian Federation, in his address to the Federal Assembly, set the task for the executive bodies of state power: to carry out work to determine and consolidate the boundaries of protected areas, and enter information about them into the Unified State Register of Real Estate (The President of the Russian Federation, 2019).

Figure 5. Summary of the goals and indicators of the project 'Conservation of biological diversity and development of eco-tourism'

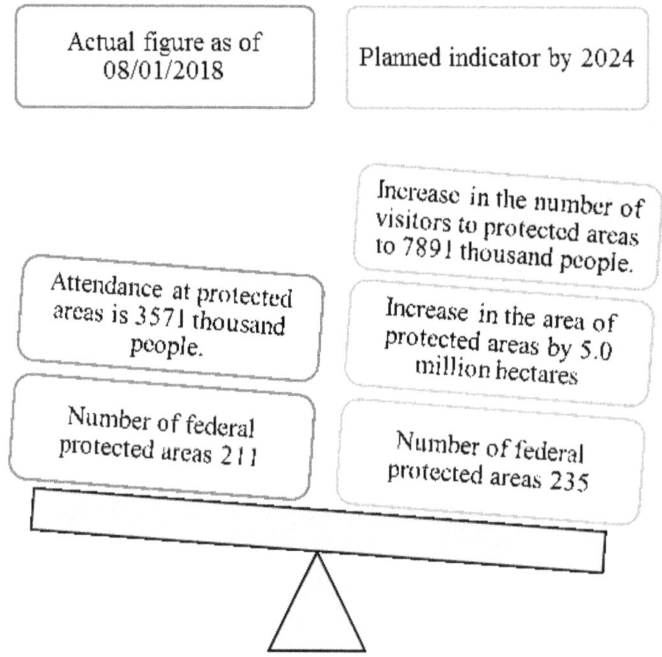

Source: National projects, 2023.

In modern conditions, the system of protected areas in the Russian Federation is actively growing and developing: comprehensive work is being carried out to form new protected areas, restore existing protected areas, and systematize data. It is important to comply with the regimes for the protection of natural objects; therefore, protected areas are of paramount importance for the ecology of the country and the world. Timely state cadastral registration of protected areas helps ensure and maintain conservation measures in such territories. It is necessary to note the key role of state environmental projects in the process of formation and expansion of the system of protected areas in Russia. Such projects contribute to the creation of new protected areas, restoration of unique ecosystems and conservation of rare species of animals and plants.

One of the significant problems that protected areas face today is the lack of information about the location of the boundaries of such territories in the Unified State Register of Real Estate. This problem has a significant impact on the protection and functioning of protected areas, posing a threat to the safety of protected natural systems. Currently, work is actively underway to record and systematize information about existing protected areas and create new protected areas. Most new protected areas are created at the regional level, which makes it possible to effectively preserve and restore biodiversity in different regions of the country. The Ministry of Natural Resources of the Russian Federation and its regional divisions are taking active measures to determine the location of the boundaries of protected areas and enter information about them into the Unified State Register of Real Estate.

When determining the location of the boundaries of protected areas, it is necessary to take into account their status. It is necessary to determine whether a specially protected natural area already exists, whether there is a regulation on this territory and whether it has been approved. The presence of approved documents on protected areas means that the necessary research has been carried out in this territory and measures for its protection have been approved. However, protected areas may be in the process of creation; the regulations on such a territory have not yet been approved, which requires the development of the boundaries of such a territory as part of the process of creating a protected area. It should be noted that the development of the boundaries of protected areas will be required in the event of changes to the boundaries of an existing territory. This may occur in connection with the expansion of the territories of existing protected areas or the creation of new ones based on unique natural objects.

Let's consider both cases. If a protected area already exists, then to establish the boundary of the territory, the specialist will have to work with a text description of its passage, which is approved in the Regulations on such a territory. This process requires careful study and understanding of the contents of the Regulations, since it is necessary to accurately display information about the boundaries of protected areas on cartographic material.

Figure 6. Fragment of the boundary of the protected area Kuloisky State Natural Biological Reserve of Regional Importance (Pinezhsky District, Arkhangelsk Region)

Source: Author's development.

On the other hand, within the framework of the organization of a previously unformed protected area, its environmental status is established - specialists prepare materials for a comprehensive environmental survey (MCES) of the territory. In these materials, the researchers confirm the ecological significance of the study area and argue for the need to assign it the status of a specially protected natural site. As part of the development of these documents, the boundaries of protected areas are determined. However, the materials of a comprehensive environmental survey do not always contain accurate information about the boundaries of protected areas or their correct description, which poses a serious problem for further work with such a territory, especially for taking into account information about the boundaries of protected areas in the Unified State Register of Real Estate. The boundaries of protected areas are drawn based on the possibility of their visual detection on the ground - along the natural boundaries of water and forest bodies, the boundaries of real estate, taking into account the cadastral situation of the territory as a whole (Fig.6) (Mnatsekanov et al, 2021).

The preparatory stage, which includes the collection and preparation of initial information, is one of the most expensive stages in all types of cadastral work. So, for example, in order to effectively spend budget funds, a polygonal metric grid method has been developed and justified in cadastral valuation in determining the pricing factors of cadastral value (Raguzin et al, 2023). Currently, new methods of research and information processing are increasingly being used in cadastral activities.

In modern conditions, cadastral research methods are rapidly changing due to the intensive development of technologies (photogrammetry, remote sensing, 3D, etc.), which leads to their use in production and research activities (Choi, 2020). The possibility of determining the elevation of terrain using satellite measurements and leveling methods is being explored to calculate the deviation of the plumb line, as this approach is especially important in difficult-to-reach areas and in areas where the local geodetic network is not available (Mustafin & Moussa, 2023). Modern methods of obtaining information about a certain territory, such as the use of unmanned aerial vehicles and laser scanning of territories, make it possible to quickly and effectively identify all objects located in the study area. Laser scanning methods are actively developing and are used as an important tool for obtaining the most complete and reliable information about an object, such as cultural heritage objects (Choker & Mustafin, 2021). Modern researchers are developing algorithms to automate the analysis of data from UAVs, which can independently determine the boundaries of objects on the ground, which is important for cadastral registration of real estate and land monitoring, since many boundaries coincide with visible boundaries and can be automatically identified by image analysis (Crommelinck et al, 2017).Among other things, modern researchers focus on developing and improving semi-automatic and automatic methods for extracting the boundaries of existing objects from point clouds obtained in the field after processing airborne laser scanning materials (Luo et al, 2017). Experts are improving aerial photography methods using laser scanning of forest plantations in order to determine their boundaries and monitor those (Sivák et al, 2023). Terrestrial laser scanning methods have now been developed to study the structure of forest stands using LiDAR systems (Kabonen & Ivanova, 2023). Modern researchers analyze percentage trends in the state of vegetation, offer recommendations for increasing the reliability of the results of the analysis of the vegetation index and other vegetation indicators, and also develop ways to improve the quality of initial remote sensing data (Skachkova & Guryeva, 2023). In other words, monitoring and research of the terrain of specially protected natural areas is possible taking into account the definition of objects, including taking into account each tree and the characteristics of other objects, which is important for the conservation of plant species and landscapes in protected areas, including this is important for the correct determination of the boundaries of protected areas. For example, methods for interpreting aerial photography data from unmanned aerial vehicles (DJI Phantom 4) in combination with the use of Earth remote sensing data (Sentinel-2) were tested - studies of the territory of protected areas were carried out and a landscape map was compiled (Ostroukhov et al, 2020).

Description of the location of the boundaries of the formed protected areas implies the implementation of work to determine the coordinates of the characteristic points of the boundaries of such territories. In most cases, existing protected areas are located on lands of forest and water funds, the boundaries of which have not yet been included in the unified state register of real estate, which creates an additional burden when carrying out work to prepare a description of the location of the boundaries of protected areas. However, not only the absence of boundaries of lands of these categories leads to difficulties in the work performed: the lack of information about any real estate objects – buildings, structures, land plots, boundaries between municipalities, the boundaries of constituent entities of the RF and other objects - leads to serious problems when performing work to prepare a description of the location of the boundaries of protected areas.

Currently, most forest management materials have lost their relevance and do not meet modern requirements, especially graphic forest management materials. However, today the state is taking measures to preserve and protect forests, develop their infrastructure, record and systematize information about them, realizing the importance of forests for the planet: they play a significant role in purifying the air, conserving water resources and providing housing for animals and plants (Kovyazin et al., 2021). In the Russian Federation, the digitalization and updating of forest management materials is currently being actively carried out, and the boundaries of forest fund lands are being established. An important aspect when describing the location of the boundaries of protected areas is the availability of approved urban planning documentation, which is an important factor in the planning, creation and development of the system of protected areas in general, and determining the location of the boundaries of such territories, in particular. And although the Russian Federation has embarked on the path of digitalization in the field of urban planning, as studies show, in practical activities we have to face the problems of implementing measures to create comprehensive accessible information support for urban planning. (Skachkova & Gureva, 2022).

Figure 7. Fragments of the boundaries of protected areas: (a) Primorsky State Natural Landscape Reserve of regional significance (Primorsky district, Arkhangelsk region); (b) Kuloisky state natural biological reserve of regional significance (Pinega district, Arkhangelsk region); (c) Solvychegodsk State Natural Biological Reserve of regional significance (Kotlas district, Arkhangelsk region)

Source: Author's development.

The development and formation of the boundaries of protected areas are often associated with the problem of the mandatory exclusion of certain categories of land from protected areas. Experts encounter this problem both when describing the location of the boundaries of existing protected areas and when developing the boundaries of new territories. For example, we can cite the boundaries of the Mudyugsky State Natural Landscape Reserve of regional significance. According to the Regulations (The Government of the Arkhangelsk region, 2016), the reserve is located within the boundaries of Mudyug Island, Primorsky District, Arkhangelsk Region, however, lands of certain categories must be excluded from its composition. When studying the territory of the protected area «Mudyugsky State

Natural Landscape Reserve of Regional Importance», inclusions of lands «industry, energy, transport, communications, radio broadcasting, television, computer science, lands for space activities, defense lands, security lands and lands for other special purposes» are observed in the composition single land use with the permitted type of use «for the operation of the 10 kV overhead power line to the NAVTEX station». These territories are excluded from the composition of protected areas in accordance with the Regulations. However, the result of these actions is often a violation of the integrity of the boundaries of protected areas - peculiar «holes» appear. Thus, the exclusion of certain categories of land from protected areas leads to the formation of «holes» within the boundaries of such territories, their fracture, as well as the division of the territory into several clusters (multiple clusters) (Fig. 7).

In protected areas, special zones can be allocated according to their functional significance (with a differentiated regime of special protection). Established functional zones contribute to the distribution of human activities in a certain territory and allow the implementation of limited economic activities in specially designated areas. If functional zones of protected areas are allocated, information about the boundaries of these zones is included in the Regulations on such territory. However, at present, the boundaries of functional zones of protected areas are not recorded in the Unified State Register of Real Estate. In particular, the Unified State Register does not contain information about zones of limited economic use of protected areas, where economic activity is permitted in accordance with the Regulations on the territory.

In order to prevent adverse anthropogenic impacts, protective zones are established for some categories of protected areas: nature reserves, national and natural parks, as well as natural monuments (The State Duma, 1995). It should be noted that the legislation of the Russian Federation does not prohibit the formation of protective zones in protected areas of other categories. The formation of protective zones in protected areas ensures the introduction of restrictions on adjacent territories, which makes it possible to preserve natural ecosystems untouched and reduce the anthropogenic impact on the protected environment. However, at present, protected areas have not been established (except for isolated cases). Approaches and methods for determining the boundaries of protected areas have not been defined, which makes it impossible to objectively form protective zones.

It is important to note that the MCES justifying the granting of environmental status to a natural area are currently not regulated at the state level; there are differences in form and content in different subjects of the Russian Federation. In some constituent entities of the Russian Federation there are local regulations regulating the content of MCES, justifying the granting of protected area status, as, for example, in the Murmansk region (The Government of the Murmansk region, 2020), however, not in all regions they exist or do not reflect a clear position on the content of materials from a comprehensive environmental survey. There are no general regulations for materials from comprehensive environmental surveys on the territory of the Russian Federation.

Currently, MCES are transferred to the relevant authorities of such territories in the format of a text document on paper and electronic media in *.doc and *.pdf formats in several copies, unless otherwise specified in the service agreement. The format and composition of materials for a comprehensive environmental survey are regulated by regional regulatory documents, if there are any, and by the state contract for the development of MCES of areas of the territory, justifying giving the territories the legal status of protected areas.

To effectively manage existing and newly created protected areas, it is necessary to develop a unified methodology that will allow the formation and consideration of their boundaries. The main idea for effective management of the protected areas system is the creation of a unified cadastre of SPNA in the RF, which will contain all the information about such territories of different levels of ownership.

SOLUTIONS AND RECOMMENDATIONS

Currently, in the Russian Federation there is no unified cadastre of protected areas, and information about them is scattered and stored only in subordinate institutions: in federal, regional or local authorities of the RF.

Among other things, information about protected areas is taken into account in the Unified State Register of Real Estate as information about the boundaries of such territories. However, information about the boundaries of protected areas in the Unified State Register of Real Estate is scattered; many protected areas are included in the Unified State Register of Real Estate in different ways: as land plots, zones (the boundaries of protected areas) and as zones with special conditions for the use of the territory. To correctly record the boundaries of protected areas, it is necessary to adjust the entry of information about such territories in the unified state register of real estate: unify the entry of the boundaries of protected areas into the unified state register of real estate, enter information about the boundaries of all existing protected areas, and also ensure the implementation of a single general section of protected areas in the register of boundaries with the entry of data about such territories on the Public Cadastral Map in a separate section.

Works to describe the location of the boundaries of protected areas are often faced with the problem of the lack of information in the Unified State Register of Real Estate on the boundaries of real estate from land plots, capital construction projects to the boundaries between municipalities of the RF, water and forest facilities, which undoubtedly leads to difficulties in determining the boundaries of protected areas. This problem also correlates with the problem of excluding certain categories of lands from the boundaries of protected areas. Exclusion of land is possible only within established boundaries. The problem of excluding lands from the boundaries of protected areas also negatively affects the configuration of protected areas and the concept of protecting the natural biodiversity of a given territory.

It should be noted that there is no possibility of cadastral registration of the boundaries of functional zones of protected areas and the boundaries of zones of limited economic use of protected areas. These zones play an important role in the organization and management of protected areas; they regulate the environmental regime and economic activities in different parts of the protected areas, which is undoubtedly important information that requires accounting.

Attention should be paid to the need to establish protected areas of protected areas. The established security zone contributes to additional protection of the protected object. Due to the lack of established protected areas, it is necessary to develop a methodology for determining the boundaries of protected areas.

Currently, there is no unified methodology for the creation and formation of protected areas. There is no general regulation for the development of MCES on the territory of the Russian Federation. The format for transmitting and storing data on protected areas is limited to paper and electronic document carriers.

In order to effectively manage and record specially protected natural areas, we consider it necessary to develop and establish a unified cadastre of protected areas, a publicly accessible and informative resource for various purposes of managing the territories of the Russian Federation. As part of the cadastre of protected areas, it is necessary to develop a unified methodology for organizing and accounting for protected areas, which will create a unified database of protected areas in the Russian Federation.

To effectively create a modern cadastre of protected areas, it is necessary to adjust the general approach to the formation and recording of protected areas. The unified cadastre of protected areas should become an open information platform that combines all the necessary data on such territories. The system should include tools convenient for working with the map, the ability to obtain information about protected areas and search for information about such territories. In particular, it is necessary to ensure the possibility of publishing the «Map of Protected Areas» system on the public Internet, which will provide all interested parties with access to the data of the state cadastre of protected areas.

The «Map of Protected Natural Areas» system should display information in vector format: about the boundaries of protected natural areas, security zones, functional zoning of protected natural areas, existing zones of limited economic use. Each protected area must have an individual cadastral registration file in electronic form with all the data from the environmental survey of the specified territory, documents approving the creation of such a territory, permitted types of activities and restrictions.

In order to effectively manage protected areas, it is necessary to develop a unified methodology for developing materials for environmental justification for giving a territory the status of a protected area. Differences in the form of data presented by research organizations and differences in requirements for these materials in the constituent entities of the Russian Federation lead to confusion in recording information about such territories.

Currently, the most used format for storing and exchanging data is XML. Data exchange via the XML format has been actively used for many years in the state real estate cadastre; transfer of data to the registration authority is possible only in the specified format.

As part of the development of a modern cadastre of protected areas, it is necessary to develop an XML schema that meets all the requirements of a modern cadastre. Using the XML format for storing and transmitting data on protected areas will provide a fast, high-quality and unified approach to processing the information received.

Thus, in order to solve the identified problems in the field of accounting for protected areas, it is important to reconsider the methods and methods of accounting for protected areas. The creation of a new modern unified cadastre of protected areas will allow for high-quality and efficient management of these territories.

CONCLUSION

As part of the ongoing research, information about the protected areas of the Russian Arctic was collected from open sources of information, and a comprehensive and up-to-date study of the state of the Arctic protected areas system was carried out. During the study of the state of the PA system of the Russian Arctic, the following problems were identified:

1. Fragmentation of data storage on protected areas based on territorial affiliation;
2. Lack of a unified regulation of MCES, justifying the granting of protected area status;
3. Elimination of «interspersed» land plots by cutting them out from the boundaries of protected areas - the emergence of «holes» within the boundaries of protected areas;
4. Errors in entering data on the boundaries of protected areas into the Unified State Register of Real Estate;

5. The absence of boundaries of protective and functional zones of protected areas provided for by the legislation of the Russian Federation, with the help of which more effective and rational management of such territories is carried out.

As a result of the research, the named current problems of protected areas of the AZ of the RF were identified and a conclusion was made about the need to revise the ways and methods of recording specially protected natural areas and develop GIS technology for their cadastral registration. The creation of a unified methodology for organizing and recording objects and areas of PA, the formation of a modern digital unified cadastre of these territories will allow for high-quality and efficient management of these land plots and will ensure high-quality protection of the biological and landscape diversity of the entire territory of the Russian Arctic. The unified digital cadastre of protected areas should be a multifunctional publicly accessible information system of protected areas containing information about such territories in electronic form, including tools for working with the service and displaying information in graphical form.

The unified cadastre of protected areas in the Arctic zone proposed by the authors will allow us to combine information that is scattered according to jurisdiction and will provide the opportunity to create a public map of natural territories. The map must display not only the boundaries of protected areas, but also existing functional zones and zones of limited economic use. The developed unified cadastre of protected areas in the Arctic zone will become a platform for effective and high-quality management of natural areas; will create a convenient and effective basis for eco-tourism at some natural sites where this type of activity is possible.

REFERENCES

Aleynikova, A., Byashkin, M., Gaivoron, T., Mainasheva, G., & Marsheva, N. V. (2020). The role of specially protected natural areas in solving environmental problems of the Russian arctic (using the example of Vaygach Island). *E3S Web of Conferences, 169*. 10.1051/e3sconf/202016902019

Baltyzhakova, T. I., & Romanchikov, A. (2021). SPATIAL ANALYSIS OF SUBWAY PASSENGER TRAFFIC IN SAINT PETERSBURG. *Geodesy and Cartography (Vilnius)*, 47(1), 10–20. 10.3846/gac.2021.11980

Berezikov, S. A. (2019). Structural changes and innovation economic development of the Arctic regions of Russia. *Journal of Mining Institute*, 240(6), 716–723. 10.31897/PMI.2019.6.716

Bogachev, V. F., Gorenburgov, M. A., & Alekseeva, M. B. (2019). Systemic diagnostics of the Arctic Industry Development Strategy. *Journal of Mining Institute*, 238(4), 450–458. 10.31897/pmi.2019.4.450

Bykova, E. (2021). Assessment of negative infrastructural externalities when determining the land value. *Journal of Mining Institute*, 247, 154–170. 10.31897/PMI.2021.1.16

Choi, H. O. (2020). An Evolutionary Approach to Technology Innovation of Cadastre for Smart Land Management Policy. *Land (Basel)*, 9(2), 50. 10.3390/land9020050

Choker, H. M., & Mustafin, M. (2021). Geodesic support of laser scan technology use for fixing cultural heritage objects. *Geodeziâ I Kartografiâ, 968*(2), 2–10. 10.22389/0016-7126-2021-968-2-2-10

Crommelinck, S., Bennett, R., Gerke, M., Yang, M. Y., & Vosselman, G. (2017). Contour detection for UAV-Based cadastral mapping. *Remote Sensing (Basel)*, 9(2), 171. 10.3390/rs9020171

Dolgikh, N. U., & Kusticova, M. A. (2019). Ecological monitoring of specially protected natural areas of the Russian Arctic. *IOP Conference Series. Earth and Environmental Science*, 272(2), 022171. Advance online publication. 10.1088/1755-1315/272/2/022171

Gusev, E. A. (2022). Results and prospects of geological mapping of the Arctic shelf of Russia. *Journal of Mining Institute*, 255, 290–298. 10.31897/PMI.2022.50

Il Tumen's press service. (2016). *The Government of the Russian Federation has begun work on clarifying the composition of the Arctic zone of the country*. The website of the State Assembly (Il Tumen). http://old.iltumen.ru

Kabonen, A.V., & Ivanova, N.V. (2023). Assessment of biometric characteristics of trees according to ground LiDAR and multi-season aerial photography in artificial plantations. *Nature Conservation Research. Reserved science, 8*(1), 64-83. 10.24189/ncr.2023.005

Kovyazin, V., Kitcenko, A., & Shobairi, S. O. R. (2021). Cadastral valuation of forest lands, taking into account the degree of development of their infrastructure. *Journal of Mining Institute*, 249, 449–462. 10.31897/PMI.2021.3.14

Lipski, S. A. (2018). Arctic lands: arrangements for their use and monitoring. *Izvestiya vuzov «Geodeziya i aerofotosyemka», 62*(2), 196–203. . EDN YWZHZR10.30533/0536-101X-2018-62-2-196-203

Luo, X., Bennett, R., Koeva, M., & Lemmen, C. (2017). Investigating Semi-Automated Cadastral Boundaries Extraction from Airborne Laser Scanned Data. *Land (Basel)*, 6(3), 60. 10.3390/land6030060

Matveeva, V. A., Bech, J., & Danilov, A. S. (2023). Prospects of nitrogen removal from mine drainage by micro-algae in northern environments. *Sustainable Development of Mountain Territories*, 15(1), 134–142. 10.21177/1998-4502-2023-15-1-134-142

Mnatsekanov, R. A., Sergeeva, M. I., & Filippova, E. V. (2021). *Methodological recommendations for the creation of specially protected natural territories of federal and regional significance*. World Wildlife Fund (WWF).

Mustafin, M., & Moussa, H. (2023). The determination of plumb line deviation using Satellite/Levelling Technique. In *Lecture notes in civil engineering* (pp. 37–47). 10.1007/978-981-99-4229-9_4

National projects. (2022). *Ecology*. National projects. https://национальныепроекты.рф/projects/ekologiya

National projects. (2023). *Passport of the project «Conservation of biological diversity and development of ecological tourism»*. National projects. https://www.mnr.gov.ru/activity/np_ecology

Ostroukhov, A.V., Klimina, E.M., & Kuptsova V.A. (2020). Landscape mapping of hard-to-reach territories on the example of the Bolonsky State Nature Reserve (Russia). *Nature Conservation Research. Reserved science, 5*(2), 47-63. 10.24189/ncr.2020.015

Pristupa, A., Tysiachniouk, M., Mol, A. P., Leemans, R., Minayeva, T., & Markina, A. V. (2017). Can zoning resolve nature use conflicts? The case of the Numto Nature Park in the Russian Arctic. *Journal of Environmental Planning and Management*, 61(10), 1674–1700. 10.1080/09640568.2017.1370365

Raguzin, I., Bykova, E. N., & Lepikhina, O. Y. (2023). POLYGONAL METRIC GRID METHOD FOR ESTIMATING THE CADASTRAL VALUE OF LAND PLOTS. *Vestnik Moskovskogo Universiteta. Seriia V, Geografiia*, 78(3), 92–103. 10.55959/MSU0579-9414.5.78.3.8

Sivák, M., Kardoš, M., Kadlečík, R., Chudá, J., Tomaštík, J., & Tuček, J. (2023). Identification of the Forest Cover Growth on Landscape Level from Aerial Laser Scanning Data. *Land (Basel)*, 12(5), 1074. 10.3390/land12051074

Skachkova, M., & Gureva, O. S. (2022). Information support of urban planning activities in Russia. *Geodeziâ I Kartografiâ, 986*(8), 45–55. 10.22389/0016-7126-2022-986-8-45-55

Skachkova, M., & Guryeva, O. (2023). Monitoring of the State of Saint Petersburg green spaces by remote sensing data. *Экология И Промышленность России, 27*(5), 51–57. 10.18412/1816-0395-2023-5-51-57

The Government of the Arkhangelsk region. (2016). *On approval of the Regulations on the Mudyug State Natural Landscape Reserve of regional significance* (Resolution of the Government of the Arkhangelsk region 363-pp). The official Internet portal of legal information http://www.pravo.gov.ru

The Government of the Murmansk region. (2020). *On approval of the Procedure for the preparation and composition of materials justifying the creation of specially protected natural territories of regional significance, changing the regime of their special protection* (Resolution of the Government of the Murmansk Region 701-pp).

The Government of the Russian Federation. (2002). *On the approval of the Environmental Doctrine of the Russian Federation* (Decree of the Government of the Russian Federation 1225-p). Collection of legislation of the Russian Federation The President of the Russian Federation. (2019). *Message from the President of the Russian Federation Vladimir Putin to the Federal Assembly of the Russian Federation* (38). Rossiyskaya gazeta.

The State Duma. (1995). *On Specially protected natural territories* (Federal Law 33-FZ). Collection of legislation of the Russian Federation.

Tishkov, A. (2012). CONSERVATION OF RUSSIAN ARCTIC BIODIVERSITY. *Geography, Environment. Sustainability*, 5(3), 48–63. 10.24057/2071-9388-2012-5-3-48-63

Tsvetkov, P., Andreychik, A., & Kosarev, O. (2024). The impact of economic development of primary and secondary industries on national CO2 emissions: The case of Russian regions. *Journal of Environmental Management*, 351, 119881. 10.1016/j.jenvman.2023.11988138150925

Vasilev, Y., Tsvetkova, A., & Stroykov, G. (2020). SUSTAINABLE DEVELOPMENT IN THE ARCTIC REGION OF THE RUSSIAN FEDERATION. *International Multidisciplinary Scientific GeoConference SGEM*, 763–770. 10.5593/sgem2020/5.1/s20.096

Zuckerman, V. A. (Ed.). (2022). *Innovative development of industry in the Arctic regions: problems and prospects*. FRC KSC RAS. 10.37614/978.5.91137.462.4

Chapter 14
Virgin Lands Development in Kazakhstan:
Causes, Implementation, Consequences

Laila Seysembekovna Akhmetova
http://orcid.org/0000-0002-3607-3688
al-Farabi Kazakh National University, Kazakhstan

Sergey Sergeevich Akhmetov
al-Farabi Kazakh National University, Kazakhstan

Usembayeva Aygerim
al-Farabi Kazakh National University, Kazakhstan

ABSTRACT

The consequences of the virgin lands development are still felt in the Republic of Kazakhstan. In addition to the environmental burden caused by the expansionary cultivation in the north of the Kazakh Soviet Socialist Republic (KazSSR), modern Kazakhstan has perceived the social and demographic effects of the all-Union Virgin Land program since the early 1990s. This chapter examines the motives of the Soviet leadership, initiators and opponents of the Virgin Land program (1953-1958), and critically analyzes the plans and implementation based on the experience gained. The impact on economic and social conditions is subjected to historical analysis. Further environmental and demographic results of the program, factors influencing the current soils and agricultural conditions in the Republic of Kazakhstan are reviewed and demonstrated. This allows giving an objective, scientifically based assessment of the Virgin Land program, taking into account multi-vector economic and functional influences.

INTRODUCTION

The program for the development of virgin lands (or Tselina) has received many assessments after the collapse of the Soviet Union, where critical trends can be easily traced. At the same time, a number of researchers come to conclusions about the need for the program in a specific historical period, emphasizing the poor quality of implementation (Saktaganova & Abylkhozhin, 2018, p. 104). There is a

DOI: 10.4018/979-8-3693-3374-7.ch014

consensus in literature when it comes to environmental consequences. However, some authors came to the predictable conclusion that it was impossible to consider, much less evaluate one aspect of the Virgin Land program, for example, related to negative impacts on the environment, in isolation from others: the political motives of the initiators and economic necessity.

In this regard, the authors decided to focus on the political and economic motives of leaders at different levels, and the results of the study, apparently, put an end to thoughts about the expediency of the Virgin Land program. Outlining the psychological portrait of the General Secretary N. Khrushchev researchers note his restless temperament, and the plan for the development of virgin lands is equated with such foreign policy maneuvers as ultimatums to the West regarding Berlin and the deployment of nuclear missiles in Cuba (Easter, 2021, p. 460). This work shows that the motives and goals of the economic initiative under study are much deeper, despite the fact that the so-called "Gigantism" of the era, as well as the personal decisions of the General Secretary to increase and unreasonably maximize crop areas and excessive expected results, were recognized as erroneous back in Soviet times.

Criticism of the program in certain aspects, within specific farms or geographic locations, was expressed already at the end of the 50s on the pages of all-Union magazines (Verkhovskiy, 1969; Cherniyenko, 1964). The analysis shows that at different stages and in different areas, all the participants in the decision-making processes had their reasons: N. Khrushchev, who faced the need to solve the food crisis, V. Molotov, who agreed with the Tselina project on a limited scale (Chuyev, 1991) (which, as the research has shown, could be the most effective solution), Zh. Shayakhmetov, who feared demographic shifts in the republic and criticized the scale and methods of land development (Akhmetova & Grigoryev, 2021), leaders and implementers of the program, scientists, incl. the second Secretary of the Central Committee of the Communist Party of Kazakhstan [CC CP] L. Brezhnev (The Resolution of the VIII Congress of the Communist Party of Kazakhstan on the Report of the CC CP of Kazakhstan, 1956) and the President of the Academy of Sciences of the Kazakh Soviet Socialistic Republic [KazSSR] K. Satpayev (Degitayeva (Ed.), 2018). Today's knowledge allows to recognize the fallacy of the actions and statements of each of them for different reasons: from lack of information, from ideological restrictions, from personal political or social tensions.

Next, the demographic and environmental consequences of the program implementation were considered. Demographic effects were sharply critisized during the discussion of the program and in the first stages of virgin lands development, social risks were declared, and unfavorable forecasts were proposed (Khrushchev, 1999). Arguments against the program were based on deterioration of local residents' life conditions, however, it was demographic shifts and related processes that had a positive impact on the social situation and continue to make a significant contribution to the economy of contemporary independent Kazakhstan.

The second consequence - environmental, perhaps the most pressing after economic necessity, was inherited by the Republic of Kazakhstan from the Union of Soviet Socialist Republics [USSR]. In general, critically examining the environmental results of lands development, Kazakh scientists conlude that positive aspects prevailed for the KazSSR, at least in socio-economic terms (Mazhitova et al., 2021). There are a number of researches on the environmental consequences of the program, containing criticism, comparative analyzes and accumulated experience exstracts (Kozybayeva (Ed.), 2022). However, providing the state with grain (both for own use and for export) is almost the same with the situation in the early 50s, remained, if not a priority, then at least a particularly important task even in the 90s and at the present time. This theory determines the relevance of multifaceted and interdisciplinary research, including land management, sociological and historical research that will concern the four regions of

Kazakhstan affected: Akmolinsk (North Kazakhstan region in Soviet period), Kostanay, North Kazakhstan (Kokshetau region in Soviet period) and Pavlodar regions.

Considering the Virgin Land campaign as a certain factual reality, regardless of accumulated experience, incl. Soviet period and the first years of independence of the Republic of Kazakhstan seems to be a big omission. Such challenges as food security, global demographic processes, agricultural prices (especially in regions, that suffer from shortages), as well as environmental problems and land protection issues remain relevant today and, perhaps, have never lost importance. In this regard, the political, economic and social factors of the topic under study were taken into account as a whole, since they allow us to most fully analyze the current state of the regions affected by the Virgin Land program and concentrate knowledge for further work.

BACKGROUND

Late 1940s - early 1950s post-war era demanded restoration and further redistribution of labor. At the same time, agricultural sector requested government support, because could no longer ensure the food supply of cities as planned in the official guidelines of the fifth five-year plan. The volume of gross agricultural output in the first post-war year was approx. 40% less than the pre-war figure (Central Statistical Directorate under the Council of Ministers of the USSR, 1967, p. 30).

The course of first party leader N. Khrushchev was marked by acute shifts in internal policy. The dismantling and further rethinking of a number of fundamental ideological and political guidelines of the Stalin era began. The new leader criticized the agricultural sector and the course of I. Stalin due to low production indicators, as well as for incorrect guidelines for the peasantry (Egorov & Bogolyubov (Eds.), 1985, pp. 305-307).

In the 1950s, the issue of agricultural products supply became particularly intensified. N. Khrushchev and his associates chose a way to solve existing problems with the massive development of virgin and fallow lands. The bulk of such lands were supposed to be developed in KazSSR.

N. Khrushchev indicated his political and economic course during the February-March Plenum of the Central Committee in 1954, devoted to the problems of virgin lands development (as cited in Akhmetova, 1996, p. 54). This new course had a number of key differences from the previous, post-war development plan and Stalinist doctrines, where, according to Volin (1970) "Agriculture had to play a second fiddle to industrialization, with underproduction the inexorable outcome" (p. 308).

The lands development, according to N. Khrushchev's plan, was supposed to solve a number of interrelated problems. Firstly, to create a zone for obtaining a large commercial mass of grain, which removed the problem of bread supply from the agenda. Secondly, providing the state with surplus grain, which allowed to enter international markets. Accordingly, the USSR could become a competitor to the USA, Canada, Australia, Argentina and a supplier to India, China and African states. This stimules where not only about export increase, but also enabled to gain additional geopolitical leverage, given the strategic importance of the product.

This turning point decision radically changed the fate of Kazakhstan. In the speech at the September plenum of the Central Committee of Communist Party in 1953, N. Khrushchev sharply criticized the entire leadership of the KazSSR for the state of affairs in agriculture (Akhmetova & Grigoryev, 2021, p. 191). The main complaint was that the Republic had a huge plot of virgin land and did not use it for grain farming, despite the fact that the country was in desperate need of bread. Secondly, Kazakhstan produced insufficient amount of feed grain, and this seriously hampered the development of transhumance livestock farming (a traditional type of cattle breeding in the north of Kazakhstan, where such grain served as the basis for livestock nutrition). At the plenum, the Bureau of the CC CP of Kazakhstan was instructed to prepare a justification for expanding the acreage in the republic with the aim of massively increasing grain production and livestock farming indicators.

However, the almost complete lack of infrastructure, unprepared research base, poor planning and other factors have led to dire consequences in the environmental sphere. At the same time, the reasonable goals set during the discussion of the Virgin Land program, the combination of the efforts of human and material resources of the entire country led to the solution of the problem and the implementation of the assigned tasks at the beginning. These factors subsequently reveal unexpected, including positive, effects on the social and economic environment. N. Khrushchev's internal economic and agricultural reforms, in comparison with changes in the other communist countries, demonstrate as one of the driving factors of improving the viability of the collective farm (kolkhoz) (Volin, 1970, p. 563).

The opposition to Khrushchev's plan on the part of the Kazakh party elite should be noted. It was discussed in the biographical work of W. Taibman (Taubman, 2003, p. 262), although the motives of the leader of the KazSSR Zh. Shayakhmetov, were most likely not of an economic nature. The leadership of the republic had reason to fear the further (after the resettlement policy of the times of the Russian Empire, as well as the war and post-war deportation policy (Alpyspaeva & Nurmukhanova, 2023, pp. 327-341), settlement of the country by representatives of different ethnic groups to the detriment of the local population. The study shows the paradoxical validity of these fears and the strategic effectiveness of Khrushchev's decision to ignore the arguments of Zh. Shayakhmetov.

RESULTS AND DISCUSSIONS

Political and Economic Reasons

The Virgin Land program was associated with N. Khrushchev's desire to change the situation in the Communist Party of the Soviet Union, to end the dominance of the city over the countryside and equalize them in contrast to Stalin's programs of total industrialization. The General Secretary has repeatedly criticized the course and agricultural policy of Stalin, as well as the well-founded drop in grain yields, in particular, in the regions under study (modern Akmolinsk - previously North Kazakhstan, Kostanay, modern North Kazakhstan – previously Kokshetau, and Pavlodar regions) in the post-war years relative to previous periods. However, back in 1946, the Council of Ministers of the USSR adopted a resolution "On expanding sown areas and increasing the yield of grain crops," which planned to expand crops in Kazakhstan, Siberia and the Southern Urals by 6.5 million hectares. In addition, the agricultural and botanical solutions introduction was planned even during the implementation of Stalin's course. For example, it was recommended to create forest shelter belts around the fields, and forced grass and forest planting in order to strengthen the soil (Kozybayeva, 2022, p. 140). But these efforts produced minimal results and did not satisfy either the real requirements of the economy or the calculations of the pre-Khrushchev leadership.

A new method for problems solving was the strategy of extensiveness. As a result, the modernization of agricultural production, its inclusion in the global flow of technological change was postponed for a long time, however, this became obvious only decades later.

N. Khrushchev was based on materials prepared by the Ministry of Agriculture back in the Witte-Stolypin era, where such factor as dust erosion were not considered at all, and the emphasis was placed on expanding the acreage of crops in new peasant farms (with an understandable difference in the form of ownership and rights to land use). This decision, being essentially a continuation of conceptual approaches to the quantitative expansion of cultivated areas and production volumes, contained serious errors and miscalculations that negatively affected the agricultural and environmental situation in Kazakhstan.

Agriculture in the entire country was in crisis and could not provide necessary amount of food before virgin lands development. This reason for the program initiation should not be forgotten, believing that the General Secretary and local leaders (with the exception of Zh. Shayakhmetov and some members of his team) acted solely out of personal ambitions or under the influence of political course. It was the food shortage that forced the country's leadership to solve the problem in an extensive way (Kozybayeva (Ed.), 2022, p. 80). Other reasons include the weak technical equipment of collective farms (including the technological lag behind capitalist countries), outdated methods of farming, and the difficult post-war demographic and economic situation, which, in total, led to weak and ineffective productivity in agriculture.

The main, large-scale and most pressing reason left no room for discussion. Even though N. Khrushchev certainly understood the degree of economic burden associated with such a huge project as the development of virgin lands, necessity forced him to act despite the counter-arguments of the Kazakh leadership. The development of lands required colossal funds and strained both the union and republican budgets. The Irish historian McCauley (1976) stated, "Taking into account economies of scale in Kazakhstan, it is still difficult to resist the temptation to attribute all virgin soil to a very expensive method of grain production, without even trying to estimate the cost of the operation," speaking about the cost and, as a result, questioning the effectiveness of the project from the capitalist Western European economies point of view (p. 147). However, according to reports from the March 24-26, 1965 Plenum of the CP, public investment in agriculture declined from 1953 to 1958 (Chernenko & Smirtyukov (Eds.), 1968). Precisely because the overwhelming scale and the economic impossibility of the project, as it turned out, the development of virgin lands became a failure as a whole, without solving its strategic objectives in the next ten years.

At the same time, in the first productive years, the Virgin Land campaign allowed to avoid food shortages and possible famine (which is predictable with extensive farming methods). The initial results of the program implementation were impressive, despite the lack of detailed planning and sufficient soil and climate characteristics awareness deficite (Edygenov et al. (Eds.), 1959, p. 159), so that the president of the KazSSR Academy of Sciences K. Satpayev even advocated expanding the land development further south – into the "lifeless" spaces of Central Kazakhstan, given the water resources availability (Degitayeva (Ed.), 2018, p. 158). However, the project lacked further fundamental reforms, technical support and capital investments to introduce the latest methods and technologies.

Demographic Consequences

February 22, 1954 N. Khrushchev makes a call at a rally of the Moscow region members of Komsomol (Youth division of the CP). He encouraged the youth to take part in Kazakhstan and Siberia Virgin Land cultivation project (Khrushchev (Ed.), 2007 p. 987). This gave impetus to the beginning of the process that Zh. Shayakhmetov feared.

Even during the period of the first globalization (late 19th century), according to Sanchez-Sibony (2014) "Rising prices for most agricultural produce encouraged the settlement of uncultivated land in places as disparate as Canada, Australia, and Argentina." (p. 30) - countries that later became the world's main grain suppliers. The reduction in the proportion of the indigenous population was the main motive for the opposition of the republic's leadership, as Khrushchev himself and some regional committees' leaders believed (Khrushchev, 1999, p. 75).

However, Zh. Shayakhmetov had doubts and justifications not only of a demographic or, even more so, "nationalistic" nature. During a personal meeting with Khrushchev, he reported about the consequences of the loss of lands located in the zone of risky agriculture, a thin layer of humus soil that would dissipate during plowing, the need to develop livestock farming in the region and, importantly from a political point of view, about the already existing agreement on these issues with I. Stalin (Akhmetova & Grigoryev, 2021, p. 192-193.). This disagreement, references to the previous leadership of the country, and a different, more pragmatic agricultural program from that proposed by Khrushchev, led to a change in the leadership of the republic and the implementation of an extensive program.

The purposeful or more precisely traditional use of steppe and forest-steppe zones (illustrated in Figure 1), which Zh. Shayakhmetov spoke about in the early 50s, in the context of agricultural megaprojects, was considered by Russian and Kazakh scientists from the environmental consequences point of view (Levykin et al., 2021). At the same time, the problem of fodder areas for livestock, which arose with particular urgency after the land was given over to crops, did not remain completely unaccounted. Some state farms successfully grew fodder (Mazhitova et al., 2023) using new methods instead of archaic ones. That is, these arguments of Zh. Shayakhmetov, as individual practice has shown, could be refuted. At the end of 1960, the total number of livestock increased by 52.3% compared to 1955 due to the expansion of cultivated areas, mechanization, and the introduction of new methods of housing and feeding (Abisheva et al. (Eds.), 1980, pp. 278, 281).

The directive redistribution of funds in favor of the program significantly complicated the situation in the economy and social situation. On the one hand, there was electrification of rural areas, new state farms were created and, as a result, new settlements and system-forming industries, water pipelines, highways and railways were built at an accelerated pace. On the other hand, there were painful cultural processes that affected the Kazakh village, language and culture, as well as changes in the ethnic structure of the population to the detriment of the indigenous Kazakh population, which, however, strategically turned out to be beneficial. As a result of the Virgin Land program and flanking social policy decisions to send new settlers to these lands, Kazakhstan has become a region of national importance and a key one in the agricultural sector. The Tselinniks (the workers of Tselina) and the internationalists were opposed to the national or local employees (Akhmetova, 1996, p. 56). However, the demographic imbalance was leveled off already in the 70s (Alexeenko, 1994, pp. 34-39).

The research materials show the influx of approximately 2.5 million people into the regions of Kazakhstan for the development of virgin lands and accompanying industrial construction projects. This led to a radical change in the demographic structure of the region. As a result of population movements due to post-war policies and the program, the demographic situation in the villages of Kazakhstan, on the one hand, was characterized by multi-ethnicity, and on the other, a significant reduction in the local Kazakh population (Kozybayeva, 2022, p. 285). However, the new cities and settlements, jobs, an increase in the population of Central and Northern Kazakhstan, industrialization, highly qualified workers, technicians and machine operators and, importantly, young people incoming (Eleuov (Ed.), 1960, pp. 11-12), can be identified as a strategically positive factor. And although initially, as Talamani (2023) stated "the economy of the small settlement was based on trade, while in the following two phases, the economy was sustained by organized exploitation of natural resources – through agriculture and materials extraction, respectively" (p. 29), considering the fact of a rapid increase in the number of foreigners was quite understandable. However still was perceived painfully by contemporaries. By 1959, the ethnic composition of the population of the KazSSR had become extremely unfavorable for the indigenous population. According to the population census, the number of Kazakhs was 30% (Central Statistical Office under the Council of Ministers of the KazSSR, 1980).

Today multi-ethnicity in villages is not considered as a problematic or potentially risky fact, while the descendants of Virgin Land campaign settlers have been living on the territory of the KazSSR and the Republic of Kazakhstan for several generations, being full citizens of the country today. It was an ideologically justified process, and the results contribute to the economy and demography of modern Kazakhstan. This is confirmed by the ethnic composition aligned in favor of the indigenous population. By 2023, the share of the Kazakh population in the republic will be 70.7%. For four regions affected by the program, the latest population census showed the following results: the share of Kazakhs in the Akmolinsk region was 56.2%, in Kostanay 44.4%, in Pavlodar 56.5% and in North Kazakhstan 37.8%. (Bureau of national statistics of the agency for strategic planning and reform of the republic of kazakhstan, 2023). These data allow us to conclude that the situation regarding the decrease in the share of the indigenous population in these regions has leveled off, taking into account the fact that the share of representatives of other ethnic groups is not always a direct consequence of the Virgin Land program, but includes the policy of resettlement of peoples in Soviet times, the movement within the country and migration policy after independence.

Today, the multi-ethnicity of the Republic of Kazakhstan, the inadmissibility of ethnic infringements and freedom of self-determination are secured by the 19, 20, 39, 44, etc. articles of the Republic of Kazakhstan Constitution (Institute of Legislation and Legal Information of the Republic of Kazakhstan, 2012), while tolerant coexistence is regulated by national legislation (Resolution of the Government of the Republic of Kazakhstan, 2019).

Environmental Consequences

Humus layer destruction. The result of the development of the total estimated at approx. 42 million hectares of steppe lands, was an increase in the area of surface soil erosion and destruction of the humus layer (which is currently regulated by the Land Code of the Republic of Kazakhstan) (Ministry of Justice of the Republic of Kazakhstan, June 20, 2003). Monoculture cultivation of grain and industrial crops on low agrotechnical conditions causes of dehumidification process and loss of soil fertility by 20-30% (as cited in Issanova et al., 2020, p. 99).

The long-term non-renewable use of soil fertility in the conditions of the arid steppe intensified the processes of anthropogenic desertification; insufficient use of fertilizers and ignoring the agro-ecological potential of the territory led to a significant decrease in soil fertility. Thus, the chernozems (the black soils) lost 25-30% of their natural humus content - the main indicator of soil fertility, together with moisture deficiency, destruction of the soil composition and high susceptibility to erosion (Gindemit, A. & Aksenova, Yu., 2021).

Figure 1. The natural areas and the virgin lands regions of Kazakhstan

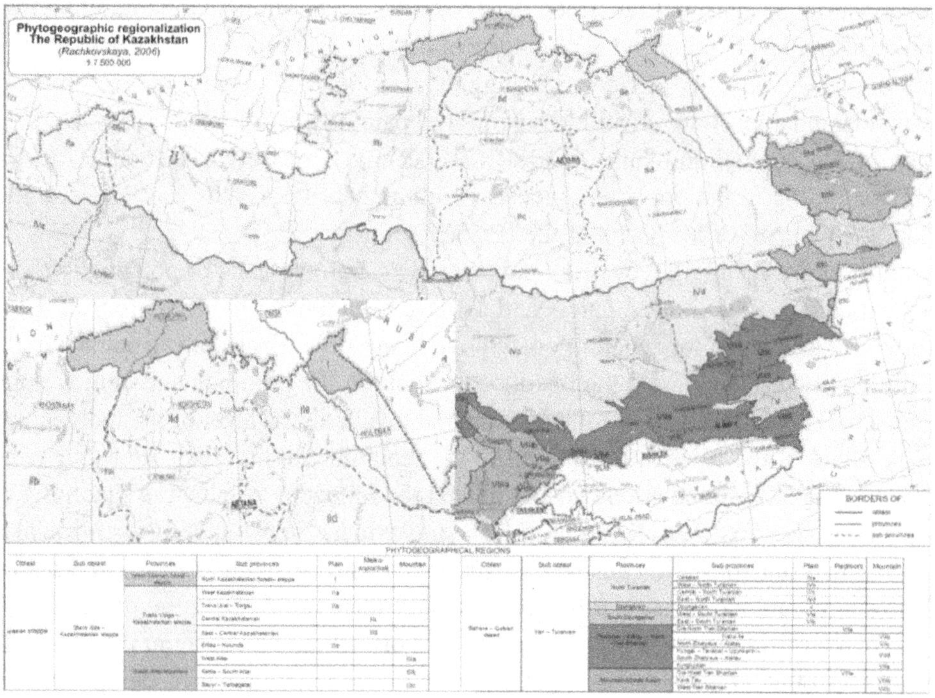

Taken from Medeu, A.R. (Ed.), 2010.
Note. Figure 1 supplemented by the authors.

Figure 1 shows phytogeographic regions of contemporary Republic of Kazakhstan, with steppe and forest steppe zones marked (those affected by the program). A close-up map shows the regions in which the Virgin Land program was implemented. According to the modern index, the territories included: Akmola region with the city of republican significance Astana and the city of Kokshetau (Kokshetau was a region in Soviet period), Kostanay region (Kostanay city), North Kazakhstan region (Petropavlovsk city), Pavlodar region (Pavlodar city) (Bureau of National Statistics of the Agency for Strategic Planning and Reform of the Republic of Kazakhstan, 2023).

Due to significant losses of humus, plants are not provided with nutrients in the required quantities, and the physical condition of the soil deteriorates. As a result, the structure becomes less durable and collapses (sprays) under the influence of annual cultivation. The porosity favorable for plant growth and, as a consequence, the water-air regime was disrupted. In addition, the lack of crop rotation (only grain crops) and monoculture provoked the massive spread of weeds, especially wild oats.

In addition, there is a constant estragement of carbon, nitrogen and ash elements with the grain harvest. This happened even though the expansion of cropland during the program mainly took place on the most suitable soils, although not always. However, the maximum negative effect was probably observed after the Virgin Land program, "While nearly 60% of the expansion after the Campaign, from 1962 to 1990, were found on soils of lower suitability", as mentioned by Kraemer (2015). In cultivated soils against the backdrop of grain monoculture, low biogenicity and insignificant accumulation of amino acids and proteins are observed, as well as a general decline in the rate of fiber decomposition.

Humus content in soils decreases most strongly in the first 10-15 years after plowing, especially its labile forms. The soils are transformed into a lower rank in terms of their suitability (Almanova et al., 2023), which, as a result of many years of improper use, affects the agricultural situation in the region and can be attributed to the worst legacy of the Virgin Land program in the strategic economic aspect.

The results of outdated methods use. Considering the main negative (ineffective) factors of the Virgin Land program that cause environmental damage (as mentioned in Figure 2), we find inappropriate use of land and outdated methods of cultivation.

Figure 2. Main negative (ineffective) factors of the Virgin Land program causing environmental damage

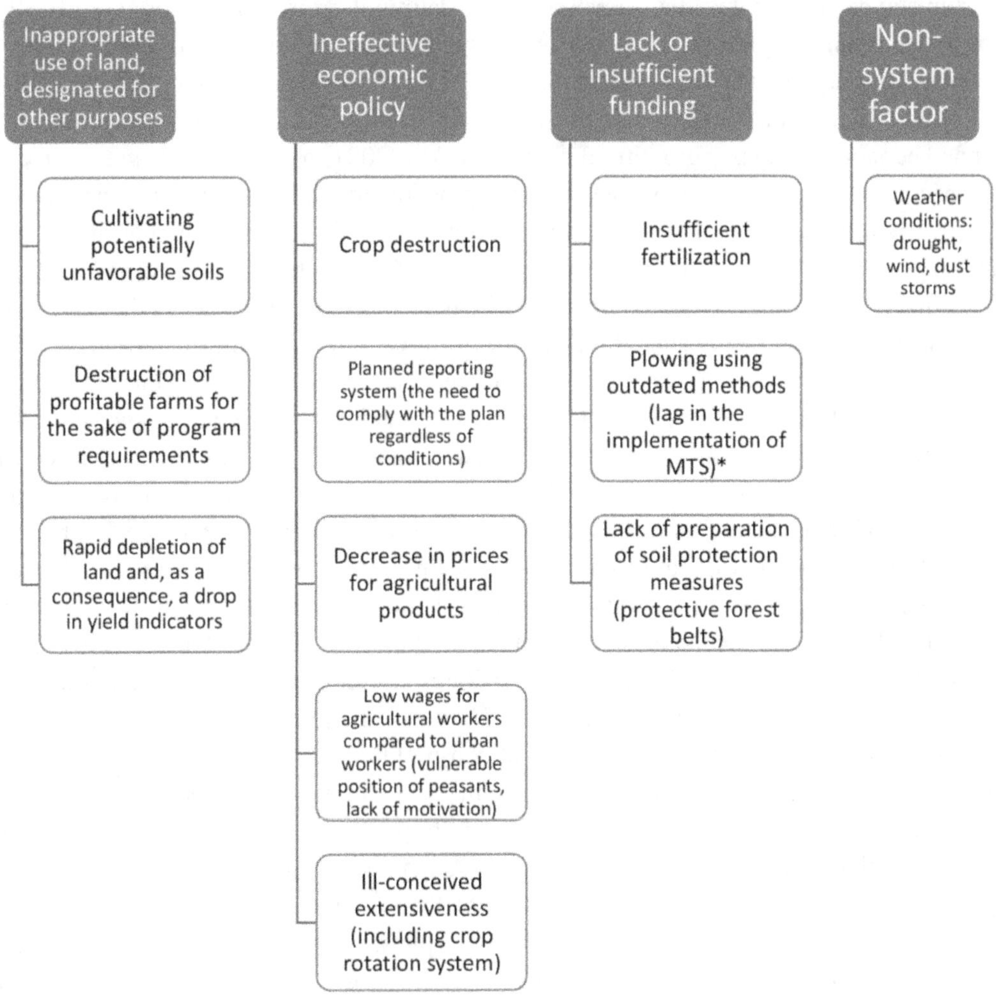

Note. MTS: the machine tractor station, providing technical support to state agricultural producers.

Despite the mentioned resolutions of the Council of Ministers of the USSR and the introduction of new methods into the agricultural sector under I. Stalin, including the forced fencing of cultivated areas with trees, during the implementation of the project (since 1958), seemingly predictable negative environmental consequences began to arise. Along with a decrease in yield after the first four years, the soil was subject to wind erosion (due to the lack of natural protection) and massive growth of weeds. In addition, the unrealistic planning, the pursuit of fulfilling an economic plan issued from above and often overestimated, leveling without regard to the specifics of geography and soil types led to absurd situations when previously profitable livestock farms had been plowed (which had been operating since the time of Stolypin) and became unusable even though the planned grain harvest still could not be collected. It was necessary to increase the arable land of the state farm at the expense of unsuitable grass and sand manes, which already in the early 60s led to erosion (Chernienko, 1964, p. 168-169). "According to its geographic and climatic location, north Kazakhstan suffers from highly variable plant growing conditions due to the permanent risk of drought and both late and early frost. Wind erosion frequently attacks the newly cultivated soil", - stated Petrick (2013, p. 3). Moreover, the features of the northern and central Kazakhstan steppe zone within the framework of agricultural projects and more specifically in grain growing are determined by climatic characteristics, such as warmth, rainfall, soil moisture retention and seasonal moisture supply in general (Rau et al., 2023). Local climatic characteristics were not taken into account. This brings us back to the "Gigantism" of the country's leadership and the ineffectiveness of the planned economy in some agricultural areas. Levykin (2021) asserts, "All negative processes, first of all agroecological, were caused by not the idea of plowing but threefold excess of the grounded plane. Potentially low-productive lands were plowed, and they are officially reckoned as affordable lands at present" (p. 4). The plowing of the Western Siberian and Kazakh steppes, as well as forest-steppes, led to a disruption of the ecological natural balance, which manifested itself later - in the early sixties, when numerous sandstorms began to occur (M. A. Uvarkina & O. Yu. Uvarkina, 2013).

Beginning in 1961, a new system of crop rotation, without using convertible husbandry method, was forcibly promoted (Boffa, 1990, p. 491). The need for up-and-down husbandry (plowed fields that are left unseeded in the summer) in the North Kazakhstan region led to soil oxidation. Deep moldboard plowing today considered as a Tselina seeding method, outdated and shown to be environmentally harmful. Although, this method is still used in old arable areas. This is despite the fact that even at the end of the 19th century, the Russian Empire scientists drew attention to the need to develop own unique method of plowing steppe soil. Carrying out an analysis in search of the reasons for the better productivity of arable land in the Canadian steppes compared to those in Kazakhstan, farmers noticed that for the steppe regions of Kazakhstan, where wind erosion is common, Canadian surface soil development tools are better suited (Verkhovskiy, 1969, p. 118). A long-term negative effect on the state of soils was predicted and discussed even then, and a system of non-moldboard tillage was developed by the Institute of Grain Farming: soil protection farming systems, in particular, non-moldboard tillage (the moldboard plow was replaced with a flat-cut one, which allowed to preserve stubble and other organic residues), sowing with stubble seeders, grain crop rotations (Saktaganova & Abylkhozhin, 2018, p. 104). A monitoring study of the influence of various cultivation methods on the fertility of chernozems showed that the best results are achieved by alternating moldboard plowing (once every four years) with moldless loosening. Today the dominant theory is about creating a cultural arable layer of steppe and dry-steppe soils through periodically repeated deep cultivation - at least 30 cm (Kogut, 2012).

However, the mass introduction of such methods in the virgin lands never occured, and the state's response to the problem of post-virgin lands was reduced mainly to soil protection measures. Levykin (2021) stated that, "The state response at the post-virgin land challenge was mainly soil-protecting measures. Series of state resolutions on immediate measures to protect soils from wind and water erosion, on prospects of lands melioration and redistribution of river's flow was approved from 1966 to 1970" (p. 5). That is even during the program implementation period they tried to find ways to both maximize the efficiency of agricultural methods and protect soils, although the comprehensive introduction of technologies did not happen, because of economic system features includingly.

The soil erosion. According to summary analytical report on the state and use of the lands of the Republic of Kazakhstan for 2015, there are more than 90 million hectares of eroded and erosion-exposed lands with 29.3 million hectares actually eroded in Kazakhstan (Tokbergenova, 2016, p. 163). At the same time, the studied areas (directly subject to the Virgin Land program), excluding Pavlodar region, are the carriers of the smallest proportion of eroded agricultural lands. The data compared and updated according to the recent Consolidated analytical report on the status and use of land of the Republic of Kazakhstan for 2022, and in fact, have not undergone significant changes since previous report.

Figure 3. The share of eroded farmland from the total area in the region in 2022

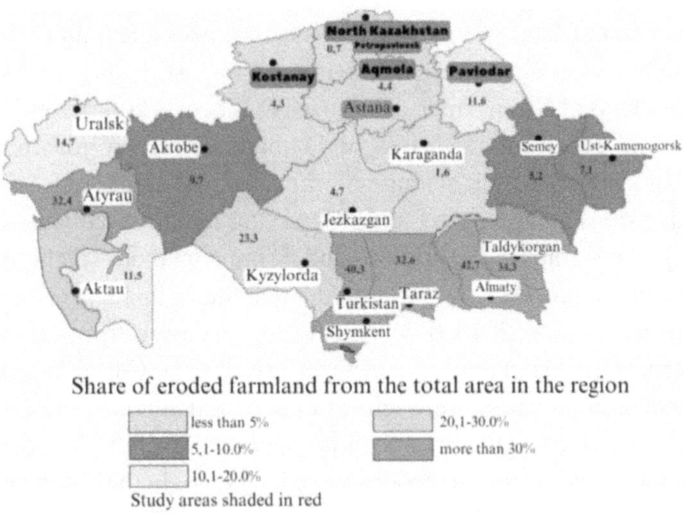

Taken from the Ministry of Agriculture of the Republic of Kazakhstan Land Management Committee, 2022.
Note. Translated and compiled by the authors according to the research questions.

The share of agricultural land and arable land that is subject to wind and/or water erosion in the studied regions was not so high in comparison with indicators in other regions of Kazakhstan. In general, in terms of erosion of agricultural land, the situation in the studied regions does not seem to be as critical in relation to other regions of the country, however, Table 1 shows specific data on arable lands, which allows to draw conclusions regarding the contribution of the virgin land development and specifically plowing.

Table 1. Area of eroded agricultural land in the studied regions as of November 1, 2022 (in thousand of hectares)

Name of the region	Total eroded agricultural lands	Total eroded arable lands	% out of total areas of eroded agricultural lands	% out of total areas of eroded arable lands
Akmolinsk	571,6	352,2	1,95	20,03
Kostanay	769,9	93,5	2,63	5,32
North Kazakhstan	56	28	0,19	1,59
Pavlodar	1297,2	334,3	4,42	19,01
Other	26625,4	950,4	90,81	54,05
Total	**29320,1**	**1758,4**	**100**	**100**

Taken from the Ministry of Agriculture of the Republic of Kazakhstan Land Management Committee, 2022.

Note. Data obtained from The Consolidated Analytical Report on the Condition and Use of Land in the Republic of Kazakhstan for 2022, processed and analyzed by the authors in accordance with the research questions.

Table 1 indicates the difference between the shares of agricultural lands subject to wind and water erosion in the studied regions and the corresponding shares of arable lands. The final 45.95% of arable lands in Kazakhstan subject to soil erosion shows the impact of the Virgin Land program in the northern regions. At the same time, the difference between agricultural land in general, including pasture land, and arable land specifically is indicative.

The direct relationship of land use methods in specific geological and climatic conditions to the sources of wind erosion is shown by American scientists using materials from a study of dust storms and droughts on the Great Plains of North America (Hansen & Libecap, 2004). Environmental consequences led to severe agricultural and economic crises, and stimulated further government intervention in the agricultural sector and a variety of soil-saving programs. Here we can draw a direct parallel with some of the excesses and decisions in the Virgin Land program for intensive cultivation of inappropriate lands - due to the similarity of geographical conditions and farming methods (including the intensification of land use and mechanization process). By 1927 the United States had reached to second place in wheat exports after Canada (Marchildon, 2010, p. 3), and during the desert storms of the mid-30s fell to record lows for obvious reasons. That is, the socio-economic crisis in the United States occurred immediately after the environmental consequences were felt (although, undoubtedly, the causes of the crisis of the Great Depression were much more than just the improper land use). In the Soviet planned economy of the 50-60s, such a quick reaction and corrections were incredible in the absence of political will and the impossibility of market economic tools employment.

Here we can state that neither the US experience nor own pre-revolutionary developments were sufficiently taken into account during the Virgin Land campaign.

Table 2. Considered practices affecting the environmental situation

	Cause	Consequence	Consequence	Solution
Planning	Plowing grass, sandy ridges and light steppe soils		Wind erosion	Purposeful land use
	Plowing potentially unproductive land		Water and wind erosion	Consideration of local climatic conditions, preliminary soil analysis
Farming activities	Fallow-free plowing		Soil acidification	Consideration of local climatic conditions, preliminary soil analysis Non-mouldboard plowing, alternating moldboard plowing with non-mouldboard loosening, the use of grain crop rotations, taking into account the specifics of steppe soil
	All of the above	Destruction of the humus layer	Desertification	Consider the agroecological potential of soils
	No crop rotations	Weed spread	Soil fertility decrease	Crop rotation
	Monoculture of grains	Violation of soil porosity	Reduced soil suitability rank	Replenishment of humus reserves, compensating fertilizers
	Long-term use of soils in arable land	Loss of humus layer	Deterioration of soil structure and water stability	Balanced application of fertilizers (organic and chemical)
	Continuous and full plowing	Natural balance disturbance	Disappearance of endemics	Optimization of livestock and crop production ratio of farms, creation of protected areas
Natural and geographical factor	Drought, wind, soils lightness, poor soil erosion resistance (lack of natural or planted fences)	Wind erosion, water erosion	Changes in soil structure, high susceptibility to changes in water regime	Protective forest belts planting
	Sand and dust storms	Wind erosion	Violation of soil stability	Consider genetic type of soils

The erosion, the content of humus and other elements of fertility decreases, the structure and water stability of soil aggregates deteriorates, which inevitably leads to changes in water and air regimes, a violation of stability soils to technogenic pressure, prolonged use of soils in arable lands - are the main further causes for the loss of fertility resources and a decrease in agricultural crop yields. These consequences were fully felt by northern regions of Kazakhstan, as a study of desertification in Central Asian regions shows (Jiang et al., 2019).

During the period of development of virgin lands, soils of light mechanical composition were plowed up with rotation, potentially dangerous to wind erosion. In developed soils, the processes of nitrification and ammonification are activated, and mobile forms of nitrogen, phosphorus and potassium that are easily absorbed by plants are mobilized. As a result, erosion-deflation processes were provoked in large areas of Pavlodar and Kostanay regions, which led not only to the loss of humus and deterioration in the physical condition of soils, but also sometimes to the death of crops (Shishov, 1987). In each genetic type (regular, carbonate, solonetzic) soils are found with altered (downwards) humus content with a deteriorated structural state, strong compaction (especially carbonate soils), contaminated with heavy metals near industrial centers.

Natural factors also contribute to the manifestation of soil erosion: the presence of open spaces (naturally in the steppe conditions of studied areas), frequent recurrence of droughts (also considered as a climatic feature of the region), increased wind activity, weak anti-erosion resistance of light soils, etc. The area eroded to varying degrees soils in northern Kazakhstan is about 20 million hectares. In addition, water erosion occurred on slope soils. It is currently manifesting itself mainly during spring snowmelt. It is associated, first of all, with a significant loss of humus since the development of virgin lands and a deterioration in the structural water stability. Considering that the conditions for growing grain crops in the territories of northern Kazakhstan directly depend on the amount of precipitation, the timing of its fall relative to the phases of crop development and seasonality (Kusainova et al., 2024), the issues of the occurrence of water erosion were a matter of a short time after the extensive plowing undertaken in the first years of the program. More than a half of the arable land in the hilly area is located on slope surfaces, which is another negative phenomenon of water erosion due to disturbance of the soil layer.

Thus, as a result of farming activity, the soil cover of northern Kazakhstan has undergone significant changes, depending on the level of farming culture, different duration and intensity of use, a decrease in biomass reserves, and varying degrees of soil compliance to anthropogenic pressures. These changes have a very different character and are mainly negative. Additionally, continuous plowing is approx. 18.2 million hectares of land led to the disappearance of a number of steppe endemics. Traditional game species were either displaced or destroyed. Among them: little bustard, gray partridge, quail, lapwing, marmot, saiga (Chibilev, 2014, pp. 93-94).

These and other consequences are considered (as collected in Table 2) as an additional result, which allows to assess more accurately the ineffectiveness and poor quality of the implementation and management of the Virgin Land program given the stipulated economic necessity and political setting of the time.

CONCLUSION

Numerous causes of land degradation, environmental impacts and problems of modern management in agriculture are considered. At the same time, Mirzabaev (2016) emphasized that, "The main areas affected by land degradation is concentrated in the north of Kazakhstan" (p. 286), which also justifies the ecological impact of the Virgin Land campaign geographically, as shown in Figure 3.

Overall negative agricultural and economic consequences of the virgin lands development are associated not only with N. Khrushchev's program, but also with the collapse of the USSR and further fluctuations in the economy of the independent Kazakhstan. Transition to a new economic model, the lack of government funding and investment inflows led to the redistribution of former arable lands. That is, the potential of Virgin Land program in 1990 was abandoned. The same thing happened in the traditionally sedentary agricultural regions of Russia and Ukraine after the collapse of the USSR. According to Schierhorn (2013), "Almost 70% of cropland abandonment occurred within the first 10 years of the transition from a state command to a market-driven economy" (p. 1182). That is, even a huge tract of land that was directly intended or favorable in terms of soil quality and climatic conditions, which was also sufficient in the planning of the Virgin Lands program (with the exception of the described unfavorable and unproductive lands), was abandoned already in post-Soviet times. The abandonment of agricultural lands that occurred in the 90s, in turn, led to the transformation of these lands into pastures. This contributes to the further spread of wind erosion and soil degradation, as follows from a recent study by colleagues from Germany and Kazakhstan (Koza et al., 2024).

Even though a significant portion of abandoned cropland remains unusable, "re-cultivation of currently unused cropland may contribute to regional and global food security" (Krämer et al., 2015, p. 14). At the same time, researchers to this day propose a reduction in the area of arable land by 40-45% in the North Kazakhstan region and an increase in the area of protective forest belts on arable lands to 7-10% - a measure proposed back in the late 40s - in including recommendations for optimizing the existing land use structure (Shogelova & Sartin, 2021, p. 139).

It is now almost impossible to restore the amount of humus in soils to the pre-virgin level, and yet this issue needs to be resolved. To replenish and qualitative changes in the physicochemical and water-physical properties of soils, it is necessary to ensure an increase in biological productivity and effective fertility, replenishment of humus reserves in old arable soils through compensating doses of fertilizers, to protect soils from wind and water erosion. At the same time, introduction of preventive measures, such as: rational use of arable land, development of grass crop rotations with an optimal ratio of crop and livestock products, balanced application of fertilizers, especially organic ones (as mentioned in Table 1), are considered to be the main conditions for maintaining soil fertility.

However, such environmental measures, apparently, create a contradiction with economic demands, which once again emphasizes the relevance of the issues under consideration. "Aided by new technologies of moisture conservation, Kazakhstan's farming sector is now among the world's ten largest producers and five largest exporters of wheat" (Organization for Economic Cooperation and Development (OECD), 2011), while the Republic of Kazakhstan remains one of the major wheat producing countries (Reynolds & Braun (Eds.), 2022, p. 241), which, of course, should be considered as overcoming the crisis of the early 90s, and also as a direct legacy of the Virgin Land program (Nazarbayev, 2023, pp. 385-386). If we consider the agricultural sector more specifically, then by 2021 there was an increase in wheat exports by 24.2% (from 971.8 million US dollars to 1 billion US dollars) (Ministry of Agriculture of the Republic of Kazakhstan, 2021).

Today it is still possible to draw parallels with the motives and goals set by the initiators of the Virgin Land program. If we ignore other consequences, and concern only environmental and partly economic ones, then when arable land is abandoned, food production stops, the risk of fire and loss of biodiversity of agricultural land increases. At the same time, abandonment or lack of agricultural exploitation makes a positive contribution to the environment, reduces soil erosion, improves water quality, etc. (Alcantara et al., 2013). That is, even in one environmental aspect, with today's experience and knowledge, including the practice of the Soviet project, the level of technology and different state and social ideology, there will be arguments for plowing (undoubtedly, not as extensive as was observed under N. Khrushchev) and against. However, at the present time there is no food issue that could lead to famine within the country, as was the case in the post-war decade in the USSR. That is, the economic and social aspects are not so sharply opposed to the environmental.

The accumulated experience, the geographical and biological research allow to partially predict the consequences of farming activities and assess the likely impact of political intervention in such a strategic sector as agriculture (using the examples of the USA in the 20-30s and the USSR in the 50-60s). This determines the relevance of further historical and cultural research, botanical examinations, economic and agricultural projects aimed at forming an objective attitude to the Virgin Land campaign experience, developing methods for effective and at the same time environment-friendly use of the land. Today, the fight against desertification and soil erosion is one of the principles of the Concept for the development of the agro-industrial complex of the Republic of Kazakhstan for 2021 – 2030, which demonstrates at least a position of acceptance and consideration of the facts and realities in the field of ecology and land conservation. However, a comprehensive and interdisciplinary study of the contribution of the Virgin Land program to the current state of agriculture, economics and ecology is applicable not only in the regions considered, but throughout the country.

A study of the political and economic motives, demographic and environmental consequences of the Virgin Land program in Kazakhstan shows a complex multifactorial picture of this historical process. One-sided critical interpretations and conclusions seem exaggerated. It is impossible to construct an unambiguous positive or negative assessment taking into account the socio-economic, political factors of influence and consequences, which also need to include the post-Soviet results, where there are both extremely unfavorable environmental consequences, and largely salutary socio-economic effects and demographic results.

REFERENCES

Abisheva, B. N., Balakayev, T. B., & Suleymenov, R. B. (1980). *Istoriya Kazahskoj SSR (s drevnejshih vremen do nashih dnej). Tom 5 [The history of the KazSSR (from ancient times to the present day), V. 5].* Nauka.

Alcantara, C., Kuemmerle, T., Baumann, M., Bragina, E., Griffiths, P., Hostert, P., Knorn, J., Müller, D., Prishchepov, A., Schierhorn, F., Sieber, A., & Radeloff, V. (2013). Mapping the extent of abandoned farmland in Central and Eastern Europe using MODIS time series satellite data. *Environmental Research Letters*, 8(3), 035035. 10.1088/1748-9326/8/3/035035

Alexeenko, A. (1994). *Selskoe naselenie Kazahstana. 1920-1990 gg.* [Rural Population of Kazakhstan. 1920-1990]. Institut istorii i etnologii im. Ch.Ch. Valihanova.

Almanova, Z., Kenzhegulova, S., Kashkarov, A., Kekilbayeva, G., Ussalinov, E., Yerzhan, D., Zhakenova, A., & Zvyagin, G. (2023). Changes in soil fertility indicators after long-term agricultural use in Northern Kazakhstan. *International Journal of Design & Nature and Ecodynamics*, 18(5), 1045–1053. 10.18280/ijdne.180504

Alpyspaeva, G.A., & Nurmukhanova, R.Zh. (2023). Participation of ethnic special settlers in the development of virgin lands in Kazakhstan. *Herald of Science of S. Seifullin Kazakh Agrotechnical Research University.*

Boffa, J. (1990). *Istoriya Sovetskogo Soyuza. T. 2. Ot Otechestvennoj vojny do polozheniya vtoroj mirovoj derzhavy. Stalin i Hrushev. 1941-1964 gg[The History of the Soviet Union. V. 2. From the Patriotic War to the Position of the Second World Power. Stalin and Khrushchev. 1941-1964].* Mezhdunarodniye Otnosheniya.

Bureau of National Statistics of the Agency for Strategic Planning and Reform of the Republic of Kazakhstan. (February 17, 2023). *Administrativno-territorialniye edinitsy Respibliki Kazakhstan (na nachalo 2023).[Administrative-territorial units of the Republic of Kazakhstan (beginning of 2023)].*https://stat.gov.kz/ru/industries/economy/national-/publications/6381/

Bureau of national statistics of the agency for strategic planning and reform of the Republic of Kazakhstan. (July 19, 2023). *Chislennost naseleniya Respubliki Kazakhstan (na nachalo 2023.[Population of the Republic of Kazakhstan at the beginning of 2023].*https://stat.gov.kz/ru/industries/economy/national-/publications/6373/

Central Committee of the Communist Party. (1956). *The Resolution of the VIII Congress of the Communist Party of Kazakhstan on the Report of the Central Committee of the Communist Party of Kazakhstan.* CCCP.

Central Statistical Directorate under the Council of Ministers of the USSR. (1967). *Strana Sovetov za 50-let. Sbornik statisticheskih materialov[The Country of Soviets for 50 Years. Collection of Statistical Materials*].

Central Statistical Office under the Council of Ministers of the Kazakh SSR. (1980). *Narodnoe hozyajstvo Kazahstana za 60 let: Stat. sb. [National economy of Kazakhstan for 60 years: Statistical collection].* CSO of the KazSSR.

Chernenko, K. U., & Smirtyukov, M. S. (Eds.). (1968). *Resheniya partii i pravitelstva po hozyajstvennym voprosam: Sb. dok. za 50 let. T. 5. 1962-1965 gg[Decisions of the Party and government on economic issues: collection of documents for 50 years. Volume 5. 1962-1965].* Politizdat.

Chibilev, A.A. (2014). Ekologicheskie uroki celiny [The ecological lessons of Tselina]. *Voprosy stepevedeniya,* 11, 91-95.

Chuyev, F. (1991). *140 besed s Molotovym* [140 Conversations with Molotov].

Easter, D. (2021). Nikita Khrushchev and the Compromise of Soviet Secret Intelligence Sources. *International Journal of Intelligence and CounterIntelligence*, 35(3), 1–23. 10.1080/08850607.2021.1874191

Edygenov, N. E., Kolodin, F. I., & Mihajlov, F. K. (Eds.). (1959). *Narodnoe dvizhenie za osvoenie celinnyh zemel v Kazahstane. Sb. materialov i dokumentov[People's Movement for the Development of Virgin Lands in Kazakhstan. Collection of Materials and Documents.].* Gospolitizdat.

Eleuov, T. (Ed.). (1960). *Iz istorii socialisticheskogo stroitelstva v Kazahstane. T.9[From the History of Socialism Construction in Kazakhstan. V.9.].* Izdatelstvo Akademii Nauk KazSSR.

Gindemit, A., & Aksenova, Yu. (2021). Evaluation of soils of agrocenoses in the forest-steppe zone of the Omsk Irtysh region by the level of cultivation and fertility. *IOP Conference Series. Earth and Environmental Science*, 659(1), 012102. 10.1088/1755-1315/659/1/012102

Hansen, Z., & Libecap, G. (2004). Small Farms, Externalities, and the Dust Bowl of the 1930s. *Journal of Political Economy*, 112(3), 665–694. 10.1086/383102

Issanova, G., Saduakhas, A., Abuduwaili, J., Tynybayeva, K., & Tanirbergenov, S. (2020). Desertification and land degradation in Kazakhstan. *Bulletin of National Academy of Sciences of the Republic of Kazakhstan.*, 5(387), 387. 10.32014/2020.2518-1467.148

Jiang, L., Jiapaer, G., Bao, A., Kurban, A., Guo, H., Zheng, G., & De Maeyer, P. (2019). Monitoring the long-term desertification process and assessing the relative roles of its drivers in Central Asia. *Ecological Indicators*, 104, 195–208. 10.1016/j.ecolind.2019.04.067

Khrushchev, N.S. (1999). *Vremya. Lyudi. Vlast. Vospominaniya v 4 kn.[Time. People. Power. Memoirs in 4 books].* Moskovskiye Novosti.

Khrushchev, S. (Ed.). (2007). *Memoirs of Nikita Khrushchev: Statesman, 1953-1964*Shriver, G., & Shenfield, S., Trans.). Brown University.

Kogut, B. (2012). Assessment of the humus content in arable soils of Russia. *Eurasian Soil Science*, 45(9), 843–851. 10.1134/S1064229312090062

Koza, M., Funk, R., Pöhlitz, J., Conrad, Ch., Shibistova, O., Meinel, T., Akshalov, K., & Schmidt, G. (2024). Wind erosion after steppe conversion in Kazakhstan. *Soil & Tillage Research*, 236, 105941. 10.1016/j.still.2023.105941

Kozybayeva, M. M. (Ed.). (2022). Krestyanstvo Kazahstana v poslevoennoe desyatiletie: socialnaya transformaciya i povsednevnost. *Kollektivnaya monografiya* [The Peasantry of Kazakhstan in the Post-war Decade: Social Transformation and Everyday Life. Collective Monograph]. IP "Madiyar".

Kraemer, R., Prishchepov, A., Müller, D., Kuemmerle, T., Radeloff, V., Dara, A., Terekhov, A., & Fruehauf, M. (2015). Long-term agricultural land-cover change and potential for cropland expansion in the former virgin lands area of Kazakhstan. *Environmental Research Letters*, 10(5), 054012. 10.1088/1748-9326/10/5/054012

Kusainova, A., Mezentseva, O., & Tusupbekov, Zh. (2024). Trends of long-term fluctuations of atmospheric humidification of the growing season in the steppe zone of Northern Kazakhstan. *E3S Web of Conferences*. 10.1051/e3sconf/202449803009

Levykin, S. V., Chibilev, A. A., Gulyanov, Yu. A., Kazachkov, G. V., Yakovlev, I. G., & Nurushev, M. Zh. (2021). The Virgin Land Megaproject and the Land Reform as the Global Experiment of Steppe Self-Restoration in North Eurasia. *IOP Conference Series: Earth and Environmental Science,* Volume 817.10.1088/1755-1315/817/1/012058

Marchildon, G. (2010). The Impact of the Great Depression on the Global Wheat Trade. *Johnson-Shoyama Graduate School of Public Policy Working Paper*, 4. https://www.researchgate.net/publication/265907875_The_Impact_of_the_Great_Depression_on_the_Global_Wheat_Trade

Mazhitova, Zh. S., Khassenova, Zh. O., & Issayev, A. U. (2023). Cultivation of Fodder Beans at the Zarechny State Farm during the Development of Virgin Lands in Kazakhstan. *IOP Conference Series: Earth and Environmental Science,* Volume 1154.10.1088/1755-1315/1154/1/012057

Mazhitova, Zh., Zhalmurzina, A., Kolganatova, S., Orazbakov, A., & Satbai, T. (2021). *Environmental consequences of Khrushchev's Virgin Land Campaign in Kazakhstan (1950s–1960s).* E3S Web of Conferences. 258. 10.1051/e3sconf/202125805036

McCauley, M. (1976). *Khrushchev and the Development of Soviet Agriculture: The Virgin Land Program, 1953–1964*. Holmes & Meier. 10.1007/978-1-349-03059-0

Medeu, A. R. (Ed.). (2010). *Natcionalniy atlas Respubliki Kazakhstan. T. 1: Prirodniye usloviya i resursy[National atlas of the Republic of Kazakhstan. V. 1: natural conditions and resources].* TOO "Vit Brand".

Ministry of Agriculture of the Republic of Kazakhstan. (December 30, 2021). *Ob utverzhdenii Koncepcii razvitiya agropromyshlennogo kompleksa Respubliki Kazahstan na 2021 – 2030 gody* [*On approval of the Concept for the development of the agro-industrial complex of the Republic of Kazakhstan for 2021 – 2030].* Decree of the Government of the Republic of Kazakhstan No. 960. https://www.gov.kz/memleket/entities/moa/documents/details/274531?lang=ru

Ministry of Agriculture of the Republic of Kazakhstan Land Management Committee. (2022). *Svodnyj analiticheskij otchet o sostoyanii i ispol'zovanii zemel' Respubliki Kazahstan za 2022 god[The consolidated analytical report on the condition and use of land in the Republic of Kazakhstan for 2022].*https://www.gov.kz/memleket/entities/land/documents/details/579164?lang=ru

Ministry of Justice of the Republic of Kazakhstan. (June 20, 2003). *Zemelniy kodeks Respubliki Kazakhstan.[The Land Code of the Republic of Kazakhstan] (No. 442).* Available from: https://adilet.zan.kz/rus/docs/K030000442_

Ministry of Justice of the Republic of Kazakhstan. (2012). *Konstituciya Respubliki Kazahstan[Constitution of the Republic of Kazakhstan]. Institute of legislation and legal information of the Republic of Kazakhstan].* https://adilet.zan.kz/rus/docs/K950001000_

Mirzabaev, A., Goedecke, J., Dubovyk, O., Djanibekov, U., Le, Q. B., & Aw-Hassan, A. (2016). Economics of Land Degradation in Central Asia. In Nkonya, E., Mirzabaev, A., & von Braun, J. (Eds.), *Economics of Land Degradation and Improvement – A Global Assessment for Sustainable Development.* Springer. 10.1007/978-3-319-19168-3_10

Nazarbayev, N. (2023). *Moya zhizn. Ot zavismosti k svobode[My Life. From Dependence to Freedom].* Foliant.

Organisation for Economic Cooperation and Development (OECD). (2011). Competitiveness and Private Sector Development: Kazakhstan 2010: Sector Competitiveness Strategy. *Competitiveness and Private Sector Development.* 10.1787/9789264089792-en

Petrick, M., Wandel, J., & Karsten, K. (2013). Rediscovering the virgin lands: Agricultural investment and rural livelihoods in a Eurasian frontier area. *World Development*, 43, 164–179. 10.1016/j.worlddev.2012.09.015

Politizdat. Akhmetova, L.S. (1996). *Patriotizm v rakurse istorii. Uchebnoe posobie[Patriotism from the Perspective of History. Tutorial].* Sanat.

Rau, A., Koibakova, Y., Nurlan, B., Nabiollina, M., Kurmanbek, Z., Issakov, Y., Zhu, K., & David, L. D. (2023). Increase in Productivity of Chestnut Soils on Irrigated Lands of Northern and Central Kazakhstan. *Land (Basel)*, 12(3), 672. 10.3390/land12030672

Resolution of the Government of the Republic of Kazakhstan. (August 27, 2019). *Ob utverzhdenii vosmogo-desyatogo svodnogo periodicheskogo doklada o vypolnenii Respublikoj Kazahstan Mezhdunarodnoj konvencii o likvidacii vseh form rasovoj diskriminacii[On approval of the eighth to tenth consolidated periodic report on the implementation by the Republic of Kazakhstan of the International Convention on the Elimination of All Forms of Racial Discrimination] (No. 630).*https://adilet.zan.kz/rus/docs/P1900000630

Reynolds, M. P., & Braun, H.-J. (Eds.), *Wheat Improvement. Food Security in a Changing Climate.* Springer., 10.1007/978-3-030-90673-3

Saktaganova, Z.G., & Abilkhozhin, Zh.B. (2018). "Tselinniy proekt" dostizheniya i problemy (socialno-ekonomicheskie i ekologicheskie aspekty osvoeniya celiny) [Virgin Land Project": achievements and problems (socio-economic and environmental aspects of virgin lands development)]. *Bulletin of Karaganda University. Series History. Philosophy, 4*(92).

Sanchez-Sibony, O. (2014). *Red Globalization. The Political Economy of the Soviet Cold War from Stalin to Khrushchev.* University Printing House. 10.1017/CBO9781139628778

Schierhorn, F., Muller, D., Beringer, T., Prishchepov, A. V., Kuemmerle, T., & Balmann, A. (2013). Post-Soviet cropland abandonment and carbon sequestration in European Russia, Ukraine, and Belarus. *Global Biogeochemical Cycles*, 27(4), 1175–1185. 10.1002/2013GB004654

Shishov, L. L. (1987). *Kriterii i modeli plodorodiya pochv [Soil fertility criteria and models].* Agropromizdat.

Shogelova, N.T., & Sartin, S.A. (2021). Application of Landscape-Geochemical Approach in the Study of the Territory of the North Kazakhstan Region. *QazBSQA Habarshysy. urylys konstrukciyalary zhane materialdary. 4*(81). 10.51488/1680-080X/2021.4-02

Statistika. Egorov, A. G., & Bogolyubov, K. M. (Eds.). (1985). *KPSS v rezolyuciyah i resheniyah sezdov, konferencij i Plenumov CK (1898-1988)[The Communist Party of the Soviet Union in Resolutions and Decisions of Congresses, Conferences and Plenums of the Central Committee (1898-1988)].*

Talamini, G. (2023). Urbanising the Virgin Lands. In a Mariotti, J. & Leetmaa, K. (Eds.). *Urban Planning During Socialism. Views from the Periphery* (15-30). Routledge. 10.4324/9781003327592-3

Taubman, W. (2003). *Khruschev: The Man and His Era.* Free Press.

Terra. Akhmetova, L., & Grigoryev, V. (2021). *Zhumabaj Shayahmetov i ego vremya[Zhumabai Shayakhmetov and his time].* Foliant.

Tokbergenova, A., & Nyussupova, G. Kairova, & Sh., Kiyassova L. (2016). Causes and Impacts of Land Degradation and Desrtification: Case Study of the Republic of Kazakhstan. *4th International Geography Symposium.* Turkey. https://pps.kaznu.kz/kz/Main/FileShow2/77365/79/3/4080/0/

Uzarkina, M.A., & Uzarkina, O.Yu. (2013). *«Celinnaya epopeya»: istoriko-politologicheskij analiz* ["The Tselina Epic": Historical and Political Analysis]. *Nauka o cheloveke: gumanitarnye issledovaniya.1* (11), 180-189.

Verkhovskiy, N. (1969). *Tretij zvonok* [The Third Call]. *Sibirskie ogni, 3.* Chernienko,

Volin, L. A. (1970). *Century of Russian Agriculture: From Alexander II to Khrushchev.* Harvard University Press. 10.4159/harvard.9780674366350

Yu. (1964). *Celinnaya doroga* [The road of Tselina]. *Novyj mir, 1.*

Chapter 15
ESP in the Educational Process of Master's Degree Students and Majoring in Ecology and Land Management

Nadezhda Anatolievna Lebedeva
http://orcid.org/0000-0003-4095-2631
International Personnel Academy, Germany

ABSTRACT

In modern language education, when teaching foreign languages to master's students in environmental specialization and land management, as well as for students of other non-linguistic universities, the term ESP (English for specific purposes) is used. A competency-based approach should be implemented, in line with which the acquisition of knowledge is combined with the experience of using the acquired knowledge in situations close to real life. Lexical, grammatical, socio-cultural, and other competencies formed in the process of learning a foreign language will help the graduate to cooperate effectively with native speakers and feel free in the country of the language being studied. The purpose of this chapter is to describe foreign language teaching in the educational process of master's degree students majoring in ecology and land management and improving their communicational skills.

INTRODUCTION

In modern language education, when teaching foreign languages to Master's students in environmental specialization and land management, as well as for students of other non-linguistic universities, the term ESP (English for Specific Purposes) is used. The main task for such foreign language teaching is that the educational process takes place within the framework of the material necessary for students to foreign language communication in their professional activities. ESP involves specialization of the learning process.

Foreign language for professional communication programs allows the development of training modules that will meet the requirements of the national credit-module system, which is introduced in higher education institutions. The analysis of scientific research gives grounds to assert that for the

DOI: 10.4018/979-8-3693-3374-7.ch015

implementation of new methodological principles in the practice of foreign language teaching and the introduction of modern approaches to the formation of students' foreign language competence, it is necessary to create a positive emotional atmosphere in a foreign language during the classes and master their linguistic skills. A competency-based approach should be implemented, in line with the acquisition of knowledge combined with the experience of using the acquired knowledge in situations close to real life. Lexical, grammatical, socio-cultural and other competencies formed in the process of learning a foreign language will help the graduate to cooperate effectively with native speakers and feel free in the country of the language being studied. The activity of lecturers should be based on taking into account the personality of each student, the development of his abilities, the formation of worldview, the development of cognitive motivation, and independence in solving complex problems. Successful professional activity requires the possession of foreign language professional communicative competence, which involves not only the amount of knowledge in the speciality but also the ability to communicate in foreign languages with professionals from other countries, as well as perform professional activities in a foreign language professionally-oriented environment (Lebedeva, 2022).

The purpose of this chapter is to describe foreign language teaching in the educational process of Master's degree students majoring in ecology and land management and improving their communicational skills.

Achieving this goal involves the **following tasks** to solve:

1. Consider the main methods and techniques of teaching a foreign language to students of the specialities "Ecology" and "Land Management (Land Resources)" and outline the lingua didactic foundations of teaching English-language communication in ecology and land management.
2. Determine the content of teaching the English language and the criteria for selecting educational material for the formation of ecological and land management competence.
3. The chapter will present both the program and theoretical materials.

The object of research is the process of teaching professionally oriented English-language communication on ecology and land management in agricultural institutions of Higher Education.

The research subject is the methodology of teaching professionally oriented English–language communication about ecology and land management.

The research hypothesis is that the effectiveness of teaching professionally oriented foreign language communication with ecology and land management can be improved by increasing the lexical and grammatical skills of students, writing, independent work, project activities, etc., which provide an expansion of the communicative and linguistic components of professional competence in foreign language communication.

To solve the tasks set, the following methods were used: a theoretical and critical analysis of different scientific literature in the pedagogical field and methods of teaching foreign languages, which made it possible to study and generalize modern methods of teaching professional foreign language writing, systematize and clarify the theoretical foundations of solving the problem; analysis of textbooks and programs in a foreign language of professional direction; forms and methods of teaching professionally oriented English-language written communication; empirical observations of the educational process, students' activities, means of checking the effectiveness of written tasks on ecology and land management.

The scientific novelty of the obtained results lies in the fact that the chapter may be interesting to exchange experiences with colleagues from other countries. The results of the study can be used to develop methodological manuals, innovative means of teaching foreign languages for professional communication, and advanced training for research and teaching staff.

BACKGROUND

The Problem of Interpreting Terminological Concepts as a Basic ESP Platform

In his work, Bonkovsky O.A. (2023) writes: "The adoption of foreign agrarian terminology has many different officials and features. Agricultural terminology can be very specific and difficult to understand through a large number of special terms. These terms may be unique to the agricultural sector. The acquisition of foreign agricultural terminology also lies in the foreign region of the students. Understanding is more consistent with the correct use of foreign terms. In different countries and some parts of the rural kingdom, the terminology may differ. The lack of standardized agricultural terminology can be confusing for students and farmers. These terms may be important to understand because there is a lack of practical knowledge in the agricultural sector. As it is possible, it is important to obtain experts in agricultural fields. Their knowledge and evidence will add value to the course. The development of interactive online resources and mobile apps for learning agricultural terminology can significantly facilitate access before the start and development of the process in a straightforward and user-friendly manner for the audience of farmers. Foreign terminology is constantly developing, so it is important to gradually update the initial platforms, adding new ones and responding to the needs of customers. The development of foreign terminology, including agricultural terminology, requires a special approach and strategy. With the imported foreign terminology, it is necessary to reach the target audience, benefiting from educational programs and the level of student training" (Bonkovsky 2023, p. 24-25).

In agricultural universities, at the faculties of agronomy and construction, the specialities "Ecology" and "Land Management" are being actively introduced as part of the preparation for bachelor's and master's. Ecology as an academic discipline is increasingly included in the educational programs of high agricultural institutions. This discipline is studied in agricultural higher education institutions. There are university educational programs in ecology and environmental management. Complete and detailed textbooks in national languages provide sufficient information about the terminology accepted in environmental science and about the main problems of world ecology. For example, a textbook on the discipline "Public Administration in the Field of Protection of Land Resources" by Safina G.R., Fedorova V.A. (2020) is intended for masters studying in the field of "Land Management and Cadastres" (Land Assessment and Monitoring). The manual introduces students to the features of state management of land resources and their protection: the theoretical foundations and organizational mechanism of land management at various levels, the effectiveness of the land management system and state control over the rational use and protection of land.

At the same time, a huge layer of the most up-to-date information on environmental issues remains inaccessible to the student, since it is found in scientific and popular science publications published abroad and written in English. To use this material, they must have a reasonable knowledge of the English language and English terminology. The material in this chapter corresponds to the topics of environmental education courses at an agronomic university in Master's specialities: "Ecology", and "Land Management and Cadastres" within the framework of the relevant educational programs.

According to (Giniyatov, 2022), any science should have at its disposal a special terminology that represents its conceptual apparatus. This logically constructed system of special terms allows us to unify the processes of understanding and their interpretation, phenomena and various kinds of relationships that arise and develop in science. The scientific disciplines of master's programs in ecology, land management and cadastre also have a terminological arsenal, which is characterized by the general term "land". As Giniyatov writes (2022):

> "Land is the main natural resource, the spatial basis for the location of all economic sectors of the economy, the main means of production in agriculture and forestry and the main component in the food production system.

In other words, land is a necessary material condition for the life of people living on it" (Giniyatov, 2022, p. 152).

The speciality "Land Management and Cadastres" refers to the field of geodesy and land management, which deals with the organization and management of land resources. Cadastres play an important role, representing systems for recording and registering land plots, their owners and their rights to them. For training in this speciality, there are also various programs and directions in higher educational institutions, in particular agricultural universities. Based on the specifics of this speciality, let's turn to the problem of interpreting the terminology of land resources, described in detail by Giniyatov (2022), which he cites, referring to documents of the state standard of the Russian Federation.

> *"Land is the most important part of the natural environment, characterized by space, relief, climate, soil cover, vegetation, subsoil, water, and is the main means of production in agriculture and forestry, as well as the spatial basis for the location of enterprises and organizations of all sectors of the national economy*" (Giniyatov, 2022, p. 152).

"Land" here is considered a natural resource, the main characteristic of which is the territory where people exist and develop it according to their interests. From such a definition, it follows: the earth is a complex multicomponent ecological system consisting of natural components. Thus, the concept of "land" can be interpreted as follows:

Figure 1. Land is a complex multicomponent ecological system

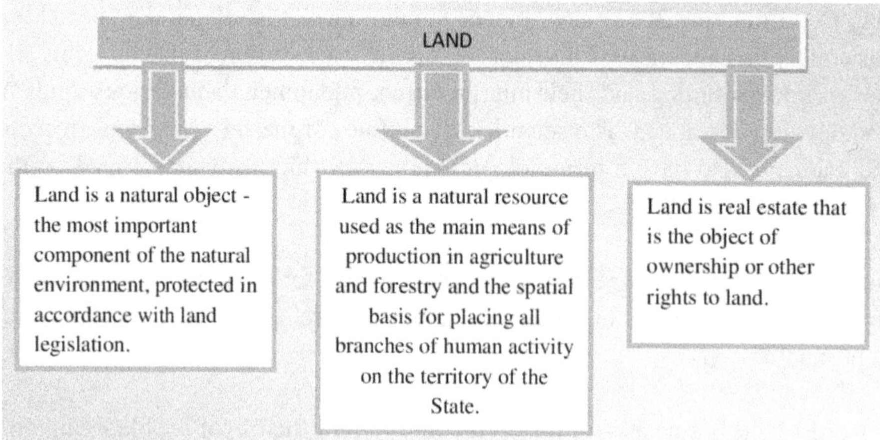

(summarized by the author)

"The characteristics of the state's land resources as lands, for example, which are used or can be used in sectors of the national economy, are carried out using several indicators that determine the distribution of lands according to their intended purpose, intended use and forms of land ownership" (Giniyatov, 2022, p. 152).

Federal executive authorities carry out state accounting of the availability and use of land in the Russian Federation according to the following categories:

- lands
- land
- forms of ownership.

The current land legislation divides all state lands, depending on their intended purpose, into seven categories, which can be represented by Figure 2:

Figure 2. State lands depending on their intended purpose

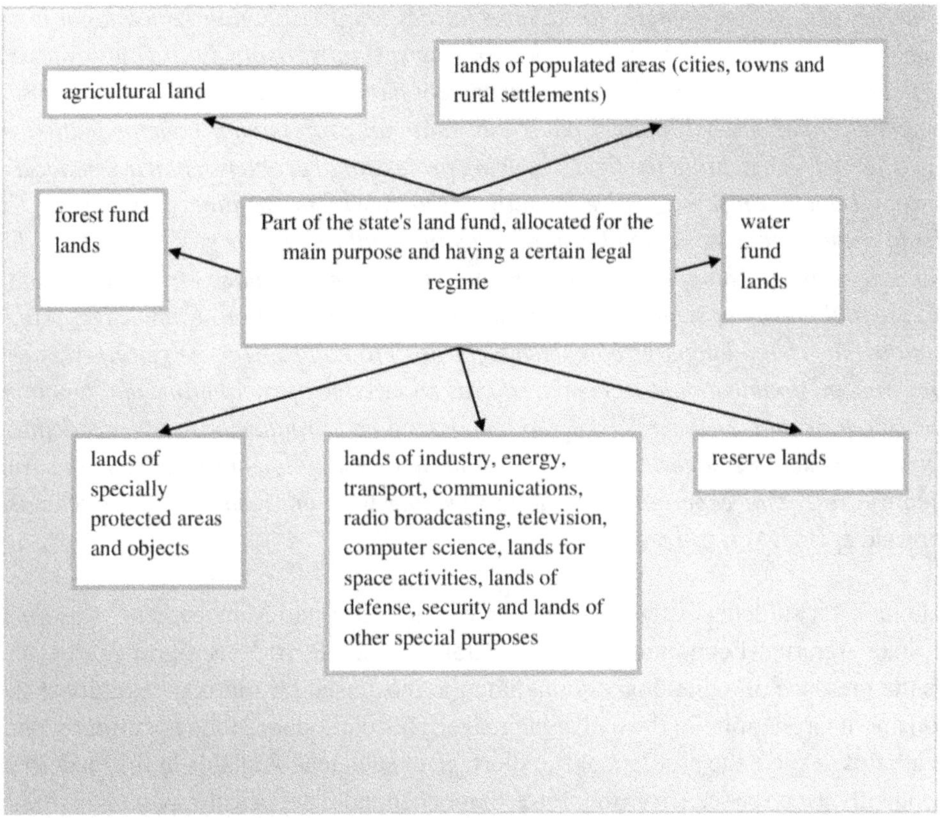

(summarized by the author according to Giniyatov's definitions)

According to intended use, as noted by Giniyatov (2022, p. 153), all lands are divided into land, which is understood as territory that is systematically used or suitable for use for specific economic purposes and differs in its natural and historical characteristics.

The land is usually divided into two large groups:

- agricultural grounds; arable lands, pastures, hayfields, perennial plantings and fallow lands.
- non-agricultural land; forest lands, tree and shrub plantations, swamps, lands under water, lands under roads and other non-agricultural lands.

Today, all lands, for example, in the Russian Federation, can be in one of three forms of ownership: state, municipal or private (Giniyatov 2022, p. 153).

Thus, from one concept "land" a fairly large conceptual apparatus of land management, real estate cadastre and land monitoring has been derived, which represents a significant terminological base of ESP.

> *«According to Russian state higher educational standards students are supposed to acquire key and professional competencies specific to their speciality. Foreign language competence is included in the list of key competencies for all specialities taught at higher educational institutions. The relevance of foreign language competence for any specialist aspiring to gain professional success in a career (among them ecology and environment managers) is evident. Lack of foreign language competence of a specialist hinders one's international professional communication and studying cutting-edge achievements in the field of solving professional problems, making impossible academic mobility and conducting cooperative scientific research with international colleagues. Consequently, the list of competencies specified by state education authorities for postgraduates, who are getting Master's degrees in ecology and environment management, includes foreign language competence as an essential means of scientific and professional communication. Knowledge, skills and experience related to foreign language competence for postgraduates, who are getting Master's degrees in ecology and environment management, requires knowledge of vocabulary and functional grammar of scientific texts and documents; skills to understand and produce business, academic or scientific written and spoken messages, to initiate, sustain and conclude conversations concerning the field of scientific research; experience in public speaking with the results of their scientific research»* (Greenwald et al., 2017, p. 1).

A dictionary for students of the speciality "Geodesy and Land Management" was published at the Kherson State Agrarian-Economic University (Bokshan, 2019, p. 5) A significant advantage of this edition is the presence of a dictionary containing the most used terminological units of the field, their translation and interpretation in the native language. This dictionary guide is currently particularly relevant, as it helps to solve the problem of the shortage of such teaching aids in the modern book market.

Consequently, as can be seen from the above, the conceptual-categorical apparatus of the major subjects of master's programs forms the lexical base of ESP, which is reflected in curricula and textbooks, and also represents an extensive layer of knowledge in the field of ecology and land management, which is taught to students of agricultural universities, which will be discussed speech later in this chapter, where the program and educational materials that were created and used by the author in foreign language classes with students will also be presented. Thus, the educational process contributes to the development of professional competence on a global scale, since it contains professional knowledge in a foreign language.

MAIN FOCUS OF THE CHAPTER

Literature Review

Quite a lot of scientific work has been devoted to teaching ESP in ecology. In particular, the scientific article "Mastering Foreign Language Competence of Ecology and Environment Managers for Mining Industry of Kuzbass» by Oksana Greenwald, Roman Islamov, and Tatyana Sergeychick (2017). Scientists write: *«The European Council identified key competencies that are fundamental for each individual as a part of one's lifelong learning strategies. The second key competence stipulated by the European Council is competence in foreign languages which is defined as "communicating in a foreign language: ability to express and interpret concepts, thoughts, feelings, facts and opinions both orally and in writing, also includes mediation skills (i.e. summarizing, paraphrasing, interpreting or translating) and intercultural understanding. According to Russian state higher educational standards students are supposed to acquire key and professional competencies specific to their specialty. Foreign language competence is included in the list of key competencies for all specialities taught at higher educational institutions"* (Greenwald et al., 2017, p. 2).

Therefore, foreign language proficiency for any specialist in the field of ecology and land management is an urgent goal for achieving career success, and "*insufficient language competence of a specialist impedes international professional communication and the study of advanced achievements*" (Greenwald et al., 2017, p. 2) "*makes academic mobility and joint research with foreign colleagues impossible*" (Greenwald et al., 2017, p. 2). «*Knowledge, skills and experience related to foreign language competence for postgraduates, who are getting Master's degree in ecology and environment management, requires knowledge of vocabulary and functional grammar of scientific texts and documents; skills to understand and produce business, academic or scientific written and spoken messages, to initiate, sustain and conclude conversations concerning the field of scientific research; experience in public speaking with the results of their scientific research*» (Greenwald et al., 2017, p. 2).

When describing their program (Greenwald et al., 2017), the scientists emphasize academic reading skills.

Figure 3. Academic reading (summarized by the author based on definitions by Greenwald et al., 2017)

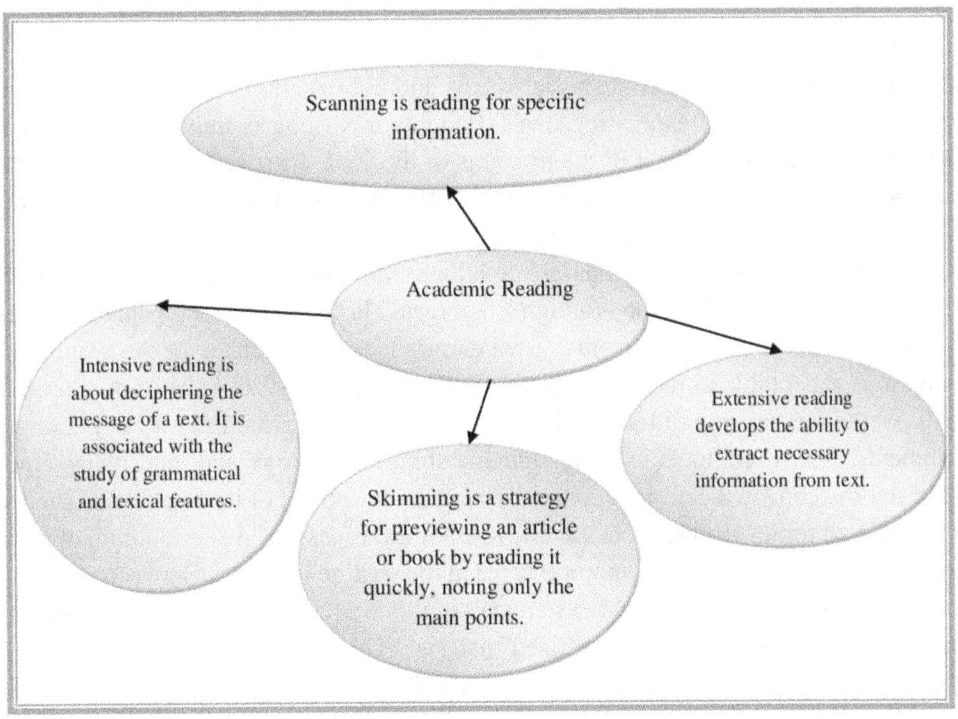

Thus, students learn to see the idea of each paragraph and can turn to explanatory or defining dictionaries and encyclopedias. At the same time, there is no need to look up every word in dictionaries, since the meaning of some words is inferred from the context. "The requirement to develop academic reading skills is caused by the need to work with scientific texts, which are characterized by general difficulty of perception, long paragraphs, special vocabulary and complex sentences.

The next important step for master's students studying ecology and land management, in our opinion, and according to Giniyatov's definitions is the ability to present their research results in a foreign language, which is possible with the help of academic writing. The collective monograph

"Cases on Error Analysis in Foreign Language Technical Writing", edited by the author of this chapter, is devoted to current issues of academic writing and its significance in various fields (Lebedeva, 2023). Special attention to writing is also emphasised by Greenwald and her colleague's work (2017).

The main theses of teaching a foreign language to master's students in the field of land management and ecology in the context of writing activities are determined by step-by-step activities:

- read the text to write an article;
- make a list of recommendations for writing a scientific article;
- read the text of writing a report at the conference;
- make a list of recommendations for writing a report for the conference;
- determine the difference between writing a scientific article and writing a report for a conference;
- familiarize yourself with the text of writing a thesis and make a list of recommendations for writing a thesis.

As a result of regular writing classes, master's degree students of ecology and land management gain knowledge. The process of mastering special words and constructions is faster, and writing skills are strengthened. The memory of expressing one's thoughts in writing is important not only in itself but also as a tribute to the advancement of the thoughts of the soul. The use of auditory images of internal movement and visual-motor images of hands in the writing process provides a comprehensive acquisition of knowledge on ecology and land management.

Students are invited to conduct a comparative evaluation analysis (independently or under the guidance of a teacher) of texts in their native and foreign languages. The texts are accompanied by lexical-stylistic and translation-methodological commentary, which provides the rationale for the proposed translation, provides options acceptable in the given text, and contains appropriate links to the theoretical part (if necessary). Written translation of technical text is a conscious act from the very beginning.

All elements, forms and tools used by students studying ecology and land management, through written translation, convey the desired content, which is the subject of its awareness. The use of lexical units of specialized topics further increases the level of awareness. The combination of auditory-motor images of internal speech and visual-motor images in the writing process ensures comprehensive assimilation of knowledge. The association racks created in this way contribute to more effective perception, memorization and reproduction of specialized material. The strength of memorization is ensured by the participation of all analysts, the interaction of analysis and synthesis, as well as the awareness and involuntary nature of the translation process. Graphic representations of words and structural patterns of sentences learned through written exercises are more easily recognized visually. In written speech, the verbal nature of a coherent utterance, which usually consists of several logically and structurally interrelated sentences, is more clearly revealed.

So, in the process of teaching foreign language communication to students on ecology and land management, an important function of written speech is that it can serve as an effective means of control. Written work, both current and final, is used to test knowledge of language material, as well as practical skills in listening, speaking and reading. Current tests using technical translation are carried out after mastering a certain part of the language material or oral topic - at least once every two weeks. They include one or two tasks and last up to 15 minutes. Final written tests are conducted three times a semester by the curriculum. This work involves testing the proficiency of various types of speech activities of students studying hydraulic engineering and water resources and can last 90 minutes. Thus, writing is an important means of improving the lexical and grammatical skills of students specializing in the field of ecology and land management. During the learning process, special texts and exercises were used, which were approved by students as useful and interesting (Lebedeva, 2023).

Valuable information on methods of teaching a foreign language can be found in the following literary sources. For example, Anisimova (2023) studied the experience of introducing distance learning for undergraduates in the environmental field at an agricultural university. Borodych, & Hrytsai (2023) investigated the peculiarities of verbal methods of teaching ecology to students in institutions of higher education. The experience of teaching English as a foreign language to students in the fields of "Geoecology", "Biology", and "Ecology and Nature Management" is described by Islamov, Resenchuk, and Perevalova (2018). Ivanova, Priss, Taranenko, Shleina, Chebanova, Yevstafieva, Krasulya, Kolomoyets-Harkusha (2020) indicate the development trajectories of "Soft Skills" among students of the Faculty of Agricultural Technologies and Ecology as one of the important factors in the formation of professional skills of young people. The work of Karychkovska and Chuchmii (2021) is devoted to the peculiarities of studying Ukrainian and foreign languages in agricultural institutions of higher education in Ukraine. Laver (2022) presented interesting final test papers for students of higher education in the field of knowledge 19 "Architecture and Construction" speciality 193 "Geodesy and land management" of the educational and professional program "Geodesy and Land Management" for linguists in the business Ukrainian language. The work program on the study discipline "Foreign Language" for the specializations of land management of the Kherson Agrarian University OKR "Master" (according to the credit-module system of education) is also a valuable development (Lebid O.M., Chekanovych V.G. 2011). Lemeshchenko-Lagoda V.V. (2020) methodological instructions for independent work for students of the speciality 193 "Geodesy and Land Management" (based on full general secondary education) full-time education contains important professionally oriented information. Malyshev, Lukashenko, & Sluchai (2020) conducted a pedagogical study of the formation of environmental competence. Nikolayenko S.M. and Nikolayenko M.S. (2021) described in detail higher agricultural education in Ukraine. The content of teaching a foreign language to future specialists in the field of tourism and environmental management was considered in her work by Rogova (2010). Semenova (2009) introduced training in summarizing foreign language texts of a socio-political nature in practical classes at a university (using the example of the German language. Modern problems of land management and cadastres were described by Sizov (2012). The work of Stolbetska (2023), which describes the role of authentic documents in the process of learning a specialized French language, turned out to be interesting. Trotsyuk, Kirikylytsia (2023) created the Syllabus of the normative educational component "Scientific communication in a foreign (English) language" of the second (Master's) level of higher education training, fields of knowledge: 09 - Biology, 01 - Education/Pedagogy, 05 - Social and Behavioural Sciences, 10 - Natural Sciences, 19 - Architecture and Construction, 20 - Agricultural Sciences and Food; specialities: 091 Biology and biochemistry, 014 Secondary education (Biology and Human Health), 053 Psychology, 101 Ecology, 102 Chemistry, 104 Physics and astronomy, 193 Geodesy and land management, 205 Forestry under educational and professional programs: "Biology"; "Secondary education. Biology, natural science, human health"; "Laboratory diagnostics"; "Psychology"; "Ecology"; "Chemistry"; "Physics Astronomy"; "Geodesy Land Management"; and forestry"; the form of education is full-time. The work "A Special Course in a Foreign Language (based on the MOOC "Sustainable Development of Mountain Territories in the Context of the Concept of Ecosystem Services") Erasmus+ CBHE project Sustainable Natural Resource Use in Arctic and High Mountainous Areas" by Lurkova (2021) turned out to be very useful. Lazariev, Fernos, Movchan, Komisarenko, Tymchuk and Neshchadym "Definition of a Process of Forming Professional Communicative Competence of the Future Agrarian Experts" (Lazarev, et al. 2019). Information and communication aspects in the context of innovative means of teaching a foreign language are studied by Lebedeva (Lebedeva, 2017, 2019). She also writes about the peculiarities of ESP in agrarian universities (Lebedeva 2018, 2023a, 2023b, 2023c).

Another scientist Zinaida Lebiedieva (2023) studied ESP writing as a practical means of cultural communication. Pasko (2023) presented valuable experience and a work program in the scientific chapter "Features of Teaching a Foreign Language to Lecturers of Higher Education Institutions".

Thus, there are quite a lot of valuable literary sources that can be used in one way or another when teaching master's students a foreign language with a professional focus on ecology and land management.

Methods and Techniques of Teaching a Foreign Language to Students of Master's Programs in Ecology and Land Management

Greenwald with colleagues (2017) note that the total volume of research of all master's students is related to environmental problems of the native region, including global environmental problems. There is practical experience, and the presence of extensive theoretical material of lectures and seminars, which allows for creating "harmonious conditions for including the process of mastering foreign language competence in the general course of the master's degree educational program. In addition, the bilingual vision of the scientific project allows students to better understand the subject of their research" (Greenwald et al., 2017, p. 156).

Kartashova (2009, p. 46) also notes the importance of intersubjective connections. According to her, they dictate the need to include specialized topics in the English course.

> "On the other hand, ESP training should not be limited to a chaotic set of specialized texts followed by the elaboration of the lexical and grammatical material contained in these texts. ESP training is of particular importance in the language education of future ecologists, since one of the primary tasks facing the system of professional environmental education at the current stage of society's development is the preparation of a specialist with a high level of environmental culture who can work successfully in the conditions of strengthening world integration processes" (Kartashova, 2006).

The formation of a comprehensively developed specialist in the ecological sphere presupposes mastery of a foreign language as a means of:

- professional communication;
- direct implementation of one's activities in a modern multicultural and multinational society;
- development of cognitive abilities;
- understanding and analysis of the manifestations of another culture, another way of thinking and the ability to correlate them with one's cultural background for a critical understanding of what is happening;

- formation and transfer of ecological culture.

According to Kartashov (2009, p. 46), it is no longer relevant to talk about the separate training of ecologists and translators in the field of professional communication, a synthesis of these two specialties is necessary for the full-fledged work of a modern employee. Among the important circumstances, Kartashova (2009, p. 47) cites the demand for specialists with knowledge of foreign languages, and, in particular, languages of international importance, interstate integration in the field of ecology, as well as access to world experience and knowledge, to a large information base abroad.

One of the main conditions determining the effectiveness of language education is its content. According to Blyev (2016), and Karychkovska (2021), the integration processes currently taking place in the world community have a significant impact on the higher education system as a whole.

> "The process of teaching language disciplines should be focused on revealing the intellectual and creative personal qualities and reserves of future specialists, taking into account professional communication skills and abilities specific to specific specialities" (Lazaryev et al., 2019).
>
> "The basis of the formation of the content of the training of future specialists is the provision of high-quality modern education, which meets the requirements of the domestic and international labour markets, as well as the integration of educational, research and innovation activities. The basis of the educational program, which ensures program learning results, are mandatory and optional components, which include the disciplines "Ukrainian language" and "Foreign language" (Karychkovska, 2021, p. 93).

The content of language education should correspond to the possibilities of an open society, in which ESP as a subject will be "a tool for both linguistic and multicultural personality development of students, contributing to self-awareness as a cultural and historical subject" (Kartashova, 2009, p. 47)

The content of learning foreign languages is not permanent.

The main principles of the selection of educational material within the framework of the development of the course "ESP for the Master's Specialities "Ecology" and "Land Planning and Cadasters" are considered to be:

Figure 4. Basic principles for selecting ESP educational material for the master's specialties "Ecology" and "Land Management and Cadasters"

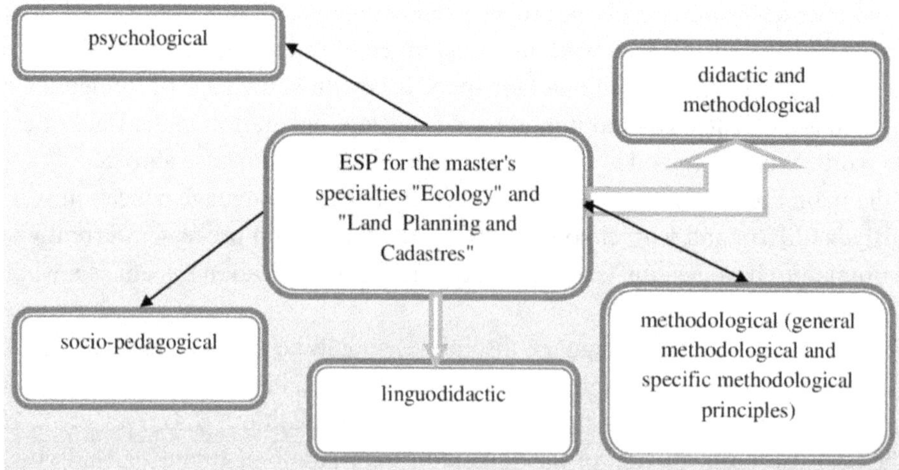

(summarized by the authors of the chapter)

The main criteria for selecting educational material can be depicted as follows:

Figure 5. Main criteria for selecting educational material

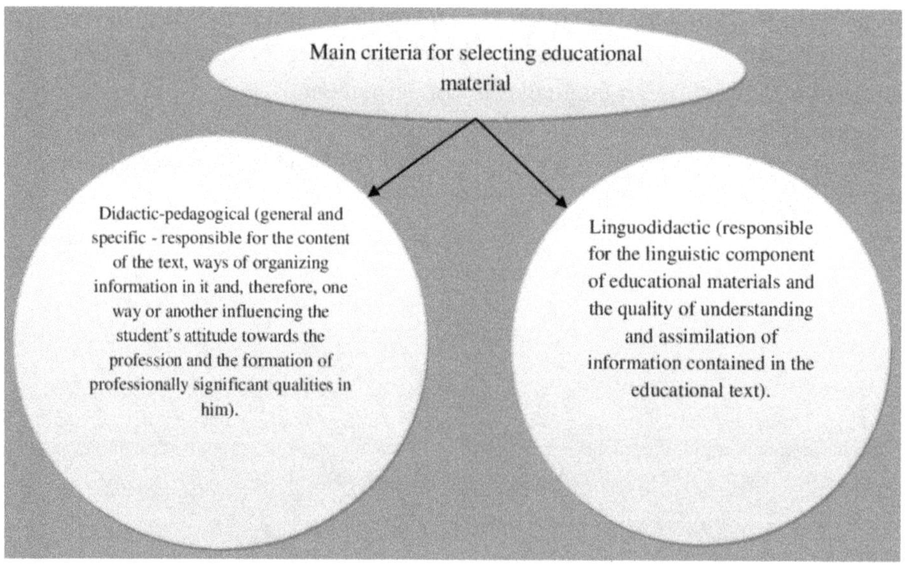

According to (Kartashova, 2009, p. 47), having determined what educational material will be used when teaching ecologists a foreign language, it is necessary:

- structure it thematically and modularly by the requirements of a systematic approach;
- organize foreign language teaching in a system of critically focused education;
- develop and apply a system of both pedagogical and linguodidactics methods and techniques for teaching foreign languages;
- actively introduce interactive forms and teaching methods (discussion, games, training) into the language learning process.

Teaching foreign languages to future environmental specialists, as proven by Kartashova (2009), must be formed taking into account the specifics of their professional activities, which was applied in teaching practice by the author of this chapter, "namely, its involvement in global integration processes and the high level of formation of environmental culture. Only in this case will the language teaching process be effective and be able to meet the requirements for language education by the modern world community" (Kartashova, 2009, p. 47).

According to the assertions of Bonkovsky (2023), it is very interesting to get students into international practice, including learning from the language through participation in agricultural forums, negotiations and projects.

The work of Greenwald et al., (2017) notes that public speaking helps undergraduates practice their acquired English language knowledge in a professional field. "Students are expected to study the recommendations and answer questions grouped according to the sections of the thesis: field of science and research, research problem, historical background of the research problem, purpose and methods, results and conclusion. Based on their answers, they make a presentation "My scientific research, carried out by the proposed recommendations," and also answer questions from the audience on the topic of their research. The presentation "My Scientific Research" gives students an initial experience of future professional dialogue and interaction with colleagues and partners. She teaches how to wisely select material for discussion and express your thoughts, how to correctly formulate your point of view and defend your interests, as well as the interests of your company. In addition, such a lesson, if properly organized by the teacher, will eliminate psychological barriers and typical fears of contact with the audience" (Greenwald et al., 2017, p. 155).

Bonkovskyi (2023) writes that it is important to achieve linguistic and cultural immersion, i.e. studying foreign terminology in the linguistic and cultural context of the country where it is used. To improve the process of learning foreign terminology, it is necessary to use new approaches, methods and technologies, for example, interactive mobile applications, online platforms, video lessons and web resources. Interactive applications may contain exercises to study words and expressions in context, as well as tests to assess knowledge. The use of virtual reality and augmented reality will help to build an immersive study of terminology, which will allow the creation of interactive scenarios and situations where students can use the terms in real conditions. The use of adaptive learning platforms will help to adapt to the individual needs and learning levels of students. This allows you to optimize learning and provide more material to students who move faster or provide extra support to those who need more time. Today, artificial intelligence technologies are increasingly used to analyze students' needs and progress, as well as to recommend additional materials and exercises. The involvement of students in joint study and cooperation in an online environment (collaborative learning) can be used to jointly discuss terms and solve tasks. Another important tool for student motivation is gamification, which allows to stimulate students to learn foreign terminology through role-playing games and competitions. Enabling students to choose materials and tasks that meet their individual needs and goals (personalization of learning) also stimulates their desire to learn the language material. Of course, a multimedia approach in the learning process is also indispensable: different types of materials, such as audio, video, interactive simulations and illustrations, to facilitate understanding and memorization of terms are a proven way to increase the effectiveness of learning. Innovations in teaching foreign terminology help to make the process more effective, interesting and convenient for students, providing them with tools and resources for successful learning (Bonkovskyi, 2023, p. 24-25).

Balieva (2016) notes that one of the oldest taught disciplines, which is an important component of general culture and one of the best ways to form a multicultural worldview, is speech and language, and therefore languages should be studied in inseparable unity with the world and culture of peoples who speak these languages. The ideas of lingua-cultural education using foreign languages are beginning to be embodied in educational and methodological complexes (UMK) in higher schools for students of non-language majors. Such complexes can be used instead of official textbooks by the concept of modular training successfully applied in European colleges and the corresponding general trend - an integrative approach to language learning based on the interrelated formation of skills in all types of speech activity: listening, speaking, reading, and writing. This approach to learning is opposed to sequential learning (listening – speaking – reading – writing).

> "Designing an English course for special purposes is a complex process. The teacher is directly involved in it because the purpose of such a course is to satisfy not only the needs of students of a certain training direction and profile but also the requirements of the educational institution of educational standards" (Blyeva, 2016, p. 74).

Blyeva (2016, p. 74) also notes the abundance of textbooks in English, indicating the problem that a universal textbook has not yet been created. Therefore, programs and manuals containing information developed by a teacher for students are the quintessence of knowledge for studying at a particular university.

The concepts listed above are reflected in the educational and methodological package, taking into account the national-regional component, for students of the master's degree in "Ecology" and "Land planning and cadastre" of the Kherson Agrarian University (Lebedeva, 2022)

As V. I. Gorbatov (2023, c. 5) writes in his scientific work, the topics of the program in a foreign language can be ecological problems of Russia and the countries of the language being studied, the problems of urbanization, global warming, the use of resource-saving technologies are discussed, and the activities of environmental organizations are discussed. Depending on the age of the students and the level of their language training, teaching methods and methodical methods are selected: discussion aquarium, round table, presentation, project method, reports, ecological forum, role-playing and business games on ecological topics. These methods require preliminary preparation, namely: preliminary study of new vocabulary and repetition of previously learned ones, mastery of grammatical structures necessary for pronouncing a statement, as well as understanding the essence of the problem under discussion.

SELF-GUIDED WORK OF STUDENTS

The importance of students' self-study work is reflected in the program which the author of this chapter used. In this section, similar statements from scientists who incorporated self-study of students into their work will be presented.

According to Lemeshchenko-Lagoda (2020), the main goal of the individual work of higher education students of agricultural technologies and ecology majoring in "Geodesy and Land Management" at all stages of education is the formation of professional competence of higher education students, the components of which are "independence of thinking, the ability to organize their educational activities without the help of a teacher, the ability to systematize, consolidate and apply acquired knowledge to solve tasks of a professional nature. During the performance of tasks of self-study work, the student of higher education can learn at a comfortable pace, choosing the optimal pace and conditions that will contribute to the deepening of acquired knowledge and preparation for active interpersonal interaction during practical classes, as well as in everyday and professional activities" (Lemeshchenko-Lagoda, 2020, p. 4).

According to the structure of the educational process of the agricultural higher educational institution, the master's students individual work during the semester consists of four parts: - performance of independent work tasks during preparation for practical classes; - performance of independent out-of-classroom reading tasks; - performance of an individual creative task (presentation); - current and final modular control.

> "Methodical guidelines for independent work for students of the faculty of agricultural technologies and ecology, major 193 "Geodesy and Land Management" consist of general provisions on the organization of self-study work, the structure of the educational discipline, recommendations for preparing for practical classes and final module control, recommendations for preparing tasks for independent extracurricular reading, recommendations for the preparation of individual creative tasks (presentations), an outline of the basic norms and rules for observing academic integrity, and general recommendations for searching for information on the Internet; and offer educational strategies for completing tasks and criteria for their evaluation" (Lemeshchenko-Lagoda, 2020, p. 4)

In addition to individual work, they use the creation of projects that play an important integrative and communicative role. Scientists believe that the application of the project method of learning in the process of forming intercultural competence is very effective. So, for example, Kryvulchak (2020) considers the formation of students' cognitive skills, the ability to independently build their knowledge, the ability to navigate in the information space, and the development of critical and creative thinking to be the basis of project activity. "The project method is based on an idea that is the essence of the concept of "project", its pragmatic focus on the result that can be obtained by solving one or another practically or theoretically significant problem. This result can be seen, understood, and applied in real practice. To achieve such a result, it is necessary to develop students' independent thinking skills, the ability to identify and solve problems, with the help of the ability to predict the results and possible consequences of various decision options, the ability to determine cause and effect relationships" (Kryvulchak, p. 345).

Project activities are always focused on the independent work of students, which they perform during the semester, regardless of whether it is individual, paired, or group. This type of work means a permanent solution to any problem. On the one hand, solving the problem involves the use of a complex of different methods and means of learning, but on the other hand - the need to integrate knowledge and the ability to apply knowledge from various professional and creative fields. Before starting the project activity, you should decide on the type of the intended project. The creative project involves the appropriate design of the results. As a rule, it does not have a detailed structure of the joint activities of the participants. It is only planned and continues to develop, obeying the logic of joint activity adopted by the group, and the interests of the project participants. In this case, it is necessary to agree on the planned results and the form of their presentation (in a newspaper, a work, a video film, a dramatization, a role-playing game, etc.)

Creative activity confirms that a foreign language is not only the goal of learning but also a means for further self-education, in the process of which students become creative individuals, and create and present their understanding of the task. The process of working on the project makes it possible to combine the knowledge obtained during the study of various disciplines into one whole, to bring them to a new level of understanding and application. Kryvulchak (2020) also describes the presence of an algorithm that includes 5 stages: preparation, planning, research, conclusions, and presentation (defence).

According to the scientists designing an English language course for special purposes is a special type of professional activity of a teacher based on the creation of an educational process that contributes not only to learning a foreign language but also to the education and development of a student's personality. The design creates the basis for the processes that follow it - drawing up the own program, formulating goals, selecting content, defining principles, developing or choosing educational materials, actually teaching students the language and monitoring and evaluating its results. The questions listed above are interrelated and affect the design of the English language course for special purposes. In this aspect, a foreign language for ecologists and land management specialists acquires additional professionally oriented functions:

Figure 6. Professionally oriented functions of a foreign language for ecologists and land management specialists

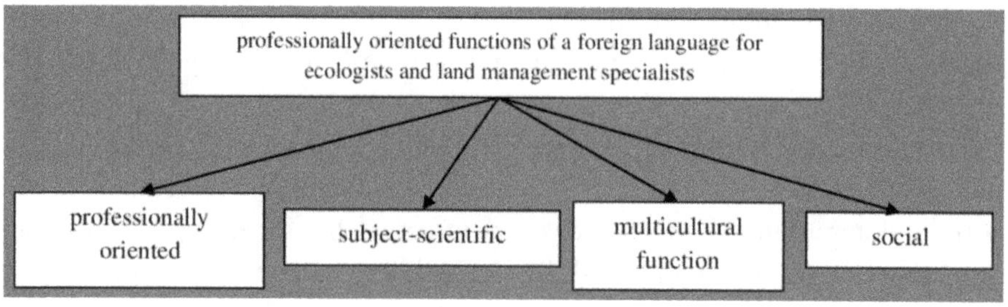

RESULTS AND DISCUSSION

In the dissertation of Lakatosh (2021, p. 109) it is said that the model of professional training of the future specialist in the agricultural sector is an open and dynamic integrative integrity, which, based on scientific and research activities, provides for the variability of the professional qualification structure and scientifically based didactic complexes. The main aspect of the methodology of this model is the integration of training and production based on the professionally oriented content of training on the formation of a system of functional professional skills. Describing the peculiarities of the training of students of agricultural specialities, the scientist characterizes the model of professional training as one that will become suitable for professional communication in a foreign language environment. She proposed to interpret the educational process as a complex of interrelated and interdependent components: target (approach and purpose of training), content (principles and content of training), organizational methodical (technology of training), evaluation and result (result of training). Investigating the professional training of future agronomists based on the competence approach, in his work Lakatosh (2021, p. 109) describes a model of process implementation in the context of five interrelated blocks: target (formation of a complex of professional knowledge, abilities and skills; consolidation of practical experience; development of professionally significant personality qualities); content (general didactic and specific principles of educational activity); operational (forms, methods, teaching aids describing the methodology of organizing the educational process); diagnostic (criteria, indicators and levels of formation of professional competence of future agronomists); effective (formation of professional competence).

At the Kherson Agrarian University, a foreign language program was developed with a professional direction for the master's degree in the specialities "Ecology" and "Landscape and Cadastres", which was used during the hostilities of the 2022/2023 academic year.

WORKING PROGRAM ON EDUCATIONAL DISCIPLINE

"Foreign Language" for the Specialties of Land Surveying and Cadastry, Ecology of the Kherson University of Agronomy Education and Qualification Level "Master" (According to the Credit-Module Education System) Lebedeva N.A. (2022)

The purpose of the foreign language course is for students of such a level of knowledge, skills and abilities that will provide the necessary communicative ability for a specialist in the areas of professional and situational communication in oral and written forms.

The task of the course is to improve and further develop knowledge, skills and abilities in a foreign language, acquired in the scope of the university program, in various types of speech activity.

Listening. The task is to achieve the skills of perception and understanding of monologic and dialogic expressions of native speakers within the limits of professional and everyday-situational topics at the level of 85-90% of information at the rate of its presentation of 100-120 words per minute.

Communication skills. Course objectives: development of naturally motivated monologic and dialogic speaking skills at a rate of 200-250 words per minute.

At the end of the course of study, students should have the skills of dialogic speech to lead a discussion and elementary communication within the limits of everyday topics.

Reading. The task is to further develop the skills of overview, informative and in-depth reading on the materials of original scientific information in the field.

Skills in review reading should provide the ability to get acquainted with the subject of scientific material and in general terms to characterize the general idea of its content. Deep reading skills are practised at the level of complete and accurate understanding of the text.

Translation. Oral and written translation from a foreign language into one's mother tongue can be used as a means of mastering a foreign language, a technique for developing reading skills and abilities, and an effective way to control the completeness and accuracy of understanding. For the formation of translation skills, some information about the peculiarities of language and style, from the theory and practice of translation is necessary.

Writing. Writing in a foreign language is also important for the development of various speaking skills and abilities. The goal is to periodically perform written exercises, make plans or summaries of what has been read, and present the content of what has been heard or read in writing. Development of skills in annotation and abstracting of scientific text.

The program takes into account the latest trends in the development of global higher education. It is focused on the training of masters; takes into account the Pan-European recommendations on language education regarding language proficiency levels that are sufficient to ensure the academic mobility of students; and is modular in its organization by the requirements of the credit-modular system.

Tasks:

1) rational division of the academic discipline into modules and checking the quality of theoretical and practical learning material of each module;
2) a wider scale of knowledge assessment;
3) improving the quality of their knowledge;

Advantages:

1) modules – logical; interconnected parts;
2) students' independent work constitutes at least half of their total educational workload;
3) self-control and self-assessment by the student of the level of assimilation of his knowledge;
4) individual training by performing separate individual tasks.

A module is a documented completed part of an educational and professional program (academic discipline, practice, state certification), which is implemented by appropriate forms of the educational process.

A content module is a system of educational elements that are combined according to the feature of compliance with a certain educational object.

The content module (chapter, subsection) contains separate modules (topics) of the student's classroom and independent work. Each content module must be evaluated and the student informed about the evaluation results.

Self-study of students is an integral and mandatory part of content modules of disciplines and is a part of knowledge assessment.

The determination of the number of points for a certain type of educational load or the performance of a certain type of independent work, which is an integral part of the content module (discipline), is proposed by the relevant department, depending on the types of educational load of a specific discipline.

During the study of a specific discipline, current, intermediate (modular) and final control of knowledge is used to stimulate effective independent work of students during the semester and ensure an objective assessment of the level of their knowledge and skills.

PROGRAM CONTENT BY TOPICS

1. The subject of ecology. Environmental pollution. Environmental protection.
2. Ecology terms.
3. Ecology problems.
4. Global Ecology: The Human Effect. Understanding Human Ecology.
5. Population Ecology.
6. Social, Economics and Ecology. Sustainability All Works Together.
7. Studying Forest Ecology. Bird Ecology and Conservation.
8. Microbial Ecology: What Is It?
9. Environmental Pollution.
10. Air Pollution. What Causes Indoor Air Pollution?
11. The problem of garbage disposal.
12. Plant peace. Plants under threat of extinction.
13. Animal world. Habitats of animals. Animals are under threat of extinction.
14. Soil. Agriculture. Agronomy.
15. Land use.
16. Land use in Russia.
17. Water Pollution. Plastic Pollution Solutions.

GENERAL EDUCATION BLOCK-MODULE

1. Pronouns. Types of pronouns
2. Verbs to be, to have, their functions
3. Noun: singular and plural forms, possessive case. Use of noun substitute words
4. Adjective, adverb. Degrees of comparison. Comparative constructions
5. Numerals: quantitative, ordinal and fractional. Time marking. Reading dates.
6. Modal verbs and their equivalents
7. Times of the Simple (Indefinite) group
8. Times of the Continuous group
9. Times of the Perfect Group
10. Passive state
11. Coordination of times
12. Direct and indirect speech
13. Participle (Participle I, Participle II)
14. Adverbial inflections. Selection and translation of adverbial inflections
15. Gerund. Peculiarities of functioning and translation
16. Infinitive. Peculiarities of functioning in professional texts
17. Infinitive. Constructions with the infinitive
18. Sentence. Types of sentences. The specifics of translating sentences in popular science style. Inversion
19. Vocabulary of a scientific text: terminological units, their use, translation techniques
20. Word formation. The most used suffixes and prefixes of scientific and technical terminology. Conversion.

The developed profile-oriented course of studying a foreign language for specialized purposes for master's students of the specialities "Ecology" and "Land Management and Cadastres" (Lebedeva Program 2022) was an attempt to reflect the specifics of their activities in the future in the fields of ecology and land management, preparing them to use a foreign language to solve practical tasks, for example, to find the necessary information on the profile of one's speciality, to describe one's developments for presenting them at the international level.

The foreign language training course for master's students of the specialities "Ecology" and "Land Management and Cadastres" is interdisciplinary in nature and is assessed by practical results: not only by the correct use of certain grammatical structures and active vocabulary but also by a profile-oriented course:

Figure 7. Foreign language training course for master's students of the specialties "Ecology" and "Land management and cadastres"

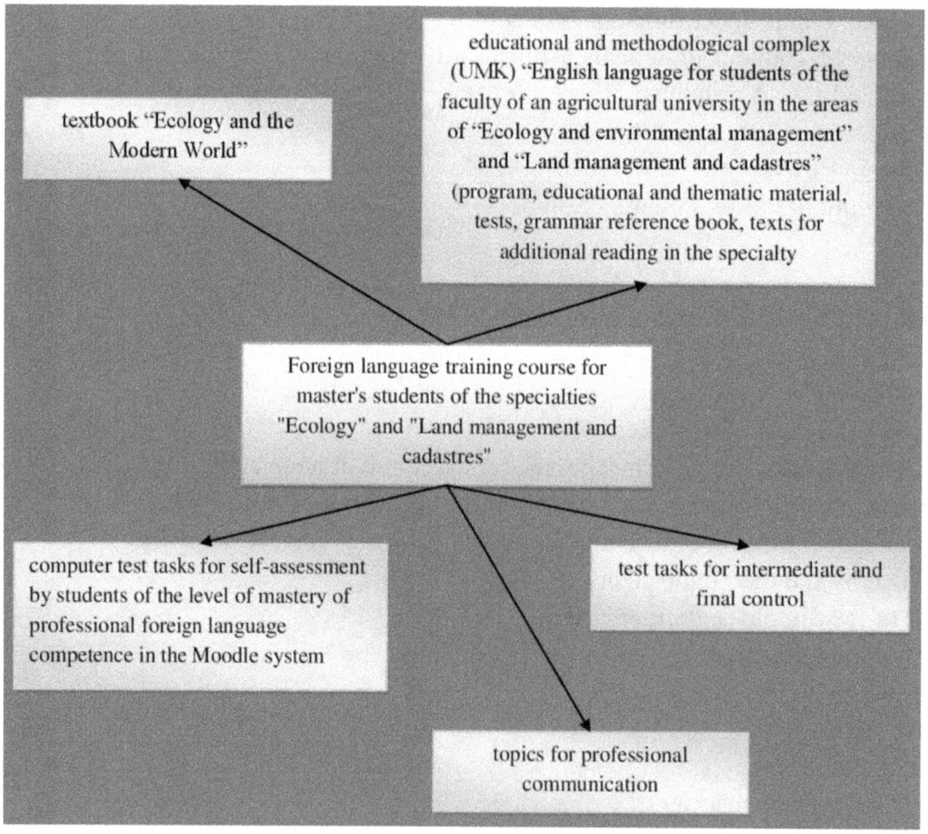

(summarized by the author of the chapter)

The textbook "Ecology and the Modern World" (Milovidov, 2013) includes 23 lessons, each of which contains texts devoted to the main problems of ecology, accompanied by exercises for consolidating active specialized vocabulary, mastering grammatical material, and also developing communication skills in conversation. In parallel, texts from the textbook in 2 volumes "English for Businessman" (Dudkina et al., 2007) related to topics for developing conversation skills on personal topics (Introducing Yourself,

Working Day, Hobbies, British Holidays, My University) were used. For additional reading, we took texts devoted to different varieties of the English language in different countries (Britain, USA, Canada, Australia), and regional studies, representing any of the countries where English is used as the state language. In addition, they practised studying additional material, which contained information about the sights of their hometown of Kherson and the region, which allowed students to get acquainted with the realities of the language they were learning and gain additional knowledge.

CONCLUSION

Consequently, as can be seen from the above, the conceptual and categorical apparatus of the main disciplines of master's programs forms the lexical base of the English language for special purposes, which is reflected in curricula and textbooks, and also represents an extensive layer of knowledge in the field of ecology and land management, which is taught to students of agricultural universities. “Land” is considered a natural resource here, the main characteristic of which is the territory where people exist and develop it following their interests. From this definition, it follows: the earth is a complex multi-component ecological system consisting of natural components. The concept of “land” is interpreted per such categories as land, land plots, and forms of ownership, with division into agricultural land and non-agricultural land. Today, all lands can be in one of three forms of ownership: state, municipal or private. Thus, a fairly extensive conceptual apparatus of land management, real estate cadastre and land monitoring was derived from one concept of “land”, which represents a significant terminological base of ESP. The use of lexical units of a special subject further increases the level of awareness.

The combination of auditory-motor images of internal speech and visual-motor images in the writing process ensures the comprehensive assimilation of students' knowledge.

As a result of the research, the purpose of this chapter has been achieved: The teaching of a foreign language in the educational process of undergraduates studying in the speciality “Ecology and land Management” is described and recommendations are given to improve their communication skills. The following tasks have been solved: the main methods and techniques of teaching a foreign language to students of the specialities “Ecology” and “Land Management (land resources)” are considered and the linguistic and didactic foundations of teaching English-speaking communication in the field of ecology and land management are outlined; the content of English language teaching and criteria for the selection of educational material for the formation of environmental and land management competence are determined, as well as the program and theoretical materials are presented.

The text of the chapter also presents both the program and educational materials that were created and used by the author in foreign language classes with students. The list of competencies established by the state educational authorities for graduate students receiving a Master's degree in ecology and environmental management includes knowledge of a foreign language as the most important means of scientific and professional communication.

The associative series created in this way contribute to a more effective perception, memorization and reproduction of specialized material. The strength of memorization is ensured by the participation of all analysts, the interaction of analysis and synthesis, as well as the conscious and involuntary nature of the translation process. Graphic images of words and block diagrams of sentences learned through written exercises are easier to recognize visually. In writing, the verbal nature of a coherent statement is more clearly manifested, which usually consists of several logically and structurally interrelated sentences.

In the process of teaching foreign language communication to students in ecology and land management, an important function of writing is that it can serve as an effective means of control. The formation of a comprehensively developed specialist in the environmental field involves mastering a foreign language as a means of carrying out one's activities in a modern multicultural and multinational society, understanding and analyzing the manifestations of another culture, another way of thinking and the transmission of environmental culture. In the learning process, the skills of all types of speech activity are formed: listening, speaking, reading and writing, including students' independent work and project activities. Thus, the educational process contributes to the development of professional competence on a global scale, since it contains professional knowledge in a foreign language.

ACKNOWLEDGMENT

This research received no grant from any organization.

REFERENCES

Anisimova, S. V. (2023). Some experience of introducing distance learning in KHNADU (opportunities and problems). *all-Ukrainian conference on problems of higher education with international participation "Environmentally oriented higher education. Methodology and practice*. Research Gate.

Blieva J. M., Khosroeva N.I. (2016). Foreign language as a subject and speech in the formation of a multicultural worldview of environmental students. *Pedagogy and psychology,4*(17). 74-78.

Bokshan, G. I. (2019). *Dictionary-a manual for the development of professionally-oriented competence of students of the specialty" Geodesy and Land Management*. KSAU.

Dudkina, G. A., Pavlova, M. V., Rey, Z. G., & Khvalnova, A. T. (2007). *English for business communication: In 2 vol.* Filomatis.

Giniyatov, I. A. (2022). On the issue of basic concepts in the field of land management, real estate cadastre and land monitoring (in the order of discussion). *Bulletin of SGUGiT*, 27(6), 152–159. 10.33764/2411-1759-2022-27-6-152-159

Gorbatov, V. I. (2022). Ecological education of students during foreign language classes. *Ecological education and ecological culture of the population: materials of the X international scientific conference on February 25–26, 2022*. Prague: Vědecko vydavatelské centrum «Sociosféra-CZ».

Greenwald, O., Islamov, R., & Sergeychick, T. (2017). Mastering Foreign Language Competence of Ecology and Environment Managers for Mining Industry of Kuzbass. *The Second International Innovative Mining Symposium. E3S Web of Conferences 21, 04012*. IEEE. 10.1051/e3sconf/20172104012

Islamov, R. S., Resenchuk, A. A., & Perevalova, A. A. (2018) From the experience of teaching English as a foreign language to students in the fields of "Geology", "Biology", "Ecology and Nature and Management". *Modern problems of science and education.*https://science-education.ru/ru/article/view?id=28417 (дата обращения: 09.03.2024).

Ivanova, I. E., Priss, O. P., Taranenko, G. G., Shleina, L. I., Chebanova, Yu. V., Evstafieva, E. S., Krasulya, Ya. G., & Kolomoets-Garkusha, O. V. (2020). *Trajectory of development of "SOFT SKILLS" in students of the Faculty of Agrotechnologies and ecology as one of the important factors of formation of professional skills of youth. Improvement of educational process in the institution of Higher Education. Issue 24, collection of scientific and methodological works. Tavrichesky State agrotechnological University named after Dmitry Motorny*. Melitopol.

Karichkovskaya S. P., Chuchmiy I. I. (2021). Features of Studying Ukrainian and Foreign Languages in Agricultural Institutions of Higher Education in Ukraine. *Academic Studios. Series "pedagogy", 4*(1), 91-98.

Kartashova, E. A. (2009). Teaching a foreign language to future ecologists as carriers of a new type of ecological culture. *Bulletin of Kostroma State University Series: Pedagogy. Psychology.Sociokinetics*, 15(3), 44–48.

Kryvulchak, L. (2020). Creative Project as a Method of Forming Cultural Competence of Students during the Lessons of the International Movement. *The modern movement of science:abstracts of the 11th International Scientific and Practical Internet Conference.* Dnipro.

Lakatosh, M. A. (2021). *Professional training of future specialists in the agricultural sector in institutions of higher education in Ukraine* [Doctoral dissertation, Mukachevo State University.] Mukachevo.

Laver O. V. (2022). *Business Ukrainian: methodological developments of final control works for applicants of higher education in the field of knowledge 19 " Architecture and Construction" specialty 193 "Geodesy and land management" of the educational and professional program "Geodesy and land management"*. Uzhgorod, 52 p.

Lazariev, O., Fernos, Yu., Movchan, L., Komisarenko, N., Tymchuk, S., & Neshchadym, L. (2019). *Definition of a Process of Forming Professional Communicative Competence of the Future Agrarian Experts.* Vision 2025. Education Excellence and Management Innovations through Sustainable Economic Competitive Advantage: Proceedings of the 34-th IBIMA Conf., Madrid, Spain.

Lebed, O. M., Bokshan, G. I., Kaminskaya, M. O., & Makukhina, S. V. (2019). *Dictionary-a manual for the development of professionally-oriented communicative competence of students of the specialty "Geodesy and Land Management"*. Kherson: HAU "Kolos".

Lebed, O. M., & Chekanovich, V. G. (2011). *Working program on the discipline "foreign language" for the specialties of Land Management and cadastre, ecology of the Kherson agronomic University of the ROC "master" (according to the credit-modular training system).* HAU.

Lebedeva, N. (2019). Information and communication technologies for teaching a foreign language regarding the modernization of public administration. *The latest information and communication technologies in the modernization of public administration: foreign and domestic experience: materials of II scientific practice seminar.* Dnipropetrovsk DRIDU NADU, 55-58.

Lebedeva N. A. (2017). Information aspects in the context of innovative means of teaching a foreign language. *All-Ukrainian science and practice internet conference "Modern management: problems and development prospects on May 19, 2017.* Kherson: KhDAU, 235-238.

Lebedeva, N. A. (2018). Teaching English in the field of knowledge "Public management and administration" in an agrarian higher educational institution. *European dimension of reforming public administration in Ukraine: materials of the International Scientific and Practical Conference.* Kyiv "Publishing House "Personal".

Lebedeva, N. A. (2022). *Work program on the academic discipline "Foreign Language" for the specialties of land management and cadastre, ecology of the Kherson Agronomic University of the Master's educational qualification level (according to the credit-modular training system).* HAU.

Lebedeva, N. A. (2022). Teaching of Foreign Language in the Educational Process of Public Administration Personnel in Ukraine. In Kulaç, O., Babaoğlu, C., & Akman, E. (Eds.), *Public Affairs Education and Training in the 21st Century* (pp. 294–310). IGI Global. 10.4018/978-1-7998-8243-5.ch019

Lebedeva, N. A. (2023). ESP in the Field of Architecture: Features of Teaching Writing and Some Aspects of Error Analysis. In Lebedeva, N. (Ed.), *Cases on Error Analysis in Foreign Language Technical Writing* (pp. 97–123). IGI Global. 10.4018/978-1-6684-6222-5.ch005

Lebedeva, N. A. (2023). Improving the Lexical and Grammatical Skills of Students Specializing in Hydraulic Engineering and Water Resources by Means of Writing. In Lebedeva, N. (Ed.), *Cases on Error Analysis in Foreign Language Technical Writing* (pp. 162–186). IGI Global. 10.4018/978-1-6684-6222-5.ch007

Lebedeva, N. A. (Ed.). (2023). *Cases on Error Analysis in Foreign Language Technical Writing*. IGI Global. 10.4018/978-1-6684-6222-5

Lebid, O. M., & Lebedeva, N. A. (2018). *Working program "Foreign language for professional purposes" for applications for higher education, studying in the educational program "Bachelor" in the specialty "6.060103. Hydraulic Engineering (Water Resources)"*. KSAU, Ukraine.

Lebiedieva, Z. (2023). ESP Writing as a Practical Mean of Cultural Communication Between Ukrainian Music Teachers and Chinese Students. In Nadezhda Anatolievna Lebedeva, I. G. I. (Ed.), *Cases on Error Analysis in Foreign Language Technical Writing* (pp. 187–206). Global. 10.4018/978-1-6684-6222-5.ch008

Lemeshchenko-Lagoda, V. V. (2020). *Foreign language II (English) guidelines for independent work for applicants for the degree of Higher Education "Bachelor" in specialty 193 "Geodesy and Land Management" according to the OPP "Geodesy and Land Management" (based on full-time general secondary education)*. Melitopol.

Lurkova, N. (2021). *A Special Course in a Foreign Language (based on the MOOC "Sustainable Development of Mountain Territories in the Context of the Concept of Ecosystem Services") Erasmus+ CBHE project Sustainable Natural Resource Use in Arctic and High Mountainous Areas*. Gorno-Altaisk State University, Department of Natural Sciences and Geography. Russia.

Malyshev, V. V., Lukashenko, T. F., & Slushna, Yu. D. (2020). Pedagogical research of environmental competence formation. *Modern movement of science: theses of the supplement. XI international scientific and practical Internet Conference, October 8-9, 2020.* Dnipro.

Milovidov, V. A. (2013). *Ecology and the modern world.* Iris-press.

Nikolaenko, S., & Nikolaenko, M. (2021). Higher agricultural education in Ukraine: challenges of the time. *Bulletin of Agrarian science, 99*(12), 5-17.

Pasko, O. A. (2023). In Lebedeva, N. A. (Ed.), *Features of Teaching a Foreign Language to Lecturers of Higher Education Institutions. Cases on Error Analysis in Foreign Language Technical Writing* (pp. 124–161). IGI Global. 10.4018/978-1-6684-6222-5.ch006

Rogova O. A. (2010). About the content of foreign language training for future specialists in the field of Tourism and nature management. *Works of BSTU. Series 6: history, philosophy,* (5), 226-228.

Safina, G. R., & Fedorova, V. A. (2020). *State management and protection of land resources of Russia at the present stage: a course of lectures. Part 1 Management in the field of land protection. Degradation of soils and lands.* Kazan. University.

Semenova, O. S. (2009). Teaching the Abstraction of Foreign-language Texts of a Socio-Political Nature in Practical Classes at a University (on the Example of the German Language). *Pedagogy. Psychology. Social work. Juvenile studies.Sociokinetics*, (3), 47–48.

Sizov, A. P. (2012). Modern problems of land management and cadastres. Part 1 Land management: a textbook for graduate students. M.: Publishing House of MIIGAiK, 69 p.

Stolbetskaya, S. B. (2023). The role of authentic documents in the process of teaching the French specialized language. *Innovative psychological-pedagogical, linguistic and legal bases of Agrarian Science in the conditions of intercultural agrarian communication: materials of the international scientific and practical conference.* Belotserkovsky NAU.

Trotsyuk, A. M., & Kirikylytsia, V. V. (2023). Scientific communication in a foreign (English) language: syllabus of normative academic discipline. *Volyn National University named after Lesya Ukrainka, Faculty of foreign philology, Department of foreign languages of natural and mathematical specialties*. Lutsk.

Compilation of References

Aasen, H., Burkart, A., Bolten, A., & Bareth, G. (2015). Generating 3D hyperspectral information with lightweight UAV snapshot cameras for vegetation monitoring: From camera calibration to quality assurance. *ISPRS Journal of Photogrammetry and Remote Sensing*, 108, 245–259. 10.1016/j.isprsjprs.2015.08.002

Abdallah, A. M., Jat, H. S., Choudhary, M., Abdelaty, E. F., Sharma, P. C., & Jat, M. L. (2021). Conservation agriculture effects on soil water holding capacity and water-saving varied with management practices and agroecological conditions: A Review. *Agronomy (Basel)*, 11(9), 1681. 10.3390/agronomy11091681

Abdalla, M. A. I., EL-WAHAB, M. A., Tawfik, M., & Khater, I. (2022). Hydrological Models Simulation in Agricultural Watershed of Wadi Kharouba in North-Western Coast Region–Egypt. *Fresenius Environmental Bulletin*, 31, 1063–1078.

Abisheva, B. N., Balakayev, T. B., & Suleymenov, R. B. (1980). *Istoriya Kazahskoj SSR (s drevnejshih vremen do nashih dnej). Tom 5 [The history of the KazSSR (from ancient times to the present day), V. 5].* Nauka.

About calendars and chronology. (n.d.) Kalendari. http://calendar.orthlib.ru/Kalendari.pdf

Ahmad, M., & Ali, A. (2023). Mapping the Future of Sustainable Development Through Cloud-Based Solutions: A Case Study of OpenStreetMap. In *Promoting Sustainable Management Through Technological Innovation* (pp. 153–176). IGI Global. 10.4018/978-1-6684-9979-5.ch011

Ahmed, N., & Al-Mutairi, K. A. (2022). Earthworms effect on microbial population and soil fertility as well as their interaction with agriculture practices. *Sustainability (Basel)*, 14(13), 7803. 10.3390/su14137803

Ahrens C. D. (2012). *Meteorology today: an introduction to weather, climate, and the environment: Cengage Learning.*

Akhmedov, Kh. M., & Karimov, Kh. S. (2003). *The possibility for produce and using the biological of gaze in Tajikistan.*

Albizua, A., Bennett, E. M., Larocque, G., Krause, R. W., & Pascual, U. (2021). Social networks influence farming practices and agrarian sustainability. *PLoS One*, 16(1), e0244619. 10.1371/journal.pone.024461933411756

Alcantara, C., Kuemmerle, T., Baumann, M., Bragina, E., Griffiths, P., Hostert, P., Knorn, J., Müller, D., Prishchepov, A., Schierhorn, F., Sieber, A., & Radeloff, V. (2013). Mapping the extent of abandoned farmland in Central and Eastern Europe using MODIS time series satellite data. *Environmental Research Letters*, 8(3), 035035. 10.1088/1748-9326/8/3/035035

Aleshina, T. A. (2010). Extraction and utilization of landfill gas to increase the efficiency of the domestic electric power industry. *Bulletin of MGSU*, 2, 140–143.

Alexeenko, A. (1994). *Selskoe naselenie Kazahstana. 1920-1990 gg.* [Rural Population of Kazakhstan. 1920-1990]. Institut istorii i etnologii im. Ch.Ch. Valihanova.

Aleynikova, A., Byashkin, M., Gaivoron, T., Mainasheva, G., & Marsheva, N. V. (2020). The role of specially protected natural areas in solving environmental problems of the Russian arctic (using the example of Vaygach Island). *E3S Web of Conferences, 169.* 10.1051/e3sconf/202016902019

Al-Hazmi, H. E., Hassan, G. K., Kurniawan, T. A., Sniatala, B., Josph, T. M., Majtacz, J., Pietchota, G., Li, X., El-Gohary, F. A., Saeb, M. R., & Makinia, J. (2024). Technological solutions to landfill management: Towards recovery of bio methane and carbon neutrality. *Journal of Environmental Management*, 354, 120414. 10.1016/j.jenvman.2024.12041438412730

Almanova, Z., Kenzhegulova, S., Kashkarov, A., Kekilbayeva, G., Ussalinov, E., Yerzhan, D., Zhakenova, A., & Zvyagin, G. (2023). Changes in soil fertility indicators after long-term agricultural use in Northern Kazakhstan. *International Journal of Design & Nature and Ecodynamics*, 18(5), 1045–1053. 10.18280/ijdne.180504

Alpyspaeva, G.A., & Nurmukhanova, R.Zh. (2023). Participation of ethnic special settlers in the development of virgin lands in Kazakhstan. *Herald of Science of S. Seifullin Kazakh Agrotechnical Research University.*

Al-Qurashi, A. (1995). *Rainfall–runoff relationships in arid areas.* [MSc thesis, Heriot-Watt University, Edinburgh, UK].

Alterra, W. (2014). Mapping Soil Erosion Prevention Using an Ecosystem Services Modeling Framework for Integrated Land Management and Policy. *Ecosystems (New York, N.Y.)*, 17(5), 878–889. 10.1007/s10021-014-9766-4

Altynov, A. E. (2010) *Spectrometry of landscapes.* Moscow: Publishing house MIIGAiK.

Altynov, A.E., Malinnikov, V.A., Popov, S.M., & Stetsenko A.F. (2010). Spectrometry of landscapes, M.: Publishing house. *MIIGAiK. UPP Reprography*, 120.

Alves, A. L., da Silva Santos, A. C., de Mattos, J. L. S., Costa, P. M. O., de Souza Motta, C. M., Melo, R. F. R., & Tiago, P. V. (2023). Diversity of filamentous fungi communities in the soils of agroecological crop polycultures and the Atlantic Rain Forest. *Archives of Agronomy and Soil Science*, 69(3), 374–386. 10.1080/03650340.2021.1995717

Ambler, J., Dearden, P. K., Wilcox, P., Hudson, M., & Tiffin, N. (2021). Including digital sequence data in the Nagoya Protocol can promote data sharing. *Trends in Biotechnology*, 39(2), 116–125. 10.1016/j.tibtech.2020.06.00932654776

Anantha, K. H., Garg, K. K., Moses, D. S., Patil, M. D., Sawargaonkar, G. L., Kamdi, P. J., Malve, S., Sudi, R., Raju, K. V., & Wani, S. P. (2021). Impact of natural resource management interventions on water resources and environmental services in different agroecological regions of India. *Groundwater for Sustainable Development*, 13, 100574. 10.1016/j.gsd.2021.100574

Anderson, J., Sarkar, D., & Palen, L. (2019). Corporate editors in the evolving landscape of OpenStreetMap. *ISPRS International Journal of Geo-Information*, 8(5), 232. 10.3390/ijgi8050232

Andreev, G. A., Bazarsky, O. V., Glauberman, A. S., Kolesnikov, A. I., & Korzhik, Yu. V. (1984). Analysis and synthesis of random spatial textures. *Foreign Radio Electronics*, (2), 3–33.

Andrianto, H. (2017). *Measurement of chlorophyll content to determine nutrition deficiency in plants: A systematic literature review.* International Conference on Information Technology Systems and Innovation (ICITSI), Bandung, Indonesia.

Androkhanov, V. A., Kulyapina, E. D., & Kurachev, V. M. (2004). *Soils of technogenic landscapes: genesis and evolution of Ros. acad.* Institute of Soil Science and Agrochemistry; Novosibirsk: Publishing house.

Andronikov, B.JI. (1979). *Aerospace methods for studying soils.* M. Kolos.

Anguluri, R., & Narayanan, P. (2017). Role of green space in urban planning: Outlook towards smart cities. *Urban Forestry & Urban Greening*, 25, 58–65. 10.1016/j.ufug.2017.04.007

Anisimova, S. V. (2023). Some experience of introducing distance learning in KHNADU (opportunities and problems). *all-Ukrainian conference on problems of higher education with international participation "Environmentally oriented higher education. Methodology and practice.* Research Gate.

Aparin, B. F. (2008). *Evolution of soils and soil cover of reclaimed lands.* St. Petersburg Publishing House.

Arinushkina, E. F. (1970). *Guide to chemical analysis of soils. M.* From Moscow State University.

Aronson, M. F. J., Lepczyk, C. A., Evans, K. L., Goddard, M. A., Lerman, S. B., MacIvor, J. S., Nilon, C. H., & Vargo, T. (2017). Biodiversity in the city: key challenges for urban green space management. In *Frontiers in Ecology and the Environment, 15*(4). 10.1002/fee.1480

Arthur, N., & Hack, J. (2022). A multiple scale, function, and type approach to determine and improve Green Infrastructure of urban watersheds. *Urban Forestry & Urban Greening*, 68, 127459. 10.1016/j.ufug.2022.127459

Asdal, Å. (2021). The role of the Svalbard Global Seed Vault in preserving crop genetic diversity. In *Plant genetic resources* (pp. 217–225). Burleigh Dodds Science Publishing.

Astell-Burt, T., & Feng, X. (2019). Association of Urban Green Space with Mental Health and General Health among Adults in Australia. *JAMA Network Open*, 2(7), e198209. 10.1001/jamanetworkopen.2019.820931348510

Awasthi, S. K., Sarsaiya, S., Kumar, V., Chaturvedi, P., Sindhu, R., Binod, P., Zhang, Z., Pandey, A., & Awasthi, M. K. (2022). Processing of municipal solid waste resources for a circular economy in China: An overview. *Fuel*, 317, 1–13. 10.1016/j.fuel.2022.123478

Ayaz, M., Ammad-Uddin, M., Baig, I., & Aggoune, E. H. M. (2018). Wireless sensor's civil applications, prototypes, and future integration possibilities: A review. *IEEE Sensors Journal*, 18(1), 4–30. 10.1109/JSEN.2017.2766364

Ayaz, M., Ammad-Uddin, M., Sharif, Z., Mansour, A., & Aggoune, E. H. M. (2019). Internet-of-Things (IoT)-based smart agriculture: Toward making the fields talk. *IEEE Access : Practical Innovations, Open Solutions*, 7, 129551–129583. 10.1109/ACCESS.2019.2932609

Ayres, E., Wall, D. H., & Bardgett, R. D. (2009). Mutualistic interactions and spatial gradients in a plant-microbial-invertebrate food web. *Ecology Letters*, 12(4), 490–499.

Aytar, E. C., & Kömpe, Y. Ö. (2023). Effect of different substrates on in vitro symbiotic seed germination for soilless production of Anacamptis laxiflora orchid. *Acta Botanica Croatica*, 82(2), 101–108. 10.37427/botcro-2023-010

Baburina, T.A. (2017). Environmental assessment of the negative impact of wastewater from a solid waste landfill (filtrate) on the environment. *Questions of science and education*, *6*(7), 177-178.

Badiu, D. L., & Ioj, . (2016). Is urban green space per capita a valuable target to achieve cities' sustainability goals? Romania as a case study. *Ecological Indicators*, 70, 53–66. 10.1016/j.ecolind.2016.05.044

BAFU (Ed.). (2017). Boden in der Schweiz, Zustand und Entwicklung, Stand 2017, Bundesamt für Umwelt, Bern. *BAFU, 2020, Bodenstrategie Schweiz, für einen nachhaltigen Umgang mit dem Boden.* BAFU. https://www.bafu.admin.ch/bafu/de/home/themen/boden/publikationenstudien/publikationen/bodenstrategie-schweiz.html

Baker, C. C., Castillo Vardaro, J. A., Doak, D. F., Pansu, J., Puissant, J., Pringle, R. M., & Tarnita, C. E. (2020). Spatial patterning of soil microbial communities created by fungus-farming termites. *Molecular Ecology*, 29(22), 4487–4501. 10.1111/mec.1558532761930

Baldoni, N., Francioni, M., Trozzo, L., Toderi, M., Fornasier, F., D'Ottavio, P., Corti, G., & Cocco, S. (2023). Effect of wood gasification biochar on soil physicochemical properties and enzyme activities, and on crop yield in a wheat-production system with sub-alkaline soil. *Biomass and Bioenergy*, 176, 106914. 10.1016/j.biombioe.2023.106914

Baltyzhakova, T. L., & Romanchikov, A. Y. (2021). Spatial analysis of subway passenger traffic in Saint Petersburg. *Geodesy and Cartography (Vilnius)*, 47(1), 10–20. 10.3846/gac.2021.11980

Banar, M., Cokaygil, Z., & Ozkan, A. (2009). Life cycle assessment of solid waste management options for Eskisehir, Turkey. *Waste Management (New York, N.Y.)*, 29(1), 54–62. 10.1016/j.wasman.2007.12.00618280731

Banerjee, S., & Van Der Heijden, M. G. (2023). Soil microbiomes and one health. *Nature Reviews. Microbiology*, 21(1), 6–20. 10.1038/s41579-022-00779-w35999468

Barnett, K. L., & Facey, S. L. (2016). Grasslands, invertebrates, and precipitation: A review of the effects of climate change. *Frontiers in Plant Science*, 7, 197607. 10.3389/fpls.2016.0119627547213

Barraqan-Escandon A., Terrados-Cepeda J., Zalamea-Leon E., & Arias-Reyes P. (2018) Electricity production using renewable resources in urban centers. *Energy*, *171*(1), 12-25 ().10.1680/jener.17.00003

Baur, J. W. R. (2018). Urban green spaces, recreation and spiritual experiences. *Leisure/ Loisir, 42*(2). 10.1080/14927713.2018.1449131

Bazzoffi P., Abbattista F., Vanino S. and Pellegrini S. (2006). Impact of land levelling for vineyard plantation on soil degradation in Italy. *Bollettino della Società Geologica Italiana. Volume speciale*(6), 191-199.

Beguería, S., López-Moreno, J. I., Lorente, A., Seeger, M., & García-Ruiz, J. M. (2003). Assessing the effect of climate oscillations and land-use changes on streamflow in the Central Spanish Pyrenees. *Ambio*, 32(4), 283–286. 10.1579/0044-7447-32.4.28312956594

Belenkov, A.I., Tyumakov, A.Yu., Sabo, Umar, & Mokicheva D.S. (2013). Agrochemical and biological characteristics of soil fertility in the experimental site of the precision farming center. *News of TSHA*, 53-62.

Bell, S. (2005). Nature for people: The importance of green spaces to communities in the east midlands of England. In *Wild Urban Woodlands*. New Perspectives for Urban Forestry. 10.1007/3-540-26859-6_5

Ben-Dor, E., Chabrillat, S., Dematte, J.A.M., Taylor, G.R., Hill, J., Whiting, M.L., Sommer, S. (2009) Using imaging spectroscopy to study soil properties. *Remote Sensing of Environment*. Research Gate.

Berezikov, S. A. (2019). Structural changes and innovation economic development of the Arctic regions of Russia. *Journal of Mining Institute*, 240(6), 716–723. 10.31897/PMI.2019.6.716

Berland, A., Shiflett, S. A., Shuster, W. D., Garmestani, A. S., Goddard, H. C., Herrmann, D. L., & Hopton, M. E. (2017). The role of trees in urban stormwater management. In *Landscape and Urban Planning, 162*. 10.1016/j.landurbplan.2017.02.017

Bertolotto, M., McArdle, G., & Schoen-Phelan, B. (2020). Volunteered and crowdsourced geographic information: The openstreetmap project. *Journal of Spatial Information Science*, 20(20). 10.5311/JOSIS.2020.20.659

Beven, K. (2001). How far can we go in distributed hydrological modelling? *Hydrology and Earth System Sciences*, 5(1), 1–12. 10.5194/hess-5-1-2001

Bhosle, K., & Musande, V. (2017). Red edge point detection for mulberry leaf. *1st International Conference on Intelligent Systems and Information Management (ICISIM)*, Aurangabad, India. 10.1109/ICISIM.2017.8122160

Biao, Z., Gaodi, X., Bin, X., & Canqiang, Z. (2012). The Effects of Public Green Spaces on Residential Property Value in Beijing. *Journal of Resources and Ecology*, 3(3), 243–252. 10.5814/j.issn.1674-764x.2012.03.007

Bispo, A. (2020). *Proposal of methodological development for the LUCAS programme in accordance with national monitoring programmes*. European Joint Programme Soil. https://ejpsoil.eu/fileadmin/projects/ejpsoil/WP6/EJP_SOIL_Deliverable_6.3_Dec_2021_final.p df) accessed 12 December 2022.

Blieva J. M., Khosroeva N.I. (2016). Foreign language as a subject and speech in the formation of a multicultural worldview of environmental students. *Pedagogy and psychology,4*(17). 74-78.

Boffa, J. (1990). *Istoriya Sovetskogo Soyuza. T. 2. Ot Otechestvennoj vojny do polozheniya vtoroj mirovoj derzhavy. Stalin i Hrushev. 1941-1964 gg[The History of the Soviet Union. V. 2. From the Patriotic War to the Position of the Second World Power. Stalin and Khrushchev. 1941-1964].* Mezhdunarodniye Otnosheniya.

Bogachev, V. F., Gorenburgov, M. A., & Alekseeva, M. B. (2019). Systemic diagnostics of the Arctic Industry Development Strategy. *Journal of Mining Institute*, 238(4), 450–458. 10.31897/pmi.2019.4.450

Bokshan, G. I. (2019). *Dictionary-a manual for the development of professionally-oriented competence of students of the specialty" Geodesy and Land Management.* KSAU.

Bolisetty, S., Peydayesh, M., & Mezzenga, R. (2019). Sustainable technologies for water purification from heavy metals: Review and analysis. *Chemical Society Reviews*, 48(2), 463–487. 10.1039/C8CS00493E30603760

Borga, M., Stoffel, M., Marchi, L., Marra, F., & Jakob, M. (2014). Hydrogeomorphic response to extreme rainfall in headwater systems: Flash floods and debris flows. *Journal of Hydrology (Amsterdam)*, 518, 194–205. 10.1016/j.jhydrol.2014.05.022

Borrelli, P., Poesen, J., Vanmaercke, M., Ballabio, C., Hervás, J., Maerker, M., Scarpa, S., & Panagos, P. (2022). Monitoring gully erosion in the European Union: A novel approach based on the Land Use/Cover Area frame survey (LUCAS)'. *International Soil and Water Conservation Research*, 10(1), 17–28. 10.1016/j.iswcr.2021.09.002

Bourke, P. M., Evers, J. B., Bijma, P., van Apeldoorn, D. F., Smulders, M. J., Kuyper, T. W., & Bonnema, G. (2021). Breeding beyond monoculture: Putting the "intercrop" into crops. *Frontiers in Plant Science*, 12, 734167. 10.3389/fpls.2021.73416734868116

Bowers, S. A., & Hanks, R. J. (1965). Reflection of radiant energy from soils. *Soil Science*, 100(2), 130–138. 10.1097/00010694-196508000-00009

Brede, B., Suomalainen, J., Bartholomeus, H., & Herold, M. (2015). Influence of solar zenith angle on the enhanced vegetation index of a Guyanese rainforest. *Remote Sensing Letters*, 6(12), 972–981. 10.1080/2150704X.2015.1089362

Bröring, J. M., Goss-Souza, D., Baretta, C. R. D. M., Sousa, J. P., Baretta, D., Oliveira Filho, L. C. I., & Klauberg-Filho, O. (2023). Soil microbial carbon and activity along with land use and geographic gradients. *Soil Science Society of America Journal*, 87(4), 856–867. 10.1002/saj2.20545

Brosi, B. J., & Briggs, H. M. (2013). Single pollinator species losses reduce floral fidelity and plant reproductive function. *Proceedings of the National Academy of Sciences of the United States of America*, 110(32), 13044–13048. Advance online publication. 10.1073/pnas.130743811023878216

Brown, C., Kovács, E., Herzon, I., Villamayor-Tomas, S., Albizua, A., Galanaki, A., & Zinngrebe, Y. (2021). Simplistic understandings of farmer motivations could undermine the environmental potential of the common agricultural policy. *Land Use Policy*, 101, 105136. 10.1016/j.landusepol.2020.105136

Bucci, G., Bentivoglio, D., & Finco, A. (2018). Precision agriculture as a driver for sustainable farming systems: State of art in literature and research. *Calitatea*, 19, 114–121.

Buendia, C., Batalla, R. J., Sabater, S., Palau, A., & Marcé, R. (2015). Runoff trends driven by climate and afforestation in a Pyrenean basin. *Land Degradation & Development.*

Bukrinskaya, E. M. (2020) Application of a logistics approach to the organization of extended producer responsibility. *News of the St. Petersburg State University of Economics*, *2*, 141-145.

Buratti-Donham, J., Venn, R., Schmutz, U., & Migliorini, P. (2023). Transforming food systems towards agroecology–a critical analysis of agroforestry and mixed farming policy in 19 European countries. *Agroecology and Sustainable Food Systems*, 47(7), 1023–1051. 10.1080/21683565.2023.2215175

Bureau of National Statistics of the Agency for Strategic Planning and Reform of the Republic of Kazakhstan. (February 17, 2023). *Administrativno-territorialniye edinitsy Respibliki Kazakhstan (na nachalo 2023). [Administrative-territorial units of the Republic of Kazakhstan (beginning of 2023)].* https://stat.gov.kz/ru/industries/economy/national-/publications/6381/

Bureau of national statistics of the agency for strategic planning and reform of the Republic of Kazakhstan. (July 19, 2023). *Chislennost naseleniya Respubliki Kazakhstan (na nachalo 2023.[Population of the Republic of Kazakhstan at the beginning of 2023].* https://stat.gov.kz/ru/industries/economy/national-/publications/6373/

Bürgi, M., Verburg, P. H., Kuemmerle, T., & Plieninger, T. (2017). Analyzing dynamics and values of cultural landscapes. *Landscape Ecology*, 32(11), 2077–2081. 10.1007/s10980-017-0573-0

Burkart, A., Hecht, V. L., Kraska, T., & Rascher, U. (2018). Phenological analysis of unmanned aerial vehicle based time series of barley imagery with high temporal resolution. *Precision Agriculture*, 19(1), 134–146. 10.1007/s11119-017-9504-y

Busch, G., Grobmann, J., Sieber, M., & Burkhardt, M. A. (2009). New and sound technology for biogas from solid waste and biomass. *Water Air and Soil Pollution Focus*, 9(1-2), 89–97. 10.1007/s11267-008-9195-5

Bykova, E. (2021). Assessment of negative infrastructural externalities when determining the land value. *Journal of Mining Institute*, 247, 154–170. 10.31897/PMI.2021.1.16

Bykowa, E. N., Khaykin, M., Shabaeva, Y. I., & Beloborodova, M. D. (2023). Development of methodology for economic evaluation of land plots for the extraction and processing of solid minerals. *Journal of Mining Institute*, 259, 52–67. 10.31897/PMI.2023.6

Bykowa, E., Skachkova, M., Raguzin, I., Dyachkova, I., & Boltov, M. (2022). Automation of Negative Infrastructural Externalities Assessment Methods to Determine the Cost of Land Resources Based on the Development of a "Thin Client" Model. *Sustainability (Basel)*, 14(15), 9383. 10.3390/su14159383

Calegari, A., de Araujo, A. G., Tiecher, T., Bartz, M. L. C., Lanillo, R. F., dos Santos, D. R., & Friedrich, T. (2020). No-till farming systems for sustainable agriculture in South America. *No-till Farming Systems for Sustainable Agriculture: Challenges and Opportunities,* 533-565.

Capolongo, D., Pennetta, L., Piccarreta, M., Fallacara, G., & Boenzi, F. (2008). Spatial and temporal variations in soil erosion and deposition due to land-levelling in a semi-arid area of Basilicata (Southern Italy). *Earth Surface Processes and Landforms*, 33(3), 364–379. 10.1002/esp.1560

Carpenter, T. M., & Georgakakos, K. P. (2004). Continuous streamflow simulation with the HRCDHM distributed hydrologic model. *Journal of Hydrology (Amsterdam)*, 298(1), 61–79. 10.1016/j.jhydrol.2004.03.032

Carranza, J. P., Piumetto, M. A., Lucca, C. M., & Da Silva, E. (2022). Mass appraisal as affordable public policy: Open data and machine learning for mapping urban land values. *Land Use Policy*, 119, 106211. 10.1016/j.landusepol.2022.106211

Central Committee of the Communist Party. (1956). *The Resolution of the VIII Congress of the Communist Party of Kazakhstan on the Report of the Central Committee of the Communist Party of Kazakhstan.* CCCP.

Central Statistical Directorate under the Council of Ministers of the USSR. (1967). *Strana Sovetov za 50-let. Sbornik statisticheskih materialov[The Country of Soviets for 50 Years. Collection of Statistical Materials*].

Central Statistical Office under the Council of Ministers of the Kazakh SSR. (1980). *Narodnoe hozyajstvo Kazahstana za 60 let: Stat. sb. [National economy of Kazakhstan for 60 years: Statistical collection].* CSO of the KazSSR.

Cheik, S., Shanbhag, R. R., Harit, A., Bottinelli, N., Sukumar, R., & Jouquet, P. (2019). Linking termite feeding preferences and soil physical functioning in Southern-Indian woodlands. *Insects*, 10(1), 4. 10.3390/insects1001000430621180

Chen, M. K., Wang, S. J., Chen, W. Q., Cao, R., Cao, Q. B., Wang, P., & Li, S. H. (2019). Effects of ant nesting on soil microbial biomass carbon and quotient in tropical forest of Xishuangbanna. *Ying Yong Sheng tai xue bao=. Journal of Applied Ecology*, 30(9), 2973–2982. 10.13287/J.1001-9332.201909.03631529872

Chen, M., Shao, M. A., Wei, X., Li, T., Zhang, Q., Sun, S., Yang, X., & Gan, M. (2022). Earthworm (Metaphire guillelmi) activities increase the risk of soil erosion–a simulation experiment. *Earth Surface Processes and Landforms*, 47(7), 1734–1743. 10.1002/esp.5343

Chen, Q. L., Hu, H. W., Yan, Z. Z., Li, C. Y., Nguyen, B. A. T., Zheng, Y., Zhu, Y.-G., & He, J. Z. (2021). Termite mounds reduce soil microbial diversity by filtering rare microbial taxa. *Environmental Microbiology*, 23(5), 2659–2668. 10.1111/1462-2920.1550733817921

Chenyin, P., Yu, W., Fenghou, S., & Yongbao, S. (2023). Review of the Current Research Progress of Seed Germination Inhibitors. *Horticulturae, 9*(4), 462. 10.3390/horticulturae9040462

Chernenko, K. U., & Smirtyukov, M. S. (Eds.). (1968). *Resheniya partii i pravitelstva po hozyajstvennym voprosam: Sb. dok. za 50 let. T. 5. 1962-1965 gg[Decisions of the Party and government on economic issues: collection of documents for 50 years. Volume 5. 1962-1965].* Politizdat.

Chibilev, A.A. (2014). Ekologicheskie uroki celiny [The ecological lessons of Tselina]. *Voprosy stepevedeniya,* 11, 91-95.

Chinnasamy, P., & Honap, V. U. (2023). Spatiotemporal variations in soil loss across the biodiversity hotspots of Western Ghats Region, India. *Journal of Earth System Science*, 132(2), 90. 10.1007/s12040-023-02098-x

Choi, H. O. (2020). An Evolutionary Approach to Technology Innovation of Cadastre for Smart Land Management Policy. *Land (Basel)*, 9(2), 50. 10.3390/land9020050

Choker, H. M., & Mustafin, M. (2021). Geodesic support of laser scan technology use for fixing cultural heritage objects. *Geodeziâ I Kartografiâ, 968*(2), 2–10. 10.22389/0016-7126-2021-968-2-2-10

Chown, S. L., Bergstrom, D. M., Houghton, M., Kiefer, K., Terauds, A., & Leihy, R. I. (2022). Invasive species impacts on sub-Antarctic Collembola support the Antarctic climate-diversity-invasion hypothesis. *Soil Biology & Biochemistry*, 166, 108579. 10.1016/j.soilbio.2022.108579

Christiansen, M., Laursen, M., Jørgensen, R., Skovsen, S., & Gislum, R. (2017). Designing and Testing a UAV Mapping System for Agricultural Field Surveying. *Sensors (Basel)*, 17(12), 2703–2715. 10.3390/s1712270329168783

Chuyev, F. (1991). *140 besed s Molotovym* [140 Conversations with Molotov].

Climate of St. Petersburg and Leningrad region. (n.d.). METEO. http://www.meteo.nw.ru/articles/index.php?id=2

Cohen, A., & Dalyot, S. (2021). Route planning for blind pedestrians using OpenStreetMap. *Environment and Planning. B, Urban Analytics and City Science*, 48(6), 1511–1526. 10.1177/2399808320933907

Colomina, I., & Molina, P. (2014). Unmanned aerial systems for photogrammetry and remote sensing: A review. *ISPRS Journal of Photogrammetry and Remote Sensing*, 92, 79–97. 10.1016/j.isprsjprs.2014.02.013

Consonni, S., Giugliano, M., & Grosso, M. (2005). Alternative strategies for energy recovery from municipal solid waste Part B: Emission and cost estimates. *Waste Management (New York, N.Y.)*, 25(2), 137–148. 10.1016/j.wasman.2004.09.00615737711

Cours, J., Bouget, C., Barsoum, N., Horák, J., Le Souchu, E., Leverkus, A. B., Pincebourde, S., Thorn, S., & Sallé, A. (2023). Surviving in changing forests: Abiotic disturbance legacy effects on arthropod communities of temperate forests. *Current Forestry Reports*, 9(4), 189–218. 10.1007/s40725-023-00187-0

Cowell, C., Paton, A., Borrell, J. S., Williams, C., Wilkin, P., Antonelli, A., Baker, W. J., Buggs, R., Fay, M. F., Gargiulo, R., Grace, O. M., Kuhnhäuser, B. G., Woudstra, Y., & Kersey, P. J. (2022). Uses and benefits of digital sequence information from plant genetic resources: Lessons learnt from botanical collections. *Plants, People, Planet*, 4(1), 33–43. 10.1002/ppp3.10216

Crommelinck, S., Bennett, R., Gerke, M., Yang, M. Y., & Vosselman, G. (2017). Contour detection for UAV-Based cadastral mapping. *Remote Sensing (Basel)*, 9(2), 171. 10.3390/rs9020171

Dao, L. T. A., Nguyen, T. A., & Chandio, A. A. (2022). Climate change and its impacts on Vietnam agriculture: A macroeconomic perspective. *Ecological Informatics*, 74, 101960. 10.1016/j.ecoinf.2022.101960

De Feo, G., & Malvano, C. (2009). The use of LCA in selecting the best MSW management system. *Waste Management (New York, N.Y.)*, 29(6), 1901–1915. 10.1016/j.wasman.2008.12.02119168344

de Kleyn, L., Mumaw, L., & Corney, H. (2020). From green spaces to vital places: Connection and expression in urban greening. *The Australian Geographer*, 51(2), 205–219. 10.1080/00049182.2019.1686195

de Mévius, J. (2022). Impact of the European Union's Seed Legislation and Intellectual Property Rights on Crop Diversity. *European Energy and Environmental Law Review, 31*(3).

De Vries, W., Schulte-Uebbing, L., Kros, H., Voogd, J. C., & Louwagie, G. (2021). Spatially explicit boundaries for agricultural nitrogen inputs in the European Union to meet air and water quality targets. *The Science of the Total Environment*, 786, 147283. 10.1016/j.scitotenv.2021.14728333958210

Decree of the Government of the Russian Federation dated 30.07.2010 No. 1292-r (ed. dated 30.05.2014) "On approval of the Concept for the development of state monitoring of agricultural lands and lands used or provided for agriculture as part of lands of other categories, and the formation of state information resources about these lands for the period up to 2020" Retrieved from https://lenoblinvest.ru

Decree of the Governor of the Leningrad Region dated October 22, 2021 N 100-pg "On approval of the scheme for the placement, use and protection of hunting grounds on the territory of the Leningrad Region. Retrieved from http://publication.pravo.gov.ru/document/4700202110260004

Decree of the President of the Russian Federation dated October 10, 2019 No. 490. On the development of artificial intelligence in the Russian Federation. https://base.garant.ru/72838946

Deguine, J. P., Aubertot, J. N., Flor, R. J., Lescourret, F., Wyckhuys, K. A., & Ratnadass, A. (2021). Integrated pest management: Good intentions, hard realities. A review. *Agronomy for Sustainable Development*, 41(3), 38. 10.1007/s13593-021-00689-w

Derbentseva, A. M., Chernovalova, A. V., Surzhik, M. M., Nesterova, O. V., Semal, V. A., Ribachuk, N. A., & Mayorova, L. P. (2015). Podgorodenka natural-technogenic soil catena: Morphological, physicomechanical, and chemical properties. *Contemporary Problems of Ecology*, 8(1), 99–111. 10.1134/S1995425515010023

Despotović, J., Rodić, V., & Caracciolo, F. (2021). Farmers' environmental awareness: Construct development, measurement, and use. *Journal of Cleaner Production*, 295, 126378. 10.1016/j.jclepro.2021.126378

Deublein, D., & Steinhauser, A. (2011). Biogas from Waste and Renewable Resources. (2nd ed.) Deggendorf: Wiley VCH.

Devaux, A., Kromann, P., & Ortiz, O. (2014). Potatoes for Sustainable Global Food Security. *Potato Research*, 57(3–4), 185–199. 10.1007/s11540-014-9265-1

Dhankhar, N., & Kumar, J. (2023). Impact of increasing pesticides and fertilizers on human health: A review. *Materials Today: Proceedings*. 10.1016/j.matpr.2023.03.766

Digital Agriculture tecnnology, *Adaption & Attitudes,* November, 2019, 68. European Soil Data Center. Coordination of the LUCAS soil module. Retrieved from https://esdac.jrc.ec.europa.eu/euso/soil-monitoring

Dobrovolsky, G. V. (2002). *Soil degradation and protection.* Moscow: Moscow State University Publishing House.

Dobrovolsky, G. V., & Urusevskaya, I. S. (2015). *Geography of soils*. Moscow State University.

Dolgikh, N. U., & Kusticova, M. A. (2019). Ecological monitoring of specially protected natural areas of the Russian Arctic. *IOP Conference Series. Earth and Environmental Science*, 272(2), 022171. Advance online publication. 10.1088/1755-1315/272/2/022171

Dos Santos, T. B., Do Nascimento, A. P. B., & De Mora Regis, M. (2019). Green areas and quality of life: Use and environment perception of an urban park in São Paulo city, Brazil. *Revista de Gestao Ambiental e Sustentabilidade*, 8(2). 10.5585/geas.v8i2.1316

Dospekhov, B.A. (1985). *Technique of field experiment*. M.-L.

Dudkina, G. A., Pavlova, M. V., Rey, Z. G., & Khvalnova, A. T. (2007). *English for business communication: In 2 vol*. Filomatis.

Duru, M., Therond, O., Martin, G., Martin-Clouaire, R., Magne, M. A., Justes, E., Journet, E. P., Aubertot, J. N., Savary, S., Bergez, J. E., & Sarthou, J. P. (2015). How to implement biodiversity-based agriculture to enhance ecosystem services: a review. *Agronomy for Sustainable Development, 35*(4), 1259–1281. 10.1007/s13593-015-0306-1

Easter, D. (2021). Nikita Khrushchev and the Compromise of Soviet Secret Intelligence Sources. *International Journal of Intelligence and CounterIntelligence*, 35(3), 1–23. 10.1080/08850607.2021.1874191

Edygenov, N. E., Kolodin, F. I., & Mihajlov, F. K. (Eds.). (1959). *Narodnoe dvizhenie za osvoenie celinnyh zemel v Kazahstane. Sb. materialov i dokumentov[People's Movement for the Development of Virgin Lands in Kazakhstan. Collection of Materials and Documents.].* Gospolitizdat.

Eikhe, E. P. (1957). *Jerusalem artichoke or earthen pear*. M.L.

Eikhe, E. P. (1957). *Jerusalem artichoke or earthen pear.* M.-L., Publishing House of the USSR Academy of Sciences.

El-Bassiony, M. F., El-Hawy, A. S., El-Lamei, A. M., & Abd-Elazem, R. A. E. (2020). Conservation Agriculture and Integrated Pest Management Practices Improve Yield and Income while Reducing Labor, Pests, Diseases and Chemical Pesticide Use in Smallholder Vegetable Farms in Nepal. *Sustainability, 12*(16), 6418. 10.3390/su12166418

Eleuov, T. (Ed.). (1960). *Iz istorii socialisticheskogo stroitelstva v Kazahstane. T.9[From the History of Socialism Construction in Kazakhstan. V.9.].* Izdatelstvo Akademii Nauk KazSSR.

Emmanuel, I., Andrieu, H., Leblois, E., Janey, N., & Payrastre, O. (2015). Influence of rainfall spatial variability on rainfall–runoff modelling: Benefit of a simulation approach? *Journal of Hydrology (Amsterdam)*, 531, 337–348. 10.1016/j.jhydrol.2015.04.058

Engels, J. M., & Ebert, A. W. (2021). A critical review of the current global ex situ conservation system for plant agrobiodiversity. I. History of the development of the global system in the context of the political/legal framework and its major conservation components. *Plants*, 10(8), 1557. 10.3390/plants1008155734451602

Engelund, F., & Hansen, E. (1967). *A monograph on sediment transport in alluvial streams. Technical University of Denmark 0stervoldgade 10*. Copenhagen K.

Ergashev, A. (1976). Intensity and dynamics of formation of photosynthesis products in Jerusalem artichoke. *Physiol. and biochem. cult. rast.*, *3*(8), 299-303.

Ergashev, A., Partoev, K., & Safarmadi, M. (2019). *Productivity of Jerusalem artichoke varieties*. International scientific and practical conference Current tasks of efficient use of land in the context of innovative development of the agricultural sector in the context of global climate change, Dushanbe.

Erhart, E., & Hartl, W. (2010). *Compost Use in Organic Farming*. Springer. 10.1007/978-90-481-8741-6_11

Erlwein, A. (2022). Exploring Ecosystems Health: Effects of Increments of Biodiversity and Trophic Complexity on the Stability of a Simple Gaian Ecosystem Model. *Agro Sur*, 50(2), 13–24. 10.4206/agrosur.2022.v50n2-02

Ertiban, S. M. (2019). Soil fauna as webmasters, engineers and bioindicators in ecosystems: Implications for conservation ecology and sustainable agriculture. *American Journal of Life Sciences*, 7(1), 17–26. 10.11648/j.ajls.20190701.14

Esquinas-Alcázar, J., Hilmi, A., & Noriega, I. L. (2012). A brief history of the negotiations on the International Treaty on Plant Genetic Resources for Food and Agriculture. In *Crop Genetic Resources as a Global Commons* (pp. 135–149). Routledge.

FAO. (2006) *World reference base for soil resources 2006: a framework for international classification, correlation and communication*. FAO, 145.

Federal Law. "On the transfer of lands or land plots from one category to another" dated December 21, 2004 N 172-FZ (latest edition). Retrieved from https://www.consultant.ru/document/cons_doc_LAW_50874

Feldman, A. D. (2000). Hydrologic modeling system HEC-HMS: technical reference manual. US Army Corps of Engineers, Hydrologic Engineering Center.

Fiener, P., & Auerswald, K. (2009). Spatial variability of rainfall on a sub-kilometre scale. *Earth Surface Processes and Landforms*, 34(6), 848–859. 10.1002/esp.1779

Fierer, N. (2017). Embracing the unknown: Disentangling the complexities of the soil microbiome. *Nature Reviews. Microbiology*, 15(10), 579–590. 10.1038/nrmicro.2017.8728824177

Filippov, M. I. (1970). Jerusalem artichoke – a valuable forage crop. *Abstracts. V symposium for new silage plants. Part 1*. L.: Kolos.

Finger, A., Smedley, M., Halley, D., Frachon, N., & Ahrends, A. (2023). Managing small populations—Using genetic data and trial translocations to help inform suitable conservation measures for the alpine blue-sowthistle (Cicerbita alpina) in Scotland. *Environmental Research. Ecology*, 2(2), 025004. 10.1088/2752-664X/acdef9

FOEN (Ed.). (2017). Soils in Switzerland, status and development, status 2017, Federal Office for the Environment, Bern. *FOEN, 2020, Swiss Soil Strategy for the Sustainable Use of Soils [Swiss Soil Strategy for the Sustainable Use of Soils]*. Federal Office for the Environment. https://www.bafu.admin.ch/bafu/de/home/themen/boden /publicationsstudies/publications/bodenstrategie-schweiz.html

Fragou, S., Kalogeropoulos, K., Stathopoulos, N., Louka, P., Srivastava, P. K., Karpouzas, S., Kalivas, D. P., & Petropoulos, G. (2020). Quantifying Land Cover Changes in a Mediterranean Environment Using Landsat TM and Support Vector Machines. *Forests*, 11(7), 750. 10.3390/f11070750

Fralenko, V.P. (2014) Methods for texture analysis of images, processing of Earth remote sensing data. *Software systems: theory and applications, 4*(22), 19–39.

Frot, E., van Wesemael, B., Benet, A., & House, M. A. (2008). Water harvesting potential in function of hillslope characteristics: A case study from the Sierra de Gador (Almeria province, south-east Spain). *Journal of Arid Environments*, 72(7), 1213–1231. 10.1016/j.jaridenv.2007.12.009

Fuentes-Peñailillo, F., Gutter, K., Vega, R., & Silva, G. C. (2024). New Generation Sustainable Technologies for Soilless Vegetable Production. *Horticulturae, 10*(1), 49. 10.3390/horticulturae10010049

Fuller, D. Q., Denham, T., & Allaby, R. (2023). Plant domestication and agricultural ecologies. *Current Biology*, 33(11), R636–R649. 10.1016/j.cub.2023.04.03837279694

Garcia-Ruiz, J. M., White, S. M., Lasanta, T., Marti, C., Gonzalez, C., Errea, M. P., Valero, B., & Ortigosa, L. (1997). Assessing the effects of land-use changes on sediment yield and channel dynamics in the central Spanish Pyrenees. *IAHS Publications-Series of Proceedings and Reports-Intern Assoc Hydrological Sciences*, 245, 151–158.

Garrett, H. E. G., Wolz, K. J., Walter, W. D., Godsey, L. D., & McGraw, R. L. (2021). Alley cropping practices. *North American agroforestry*, 163-204.

Gavrilova, O. I., & Shchukin, P. O. (2019). Characteristics of agricultural soils in the northwestern region of Russia, *Novaum.ru, 18*, 54-56. https://www.elibrary.ru/item.asp?id=38206285

Gerasimova, M.I. (2019). International classification of soils and the possibilities of its application in geographical research, *Bulletin of Moscow University, 5*, 49 – 56.

Gerlich, H. S., Holmstrup, M., Schmidt, N. M., & Høye, T. T. (2023). Phenological sensitivity of climate across taxa and local habitats in a high-Arctic arthropod community. https://doi.org/10.32942/X24590

Ghimire, U., Agarwal, A., Shrestha, N. K., Daggupati, P., Srinivasan, G., & Than, H. H. (2020). Applicability of lumped hydrological models in a data-constrained river basin of Asia. *Journal of Hydrologic Engineering*, 25(8), 05020018. 10.1061/(ASCE)HE.1943-5584.0001950

Ghinea, C., & Gavrilescu, M. (2010). Decision support models for solid waste management – An overview. *Environmental Engineering and Management Journal*, 9(6), 869–880. 10.30638/eemj.2010.116

Gibbons, W. R. (1989). Batch and continuous solid-phase fermentation of Jerusalem artichoke tubers. *Journal of Fermentation Technology*, 67(4), 258–265.

Gindemit, A., & Aksenova, Yu. (2021). Evaluation of soils of agrocenoses in the forest-steppe zone of the Omsk Irtysh region by the level of cultivation and fertility. *IOP Conference Series. Earth and Environmental Science*, 659(1), 012102. 10.1088/1755-1315/659/1/012102

Giniyatov, I. A. (2022). On the issue of basic concepts in the field of land management, real estate cadastre and land monitoring (in the order of discussion). *Bulletin of SGUGiT*, 27(6), 152–159. 10.33764/2411-1759-2022-27-6-152-159

Gliessman, S. (2018). Defining Agroecology. *Agroecology and Sustainable Food Systems*, 42(6), 599–600. 10.1080/21683565.2018.1432329

Gobattoni, F. (2009). *Hydrological models for watershed planning and water resources management: a Decision System Support for Marta River Basin.* Central Italy.

Godin, V.V., Belousova, M.N., Belousov, V.A., & Terekhova, A.E. (2020) *Agriculture in the digital era: challenges and solutions Electronic management in industries*, no. 1, 4–15

Golubev, V.N., Volkova, I.V., Kushalakov, Kh.M. (1995). *Jerusalem artichoke, properties, composition, processing methods, applications.* M.-L.

Goncharov, A. A., Leonov, V. D., Rozanova, O. L., Semenina, E. E., Tsurikov, S. M., Uvarov, A. V., Zuev, A. G., & Tiunov, A. V. (2023). A meta-analysis suggests climate change shifts structure of regional communities of soil invertebrates. *Soil Biology & Biochemistry*, 181, 109014. 10.1016/j.soilbio.2023.109014

González-Toral, C., & Cires, E. (2022). Relevance of DNA preservation for future botany and ecology. *Molecular Ecology*, 31(20), 5125–5131. 10.1111/mec.1665236214196

Goodrich, D. C., Faurès, J.-M., Woolhiser, D. A., Lane, L. J., & Sorooshian, S. (1995). Measurement and analysis of small-scale convective storm rainfall variability. *Journal of Hydrology (Amsterdam)*, 173(1), 283–308. 10.1016/0022-1694(95)02703-R

Goodrich, D., Unkrich, C., Smith, R., Guertin, D., Hernandez, M., Burns, I., Massart, J., Levick, L., Miller, S., & Semmens, D. (2010). The AGWAKINEROS2 suite of modeling tools.In: (eds). *Proc. 2010 Watershed Management Conf.* 10.1061/41143(394)116

Gorbatov, V. I. (2022). Ecological education of students during foreign language classes. *Ecological education and ecological culture of the population: materials of the X international scientific conference on February 25–26, 2022.* Prague: Vědecko vydavatelské centrum «Sociosféra-CZ».

Gorbovskaya, A. D., & Gonina, A. V. (2015). The need to use waste for energy production in the Udmurt Republic. *Eurasian Union of Scientists*, 11(20), 6–8.

Görres, J. H., & Amador, J. A. (2021). The soil fauna. In *Principles and applications of soil microbiology* (pp. 191–212). Elsevier. 10.1016/B978-0-12-820202-9.00008-3

GOST R 1.0-2004. *Standardization in the Russian Federation. Basic provisions.* https://docs.cntd.ru/document/1200038794.

GOST R 53691-2009. *Resource conservation. Waste management. Waste certificate for hazard class 1-4. Basic requirements.*

GOST R 54098-2010. Resource conservation. Secondary material resources. Terms and definitions. https://docs.cntd.ru/document/1200084875#7D20K3.

Greenbaum, N., Ben-Zvi, A., Haviv, I., & Enzel, Y. (2006). *The hydrology and paleohydrology of the Dead Sea tributaries* (Vol. 401). Geological Society of America Special Papers. 10.1130/2006.2401(05)

Greenwald, O., Islamov, R., & Sergeychick, T. (2017). Mastering Foreign Language Competence of Ecology and Environment Managers for Mining Industry of Kuzbass. *The Second International Innovative Mining Symposium. E3S Web of Conferences 21, 04012.* IEEE. 10.1051/e3sconf/20172104012

Grilli, E., Carvalho, S. C. P., Chiti, T., Coppola, E., D'Ascoli, R., La Mantia, T., Marzaioli, R., Mastrocicco, M., Pulido, F., Rutigliano, F. A., Quatrini, P., & Castaldi, S. (2021). Critical range of soil organic carbon in southern Europe lands under desertification risk. *Journal of Environmental Management*, 287, 112285. 10.1016/j.jenvman.2021.11228533725659

Guerra, C., Bardgett, R. D., Caon, L., Crowther, T. W., Delgado-Baquerizo, M., Montanarella, L., Navarro, L. M., Orgiazzi, A., Singh, B. K., Tedersoo, L., Vargas-Rojas, R., Briones, M. J. I., Buscot, F., Cameron, E. K., Cesarz, S., Chatzinotas, A., Cowan, D. A., Djukic, I., van den Hoogen, J., & Eisenhauer, N. (2021). Tracking, targeting, and conserving soil biodiversity. *Science*, 371(6526), 239–241. 10.1126/science.abd792633446546

Gulakov, V.K., Trubakov, A.O., Ogurtsov, S.N. (2011) Informative significance of texture characteristics based on the adjacency matrix of image pixel brightness levels. *Bulletin of Bryansk State Technical University, 2*(30), 81.

Gulde, S., Chung, H., Amelung, W., Chang, C., & Six, J. (2008). Soil carbon saturation controls labile and stable carbon pool dynamics. *Soil Science Society of America Journal*, 72(3), 605–612. 10.2136/sssaj2007.0251

Gunathilaka, R. P. D., James, C. R. S., & Christopher, M. F. (2018). Adaptation to climate change in perennial cropping systems: Options, barriers and policy implications. *Environmental Science & Policy*, 82, 108–116. 10.1016/j.envsci.2018.01.011

Gurevich P.A., Shavaleeva S.M., Glebov A.N., Bayanova L.N. (2013) Chemical corrosion of colleate glass and glass container waste as a factor of negative influence on the environment. *Bulletin of Kazan Technological University, 4.*

Gusev, E. A. (2022). Results and prospects of geological mapping of the Arctic shelf of Russia. *Journal of Mining Institute*, 255, 290–298. 10.31897/PMI.2022.50

Hadipour-Rokni, R., Askari Asli-Ardeh, E., Jahanbakhshi, A., Esmaili paeen-Afrakoti, I., & Sabzi, S.Ramazan Hadipour-Rokni. (2023). Intelligent detection of citrus fruit pests using machine vision system and convolutional neural network through transfer learning technique. *Computers in Biology and Medicine*, 155, 106611. 10.1016/j.compbiomed.2023.10661136774891

Hansen, Z., & Libecap, G. (2004). Small Farms, Externalities, and the Dust Bowl of the 1930s. *Journal of Political Economy*, 112(3), 665–694. 10.1086/383102

Haralik, R. M. (1979). Statistical and structural approaches to the description of textures. *TIIER*, 67(5), 98–120.

Hawes, C., Iannetta, P. P., & Squire, G. R. (2021). Agroecological practices for whole-system sustainability. *CABI Reviews*.

He, D., Yang, R., & Cao, J. (2023). The relationship between interannual variability of winter surface solar radiation over the Southeast Asian low-latitude highlands and the circumglobal teleconnection. *Atmospheric Research*, 287, 106694. 10.1016/j.atmosres.2023.106694

Henry, R. J. (2020). Innovations in plant genetics adapting agriculture to climate change. *Current Opinion in Plant Biology*, 56, 168–173. 10.1016/j.pbi.2019.11.00431836470

Hirte, J., Richner, W., Orth, B., Liebisch, F., & Flisch, R. (2021). Yield response to soil test phosphorus in Switzerland: Pedoclimatic drivers of critical concentrations for optimal crop yields using multilevel modelling. *The Science of the Total Environment*, 755, 143453. 10.1016/j.scitotenv.2020.14345333199000

Hoang, C. T., Bui, T. P., & Duong, V. H. (2021). Species biodiversity and distribution of seagrass beds in several coastal areas of central Vietnam. *Regional Studies in Marine Science*, 41, 101531. 10.1016/j.rsma.2020.101531

Hoare, R., Foggin, C., Alexander, S., & Chakombera, D. (2022). Combating the use of cyanide poison in ivory poaching. *African Journal of Wildlife Research*, 52(1), 90–98. 10.3957/056.052.0090

Honkavaara, E., Markelin, L., Hakala, T., & Peltoniemi, J. (2014). The Metrology of Directional, Spectral Reflectance Factor Measurements Based on Area Format Imaging by UAVs. *Photogrammetrie, Fernerkundung, Geoinformation*, 14, 175–188. 10.1127/1432-8364/2014/0218

Honkavaara, E., Saari, H., Kaivosoja, J., Pölönen, I., Hakala, T., Litkey, P., Makynen, J., & Pesonen, L. (2013). Processing and Assessment of Spectrometric, Stereoscopic Imagery Collected Using a Lightweight UAV Spectral Camera for Precision Agriculture. *Remote Sensing (Basel)*, 5(10), 5006–5039. 10.3390/rs5105006

Horn, R. (2022). Soil health and biodiversity: interactions with physical processes and functions. In Reyes Sanchez, L. (Eds.), *Sustainable soil management as a key to preserve soil biodiversity and stop its degradation.* International Union of Soil Sciences.

Ibrahim, A., Bartsch, K., & Sharifi, E. (2020). Green infrastructure needs green governance: Lessons from Australia's largest integrated stormwater management project, the River Torrens Linear Park. *Journal of Cleaner Production*, 261, 121202. 10.1016/j.jclepro.2020.121202

Ibrahim, Y., Arafat, H. A., Mezher, T., & AlMarzooqi, F. (2018). An integrated framework for sustainability assessment of seawater desalination. *Desalination*, 447, 1–17. 10.1016/j.desal.2018.08.019

Il Tumen's press service. (2016). *The Government of the Russian Federation has begun work on clarifying the composition of the Arctic zone of the country*. The website of the State Assembly (Il Tumen). http://old.iltumen.ru

Isakova A.V. (2020) Legal support for industrial environmental control from the investment project to operation. *Current issues of Russian law, 15*(116), 207-217.

Islamov, R. S., Resenchuk, A. A., & Perevalova, A. A. (2018) From the experience of teaching English as a foreign language to students in the fields of "Geology", "Biology", "Ecology and Nature and Management". *Modern problems of science and education*.https://science-education.ru/ru/article/view?id=28417 (дата обращения: 09.03.2024).

Issanova, G., Saduakhas, A., Abuduwaili, J., Tynybayeva, K., & Tanirbergenov, S. (2020). Desertification and land degradation in Kazakhstan. *Bulletin of National Academy of Sciences of the Republic of Kazakhstan.*, 5(387), 387. 10.32014/2020.2518-1467.148

Ityshev I.K., Potapova S.O. (2018) On the problems of fire safety of solid household waste and their storage places. *Fire safety: problems and prospects*, 9(1), 292-300.

Ivanova, I. E., Priss, O. P., Taranenko, G. G., Shleina, L. I., Chebanova, Yu. V., Evstafieva, E. S., Krasulya, Ya. G., & Kolomoets-Garkusha, O. V. (2020). *Trajectory of development of "SOFT SKILLS" in students of the Faculty of Agrotechnologies and ecology as one of the important factors of formation of professional skills of youth. Improvement of educational process in the institution of Higher Education. Issue 24, collection of scientific and methodological works. Tavrichesky State agrotechnological University named after Dmitry Motorny*. Melitopol.

Jaganmohan, M., Knapp, S., Buchmann, C. M., & Schwarz, N. (2016). The Bigger, the Better? The Influence of Urban Green Space Design on Cooling Effects for Residential Areas. *Journal of Environmental Quality*, 45(1), 134–145. 10.2134/jeq2015.01.006226828169

Jahoni, A. (2010). *The medicins and food plants*. Dushanbe, Adib.

Jeanneret, P., Aviron, S., Alignier, A., Lavigne, C., Helfenstein, J., Herzog, F., Kay, S., & Petit, S. (2021). Agroecology landscapes. *Landscape Ecology, 36*(8), 2235–2257. 10.1007/s10980-021-01248-0

Jiang, L., Jiapaer, G., Bao, A., Kurban, A., Guo, H., Zheng, G., & De Maeyer, P. (2019). Monitoring the long-term desertification process and assessing the relative roles of its drivers in Central Asia. *Ecological Indicators*, 104, 195–208. 10.1016/j.ecolind.2019.04.067

Johnson, D. L., & Miller, A. C. (1997). A spatially distributed hydrologic model utilizing raster data structures. *Computers & Geosciences*, 23(3), 267–272. 10.1016/S0098-3004(96)00084-2

Joint Research Centre (JRC) of the European Commission, Zarco-Tejada, P.J., Hubbard, N., Loudjani, P. (2014). *Precision Agriculture: An Opportunity for EU Farmers—Potential Support with the CAP 2014–2020*. European Parliament; Brussels, Belgium:. Agriculture and Rural Development; Policy Department B: Structural and Cohesion Policies European Union.

Joseph, J., van Dijk, M., & Krisztin, T. (2023). Do Large Estates Benefit Smallholder Neighbours? Evidence from Malawi. *The Journal of Development Studies*, 59(4), 570–592. 10.1080/00220388.2022.2147831

Jug, I., Brozović, B., Đurđević, B., Wilczewski, E., Vukadinović, V., Stipešević, B., & Jug, D. (2021). Response of crops to conservation tillage and nitrogen fertilization under different agroecological conditions. *Agronomy (Basel)*, 11(11), 2156. 10.3390/agronomy11112156

Kabonen, A.V., & Ivanova, N.V. (2023). Assessment of biometric characteristics of trees according to ground LiDAR and multi-season aerial photography in artificial plantations. *Nature Conservation Research. Reserved science, 8*(1), 64-83. 10.24189/ncr.2023.005

Kalin, L., & Hantush, M. H. (2006). Comparative assessment of two distributed watershed models with application to a small watershed. *Hydrological Processes*, 20(11), 2285–2307. 10.1002/hyp.6063

Karavaeva, N. A. (2008). *Agrogenic memory of soils. Soil as a memory of biosphere-theosphere-anthropogenic interactions.* M., LKI.

Karavanova, E. I. (2003). *Optical properties of soils and their nature.* M.: Moscow State University Publishing House, 187.

Karichkovskaya S. P., Chuchmiy I. I. (2021). Features of Studying Ukrainian and Foreign Languages in Agricultural Institutions of Higher Education in Ukraine. *Academic Studios. Series "pedagogy", 4*(1), 91-98.

Karpachevskii, L.O. (2005). *Ecological Soil Science.* GEOS, Moscow.

Kartashev, A. G. (2000). *Electromagnetic ecology.* TSU.

Kartashova, E. A. (2009). Teaching a foreign language to future ecologists as carriers of a new type of ecological culture. *Bulletin of Kostroma State University Series: Pedagogy. Psychology.Sociokinetics*, 15(3), 44–48.

Kaskevich, E. P., & Plekhanov, G. F. (2003). *Environmental characteristics of super- and ultra-high voltage power transmission.* International scientific and practical conference Transmission of energy by alternating current over long and ultra-long distances, Novosibirsk, Russia.

Kasten, F., & Young, A. T. (1989). Revised optical air mass tables and approximation formula. *Applied Optics*, 28(22), 4735–4738. 10.1364/AO.28.00473520555942

Kasymov, J. K. (1975). *Agricultural crops of Tajikistan.*

Kataev, M.Yu., Kartashov, E.Yu., & Smirnov, D.S. (2021). Methodology and program for atmospheric correction of images of unmanned aerial vehicles in the problem of safe vegetation location. *TUSUR Reports*, T.24, No. 4, 73–78. TUSUR.

Kataev, M.Yu., Kartashov, E.Yu., Ryabukhin, V.V., Makarov, E.V., & Pasko, O.A. (2023b) Technique for segmenting images of unmanned aerial vehicles using neural networks. *Modern problems of remote sensing of the Earth from space, 20*(1), 55–66.

Kataev, M.Yu., Kartashov, E.Yu., Ryabukhin, V.V., Makarov, E.V., Pasko, O.A. (2023b) Technique for segmenting images of unmanned aerial vehicles using neural networks. *Modern problems of remote sensing of the Earth from spac*e, *20*(1), 55–66.

Kataev, M.Yu., Skugarev, A.A., Sorokin, I.B. (2017) Possibilities of space monitoring for agricultural purposes in the Tomsk region. *TUSUR Reports,* T.20, No. 3, 186–190. TURSUR.

Kataev, M. Yu., Pasko, O. A., & Kartashov, E. Yu. (2023a). Analysis of the practical possibilities of using unmanned aerial vehicles in agriculture. *Bulletin of the Krasnoyarsk State Agrarian University*, 1(190), 54–62.

Khaled, A. Y., Aziz, S. A., Bejo, S. K., Nawi, N. M., Seman, I. A., & Izzuddin, M. A. (2018). *Dielectric constant and chlorophyll content measurements for basal stem rot (BSR) disease detection.* August 2018 International Conference on Signals and Systems (ICSigSys), Bali, Indonesia. 10.1109/ICSIGSYS.2018.8373570

Khalifaev, D. R., Khalifaev, D. R., & Popov, D. M. (2015). *Receiving and standardization of tincture of homeopathic matrix from fresh tubers of Jerusalem artichoke*. Modern aspects of the use of plant raw materials and raw materials of natural origin in medicine, Moscow.

Khalifaev, D. R., Safarzoda, R. Sh., & Gulmamadova, G. G. (2013). *About the technology for obtaining dry extract from the Helianthus tuberosus plant.* The contribution of medical sciences to practical healthcare, Dushanbe.

Khalifaev, D. R., Safarzoda, R. Sh., & Popov, D. M. (2014). *Amino acid composition of grass Jerusalem artichoke cultivated in Tajikistan*. Pharmacy development and registration of medicines, Moscow.

Khan, M. A., Ahmad, W., & Paul, B. (2018). Ecological impacts of termites. *Termites and Sustainable Management: Volume 1-Biology, Social Behaviour and Economic Importance*, 201-216. 10.1007/978-3-319-72110-1_10

Khan, N., Ray, R. L., Sargani, G. R., Ihtisham, M., Khayyam, M., & Ismail, S. (2021). Current Progress and Future Prospects of Agriculture Technology: Gateway to Sustainable Agriculture. *Sustainability, 13*(9), 4883. 10.3390/su13094883

Khan, A., & Sudheer, M. (2022). Machine learning-based monitoring and modeling for spatio-temporal urban growth of Islamabad. *The Egyptian Journal of Remote Sensing and Space Sciences*, 25(2), 541–550. 10.1016/j.ejrs.2022.03.012

Khlustov, V. K., Vasenev, V. V., & Ganokhin, A. M. (2022). The importance of climatic indicators on the productivity of pine forests. *Environmental Management*, 5, 121–131.

Kholodov, O. A. (2019). Integrated monitoring of agricultural land use in the modern period, *Bulletin of the Volgograd State University.Economy*, 3, 107–119.

Khrushchev, N.S. (1999). *Vremya. Lyudi. Vlast. Vospominaniya v 4 kn.[Time. People. Power. Memoirs in 4 books].* Moskovskiye Novosti.

Khrushchev, S. (Ed.). (2007). *Memoirs of Nikita Khrushchev: Statesman, 1953-1964*Shriver, G., & Shenfield, S., Trans.). Brown University.

Kiçaj, H. (2023). Ecology of soil animals (Diplopoda class, Myriapoda group). *Scientific Horizons.*, 26(5). Advance online publication. 10.48077/scihor5.2023.37

Killebrew, K., & Wolff, H. (2010). *Environmental impacts of agricultural technologies.* IDEAS. https://ideas.repec.org/p/udb/wpaper/uwec-2011-01.html

Király, É., Keserű, Z., Molnár, T., Szabó, O., & Borovics, A. (2023). Carbon Sequestration in the Aboveground Living Biomass of Windbreaks—Climate Change Mitigation by Means of Agroforestry in Hungary. *Forests*, 15(1), 63. 10.3390/f15010063

Kirillova, N. P., & Sileva, T. M. (2017). Colorimetric Analysis of Soils Using Digital Cameras. *Moscow University Soil Science Bulletin*, 72(1), 13–20. 10.3103/S0147687417010045

Kirsanova, A. Yu., & Filippova, T. M. (2008). Energy potential of landfill gas at the solid waste landfill in Angarsk. *AGTA Bulletin*, 1(2), 82–87.

Klausshofer, H., Abraham, B., & Leichtfried, G. (1998). Hydrolaseaktivität bei Topinambur und Zuckerind. *Zuckerind*, 113(3), 209–215.

Klimanova, O. A., & Illarionova, O. I. (2020). Green infrastructure indicators for urban planning: Applying the integrated approach for Russian largest cities. *Geography, Environment. Sustainability*, 13(1), 251–259.

Klimanova, O., Illarionova, O., Grunewald, K., & Bukvareva, E. (2021). Green infrastructure, urbanization, and ecosystem services: The main challenges for Russia's largest cities. *Land (Basel)*, 10(12), 1292. 10.3390/land10121292

Kobernichenko, V. G. (2016). *Radio-electronic systems for remote sensing of the Earth.* Ekaterinburg: Ural Publishing House.

Kochnev, N.K., & Kolinicheva, M.V. (2002). *Jerusalem artichoke - bioenergy culture of the XXI century.* M.-L.

Kogut, B. (2012). Assessment of the humus content in arable soils of Russia. *Eurasian Soil Science*, 45(9), 843–851. 10.1134/S1064229312090062

Komarov, A. A. (2018), Assessment of the state of the herbage using the vegetation index NDVI. *Proceedings of the St. Petersburg State Agrarian University,2,* 124-129.

Kondakov, A. M. (1988). Alternative energy sources. *Geography at school.* M. *Pedagogy*, 185.

Kondratenko V.V. (2008). Complex processing of Jerusalem artichoke into functional food products. *Innovative technologies in the field of refrigerated storage and food processing*, (132-133).

Kong, W., Huang, W., & Zhou, X. (2016). *Remote estimation of carotenoid/chlorophyll ratio in vertical layers using canopy multi-angle spectral data.* Fifth International Conference on Agro-Geoinformatics (Agro-Geoinformatics), Tianjin, China. 10.1109/Agro-Geoinformatics.2016.7577684

Kong, F., Yin, H., Nakagoshi, N., & Zong, Y. (2010). Urban green space network development for biodiversity conservation: Identification based on graph theory and gravity modeling. *Landscape and Urban Planning*, 95(1–2), 16–27. 10.1016/j.landurbplan.2009.11.001

Kopylova, N., & Starikov, I. P. (2021). Methods of displaying geospatial information using cartographic web technologies for the arctic region and the continental shelf. *Geodezia i Kartografia*, 971(5), 15–22. 10.22389/0016-7126-2021-971-5-15-22

Kornilov, Y. N., Romanchikov, A. Y., & Bogoliubova, A. A. (2023). Estimating deformation process through single shot close range photogrammetry method in Agisoft Metashape. *Geodezia I Kartografia*, 84(10), 2–11. 10.22389/0016-7126-2023-1000-10-2-11

Kosmas, C., Danalatos, N., Cammeraat, L. H., Chabart, M., Diamantopoulos, J., Farand, R., Gutierrez, L., Jacob, A., Marques, H., Martinez-Fernandez, J., Mizara, A., Moustakas, N., Nicolau, J. M., Oliveros, C., Pinna, G., Puddu, R., Puigdefabregas, J., Roxo, M., Simao, A., & Vacca, A. (1997). The effect of land use on runoff and soil erosion rates under Mediterranean conditions. *Catena*, 29(1), 45–59. 10.1016/S0341-8162(96)00062-8

Kostenko, I.V. (2009) Study of the optical properties of samples of soddy-steppe sandy soils in the south of Ukraine using a scanner. *Soil Science*, (9), 090–1098.

Kostyleva, N. V., & Racheva, N. L. (2016). *Characteristics of pollutants: reference book.* Federal State Budgetary Institution Ural Research Ecology Institute.

Kothencz, G., Kolcsár, R., Cabrera-Barona, P., & Szilassi, P. (2017). Urban green space perception and its contribution to well-being. *International Journal of Environmental Research and Public Health*, 14(7), 766. 10.3390/ijerph1407076628704969

Kovyazin, V. F., Nguyen, T. A., & Nguyen, T. T. (2023). Monitoring the forest fund lands of Kon Tum province, Vietnam using remote sensing data of Earth. *Geodezia I Kartografia*, 84(8), 57–64. 10.22389/0016-7126-2023-998-8-57-64

Kovyazin, V., Kitcenko, A., & Shobairi, S. O. R. (2021). Cadastral valuation of forest lands, taking into account the degree of development of their infrastructure. *Journal of Mining Institute*, 249, 449–462. 10.31897/PMI.2021.3.14

Koza, M., Funk, R., Pöhlitz, J., Conrad, Ch., Shibistova, O., Meinel, T., Akshalov, K., & Schmidt, G. (2024). Wind erosion after steppe conversion in Kazakhstan. *Soil & Tillage Research*, 236, 105941. 10.1016/j.still.2023.105941

Kozybayeva, M. M. (Ed.). (2022). Krestyanstvo Kazahstana v poslevoennoe desyatiletie: socialnaya transformaciya i povsednevnost. *Kollektivnaya monografiya* [The Peasantry of Kazakhstan in the Post-war Decade: Social Transformation and Everyday Life. Collective Monograph]. IP "Madiyar".

Kpienbaareh, D., Bezner Kerr, R., Luginaah, I., Wang, J., Lupafya, E., Dakishoni, L., & Shumba, L. (2020). Spatial and ecological farmer knowledge and decision-making about ecosystem services and biodiversity. *Land (Basel)*, 9(10), 356. 10.3390/land9100356

Kraemer, R., Prishchepov, A., Müller, D., Kuemmerle, T., Radeloff, V., Dara, A., Terekhov, A., & Fruehauf, M. (2015). Long-term agricultural land-cover change and potential for cropland expansion in the former virgin lands area of Kazakhstan. *Environmental Research Letters*, 10(5), 054012. 10.1088/1748-9326/10/5/054012

Kryvulchak, L. (2020). Creative Project as a Method of Forming Cultural Competence of Students during the Lessons of the International Movement. *The modern movement of science:abstracts of the 11th International Scientific and Practical Internet Conference.* Dnipro.

Kuhwald, M., Busche, F., Saggau, P., & Duttmann, R. (2022). Is soil loss due to crop harvesting the most disregarded soil erosion process? A review of harvest erosion'. *Soil & Tillage Research*, 215, 105213. 10.1016/j.still.2021.105213

Kulak, V., Longboat, S., Brunet, N. D., Shukla, M., & Saxena, P. (2022). In vitro technology in plant conservation: Relevance to biocultural diversity. *Plants*, 11(4), 503. 10.3390/plants1104050335214833

Kumar, V., & Vuilliomenet, A. (2021). Urban nature: Does green infrastructure relate to the cultural and creative vitality of European cities? *Sustainability (Basel)*, 13(14), 8052. 10.3390/su13148052

Kusainova, A., Mezentseva, O., & Tusupbekov, Zh. (2024). Trends of long-term fluctuations of atmospheric humidification of the growing season in the steppe zone of Northern Kazakhstan. *E3S Web of Conferences*. 10.1051/e3sconf/202449803009

Kysyydak, A. S., & Salchak, A. D. (2013). Energy potential of solid waste landfills. *Bulletin of the Tuva State Technical University*, 3, 34–39.

Lagacherie, P., McBratney, A. B., & Voltz, M. (2007). *Digital Soil Mapping: An introductory perspective*. Elsevier.

Lakatosh, M. A. (2021). *Professional training of future specialists in the agricultural sector in institutions of higher education in Ukraine* [Doctoral dissertation, Mukachevo State University.] Mukachevo.

Lakiza, N. V., & Neudachina, L. K. (2015) *Analysis of food products*. Ekaterinburg: Ural Publishing House.

Laliberte, A. S., Goforth, M. A., Steele, C. M., & Rango, A. (2011). Multispectral remote sensing from unmanned aircraft: Image processing workflows and applications for rangeland environments. *Remote Sensing (Basel)*, 3(11), 2529–2551. 10.3390/rs3112529

Lal, R. (2004). Soil carbon sequestration impacts on global climate change and food security. *Science*, 304(5677), 1623–1627. 10.1126/science.109739615192216

Lavallee, J. M., Soong, J. L., & Cotrufo, M. F. (2020). Conceptualizing soil organic matter into particulate and mineralassociated forms to address global change in the 21st century'. *Global Change Biology*, 26(1), 261–273. 10.1111/gcb.1485931587451

Laver O. V. (2022). *Business Ukrainian: methodological developments of final control works for applicants of higher education in the field of knowledge 19 " Architecture and Construction" specialty 193 "Geodesy and land management" of the educational and professional program "Geodesy and land management"*. Uzhgorod, 52 p.

Lazariev, O., Fernos, Yu., Movchan, L., Komisarenko, N., Tymchuk, S., & Neshchadym, L. (2019). *Definition of a Process of Forming Professional Communicative Competence of the Future Agrarian Experts.* Vision 2025. Education Excellence and Management Innovations through Sustainable Economic Competitive Advantage: Proceedings of the 34-th IBIMA Conf., Madrid, Spain.

Le Jing, L. Y., Hua, D., Tan, L., & Ning, C. (2011). Research on method of LED-induced chlorophyll fluorescence spectrum and image information acquisition. *IEEE 2011 10th International Conference on Electronic Measurement & Instruments.* IEEE. 10.1109/ICEMI.2011.6037913

Le Texier, M., Schiel, K., & Caruso, G. (2018). The provision of urban green space and its accessibility: Spatial data effects in Brussels. *PLoS One*, 13(10), e0204684. 10.1371/journal.pone.020468430332449

Lebed, O. M., Bokshan, G. I., Kaminskaya, M. O., & Makukhina, S. V. (2019). *Dictionary-a manual for the development of professionally-oriented communicative competence of students of the specialty "Geodesy and Land Management".* Kherson: HAU "Kolos".

Lebedeva N. A. (2017). Information aspects in the context of innovative means of teaching a foreign language. *All-Ukrainian science and practice internet conference "Modern management: problems and development prospects on May 19, 2017.* Kherson: KhDAU, 235-238.

Lebedeva, N. (2019). Information and communication technologies for teaching a foreign language regarding the modernization of public administration. *The latest information and communication technologies in the modernization of public administration: foreign and domestic experience: materials of II scientific practice seminar.* Dnipropetrovsk DRIDU NADU, 55-58.

Lebedeva, N. A. (2018). Teaching English in the field of knowledge "Public management and administration" in an agrarian higher educational institution. *European dimension of reforming public administration in Ukraine: materials of the International Scientific and Practical Conference.* Kyiv "Publishing House "Personal".

Lebedeva, N. A. (2022). Teaching of Foreign Language in the Educational Process of Public Administration Personnel in Ukraine. In Kulaç, O., Babaoğlu, C., & Akman, E. (Eds.), *Public Affairs Education and Training in the 21st Century* (pp. 294–310). IGI Global. 10.4018/978-1-7998-8243-5.ch019

Lebedeva, N. A. (2022). *Work program on the academic discipline "Foreign Language" for the specialties of land management and cadastre, ecology of the Kherson Agronomic University of the Master's educational qualification level (according to the credit-modular training system).* HAU.

Lebedeva, N. A. (2023). ESP in the Field of Architecture: Features of Teaching Writing and Some Aspects of Error Analysis. In Lebedeva, N. (Ed.), *Cases on Error Analysis in Foreign Language Technical Writing* (pp. 97–123). IGI Global. 10.4018/978-1-6684-6222-5.ch005

Lebed, O. M., & Chekanovich, V. G. (2011). *Working program on the discipline "foreign language" for the specialties of Land Management and cadastre, ecology of the Kherson agronomic University of the ROC "master" (according to the credit-modular training system).* HAU.

Lebid, O. M., & Lebedeva, N. A. (2018). *Working program "Foreign language for professional purposes" for applications for higher education, studying in the educational program "Bachelor" in the specialty "6.060103. Hydraulic Engineering (Water Resources)".* KSAU, Ukraine.

Lefevre, C., Rekik, F., Viridiana, A., & Liesl, W. (2017). *Soil Organic Carbon. The Hidden Potential, Food and Agriculture Organization of the United Nations.* FAO.

Lehmann, J., Bossio, D. A., Kögel-Knabner, I., & Rillig, M. C. (2020). The concept and future prospects of soil health'. *Nature Reviews. Earth & Environment*, 1(10), 544–553. 10.1038/s43017-020-0080-833015639

Lelong, C. C. D., Burger, P., Jubelin, G., Roux, B., Labbe, S., & Baret, F. (2008). Assessment of unmanned aerial vehicles imagery for quantitative monitoring of wheat crop in small plots. *Sensors (Basel)*, 8(5), 3557–3585. 10.3390/s805355727879893

Lemeshchenko-Lagoda, V. V. (2020). *Foreign language II (English) guidelines for independent work for applicants for the degree of Higher Education "Bachelor" in specialty 193 "Geodesy and Land Management" according to the OPP "Geodesy and Land Management" (based on full-time general secondary education).* Melitopol.

Lesaignoux, A., Fabre, S., Briottet, X., & Olioso, A. (2010) Estimation of Soil Moisture Content of bare soils from their spectral optical properties in the 0.4 – 12 μm spectral domain, *IEEE International Geoscience and Remote Sensing Symposium,* (pp. 3861-3864). IEEE. 10.1109/IGARSS.2010.5649907

Lessmann, M., Ros, G. H., Young, M. D., & de Vries, W. (2021). Global variation in soil carbon sequestration potential through improved cropland management. *Global Change Biology*, 28(3), 1162–1177. 10.1111/gcb.1595434726814

Lestari, N. P., & Pratika, Y. (2020). Tourist perception of the intention to visit green open space. *Jurnal Inovasi Ekonomi*, 5(01). 10.22219/jiko.v0i0.11018

Levick, L., Semmens, D., Guertin, D., Burns, I., Scott, S., Unkrich, C., & Goodrich, D. (2004). Adding global soils data to the automated geospatial watershed assessment tool (AGWA). *Proc. 2nd SAHRA,(Sustainability of Semi-Arid Hydrol. and Riparian Areas), U. of Ariz., Internat'l. Sym. on Transboundary Water Manage.* University of Arizona.

Levykin, S. V., Chibilev, A. A., Gulyanov, Yu. A., Kazachkov, G. V., Yakovlev, I. G., & Nurushev, M. Zh. (2021). The Virgin Land Megaproject and the Land Reform as the Global Experiment of Steppe Self-Restoration in North Eurasia. *IOP Conference Series: Earth and Environmental Science,* Volume 817.10.1088/1755-1315/817/1/012058

Lin, S., Gu, J., Yamazaki, S., & Shum, H.-Y. (2004). Radiometric calibration from a single image. In *Proc. IEEE Comp. Vision and Pattern Recognition*, (pp. 938–945). IEEE.

Lipski, S. A. (2018). Arctic lands: arrangements for their use and monitoring. *Izvestiya vuzov «Geodeziya i aerofotosyemka», 62*(2), 196–203. . EDN YWZHZR10.30533/0536-101X-2018-62-2-196-203

Li, R., Zhao, S., & Yang, B. (2023). Research on the Application Status of Machine Vision Technology in Furniture Manufacturing Process. *Applied Sciences (Basel, Switzerland)*, 13(4), 1–14. 10.3390/app13042434

Li, S., Xie, J., & Paudel, B. (2023). Do ecological restoration projects undermine economic performance? A spatially explicit empirical study in loess plateau, China. *Remote Sensing (Basel)*, 15(12), 3035. 10.3390/rs15123035

Litvinov, V. N. (1958). *Earthen pear in the Gissar Valley of the Tajik SSR.* [Dissertation... Ph.D. agricult. Sci., Stalinabad].

Litvinov, V. N. (1965). *Forage crops of Tajikistan.* Irfon.

Liu, C., Plaza-Bonilla, D., Coulter, J. A., Kutcher, H. R., Beckie, H. J., Wang, L., Floc'h, J. B., Hamel, C., Siddique, K. H. M., Li, L., & Gan, Y. (2022). Diversifying crop rotations enhances agroecosystem services and resilience. *Advances in Agronomy*, 173, 299–335. 10.1016/bs.agron.2022.02.007

Liu, X., Ma, D., Wu, Y., & Li, L. (2018). Effects of drought events on soil microbial community and activities in a temperate grassland. *Land Degradation & Development*, 29(11), 3886–3895.

Liu, Z., He, S., Men, W., & Sun, H. (2024). Impact of climate risk on financial stability: Cross-country evidence. *International Review of Financial Analysis*, 92, 103096. 10.1016/j.irfa.2024.103096

López-Tarazón, J., Batalla, R. J., Vericat, D., & Balasch, J. (2010). Rainfall, runoff and sediment transport relations in a mesoscale mountainous catchment: The River Isábena (Ebro basin). *Catena*, 82(1), 23–34. 10.1016/j.catena.2010.04.005

Lorenzo, L., Casalí, J., López, J., & Del Valle de Lersundi, J. (2002). *Long term assessment of soil erosion rates in vineyards, and its application for USLE model evaluation*. European Society of Agronomy Conference, Córdoba.

Ludwig, C., Hecht, R., Lautenbach, S., Schorcht, M., & Zipf, A. (2021). Mapping Public Urban Green Spaces Based on OpenStreetMap and Sentinel-2 Imagery Using Belief Functions. *ISPRS International Journal of Geo-Information*, 10(4), 251. 10.3390/ijgi10040251

Lu, L., Li, G., He, N., Li, H., Liu, T., Li, X., Whalen, J. K., Geisen, S., & Liu, M. (2023). Drought shifts soil nematodes to smaller size across biological scales. *Soil Biology & Biochemistry*, 184, 109099. 10.1016/j.soilbio.2023.109099

Luneva, N. N. (2009). *Technological methods for accounting and monitoring of weeds in agroecosystems, High-performance and high-precision technologies and methods of phytosanitary monitoring*. VIZR.

Luo, X., Bennett, R., Koeva, M., & Lemmen, C. (2017). Investigating Semi-Automated Cadastral Boundaries Extraction from Airborne Laser Scanned Data. *Land (Basel)*, 6(3), 60. 10.3390/land6030060

Lurkova, N. (2021). *A Special Course in a Foreign Language (based on the MOOC "Sustainable Development of Mountain Territories in the Context of the Concept of Ecosystem Services") Erasmus+ CBHE project Sustainable Natural Resource Use in Arctic and High Mountainous Areas*. Gorno-Altaisk State University, Department of Natural Sciences and Geography. Russia.

Lvov, Yu.A. (1986).Environmental changes in the construction and operation of overhead power transmission lines of high and ultra-high voltage. *Ultra-high voltage power transmission and ecology*. ENIN im. Krkikanovsky.

Ly, S., Charles, C., & Degré, A. (2013). Different methods for spatial interpolation of rainfall data for operational hydrology and hydrological modeling at watershed scale. A review. *Biotechnologie, Agronomie, Société et Environnement*, 17(2), 392.

Lyubinskaya T.V. (2010) Reducing biogas emissions from solid waste as the most important element in reducing the "greenhouse" effect. *Bulletin of RUDN University. Series: Ecology and life safety*, *1*, 76-81.

Ma, Y., Brindley, P., & Lange, E. (2023). From Modelling and Analysis of Accessibility of Urban Green Space to Green Infrastructure Planning: Guangzhou as a Case Study. In *Adaptive Urban Transformation: Urban Landscape Dynamics, Regional Design and Territorial Governance in the Pearl River Delta, China* (pp. 249–266). Springer.

Maisnam, P., Jeffries, T. C., Szejgis, J., Bristol, D., Singh, B. K., Eldridge, D. J., Horn, S., Chieppa, J., & Nielsen, U. N. (2023). Severe prolonged drought favours stress-tolerant microbes in Australian drylands. *Microbial Ecology*, 86(4), 3097–3110. 10.1007/s00248-023-02303-w37878053

Majer, J. D., & Agosti, D. (2016). Ants visible from space influence soil properties and vegetation in steppe rangelands of Iran. *Sociobiology*, 63(4), 1063–1068. 10.13102/sociobiology.v63i4.1195

Malyshev, V. V., Lukashenko, T. F., & Slushna, Yu. D. (2020). Pedagogical research of environmental competence formation. *Modern movement of science: theses of the supplement. XI international scientific and practical Internet Conference, October 8-9, 2020.* Dnipro.

Marchildon, G. (2010). The Impact of the Great Depression on the Global Wheat Trade. *Johnson-Shoyama Graduate School of Public Policy Working Paper*, 4. https://www.researchgate.net/publication/265907875_The_Impact_of_the_Great_Depression_on_the_Global_Wheat_Trade

Marcos-Pérez, M., Sánchez-Navarro, V., & Zornoza, R. (2023). Intercropping systems between broccoli and fava bean can enhance overall crop production and improve soil fertility. *Scientia Horticulturae*, 312, 111834. 10.1016/j.scienta.2023.111834

Martin, I., et al. (2022). *Derivation and use of soil screening values for assessing ecological risks (revised). Report – ShARE id26 (revised).* Environment Agency, Horizon House.

Maslikov, V. I., Chusov, A. N., Cheremisin, A. V., & Ryzhakova, M. G. (2012). Assessment of the geoecological risk of air pollution from solid waste landfill emissions for the selection of reclamation measures. *Scientific and Technical Journal of St. Petersburg State Polytechnic University.Science and Education*, 2(1), 239–243.

Maslikov, V. I., Chusov, A. N., Molodtsov, D. V., & Ryzhakova, M. G. (2012). Zonal determination of biogas emissions at a solid waste landfill to assess the geo-ecological state and substantiate the management of waste decomposition processes during reclamation. *Scientific and Technical Journal of St. Petersburg State Polytechnic University.Science and education*, 2(1), 260–265.

Matrokhina, K. V., Trofimets, V. Y., Mazakov, E. B., Makhovikov, A. B., & Khaykin, M. M. (2023). Development of methodology for scenario analysis of investment projects of enterprises of the mineral resource complex. *Journal of Mining Institute*, 259, 112–124. 10.31897/PMI.2023.3

Matveeva, V. A., Bech, J., & Danilov, A. S. (2023). Prospects of nitrogen removal from mine drainage by micro-algae in northern environments. *Sustainable Development of Mountain Territories*, 15(1), 134–142. 10.21177/1998-4502-2023-15-1-134-142

Mazhitova, Zh., Zhalmurzina, A., Kolganatova, S., Orazbakov, A., & Satbai, T. (2021). *Environmental consequences of Khrushchev's Virgin Land Campaign in Kazakhstan (1950s–1960s).* E3S Web of Conferences. 258. 10.1051/e3sconf/202125805036

Mazhitova, Zh. S., Khassenova, Zh. O., & Issayev, A. U. (2023). Cultivation of Fodder Beans at the Zarechny State Farm during the Development of Virgin Lands in Kazakhstan. *IOP Conference Series: Earth and Environmental Science,* Volume 1154.10.1088/1755-1315/1154/1/012057

McBratney, A., Mendonça, S. M., Minasny. B. (2003) On digital soil mapping. *Geoderma, 1–2* (117), 3–52.

McCauley, M. (1976). *Khrushchev and the Development of Soviet Agriculture: The Virgin Land Program, 1953–1964.* Holmes & Meier. 10.1007/978-1-349-03059-0

McCord, J., McCord, M., McCluskey, W., Davis, P. T., McIlhatton, D., & Haran, M. (2014). Effect of public green space on residential property values in Belfast metropolitan area. *Journal of Financial Management of Property and Construction*, 19(2), 117–137. 10.1108/JFMPC-04-2013-0008

Medeu, A. R. (Ed.). (2010). *Natcionalniy atlas Respubliki Kazakhstan. T. 1: Prirodniye usloviya i resursy[National atlas of the Republic of Kazakhstan. V. 1: natural conditions and resources].* TOO "Vit Brand".

Mees, J. (1988). *Astronomical formulas for microcalculators.* Jean Meyos.

Meles, M. B., Goodrich, D. C., Unkrich, C. L., Gupta, H. V., Burns, I. S., Hirpa, F. A., Razavi, S., & Guertin, D. P. (2024). Rainfall distributional properties control hydrologic model parameter importance. *Journal of Hydrology. Regional Studies*, 51, 101662. 10.1016/j.ejrh.2024.101662

Memarian, H., Balasundram, S. K., Talib, J. B., Teh Boon Sung, C., Mohd Sood, A., & Abbaspour, K. C. (2013). KINEROS2 application for land use/cover change impact analysis at the H ulu L angat B asin, M alaysia. *Water and Environment Journal : the Journal / the Chartered Institution of Water and Environmental Management*, 27(4), 549–560. 10.1111/wej.12002

Mendoza, F., Dejmek, P., & Aguilera, J. M. (2006). Calibrated color measurements of agricultural foods using image analysis. *Postharvest Biology and Technology*, 41(3), 285–295. 10.1016/j.postharvbio.2006.04.004

Meng, L., & Si, W. (2022). Pro-Environmental Behavior: Examining the Role of Ecological Value Cognition, Environmental Attitude, and Place Attachment among Rural Farmers in China. *International Journal of Environmental Research and Public Health*, 19(24), 17011. 10.3390/ijerph192417011365548 98

Meselhy, A. A., Salem, H. M., & Elhagarey, M. E. (2019). Manufacturing of machine for crop cultivating under rainfed conditions. *Bioscience Research*, 16(4), 3903–3926.

Mete, M. O., & Yomralioglu, T. (2023). A Hybrid Approach for Mass Valuation of Residential Properties through Geographic Information Systems and Machine Learning Integration. *Geographical Analysis*, 55(4), 535–559. 10.1111/gean.12350

Methodology for determining the extent of damage from soil and land degradation. Committee of the Russian Federation on Land Resources and Land Management.

Meyer, L. D., & Wischmeier, W. H. (1969). Mathematical simulation of the process of soil erosion by water. *Transactions of the ASAE. American Society of Agricultural Engineers*, 12(6), 754–0758. 10.13031/2013.38945

Miller, S. N., Semmens, D. J., Goodrich, D. C., Hernandez, M., Miller, R. C., Kepner, W. G., & Guertin, D. P. (2007). The automated geospatial watershed assessment tool. *Environmental Modelling & Software*, 22(3), 365–377. 10.1016/j.envsoft.2005.12.004

Miller, S., Semmens, D., Miller, R., Hernandez, M., Goodrich, D., Miller, W., Kepner, W., & Ebert, D. (2002). GIS-based hydrologic modeling: the automated geospatial watershed assessment tool. *Proceedings of the Second Federal Interagency Hydrologic Modeling Conference*.

Milovidov, V. A. (2013). *Ecology and the modern world*. Iris-press.

Minaei, M. (2020). Evolution, density and completeness of OpenStreetMap road networks in developing countries: The case of Iran. *Applied Geography (Sevenoaks, England)*, 119, 102246. 10.1016/j.apgeog.2020.102246

Minasny, B., McBratney, A. B., & Hartemink, A. E. (2010). Global pedodiversity, taxonomic distance, and the World Reference Base. *Geoderma*, 155(3-4), 132–139. 10.1016/j.geoderma.2009.04.024

Ministry of Agriculture of the Republic of Kazakhstan Land Management Committee. (2022). *Svodnyj analiticheskij otchet o sostoyanii i ispol'zovanii zemel' Respubliki Kazahstan za 2022 god[The consolidated analytical report on the condition and use of land in the Republic of Kazakhstan for 2022]*. https://www.gov.kz/memleket/entities/land/documents/details/579164?lang=ru

Ministry of Agriculture of the Republic of Kazakhstan. (December 30, 2021). *Ob utverzhdenii Koncepcii razvitiya agropromyshlennogo kompleksa Respubliki Kazahstan na 2021 – 2030 gody [On approval of the Concept for the development of the agro-industrial complex of the Republic of Kazakhstan for 2021 – 2030]*. Decree of the Government of the Republic of Kazakhstan No. 960. https://www.gov.kz/memleket/entities/moa/documents/details/274531?lang=ru

Ministry of Construction of Russia. (2020). *SP 131.13330.2020. SNiP 23-01-99* Construction climatology*. Ministry of Construction of Russia.

Ministry of Justice of the Republic of Kazakhstan. (2012). *Konstituciya Respubliki Kazahstan[Constitution of the Republic of Kazakhstan]. Institute of legislation and legal information of the Republic of Kazakhstan].* https://adilet.zan.kz/rus/docs/K950001000_

Ministry of Justice of the Republic of Kazakhstan. (June 20, 2003). *Zemelniy kodeks Respubliki Kazakhstan.[The Land Code of the Republic of Kazakhstan] (No. 442).* Available from: https://adilet.zan.kz/rus/docs/K030000442_

Mirzabaev, A., Goedecke, J., Dubovyk, O., Djanibekov, U., Le, Q. B., & Aw-Hassan, A. (2016). Economics of Land Degradation in Central Asia. In Nkonya, E., Mirzabaev, A., & von Braun, J. (Eds.), *Economics of Land Degradation and Improvement – A Global Assessment for Sustainable Development.* Springer. 10.1007/978-3-319-19168-3_10

Mnatsekanov, R. A., Sergeeva, M. I., & Filippova, E. V. (2021). *Methodological recommendations for the creation of specially protected natural territories of federal and regional significance.* World Wildlife Fund (WWF).

Mockshell, J., & Kamanda, J. (2018). Beyond the agroecological and sustainable agricultural intensification debate: Is blended sustainability the way forward? *International Journal of Agricultural Sustainability*, 16(2), 127–149. 10.1080/14735903.2018.1448047

Mora, F., & Iverson, L. (2002). A spatially constrained ecological classification: Rationale, methodology and implementation. *Plant Ecology*, 158(2), 153–169. 10.1023/A:1015534615415

Mor, S., & Ravindra, K. (2023). Municipal solid waste landfills in lower- and middle-income countries: Environmental impacts, challenges and sustainable management practices. *Process Safety and Environmental Protection*, 174, 510–530. 10.1016/j.psep.2023.04.014

Motoshita, M., Ono, Y., Pfister, S., Boulay, A. M., Berger, M., Nansai, K., Tahara, K., Itsubo, N., & Inaba, A. (2018). Consistent characterisation factors at midpoint and endpoint relevant to agricultural water scarcity arising from freshwater consumption. *The International Journal of Life Cycle Assessment*, 23(12), 2276–2287. 10.1007/s11367-014-0811-5

Mudrák, O., Uteseny, K., & Frouz, J. (2012). Earthworms drive succession of both plant and Collembola communities in post-mining sites. *Applied Soil Ecology*, 62, 170–177. 10.1016/j.apsoil.2012.08.004

Mudrykh, N. M. (2018). Spatial variation of properties in the agrosod-podzolic soil. *Anthropogenic Transformation of Nature*, 4, 82–84.

Mujumdar, P., & Kumar, D. N. (2012). *Floods in a changing climate: hydrologic modeling.* Cambridge University Press. 10.1017/CBO9781139088428

Muliasari, I. G. A. D., Suartika, G. A. M., & Saputra, K. E. (2021). Conversion of Public Green Open Spaces to Facilitate the Tourist Industry. *IOP Conference Series. Earth and Environmental Science*, 903(1), 012011. 10.1088/1755-1315/903/1/012011

Murguia, J. M., Ordoñez, P., Corral, L., & Navarrete-Chacón, G. (2022). *Payment for Ecosystem Services in Costa Rica: Evaluation of a Country-Wide Program.* IDB, Inter-American Development Bank. 10.18235/0004259

Murphy, J. T., Breeze, T. D., Willcox, B., Kavanagh, S., & Stout, J. C. (2022). Globalisation and pollinators: Pollinator declines are an economic threat to global food systems. *People and Nature*, 4(3), 773–785. 10.1002/pan3.10314

Mustafin, M., & Moussa, H. (2023). The determination of plumb line deviation using Satellite/Levelling Technique. In *Lecture notes in civil engineering* (pp. 37–47). 10.1007/978-981-99-4229-9_4

Mvula, R. L. S., Mundike, J., & Nguvulu, A. (2023). Spatial suitability analysis for site selection of municipal solid waste landfill using hybrid GIS and MCDA approach: The case of Kitwe, Zambia. *Scientific African*, 21, e01885. 10.1016/j.sciaf.2023.e01885

Mwendwa, P., & Giliba, R. A. (2012). Benefits and challenges of Urban green spaces. *Zhongguo Renkou Ziyuan Yu Huanjing*, 10(1), 73–79. 10.1080/10042857.2012.10685062

Nascimento, C. M., Mendes, W. S., & Silvero, N. E. Q. (2020). Soil degradation index developed by multitemporal remote sensing images, climate variables, terrain and soil atributes. *Journal of Environmental Management*, 277, 111316. 10.1016/j.jenvman.2020.11131632980636

Nath, D., Dutta, P. K., & Bhattacharya, A. K. (2021). *Detection of plant diseases and nutritional deficiencies from unhealthy plant leaves using machine learning techniques.* 4th Smart Cities Symposium (SCS 2021), Online Conference, Bahrain. 10.1049/icp.2022.0368

National projects. (2022). *Ecology*. National projects. https://национальныепроекты.рф/projects/ekologiya

National projects. (2023). *Passport of the project «Conservation of biological diversity and development of ecological tourism»*. National projects. https://www.mnr.gov.ru/activity/np_ecology

Navratil, O., Evrard, O., Esteves, M., Legout, C., Ayrault, S., Némery, J., Mate-Marin, A., Ahmadi, M., Lefèvre, I., Poirel, A., & Bonté, P. (2012). Temporal variability of suspended sediment sources in an alpine catchment combining river/rainfall monitoring and sediment fingerprinting. *Earth Surface Processes and Landforms*, 37(8), 828–846. 10.1002/esp.3201

Nazarbayev, N. (2023). *Moya zhizn. Ot zavismosti k svobode[My Life. From Dependence to Freedom].* Foliant.

Nguyen, T. X. (2020). *Monitoring of land in the Socialist Republic of Vietnam for cadastral registration in the context of changing market relations* [Dissertation, Saint-Petersburg Mining University]. Saint-Petersburg Mining University.

Nikolaenko, S., & Nikolaenko, M. (2021). Higher agricultural education in Ukraine: challenges of the time. *Bulletin of Agrarian science, 99*(12), 5-17.

Niu X.X., Wang S.Z., Niu Y.C., Wei L.F., Yu L.Y. Improvement and optimization for the first order decay model parameters at typical municipal solid waste landfills in China. *Advances in Climate Change Research*, *14*(4), 605-614 ().10.1016/j.accre.2023.07.002

Novikova, L.Yu, Novikova, S.D., & Kiru, E.V. (2017). Manifestation of economically valuable traits in potato varieties (Solanum L.) under climate change on the European territory of Russia. *Agricultural biology*, *52*(1), 75-83.

Nutsford, D., Pearson, A. L., & Kingham, S. (2013). An ecological study investigating the association between access to urban green space and mental health. *Public Health*, 127(11), 1005–1011. 10.1016/j.puhe.2013.08.01624262442

Oganesyan, A., Kurita, H., Maslov, D., & Take, A. (2005). Project for collecting landfill biogas and generating energy in Yerevan. *Alternative Energy and Ecology*, 12(32), 88–93.

Oliveira, S., Andrade, H., & Vaz, T. (2011). The cooling effect of green spaces as a contribution to the mitigation of urban heat: A case study in Lisbon. *Building and Environment*, 46(11), 2186–2194. 10.1016/j.buildenv.2011.04.034

On approval of the State Program of the Russian Federation "Environmental Protection" (2014). Decree of the Government of the Russian Federation #326 https://docs.cntd.ru/document/499091755

On Production and Consumption Waste Federal Law # 89-FZ (1998). https://legalacts.ru/doc/FZ-ob-othodah-proizvodstva-i-potreblenija/

Organisation for Economic Cooperation and Development (OECD). (2011). Competitiveness and Private Sector Development: Kazakhstan 2010: Sector Competitiveness Strategy. *Competitiveness and Private Sector Development.* 10.1787/9789264089792-en

Orlov, D. S., Sukhanova, N. I., & Rozanova, M. S. (2002). *Spectral reflectance of soils and their components.* M.: Moscow State University Publishing House.

Orlov, D. S. (1997). Color and diagnostics of soils, Soros Educational Journal. No. 4, 45-51. 23. Kataev, M.Yu., Dadonova, M.M. (2019) Vegetation recognition technique based on color and texture analysis of RGB images. *Light & Engineering*, (2), 34–39.

Ortskhanov T.A. (2010) Urban planning aspects aimed at ensuring the environmental safety of the territories where solid waste landfills are located. *Academia. Architecture and construction*, 3, 150-153.

Ostroukhov, A.V., Klimina, E.M., & Kuptsova V.A. (2020). Landscape mapping of hard-to-reach territories on the example of the Bolonsky State Nature Reserve (Russia). *Nature Conservation Research. Reserved science, 5*(2), 47-63. 10.24189/ncr.2020.015

Panagos, P., Ballabio, C., Himics, M., Scarpa, S., Matthews, F., Bogonos, M., Poesen, J., & Borrelli, P. (2021). Projections of soil loss by water erosion in Europe by 2050. *Environmental Science & Policy*, 124, 380–392. 10.1016/j.envsci.2021.07.012

Panagos, P., Ballabio, C., Poesen, J., Lugato, E., Scarpa, S., Montanarella, L., & Borrelli, P. (2020). A soil erosion indicator for supporting agricultural, environmental and climate policies in the European Union'. *Remote Sensing (Basel)*, 12(9), 1365. 10.3390/rs12091365

Pandul, I. S. (2010). *Geodetic astronomy in relation to solving engineering and geodetic problems.* Politekhnika.

Partoev, K., Saydaliyev, N.H., Yasinov, Sh.M. & Sadriddinov S. (2015).Correlation communication between signs of a sun artichoke. *News of the Orenburg state agricultural university, 6*(56), 36-37.

Pashchenko L.P. (1999). Rational aspects in processing Jerusalem artichoke. *Storage and processing of agricultural raw materials,7*, 13-17.

Pashkevich, M. A., Danilov, A. S., & Matveeva, V. A. (2021). Remote sensing of chemical anomalies in the atmosphere in influence zone of korkino open pit coal mine. *Eurasian Mining*, 35(1), 79–83. 10.17580/em.2021.01.16

Pasko, N. M. (1999). Jerusalem artichoke - biotechnical potential for food, medicinal, technical, feed and environmental purposes. *Sat. works "Innovative technologies and products"*. Novosibirsk.

Pasko, N. M. (2003). Selection and seed production of Jerusalem artichoke. Sat. scientific works, Moscow.

Pasko, O., Staurskaya, N., Zakharchenko, A., Zharnikov, V., Larionov, Y. (2022). Bio-Farming as the Basis of Environmentally-Sustainable Arable Farming at the Time of Global Warming. *Research Anthology on Strategies for Achieving Agricultural Sustainability*, (pp. 600–627). Research Gate.

Pasko, O.A., Goryachkin, V.N., Ten, A.D., & Nabieva, M.N. (2022a). Experience of agrarian reforms of the 20th century in Russia. *Economics, labor, management in agriculture,4*(86). 102-110.

Pasko, O.A., Goryachkin, V.N., Ten, A.D., Nabieva, M.N. (2022a). Experience of agrarian reforms of the 20th century in Russia. *Economics, labor, management in agriculture,4*(86). 102-110.

Pasko, O.A., Tokareva, O.S., Alshaibi, A.D.A., Chernikova, T.Yu., Kabral, P. (2019a). Assessment of the condition of near-village cedar forests of the Tomsk region using Earth remote sensing data. *Izvestiya Tomsk Polytechnic University. Georesource engineering, 330*(1), 98-109.

Pasko, O.A., Tokareva, O.S., Ibragimov, E.A. (2020a). Analysis of the ecological hazards of snow bars on the example of the city of Tomsk. *Modern problems of remote sensing of the Earth from space,17*(3),135-144.

Pasko, O.A., Zakharchenko, A.V. (2021). The Degradation of the Territories of Snow Polygons. *Ecology and industry of Russia*, *25*(11), 30-35.

Pasko, O.A., Zakharchenko, A.V., Pospelova, E.V. (2019b). Differentiation of agricultural lands by area on the example of the Tomsk region. *Bulletin of Tomsk Polytechnic University. Georesource engineering,330* (5), 100-112.

Pasko, O.A., Zakharchenko, A.V., Pospelova, E.V. (2019b). Differentiation of agricultural lands by area on the example of the Tomsk region. *Bulletin of Tomsk Polytechnic University. Georesource engineering,330*(5), 100-112.

Pasko, O. A. (2022b). Dependence of potato yield on weather conditions. *Bulletin of KrasGAU.*, 1(178), 56–61. 10.36718/1819-4036-2022-1-56-61

Pasko, O. A. (2023). In Lebedeva, N. A. (Ed.), *Features of Teaching a Foreign Language to Lecturers of Higher Education Institutions. Cases on Error Analysis in Foreign Language Technical Writing* (pp. 124–161). IGI Global. 10.4018/978-1-6684-6222-5.ch006

Pasko, O. A., & Zakharchenko, A. V. (2020b). Cadastral and geoecological characteristics of especially valuable productive agricultural land. *Bulletin of Tomsk Polytechnic University.Georesources Engineering*, 331(11), 7–19.

Patel, R. J., Rank, P. H., Vekariya, P. B., Rank, H. D., Vadar, H. R., Parmar, H. V., Tiwari, M. K., Marviya, P. B., & Baca, M. M. L. (2023). *Enhancing Wheat (Triticum aestivum L.) Crop Yield and Water Use Efficiency: A Study on Canopy Air Temperature Difference-Based Drip Irrigation Scheduling in a Semi-Arid Region of Western India.* 10.21203/rs.3.rs-3376468/v1

Pathirana, R., & Carimi, F. (2022). Management and utilization of plant genetic resources for a sustainable agriculture. *Plants*, 11(15), 2038. 10.3390/plants1115203835956515

Paul, C., Kuhn, K., Steinhoff-Knopp, B., Weißhuhn, P., & Helming, K. (2020). Towards a standardisation of soil-related ecosystem service assessments. *European Journal of Soil Science*, 72(4), 1543–1558. 10.1111/ejss.13022

Pestryakov, V.K. (1973). *Soils of the Leningrad region.* Publishing house: Lenizdat.

Petit, S., Muneret, L., Carbonne, B., Hannachi, M., Ricci, B., Rusch, A., & Lavigne, C. (2020). Landscape-scale expansion of agroecology to enhance natural pest control: A systematic review. *Advances in Ecological Research*, 63, 1–48. 10.1016/bs.aecr.2020.09.001

Petrick, M., Wandel, J., & Karsten, K. (2013). Rediscovering the virgin lands: Agricultural investment and rural livelihoods in a Eurasian frontier area. *World Development*, 43, 164–179. 10.1016/j.worlddev.2012.09.015

Peydayesh, M., & Mezzenga, R. (2021). Protein nanofibrils for next generation sustainable water purification. *Nature Communications, 12*(1), 1–17. 10.1038/s41467-021-23388-2

Pham, T. D., Yokoya, N., Bui, D. T., Yoshino, K., & Friess, D. A. (2019). Remote Sensing Approaches for Monitoring Mangrove Species, Structure, and Biomass: Opportunities and Challenges. *Remote Sensing (Basel)*, 11(3), 230. 10.3390/rs11030230

Pham, V. T., Do, T. A. T., Tran, D. H., & Do, T. N. A. (2023). Classifying forest cover and mapping forest fire susceptibility in Dak Nong province, Vietnam utilizing remote sensing and machine learning. *Ecological Informatics*, 79, 102392. 10.1016/j.ecoinf.2023.102392

Platonycheva, Yu.N., Polyakova, N.V., Berchuk, A.V., Bogomolova, Yu.A. (2013). The influence of fallow plowing methods on the mobility of organic matter in light gray forest soil. *Agricultural science of the Euro-North East*, *4* (35), 39-43.

Politizdat. Akhmetova, L.S. (1996). *Patriotizm v rakurse istorii. Uchebnoe posobie[Patriotism from the Perspective of History. Tutorial].* Sanat.

Potapov, A. A. (2003). New information technologies based on probabilistic texture and fractal features in radar detection of low-contrast targets. *Radio Engineering and Electronics*, 48(9), 1101–1119.

Poudel, S., Bansal, S., Podder, S., Paneru, B., Karki, S., Fike, J., & Kumar, S. (2022). Conversion of open pasture to hardwood silvopasture enhanced soil health of an ultisol. *Agroforestry Systems*, 96(8), 1237–1247. 10.1007/s10457-022-00783-2

Prasad, P., Loveson, V. J., Chandra, P., & Koth, M. (2022). Evaluation and comparison of the earth observing sensors in land cover/ land use studies using machine learning algorithms. *Ecological Informatics*, 68, 101522. 10.1016/j.ecoinf.2021.101522

Prasuhn, V. (2020). Twenty years of soil erosion on-farm measurement: Annual variation, spatial distribution and the impact of conservation programmes for soil loss rates in Switzerland'. *Earth Surface Processes and Landforms*, 45(7), 1539–1554. 10.1002/esp.4829

Praznik, W., & Beck, R. H. (1987). Inulin composition during growth of tubers of Helianthus tuberosus. *Agricultural and Biological Chemistry*, 51(6), 1593–1599. 10.1271/bbb1961.51.1593

Pristupa, A., Tysiachniouk, M., Mol, A. P., Leemans, R., Minayeva, T., & Markina, A. V. (2017). Can zoning resolve nature use conflicts? The case of the Numto Nature Park in the Russian Arctic. *Journal of Environmental Planning and Management*, 61(10), 1674–1700. 10.1080/09640568.2017.1370365

Prout, J. M., Shepherd, K. D., McGrath, S. P., Kirk, G. J. D., & Haefele, S. M. (2020). What is a good level of soil organic matter? An index based on organic carbon to clay ratio. *European Journal of Soil Science*, 72(6), 2493–2503. 10.1111/ejss.13012

Prudencio, L., & Null, S. E. (2018). Stormwater management and ecosystem services: A review. In *Environmental Research Letters, 13*(3). 10.1088/1748-9326/aaa81a

Pudjiastuti, A. Q., Kaluge, D., & Widowati, W. (2023). Reallocation of the use of chemical fertilizers and pesticides to increase the income of vegetable farmers and prevent land degradation. *Journal of Degraded and Mining Lands Management*, 11(1), 5095–5103. 10.15243/jdmlm.2023.111.5095

Puech, C., Brulaire, A., Paraiso, J., & Faloya, V. (2021). Collective design of innovative agroecological cropping systems for the industrial vegetable sector. *Agricultural Systems*, 191, 103153. 10.1016/j.agsy.2021.103153

Qaim, M. (2016). *Genetically modified crops and agricultural development*. https://books.google.com/books?hl=en&lr=&id=skYDAAAQBAJ&oi=fnd&pg=PP1&ots=DSYSZE-vjq&sig=mFnSsqS3Qmi3lc-Cz7d8nN11e6U

Qaim, M. (2020). Role of New Plant Breeding Technologies for Food Security and Sustainable Agricultural Development. *Applied Economic Perspectives and Policy*, 42(2), 129–150. 10.1002/aepp.13044

Raguzin, I., Bykova, E. N., & Lepikhina, O. Y. (2023). POLYGONAL METRIC GRID METHOD FOR ESTIMATING THE CADASTRAL VALUE OF LAND PLOTS. *Vestnik Moskovskogo Universiteta. Seriia V, Geografiia*, 78(3), 92–103. 10.55959/MSU0579-9414.5.78.3.8

Rahmoune, M., Radjef, M. S., Boukherroub, T., & Carvalho, M. (2024). A new integrated cooperative game and optimization model for the allocation of forest resources. *European Journal of Operational Research*, 316(1), 329–340. 10.1016/j.ejor.2024.01.018

Ramaiah, M., & Avtar, R. (2019). Urban Green Spaces and Their Need in Cities of Rapidly Urbanizing India: A Review. In *Urban Science, 3*(3). 10.3390/urbansci3030094

Ramos, M., & Martinez-Casasnovas, J. (2006). Nutrient losses by runoff in vineyards of the Mediterranean Alt Penedès region (NE Spain). *Agriculture, Ecosystems & Environment*, 113(1), 356–363. 10.1016/j.agee.2005.10.009

Ramos, M., & Martínez-Casasnovas, J. (2009). Impacts of annual precipitation extremes on soil and nutrient losses in vineyards of NE Spain. *Hydrological Processes*, 23(2), 224–235. 10.1002/hyp.7130

Ramos, M., & Martínez-Casasnovas, J. (2010). Effects of field reorganisation on the spatial variability of runoff and erosion rates in vineyards of Northeastern Spain. *Land Degradation & Development*, 21(1), 1–12. 10.1002/ldr.958

Rani, A., Kashyap, R., & Azmi, W. (2021). Conservation of Biodiversity by Biotechnology. In *Basic Concepts in Environmental Biotechnology* (pp. 149–175). CRC Press. 10.1201/9781003131427-12

Rasmussen, J., Ntakos, G., Nielsen, J., Svensgaard, J., Poulsen, R. N., & Christensen, S. (2016). Are vegetation indices derived from consumer-grade cameras mounted on UAVs sufficiently reliable for assessing experimental plots? *European Journal of Agronomy*, 74, 75–92. 10.1016/j.eja.2015.11.026

Rau, A., Koibakova, Y., Nurlan, B., Nabiollina, M., Kurmanbek, Z., Issakov, Y., Zhu, K., & David, L. D. (2023). Increase in Productivity of Chestnut Soils on Irrigated Lands of Northern and Central Kazakhstan. *Land (Basel)*, 12(3), 672. 10.3390/land12030672

Rawls, W., Brakensiek, D., & Saxtonn, K. (1982). Estimation of soil water properties. *Transactions of the ASAE. American Society of Agricultural Engineers*, 25(5), 1316–1320. 10.13031/2013.33720

Raza, M. B., Bhoi, T. K., & Samal, I. (2019). Soil arthropods as a nutrient enhancer. *International Journal of Chemical Studies*, 7, 1687–1692.

Razumnykh, A. V., Kalinina, K. V., & Vlasov, V. A. (2020). Implementation of the reform of solid waste management in the Krasnoyarsk Territory: Selected issues of theory and practice. *Bulletin of Science and Education*, 23(101), 39–42.

Reddy, C. S., Jha, C. S., & Dadhwal, V. K. (2016). Assessment and monitoring of long-term forest cover changes (1920–2013) in Western Ghats biodiversity hotspot. *Journal of Earth System Science*, 125(1), 103–114. 10.1007/s12040-015-0645-y

Reed, S., Koren, V., Smith, M., Zhang, Z., Moreda, F., Seo, D.-J., & Participants, D. (2004). Overall distributed model intercomparison project results. *Journal of Hydrology (Amsterdam)*, 298(1), 27–60. 10.1016/j.jhydrol.2004.03.031

Rees, W.G. (2006). *Fundamentals of remote sensing*. Moscow, Tekhnosphere, 336.

Refsgaard, J. C. (1997). Parameterisation, calibration and validation of distributed hydrological models. *Journal of Hydrology (Amsterdam)*, 198(1), 69–97. 10.1016/S0022-1694(96)03329-X

Refsgaard, J. C., & Knudsen, J. (1996). Operational validation and intercomparison of different types of hydrological models. *Water Resources Research*, 32(7), 2189–2202. 10.1029/96WR00896

Reganold, J. P., & Wachter, J. M. (2016). Organic agriculture in the twenty-first century. *Nature Plants, 2*(2), 1–8. 10.1038/nplants.2015.221

Reingart, E.S., Kochnev, N.K., Ponamarev, A.G., & Zvyagintsev, P.S. (2008). Prospects for the use of Jerusalem artichoke for the production of bioethanol. *Achievement of science and technology of the RPK, 1*, 38-40.

Resolution of the Government of the Republic of Kazakhstan. (August 27, 2019). *Ob utverzhdenii vosmogo-desyatogo svodnogo periodicheskogo doklada o vypolnenii Respublikoj Kazahstan Mezhdunarodnoj konvencii o likvidacii vseh form rasovoj diskriminacii[On approval of the eighth to tenth consolidated periodic report on the implementation by the Republic of Kazakhstan of the International Convention on the Elimination of All Forms of Racial Discrimination] (No. 630).*https://adilet.zan.kz/rus/docs/P1900000630

Reyes, F., Sorgonà, A., Briones, M. J., Crecchio, C., & Sofo, A. (2023). Plant Growth and Root Morphology Are Affected by Earthworm-Driven (Eisenia sp.) Changes in Soil Chemico-Physical Properties: A Mesocosm Experiment with Broccoli and Faba Bean. *Journal of Soil Science and Plant Nutrition*, 23(3), 4078–4090. 10.1007/s42729-023-01325-0

Reynolds, M. P., & Braun, H.-J. (Eds.), *Wheat Improvement. Food Security in a Changing Climate*. Springer., 10.1007/978-3-030-90673-3

Rifna, E. J., Ratish Ramanan, K., & Mahendran, R. (2019). Emerging technology applications for improving seed germination. *Trends in Food Science & Technology*, 86, 95–108. 10.1016/j.tifs.2019.02.029

Rogova O. A. (2010). About the content of foreign language training for future specialists in the field of Tourism and nature management. *Works of BSTU. Series 6: history, philosophy,* (5), 226-228.

Roitsch, D., Abruscato, S., Lovrić, M., Lindner, M., Orazio, C., & Winkel, G. (2023). Close-to-nature forestry and intensive forestry – Two response patterns of forestry professionals towards climate change adaptation. *Forest Policy and Economics*, 154, 103035. 10.1016/j.forpol.2023.103035

Rosenberg, G. V. (1963). *Twilight. M.* Fizmatgiz.

Rouland-Lefèvre, C. (2011). Termites as pests of agriculture. *Biology of termites: a modern synthesis*, 499-517. 10.1007/978-90-481-3977-4_18

Rusanov, A. M., Teslya, A. V., & Sayagfarova, A. M. (2011). Restoration of the humus state of steppe chernozems under long-term fallow land. *Bulletin of Orenburg State University*, 12(131), 132–138.

Safari, A., De Smedt, F., & Moreda, F. (2012). WetSpa model application in the distributed model intercomparison project (DMIP2). *Journal of Hydrology (Amsterdam)*, 418, 78–89. 10.1016/j.jhydrol.2009.04.001

Safarzoda, R. Sh., Khalifaev, D. R., & Popov, D. M. (2014). *Amino acid composition of grass Jerusalem artichoke cultivated in Tajikistan*. Pharmacy development and registration of medicines, Moscow.

Safarzoda, R. Sh., Khalifaev, D. R., & Popov, D. M. (2015). *Amino acid composition of grass Jerusalem artichoke cultivated in Tajikistan*. Pharmacy development and registration of medicines, Moscow, 4.

Safina, G. R., & Fedorova, V. A. (2020). *State management and protection of land resources of Russia at the present stage: a course of lectures. Part 1 Management in the field of land protection. Degradation of soils and lands*. Kazan. University.

Sahmat, S. S., Rafii, M. Y., Oladosu, Y., Jusoh, M., Hakiman, M., & Mohidin, H. (2024). Unravelling the dynamics of genotype and environment interactions on chilli (Capsicum annuum L.) yield-related attributes in soilless planting systems. *Scientific Reports, 14*(1), 1–17. 10.1038/s41598-023-50381-0

Saidaliev H., Partoev, K., Yasinov, S. (2015). *Biopotensial of topinambour in conditions of Tajikistan*. Izvestya Orenburg Government agrarian University.

Saidaliev, H. H. Kh. (2018). *Productivity of topinabure depending with cultivations in Gissar and Rasht valyes of Tajikistan* [Dissertation, Dushanbe].

Saktaganova, Z.G., & Abilkhozhin, Zh.B. (2018). "Tselinniy proekt" dostizheniya i problemy (socialno-ekonomicheskie i ekologicheskie aspekty osvoeniya celiny) [Virgin Land Project": achievements and problems (socio-economic and environmental aspects of virgin lands development)]. *Bulletin of Karaganda University. Series History. Philosophy, 4*(92).

Salem, H. M. (2014). *Effect of conservation tillage on soil physical properties, in-situ rainwater harvesting, and erosion control in arid and semi-arid regions* [Doctoral dissertation, Agronomos].

Salem, H. M., Ali, A. M., Wu, W., & Tu, Q. (2021). Initial effect of shifting from traditional to no-tillage on runoff retention and sediment reduction under rainfall simulation. *Soil Research (Collingwood, Vic.)*, 60(6), 547–560. 10.1071/SR21082

Salem, H. M., & Meselhy, A. A. (2021). A portable rainfall simulator to evaluate the factors affecting soil erosion in the northwestern coastal zone of Egypt. *Natural Hazards*, 105(3), 2937–2955. 10.1007/s11069-020-04432-8

Salem, H. M., Meselhy, A. A., Elhagarey, M. E., Abdalla, M. A., El Salem, A., Ali, A. M., & Schott, L. R. (2023). In-Situ Rainwater Harvesting Using Innovative Engineering Technologies for Rehabilitation of Degraded Lands and Climate Change Adaptation. In *Intelligent Engineering Applications and Applied Sciences for Sustainability* (pp. 92–108). IGI Global. 10.4018/979-8-3693-0044-2.ch006

Salem, H. M., Meselhy, A., Elhagarey, M., Ali, A. M., & Wu, W. (2022). Soil erosion control and wheat productivity are improved by a developed ridge-furrow and reservoir tillage systems. *Archives of Agronomy and Soil Science*, 68(2), 273–282. 10.1080/03650340.2020.1832655

Salem, H. M., Muñoz-García, M. Á., & Rodríguez, M. G. (2020). Soil physical properties and soil water tension monitoring by wireless sensor network after reservoir and minimum tillage practices. *Soil Research (Collingwood, Vic.)*, 59(3), 309–317. 10.1071/SR20129

Salem, H. M., Schott, L. R., Piaskowski, J., Chapagain, A., Yost, J. L., Brooks, E., Kahl, K., & Johnson-Maynard, J. (2024). Evaluating Intra-Field Spatial Variability for Nutrient Management Zone Delineation through Geospatial Techniques and Multivariate Analysis. *Sustainability (Basel)*, 16(2), 645. 10.3390/su16020645

Salem, H. M., Valero, C., Muñoz, M. Á., & Gil-Rodríguez, M. (2015). Effect of integrated reservoir tillage for in-situ rainwater harvesting and other tillage practices on soil physical properties. *Soil & Tillage Research*, 151, 50–60. 10.1016/j.still.2015.02.009

Salem, H. M., Valero, C., Muñoz, M. Á., Gil-Rodríguez, M., & Barreiro, P. (2014). Effect of reservoir tillage on rainwater harvesting and soil erosion control under a developed rainfall simulator. *Catena*, 113, 353–362. 10.1016/j.catena.2013.08.018

Salgotra, R. K., & Chauhan, B. S. (2023). Genetic diversity, conservation, and utilization of plant genetic resources. *Genes*, 14(1), 174. 10.3390/genes1401017436672915

Sanches, I. D., Feitosa, R. Q., Achanccaray, P., Montibeller, B., Luiz, A. J. B., Soares, M. D., Prudente, V. H. R., Vieira, D. C., & Maurano, L. E. P. (2018). Lem benchmark database for tropical agricultural remote sensing application. *The International Archives of the Photogrammetry, Remote Sensing and Spatial Information Sciences*, XLII-1, 387–392. 10.5194/isprs-archives-XLII-1-387-2018

Sanchez-Sibony, O. (2014). *Red Globalization. The Political Economy of the Soviet Cold War from Stalin to Khrushchev*. University Printing House. 10.1017/CBO9781139628778

Sangotayo, A. O., Chellappa, J., Sekaran, U., Bansal, S., Angmo, P., Jasa, P., Kumar, S., & Iqbal, J. (2023). Long-term conservation and conventional tillage systems impact physical and biochemical soil health indicators in a corn–soybean rotation. *Soil Science Society of America Journal*, 87(5), 1056–1071. 10.1002/saj2.20563

Sardorov, M. N. (1997). *Productivity and photosynthetic activity of combined crops of alfalfa with cereal crops in the conditions of Central Tajikistan* [Dissertation, Dushanbe].

Sari, Y., Maulida, M., Maulana, R., Wahyudi, J., & Shalludin, A. (2021). *Detection of Corn Leaves Nutrient Deficiency Using Support Vector Machine (SVM)*. 2021 4th International Conference of Computer and Informatics Engineering (IC2IE), Depok, Indonesia. 10.1109/IC2IE53219.2021.9649375

Sawicka, B., & Egbuna, C. (2020). Pests of agricultural crops and control measures. In *Natural remedies for pest, disease and weed control* (pp. 1–16). Academic Press., 10.1016/B978-0-12-819304-4.00001-4

Sayer, J., Sunderland, T., Ghazoul, J., Pfund, J. L., Sheil, D., Meijaard, E., Venter, M., Boedhihartono, A. K., Day, M., Garcia, C., Van Oosten, C., & Buck, L. E. (2013). Ten principles for a landscape approach to reconciling agriculture, conservation, and other competing land uses. *Proceedings of the National Academy of Sciences of the United States of America*, 110(21), 8349–8356. 10.1073/pnas.121059511023686581

Schaepman-Strub, G., Schaepman, M. E., Painter, T. H., Dangel, S., & Martonchik, J. V. (2006). Reflectance quantities in optical remote sensing—Definitions and case studies. *Remote Sensing of Environment*, 103(1), 27–42. 10.1016/j.rse.2006.03.002

Schierhorn, F., Muller, D., Beringer, T., Prishchepov, A. V., Kuemmerle, T., & Balmann, A. (2013). Post-Soviet cropland abandonment and carbon sequestration in European Russia, Ukraine, and Belarus. *Global Biogeochemical Cycles*, 27(4), 1175–1185. 10.1002/2013GB004654

Schimmel, H., & Amelung, W. (2022). *Organic soils. Encyclopedia of Soils in the Environment.* Elsevier. 10.1016/B978-0-12-822974-3.00073-2

Schjønning, P. (2021). Topsoil shear strength – measurements and predictions'. *Soil & Tillage Research*, 212, 105049. 10.1016/j.still.2021.105049

Schulte, M., Jonsson, R., Eggers, J., Hammar, T., Stendahl, J., & Hansson, P. (2023). Demand-driven climate change mitigation and trade-offs from wood product substitution: The case of Swedish multi-family housing construction. *Journal of Cleaner Production*, 421, 138487. 10.1016/j.jclepro.2023.138487

Schuol, J., & Abbaspour, K. (2007). Using monthly weather statistics to generate daily data in a SWAT model application to West Africa. *Ecological Modelling*, 201(3), 301–311. 10.1016/j.ecolmodel.2006.09.028

Seleshnikov, S. I. (1970). *History of the calendar and chronology*. Science.

Semenova, O. S. (2009). Teaching the Abstraction of Foreign-language Texts of a Socio-Political Nature in Practical Classes at a University (on the Example of the German Language). *Pedagogy. Psychology. Social work. Juvenile studies. Sociokinetics*, (3), 47–48.

Senesi, N. (1989). Composted materials as organic fertilizers. *The Science of the Total Environment*, 81–82(C), 521–542. 10.1016/0048-9697(89)90161-7

Sen, Z. (2008). *Wadi hydrology*. CRC Press. 10.1201/9781420061550

Serio, F., Manuel, R., Machado, A., Fussy, A., & Papenbrock, J. (2022). An Overview of Soil and Soilless Cultivation Techniques—Chances, Challenges and the Neglected Question of Sustainability. *Plants, 11*(9), 1153. 10.3390/plants11091153

Shah, K. K., Modi, B., Pandey, H. P., Subedi, A., Aryal, G., Pandey, M., & Shrestha, J. (2021). Diversified crop rotation: An approach for sustainable agriculture production. *Advances in Agriculture*, 2021, 1–9. 10.1155/2021/8924087

Shaimova, A. M., Nasyrova, L. A., & Faskhutdinov, R. R. (2011). Study of methane generation factors in the conditions of a solid waste landfill. *Bashkir Chemical Journal*, 18(2), 172–176.

Shain, S.S. (1931). *Ground pear is a fodder crop.* M.-L.

Shazzo, R. I., Kaishev, V. G., Gish, R. A., & Cornea, R. I. (2013). *Jerusalem Artichoke:biology, agricultural cultivation, a place in the ecosystem, processing technology (yesterday, today, tomorrow).* Krasnodar.

Shazzo, R.I., Kasyanov, G.I. (2000). Functional foods. M.-L.

Sheng, Y., He, Z., & Zhang, H. (2010). *The research on chlorophyll content in vivo-leaf based on non-destructive testing by fiber spectrometer.* April 2010 3rd International Congress on Image and Signal Processing, Yantai, China. 10.1109/CISP.2010.5646171

Sherman, L. K. (1932). The relation of hydrographs of runoff to size and character of drainage-basins. *Eos (Washington, D.C.)*, 13(1), 332–339.

Shilovskikh, P. A., & Kostarev, S. N. (2007). Conceptual approaches to modeling process management at waste disposal sites. *Bulletin of the Perm National Research Polytechnic University*, 2, 120–125.

Shishov, L. L., Tonkonogov, V. D., & Lebedeva, I. I. (2004). *Classification and diagnostics of soils in Russia.* Smolensk: Oikumena.

Shishov, L.L., Tonkonogov, V.D., Lebedeva, I.I., Gerasimova, M.I. (2004). Classification and diagnostics of soils in Russia. *Oikumena Smolensk.*

Shishov, L. L. (1987). *Kriterii i modeli plodorodiya pochv[Soil fertility criteria and models].* Agropromizdat.

Shi, X., Chen, Z., Wu, L., Wei, W., & Ni, B. J. (2023). Microplastics in municipal solid waste landfills: Detection, formation and potential environmental risks. *Current Opinion in Environmental Science & Health*, 31, 100433. 10.1016/j.coesh.2022.100433

Shmilovitz, Y., Marra, F., Wei, H., Argaman, E., Goodrich, D., Assouline, S., & Morin, E. (2023). Assessing the controlling factors on watershed soil erosion during intense rainstorm events using radar rainfall and process-based modeling. *Catena*, 231, 107282. 10.1016/j.catena.2023.107282

Shoaib, M., & Shah, B. (2023). An advanced deep learning models-based plant disease detection: A review of recent research. *Frontiers in Plant Science, 14*, 1-22.

Shogelova, N.T., & Sartin, S.A. (2021). Application of Landscape-Geochemical Approach in the Study of the Territory of the North Kazakhstan Region. *QazBSQA Habarshysy. urylys konstrukciyalary zhane materialdary. 4*(81). 10.51488/1680-080X/2021.4-02

Shrestha, D., Zinck, J., & Van Ranst, E. (2004). Modelling land degradation in the Nepalese Himalaya. *Catena*, 57(2), 135–156. 10.1016/j.catena.2003.11.003

Shultz M. J. (2007). *Comparison of distributed versus lumped hydrologic simulation models using stationary and moving storm events applied to small synthetic rectangular basins and an actual watershed basin.* ProQuest.

Shultz, M. J., & Corby, R. (2006). Development, calibration, and implementation of a distributed hydrologic model for use in real-time river forecasting.In: (eds). *Proceedings of 3rd Federal Interagency Hydrologic Modeling Conference.* Research Gate.

Shumilov, Yu.V., Luneva, N.N., Ermolenko, S.A., Savva, A.P., Zakota, T.Yu., Mysnik, E.N., & Danilov, R.Yu. (2018). Use of unmanned aerial vehicles for phytosanitary monitoring of weeds. *Plant Protection Bulletin, 4* (98), 22-27.

Singh, V. P. (1992). *Elementary hydrology.* Pearson College Division.

Sivák, M., Kardoš, M., Kadlečík, R., Chudá, J., Tomaštík, J., & Tuček, J. (2023). Identification of the Forest Cover Growth on Landscape Level from Aerial Laser Scanning Data. *Land (Basel)*, 12(5), 1074. 10.3390/land12051074

Sizov, A. P. (2012). Modern problems of land management and cadastres. Part 1 Land management: a textbook for graduate students. M.: Publishing House of MIIGAiK, 69 p.

Skachkova, M., & Gureva, O. S. (2022). Information support of urban planning activities in Russia. *Geodeziâ I Kartografiâ, 986*(8), 45–55. 10.22389/0016-7126-2022-986-8-45-55

Skachkova, M., & Guryeva, O. (2023). Monitoring of the State of Saint Petersburg Green Spaces by Remote Sensing Data. *Ecology and Industry of Russia*, 27(5), 51–57. 10.18412/1816-0395-2023-5-51-57

Smith, G. M., & Milton, E. J. (1999). The use of the empirical line method to calibrate remotely sensed data to reflectance. *International Journal of Remote Sensing*, 20(13), 2653–2662. 10.1080/014311699211994

Smith, M. B. (1993). A gis-based distributed parameter hydrologic model for urban areas. *Hydrological Processes*, 7(1), 45–61. 10.1002/hyp.3360070106

Smith, M. B., Koren, V., Finnerty, B., & Johnson, D. L. (1999). *Distributed modeling: Phase 1 results.* US Department of Commerce, National Oceanic and Atmospheric Administration, National Weather Service.

Smith, M. B., Seo, D.-J., Koren, V. I., Reed, S. M., Zhang, Z., Duan, Q., Moreda, F., & Cong, S. (2004). The distributed model intercomparison project (DMIP): Motivation and experiment design. *Journal of Hydrology (Amsterdam)*, 298(1), 4–26. 10.1016/j.jhydrol.2004.03.040

Soeprobowati, T. R., Sularto, R. B., Hadiyanto, H., Puryono, S., Rahim, A., Jumari, J., & Gell, P. (2023). The carbon stock potential of the restored mangrove ecosystem of Pasarbanggi, Rembang, Central Java. *Marine Environmental Research*, 193, 106257. 10.1016/j.marenvres.2023.10625738086137

Sofo, A., Khanghahi, M. Y., Curci, M., Reyes, F., Briones, M. J., Sarneel, J. M., Cardinale, D., & Crecchio, C. (2023). Earthworm-driven changes in soil chemico-physical properties, soil bacterial microbiota, tree/tea litter decomposition, and plant growth in a mesocosm experiment with two plant species. *Plants*, 12(6), 1216. 10.3390/plants1206121636986903

Soil Atlas of Europe. (2005). *European Soil Bureau Network.* European Commission. https://esdac.jrc.ec.europa.eu/euso/soil-monitoring

Solís, P., Anderson, J., & Rajagopalan, S. (2021). Open geospatial tools for humanitarian data creation, analysis, and learning through the global lens of YouthMappers. *Journal of Geographical Systems*, 23(4), 599–625. 10.1007/s10109-020-00339-x

Song, Y., Zhao, W., Su, Z., Guo, S., Du, Y., Song, X., Shi, X., Li, X., Liu, Y., & Liu, Z. (2024). Effect of Pulsed Electric Field Treatment on Seed Germination and Seedling Growth of Scutellaria baicalensis. *Agriculture 2024, Vol. 14, Page 158, 14*(1), 158. 10.3390/agriculture14010158

Sorokin, I. B., Kulizhsky, S. P., Pasko, O. A., & Merzlyakov, O. E. (2023). The influence of long-term use of straw and green manure on gray forest soil on the productivity of agrocenoses. *Агрохимия*, 9(9), 22–27. 10.31857/S0002188123090119

Sozina, I. D., & Danilov, A. S. (2023). Microbiological remediation of oil-contaminated soils. *Journal of Mining Institute*, 260, 297–312. 10.31897/PMI.2023.8

Spangler, K. R., Brochu, P., Nori-Sarma, A., Milechin, D., Rickles, M., Davis, B., Dukes, K. A., & Lane, K. J. (2023). Calculating access to parks and other polygonal resources: A description of open-source methodologies. *Spatial and Spatio-temporal Epidemiology*, 47, 100606. 10.1016/j.sste.2023.10060638042531

Srivastav, A. L. (2020). Chemical fertilizers and pesticides: role in groundwater contamination. *Agrochemicals Detection, Treatment and Remediation: Pesticides and Chemical Fertilizers*, 143–159. 10.1016/B978-0-08-103017-2.00006-4

Starovoitov, V.I., Starovoitova, M.B., Voronov, N.V. (2002).Cultures of the XXI century. *Technology and equipment for the village*, *8*, 9-13.

Statistika. Egorov, A. G., & Bogolyubov, K. M. (Eds.). (1985). *KPSS v rezolyuciyah i resheniyah sezdov, konferencij i Plenumov CK (1898-1988)[The Communist Party of the Soviet Union in Resolutions and Decisions of Congresses, Conferences and Plenums of the Central Committee (1898-1988)].*

Steinhoff-Knopp, B., Kuhn, T. K., & Burkhard, B. (2020). The impact of soil erosion on soil-related ecosystem services: Development and testing a scenario-based assessment approach. *Environmental Monitoring and Assessment*, 193(S1), 274. 10.1007/s10661-020-08814-033988744

Stepkin, Yu. I., & Gaidukova, E. P. (2018). Risk assessment and management in waste management. *Hygiene and Sanitation*, 8(97), 693–698. 10.18821/0016-9900-2018-97-8-693-698

Stolbetskaya, S. B. (2023). The role of authentic documents in the process of teaching the French specialized language. *Innovative psychological-pedagogical, linguistic and legal bases of Agrarian Science in the conditions of intercultural agrarian communication: materials of the international scientific and practical conference.* Belotserkovsky NAU.

Sugawara, H., Shimizu, S., Takahashi, H., Hagiwara, S., Narita, K., Mikami, T., & Hirano, T. (2016). Thermal Influence of a Large Green Space on a Hot Urban Environment. *Journal of Environmental Quality*, 45(1), 125–133. 10.2134/jeq2015.01.004926828168

Sumberg, J., & Giller, K. E. (2022). What is 'conventional' agriculture? *Global Food Security*, 32, 100617. 10.1016/j.gfs.2022.100617

Swanton, C. J., & Hamill, A. S. (2005) *Jerusalem artichoke.* BUKVI. www.bukvi.ru/stat/stat 394.html

Szcześniak, M., & Piniewski, M. (2015). Improvement of Hydrological Simulations by Applying Daily Precipitation Interpolation Schemes in Meso-Scale Catchments. *Water (Basel)*, 7(2), 747–779. 10.3390/w7020747

Tahat, M. M., Alananbeh, K. M., Othman, Y. A., & Leskovar, D. I. (2020). Soil Health and Sustainable Agriculture. *Sustainability, 12*(12), 4859. 10.3390/su12124859

Talamini, G. (2023). Urbanising the Virgin Lands. In a Mariotti, J. & Leetmaa, K. (Eds.). *Urban Planning During Socialism. Views from the Periphery* (15-30). Routledge. 10.4324/9781003327592-3

Talovina, I. V., Krikun, N. S., Yurchenko, Y. Y., & Ageev, A. S. (2022). Remote sensing techniques in the study of structural and geotectonic features of iturup island (the kuril islands). *Journal of Mining Institute*, 254, 158–172. 10.31897/PMI.2022.45

Taubman, W. (2003). *Khruschev: The Man and His Era.* Free Press.

Terra. Akhmetova, L., & Grigoryev, V. (2021). *Zhumabaj Shayahmetov i ego vremya[Zhumabai Shayakhmetov and his time].* Foliant.

Thakur, M. P., Künne, T., Unsicker, S. B., Biere, A., Ferlian, O., Pruschitzki, U., Thouvenot, L., Türke, M., & Eisenhauer, N. (2021). Invasive earthworms reduce chemical defense and increase herbivory and pathogen infection in native trees. *Journal of Ecology*, 109(2), 763–775. 10.1111/1365-2745.1350433664527

The Government of the Arkhangelsk region. (2016). *On approval of the Regulations on the Mudyug State Natural Landscape Reserve of regional significance* (Resolution of the Government of the Arkhangelsk region 363-pp). The official Internet portal of legal information http://www.pravo.gov.ru

The Government of the Murmansk region. (2020). *On approval of the Procedure for the preparation and composition of materials justifying the creation of specially protected natural territories of regional significance, changing the regime of their special protection* (Resolution of the Government of the Murmansk Region 701-pp).

The Government of the Russian Federation. (2002). *On the approval of the Environmental Doctrine of the Russian Federation* (Decree of the Government of the Russian Federation 1225-p). Collection of legislation of the Russian Federation
The President of the Russian Federation. (2019). *Message from the President of the Russian Federation Vladimir Putin to the Federal Assembly of the Russian Federation* (38). Rossiyskaya gazeta.

The State Duma. (1995). *On Specially protected natural territories* (Federal Law 33-FZ). Collection of legislation of the Russian Federation.

Threlfall, C. G., Mata, L., Mackie, J. A., Hahs, A. K., Stork, N. E., Williams, N. S. G., & Livesley, S. J. (2017). Increasing biodiversity in urban green spaces through simple vegetation interventions. *Journal of Applied Ecology*, 54(6), 1874–1883. 10.1111/1365-2664.12876

Tian, Z., Wang, J. W., Li, J., & Han, B. (2021). Designing future crops: Challenges and strategies for sustainable agriculture. *The Plant Journal*, 105(5), 1165–1178. 10.1111/tpj.1510733258137

Timofeeva, Yu. R. (2020). *Assessment of the agro-resource potential of the soils of the Leningrad region.* Materials on the study of Russian soils: collection of scientific papers, St. Petersburg.

Tishkov, A. (2012). CONSERVATION OF RUSSIAN ARCTIC BIODIVERSITY. *Geography, Environment. Sustainability*, 5(3), 48–63. 10.24057/2071-9388-2012-5-3-48-63

Titok, V. V., Rupasova, Zh. A., Kuptsov, N. S., & Popov, E. G. Duba,r D.A., Pashkevich, P.A., & Veevnik, A.A. (2018). *Jerusalem artichoke in Belarus*. Minsk.

Tittonell, P. (2014). Ecological intensification of agriculture — Sustainable by nature. *Current Opinion in Environmental Sustainability*, 8, 53–61. 10.1016/j.cosust.2014.08.006

Tokareva, O.S., Pasko, O.A., Ushakova, N.S., Makartsova, E.S., Fedorova, L.A. (2018). Comprehensive assessment of the development of vegetation degradation of snow dumps. *Modern problems of remote sensing of the Earth from space, 15*(2), 75-83. .10.21046/2070-7401-2018-15-2-75-83

Tokareva, O. S., Alshaibi, A. D. A., & Pasko, O. A. (2021). Assessment of restoration dynamics of burnt forest area vegetation using landsat satellite data. *Bulletin of the Tomsk Polytechnic University. Geo Assets Engineering*, 332(7), 191–199.

Tokareva, O. S., Pasko, O. A., Majid, S. M., & Cabral, P. (2020). Monitoring the state of vegetation cover in the territory of central Iraq using Landsat-8 satellite data. *Bulletin of Tomsk Polytechnic University.Georesources Engineering*, 331(6), 19–31.

Tokbergenova, A., & Nyussupova, G. Kairova, & Sh., Kiyassova L. (2016). Causes and Impacts of Land Degradation and Desrtification: Case Study of the Republic of Kazakhstan. *4th International Geography Symposium*. Turkey. https://pps.kaznu.kz/kz/Main/FileShow2/77365/79/3/4080/0/

Tomar, D., & Pali, N. (2023). Advantages to agroecosystem by green manures and cover crops. *Journal of Pharmacognosy and Phytochemistry*, 12(2), 146–149.

Tooming, H. G. (1977). *Solar radiation and crop formation*. Gidrometeoizdat.

Tretyakov, V. F. (2009) *Bioethanol - state and prospects in energy and petrochemicals*. Materials II int. Sci-Tech. conf. Alternative sources of raw materials and fuel, Minsk.

Tropeano, D. (1983). Soil erosion on vineyards in the Tertiary Piedmontese basin (northwestern Italy): Studies on experimental areas. *Catena Supplement*, 4, 115–127.

Trotsyuk, A. M., & Kirikylytsia, V. V. (2023). Scientific communication in a foreign (English) language: syllabus of normative academic discipline. *Volyn National University named after Lesya Ukrainka, Faculty of foreign philology, Department of foreign languages of natural and mathematical specialties*. Lutsk.

Tsvetkov, P., Andreychik, A., & Kosarev, O. (2024). The impact of economic development of primary and secondary industries on national CO2 emissions: The case of Russian regions. *Journal of Environmental Management*, 351, 119881. 10.1016/j.jenvman.2023.11988138150925

Turk, J., & Rebecca, Y. (2020). Field Conditions and the Accuracy of Visually Determined Munsell Soil Color. *Soil Science Society of America Journal*, 84(1), 163–169. 10.1002/saj2.20023

Tymchuk, A. I. (2018). On texture features in the problem of segmentation of aerial photographs based on brightness dependence matrices. *Cybernetics and Programming*, (6), 31–39.

Usanova, Z. I. (1967).Photoperiodism of Jerusalem artichoke. *Proceedings of the Tula State University. agricultural experimental station*. 1,Tula: Priokskoe book. publishing house, 161-181.

Usanova, Z. I. (2005). Programming the yield of earthen pear. *Current problems of agricultural science and practice*. Tver.

Usanova, Z. I., & Ivanov, V. A. (1993). Features of the cultivation of Jerusalem artichoke in the north-eastern part of the Central region of Russia. *Abstracts of reports V interregion. intersectoral. scientific-prod. conf.* Tver.

Usanova, Z.I., Oserbaev, A.K. (2003). Production process of the Skorospelka earthen pear in the northern part of the Central region of Russia. *Non-traditional natural resources and innovative technologies and products*. M: RANS.

Ushakov, E., Aleksandrova, T., & Romashev, A. (2021). Neural Network Modeling Methods in the Analysis of the Processing Plant's Indicators. *Advances in Intelligent Systems and Computing*, 1259, 36–45. 10.1007/978-3-030-57453-6_4

Ushakov, T. V. (1932). Earthen pear in Ingushetia. Ordzhonikidze: Publishing house. *Heart (British Cardiac Society)*, 57.

Ustimenko, G.V.(1960). *Earthen pear. M.: State*. Publishing house of agricultural literature.

Uzarkina, M.A., & Uzarkina, O.Yu. (2013). *«Celinnaya epopeya»: istoriko-politologicheskij analiz* ["The Tselina Epic": Historical and Political Analysis]. *Nauka o cheloveke: gumanitarnye issledovaniya.1* (11), 180-189.

Vadyunina, A. F., & Korchagina, Z. A. (1986). *Methods for studying the physical properties of soils* (Vol. M). Agropromizdat.

Vågen, Š. P., Daněk, P., Schaetzl, R. J., Vašíčková, I., & Valtera, M. (2016). Soil mixing and genesis as affected by tree uprooting in three temperate forests. *European Journal of Soil Science*, 66(3), 589–603. 10.1111/ejss.12245

Valladares, F. (2023). More biodiversity to improve our health: The benefits to human well-being of favouring functional and diverse ecosystems. *Metode Science Studies Journal*, (13), 111–117.

Van Steenbergen, F., Lawrence, P., Mehari, A., Salman, M., & Faurés, J. (2010). *Guidelines on spate irrigation. FAO Irrigation and Drainage Paper 65: ISBN 978-92-5-106608-09*. FAO.

Vasilev, Y., Tsvetkova, A., & Stroykov, G. (2020). SUSTAINABLE DEVELOPMENT IN THE ARCTIC REGION OF THE RUSSIAN FEDERATION. *International Multidisciplinary Scientific GeoConference SGEM*, 763–770. 10.5593/sgem2020/5.1/s20.096

Veerman, C. (2020). *Land Contamination & Reclamation*, 11, 31–36. 10.2462/09670513.617

Venter, Z. S., Shackleton, C. M., Van Staden, F., Selomane, O., & Masterson, V. A. (2020). Green Apartheid: Urban green infrastructure remains unequally distributed across income and race geographies in South Africa. *Landscape and Urban Planning*, 203, 103889. 10.1016/j.landurbplan.2020.103889

Verdú, J., Batalla, R., & Martínez-Casasnovas, J. (2006). Estudio hidrológico de la cuenca del río Isábena (Cuenca del Ebro). II: Respuesta hidrológica. *Ing Agua*, 13(4), 331–343. 10.4995/ia.2006.10498

Verkhovskiy, N. (1969). *Tretij zvonok* [The Third Call]. *Sibirskie ogni, 3*. Chernienko,

Vieux B. E. and Moreda F. G. (2003). Ordered Physics-Based Parameter Adjustment of a Distributed Model. *Calibration of Watershed Models*.

Vieux, B. E. (2001). *Distributed hydrologic modeling using GIS*. Springer. 10.1007/978-94-015-9710-4

Viscarra, R. R. A., Behrens, T., & Ben-Dor, E. (2016). A global spectral library to characterize the world's soil. *Earth-Science Reviews*, 155, 198–230. 10.1016/j.earscirev.2016.01.012

Vogel, H. J. (2020). Quantitative evaluation of soil functions: Potential and state. *Frontiers in Environmental Science*, 22. 10.3389/fenvs.2019.00164

Vohland, M., Besold, J., Hill, J., & Fründ, H.-C. (2011). Comparing different multivariate calibration methods for the determination of soil organic carbon pools with visible to near infrared spectroscopy. *Geoderma*, 166(1), 198–205. 10.1016/j.geoderma.2011.08.001

Volin, L. A. (1970). *Century of Russian Agriculture: From Alexander II to Khrushchev*. Harvard University Press. 10.4159/harvard.9780674366350

Volungevičius, S. J., & Skorupskas, R. (2011). Classification of anthropogenic soil transformation. *Geologija. Vilnius,4*(76), 165–177. 10.6001/geologija.v53i4.1904

Vonk, W. J., van Ittersum, M. K., Reidsma, P., Zavattaro, L., Bechini, L., Guzmán, G., Pronk, A., Spiegel, H., Steinmann, H. H., Ruysschaert, G., & Hijbeek, R. (2020). European survey shows poor association between soil organic matter and crop yields. *Nutrient Cycling in Agroecosystems*, 118(3), 325–334. 10.1007/s10705-020-10098-2

Wajchman-Świtalska, S., Grabowska-Chenczke, O., & Woźniak, M. (2023). Exploring the Patterns of Recreational Activity among Urban Green Spaces in Poland. *Sustainability (Basel)*, 15(6), 5425. 10.3390/su15065425

Wambugu, P. W., & Henry, R. (2022). Supporting in situ conservation of the genetic diversity of crop wild relatives using genomic technologies. *Molecular Ecology*, 31(8), 2207–2222. 10.1111/mec.1640235170117

Wang, L., Zhao, Y., Al-Kaisi, M., Yang, J., Chen, Y., & Sui, P. (2020). Effects of Seven Diversified Crop Rotations on Selected Soil Health Indicators and Wheat Productivity. *Agronomy, 10*(2), 235. 10.3390/agronomy10020235

Wang, J., Rienow, A., David, M., & Albert, C. (2022). Green infrastructure connectivity analysis across spatiotemporal scales: A transferable approach in the Ruhr Metropolitan Area, Germany. *The Science of the Total Environment*, 813, 152463. 10.1016/j.scitotenv.2021.15246334952053

Wang, Q., Robson, T. M., Pieristè, M., Kenta, T., Zhou, W., & Kurokawa, H. (2022). Canopy structure and phenology modulate the impacts of solar radiation on C and N dynamics during litter decomposition in a temperate forest. *The Science of the Total Environment*, 820, 153185. 10.1016/j.scitotenv.2022.15318535065130

Wang, X., Zhang, Q., Zhang, Z., Li, W., Liu, W., Xiao, N., Liu, H., Wang, L., Li, Z., Ma, J., Liu, Q., Ren, C., Yang, G., Zhong, Z., & Han, X. (2023). Decreased soil multifunctionality is associated with altered microbial network properties under precipitation reduction in a semiarid grassland. *iMeta*, 2(2), e106. 10.1002/imt2.106

Weindorf, D. C., Muir, J. P., & Landeros-Sánchez, C. (2011). *Organic Compost and Manufactured Fertilizers: Economics and Ecology*. Springer. 10.1007/978-94-007-1309-3_2

Wei, W., Yan, Y., Cao, J., Christie, P., Zhang, F., & Fan, M. (2016). Effects of combined application of organic amendments and fertilizers on crop yield and soil organic matter: An integrated analysis of long-term experiments. *Agriculture, Ecosystems & Environment*, 225, 86–92. 10.1016/j.agee.2016.04.004

Wezel, A., Casagrande, M., Celette, F., Vian, J. F., Ferrer, A., & Peigné, J. (2014). Agroecological practices for sustainable agriculture. A review. *Agronomy for Sustainable Development*, 34(1), 1–20. 10.1007/s13593-013-0180-7

Wezel, A., Herren, B. G., Kerr, R. B., Barrios, E., Gonçalves, A. L. R., & Sinclair, F. (2020). Agroecological principles and elements and their implications for transitioning to sustainable food systems. A review. *Agronomy for Sustainable Development*, 40(6), 1–13. 10.1007/s13593-020-00646-z

Wheater H. (2002a). Progress in and prospects for fluvial flood modelling. *Philosophical Transactions of the Royal Society of London A: Mathematical, Physical and Engineering Sciences*, *360*(1796), 1409-1431.

Wheater H. S. (2002b). Hydrological processes in arid and semiarid areas. *Hydrology of wadi systems*, 5-22.

Wheater, H., Butler, A., Stewart, E., & Hamilton, G. (1991). A multivariate spatial-temporal model of rainfall in southwest Saudi Arabia. I. Spatial rainfall characteristics and model formulation. *Journal of Hydrology (Amsterdam)*, 125(3), 175–199. 10.1016/0022-1694(91)90028-G

Wicherek, S. (1991). Viticulture and soil erosion in the north of Parisian basin. Example: The mid Aisne region. *Zeitschrift für Geomorphologie*, 83(Supplementband), 115–126.

Wieczochek, A. (1988). Helianthus tuberosus as a potential energy source. *Acta aliment.Hol*, 14(2), 115–122.

Wiegant, D., Bakx, J., Flohr, N., Oel, P., & Dewulf, A. (2022). Ecuadorian water funds' use of scale-sensitive strategies to stay on course in forest and landscape restoration governance. *Journal of Environmental Management*, 311, 0301–4797. 10.1016/j.jenvman.2022.11485035278923

Wiki, O. (2023). *About OpenStreetMap*. https://wiki.openstreetmap.org/wiki/About_OpenStreetMap

Woolhiser, D. A., Fedors, R. W., Smith, R. E., & Stothoff, S. A. (2006). Estimating infiltration in the upper split wash watershed, Yucca Mountain, Nevada. *Journal of Hydrologic Engineering*, 11(2), 123–133. 10.1061/(ASCE)1084-0699(2006)11:2(123)

Woolhiser, D. A., Smith, R., & Goodrich, D. C. (1990). *KINEROS: a kinematic runoff and erosion model: documentation and user manual*. US Department of Agriculture, Agricultural Research Service.

Wu, J., Ayres, E., Vance, G. F., Luo, Y., & Guo, Q. (2012). Effects of elevated CO2 and N deposition on litter decomposition and soil microbial community in a Tibetan alpine meadow. *Plant and Soil*, 351(1-2), 101–112.

Xie, S., Tran, H. T., Pu, M., & Zhang, T. (2023). Transformation characteristics of organic matter and phosphorus in composting processes of agricultural organic waste: Research trends. *Materials Science for Energy Technologies*, 6, 331–342. 10.1016/j.mset.2023.02.006

Xu, H., Fan, Z., Ahmad, F., & Zhang, D. (2023). Exploring the ecological protection impacts of cultivated land transfer: Explanation based on fertilizers and pesticides. *Ecological Indicators*, 154, 110681. 10.1016/j.ecolind.2023.110681

Yadav, J., Srivastva, A. K., & Singh, R. (2024). Diversity of halotolerant endophytes from wheat (Triticum aestivum) and their response to mitigate salt stress in plants. *Biocatalysis and Agricultural Biotechnology*, 56, 103000. 10.1016/j.bcab.2023.103000

Yang, C., He, X., Wang, R., Yan, F., Yu, L., Bu, K., Yang, J., Chang, L., & Zhang, S. (2017). The effect of urban green spaces on the urban thermal environment and its seasonal variations. *Forests*, 8(5), 153. 10.3390/f8050153

Yan, Z. Z., Chen, Q. L., Li, C. Y., Nguyen, B. A. T., Zhu, Y. G., He, J. Z., & Hu, H. W. (2021). Termite mound formation reduces the abundance and diversity of soil resistomes. *Environmental Microbiology*, 23(12), 7661–7670. 10.1111/1462-2920.1563134097804

Yaroshevich, M. I. (2010). Jerusalem artichoke is a promising multi-purpose crop. *Proceedings of BSU*, 4(2), 1–12.

Yaroshevich, M. I. (2014). *Jerusalem artichoke is a promising multi-purpose crop.*

Yatheendradas, S., Wagener, T., Gupta, H., Unkrich, C., Goodrich, D., Schaffner, M., & Stewart, A. (2008). Understanding uncertainty in distributed flash flood forecasting for semiarid regions. *Water Resources Research*, 44(5), 2007WR005940. 10.1029/2007WR005940

Yeboah, G., de Albuquerque, J. P., Troilo, R., Tregonning, G., Perera, S., Shifat Ahmed, S. A. K., Ajisola, M., Alam, O., Aujla, N., Azam, S. I., Azeem, K., Bakibinga, P., Chen, Y. F., Choudhury, N. N., Diggle, P. J., Fayehun, O., Gill, P., Griffiths, F., Harris, B., & Yusuf, R. (2021). Analysis of openstreetmap data quality at different stages of a participatory mapping process: Evidence from slums in Africa and Asia. *ISPRS International Journal of Geo-Information*, 10(4), 265. 10.3390/ijgi10040265

Yu. (1964). *Celinnaya doroga* [The road of Tselina]. *Novyj mir, 1.*

Yuritsina, T.N. (1998). Jerusalem artichoke is an amazing plant. *Potatoes and vegetables*, *3*, 17-21.

Zakharchenko A.V, Moskovchenko D.V., Tigeev A.A. (2022). Analysis of the heterogeneity of the chemical composition of snow cover using clustering methods (k-means) and GIS-technology. *Ice and snow, 62*, 2. 227-240. 10.31857/S2076673422020128

Zakharchenko, A.V. (2005a). Morphological features of the surface nanorelief of soddy-podzolic soils of forest biogenocenoses. *Bulletin of Tomsk State Pedagogical University, Series: Natural and exact sciences (information technologies)*, 7(51), 100-108.

Zakharchenko, A.V. (2005b). Natural restoration of anthropogenically modified soils. *Bulletin of Tomsk State Pedagogical University, Series: Natural and exact sciences (information technologies)*, 136-144.

Zakharchenko, A.V. (2007). Spatial conjugacy of morphological soil surfaces. *Bulletin of Tomsk State Pedagogical University, Series: Natural and exact sciences (information technologies)*, 146-152.

Zakharchenko, A. V., Alekseev, V. I., & Ipatova, D. V. (2016). Hierarchical concept of soil heterogeneity and planning of study scale. *Bulletin of Tomsk Polytechnic University.Georesources Engineering*, 327(4), 149–163.

Zakharchenko, A. V., & Pasko, O. A. (2022). *Plants, soils and sustainable farming.*

Zakharchenko, A. V., Pasko, O. A., & Sorokin, I. B. (2021). The dynamics of the reduction of arable land areas according to their long-term monitoring in the Tomsk region. *Ecology and Industry of Russia*, 25(11), 30–35. 10.18412/1816-0395-2021-7-54-59

Zakharchenko, A. V., & Zakharchenko, N. V. (2006). Three-Dimensional Surface Morphometry of Soil Horizons in Field Studies. *Eurasian Soil Science*, 39(2), 134–140. 10.1134/S1064229306020037

Zakharchenko, A., Pasko, O., Tigeev, A., & Moskovchenko, D. (2023). On the Background Values of the Geochemical Spectra of Microelements of the Solid Phase of Snow During their Comparative Analysis in the Cities of Tobolsk And Tyumen. *Ecology and Industry of Russia*, 27(4), 61–65. 10.18412/1816-0395-2023-4-61-65

Zakharov, S. A. (1931) *Soil science course*. Moscow, Moscow State University.

Zhang, L., Qi, S., Li, P., & Zhou, P. (2024). Influence of stand and environmental factors on forest productivity of Platycladus orientalis plantations in Beijing's mountainous areas. *Ecological Indicators*, 158, 111385. 10.1016/j.ecolind.2023.111385

Zhang, Q. (2015). *Precision Agriculture Technology for Crop Farming*. CRC Press. 10.1201/b19336

Zhang, Y., Li, X., Wang, A., Bao, T., & Tian, S. (2015). Density and diversity of OpenStreetMap road networks in China. *Journal of Urban Management*, 4(2), 135–146. 10.1016/j.jum.2015.10.001

Zhang, Y., Wu, Q., Wu, L., & Li, Y. (2021). Measuring community green inequity: A fine-scale assessment of beijing urban area. *Land (Basel)*, 10(11), 1197. 10.3390/land10111197

Zhang, Z. Q. (2011). *Animal biodiversity: An outline of higher-level classification and survey of taxonomic richness*. Magnolia press.

Zhao, J. (2020). The role of international organizations in preventing conflicts between the SPS Agreement and the Cartagena Protocol on Biosafety. *Review of European, Comparative & International Environmental Law*, 29(2), 271–281. 10.1111/reel.12326

Zharnikov, V. B., Larionov, Yu. S., & Pasko, O. A. (2019). On traditions and innovations in domestic land management and cadaster. *Interexpo Geo-Siberia*, 3(2), 73–79. 10.33764/2618-981X-2019-3-2-73-79

Zhou, P., Zhang, H., Huang, B., Ji, Y., Peng, S., & Zhou, T. (2022). Are productivity and biodiversity adequate predictors for rapid assessment of forest ecosystem services values? *Ecosystem Services*, 57, 101466. 10.1016/j.ecoser.2022.101466

Zhou, S., Jia, P., Xu, W., Shane Alam, S., & Zhang, Z. (2023). A novel composting system for mitigating ammonia emissions and producing nitrogen-rich organic fertilizer. *Bioresource Technology*, 386, 129455. 10.1016/j.biortech.2023.12945537419288

Zhu, Y., Mao, H., Li, P., & Wu, Y. (2011). *Measurement of lettuce leaf chlorophyll content by means of VIS-NIR spectroscopy*. International Conference on Remote Sensing, Environment and Transportation Engineering, Nanjing.

Zhukova, M. A., & Ulezko, A. V. (2021). *Prospects for the digital transformation of agriculture*. Voronezh: Voronezh State Agrarian University.

Zhu, Q., Liu, X., Hao, T., Zeng, M., Shen, J., Zhang, F., & de Vries, W. (2020). Cropland acidification increases risk of yield losses and food insecurity in China'. *Environmental Pollution*, 256, 113145. 10.1016/j.envpol.2019.11314531662249

Zinck, J. A., Metternicht, G., & Bocco, G. (2016). *Geopedology. An Integration of Geomorphology and Pedology for Soil and Landscape Studies*. Springer International publishing.s 10.1007/978-3-319-19159-1_1

Zink, A., Fleige, H., & Horn, R. (2011). Verification of harmful subsoil compaction in loess soils'. *Soil & Tillage Research*, 114(2), 127–134. 10.1016/j.still.2011.04.004

Zowada, C., Frerichs, N., Zuin, V. G., & Eilks, I. (2020). Developing a lesson plan on conventional and green pesticides in chemistry education – a project of participatory action research. *Chemistry Education Research and Practice*, 21(1), 141–153. 10.1039/C9RP00128J

Zubkova, T. A., & Karpachevsky, L. O. (2001). *Matrix organization of soils*. State University named after M. V. Lomonosov.

Zuckerman, V. A. (Ed.). (2022). *Innovative development of industry in the Arctic regions: problems and prospects*. FRC KSC RAS. 10.37614/978.5.91137.462.4

Zurbaran, M. A., Wightman, P., & Brovelli, M. A. (2019). A MACHINE LEARNING PIPELINE ARTICULATING SATELLITE IMAGERY and OPENSTREETMAP for ROAD DETECTION. *International Archives of the Photogrammetry, Remote Sensing and Spatial Information Sciences - ISPRS Archives, 42*(4/W14). 10.5194/isprs-archives-XLII-4-W14-255-2019

About the Contributors

Olga Pasko worked at the National Research Tomsk Polytechnic University, Engineering School of Natural Resources, Department of Geology, Professor. Currently, he is Vice-Rector for Research at the National Open Institute St. Petersburg. She has a work experience in the "Agrophysical Research Institute", St. Petersburg as the head of the experimental research sector. Main responsibilities: management of profiles in the areas of "Land Management" (master's degree) and "Land management, cadastre and land monitoring" (postgraduate studies); preparation and implementation of accreditation for these specialities in 2016 and 2021; planning, organization and control of educational, educational and methodological work in the disciplines provided, taking into account the specifics of the contingent of students. Olga Anatolievna created some working curricula for the provided disciplines and management of their development by other teachers. She also conducted all types of training sessions, managing course and diploma projects and research work for bachelor's and masters. There is her control of the methodological support of the provided disciplines, management of the textbooks preparation, educational and teaching aids, lecture notes and other methodological material on the disciplines provided, direct participation in their development and in preparing them for publication. Her experience is the management of research work in the scientific direction of the department; involvement in its implementation of teachers, teaching and support staff of the department; graduate students and students of the department, as well as specialists from other structural divisions of the university; management of the preparation of graduate students and applicants at the department. She is also known for her teaching experience at universities in Tomsk: National Research Tomsk Polytechnic University, National Research Tomsk State University, Tomsk University of Architecture and Civil Engineering, Tomsk Agricultural Institute – a branch of the Novosibirsk Agrarian University. More than 200 publications have been published, including in journals included in the Scopus and WoS databases – 20. Hirsch Scopus Index – 4, WoS – 3, Hirsch RSCI – 10. Awarded with thanks from the Ministry of Science and Higher Education of the Russian Federation, the Administration of the Tomsk Region, State Duma of the Tomsk Region, Administration of the Novosibirsk Region, Government of the Novosibirsk Region, Omsk State University, Tomsk State University, Tomsk Polytechnic University. Olga Pasko is a well-known screenwriter and film director of documentaries, a full member of the Eurasian Academy of Television and Radio, a member of the International Union of Journalists, and the winner of numerous film and television festivals.

Nadezhda Anatolievna Lebedeva is a Doctor of Philosophy in the field of Cultural Studies and works at the Kherson State Agrarian University (Kherson, Ukraine) as the department head of Humanitarian Disciplines Chair, and International Personnel Academy (Kyiv, Ukraine), professor, and a full member of the Eurasian Academy of Television and Radio. She is the main editor of the international journal "Universum: Philology & Arts", a reviewer and member of the editorial board of scientific collections of international conferences and scientific journals ("Journal of Social, Humanity, and Education", "Journal of Sustainable Tourism and Entrepreneurship" Goodwood Publishing), the author of more than a hundred publications, including Scopus and WoS, has recommendations and certificates of honour for successful scientific activity). Experienced in project cinematography management and developed lecture courses in a foreign language for students of state administration. She is the author of more than 100 scientific works. The focus of her research is the cultural and social aspects of regional development. N. Lebedeva graduated from the Kherson State Pedagogical Institute in 1996. After graduating from the post-graduate course at the Kherson State University, became a lecturer at the History of World Culture and Literature Department of the Kherson State University. In June 2008 was enrolled in the doctoral program of the Interregional Academy of Personnel Management in Kyiv, specializing in Cultural Studies. She was awarded the medal "For Civil Courage" of the Afghan Veterans' Ukrainian Union for the book "My Afghan" by Nikolay Vishnyak". Participation in the creation of documentary films and scientific articles is marked by membership in the Eurasian Academy of Television and Radio (November 2012). The monograph "Cultural Aspect of Alexander Elagin's Creativeness" was awarded 1st place in scientific direction "Cultural Studies" in 2017 and became the winner of the competition for educational projects "Interclover – 2017" in the nomination – "Research Project". She also was invited to be a member of the Ukrainian Assembly of Doctors of Sciences in the Field of State Administration (February 2018). Scope of her scientific interests: state administration, philosophy, pedagogy, Russian, Ukrainian and English philology, cultural studies and ethnology. Currently a reviewer for Indonesian journals.

***

About the Contributors

Mohamed Abdalla is a researcher at the Department of soil and water conservation, water resources and desert Lands division, Desert Research Center. His interests are hydropolitics and resilient use of water resources as a driver for sustainable development. His field of expertise is mainly water resources engineering, hydrological modeling. Remote Sensing - GIS Climate change- Climate Smart Agriculture - Water harvesting - Soil and water conservation Hydrological modeling (KINEROS2 - SWAT)

Munir Ahmad, Ph.D. in Computer Science, brings over 24 years of invaluable expertise in the realm of spatial data development, management, processing, visualization, and quality assurance. His unwavering commitment to open data, big data, crowdsourced data, volunteered geographic information, and spatial data infrastructure has solidified him as a seasoned professional and a trusted trainer in cutting-edge spatial technologies. With a profound passion for research, Munir has authored more than 30 publications in his field, culminating in the award of his Ph.D. in Computer Science from Preston University Pakistan in 2022. His dedication to propelling the industry forward and sharing his extensive knowledge defines his mission.

Sergey Sergeevich Akhmetov, UNESCO Department of Journalism and Communication, al-Farabi Kazakh National University, Kazakhstan. Sergey Akhmetov - born in Almaty, Kazakhstan. Education: Kazakhstan Institute of Economics, Management and Strategic Research (Bachelor's dergree: economics of public policy and business economics); al-Farabi Kazakh National University (master's degree: public relations). Coauthor of the books: "Nepravitelstvenniy sector Almaty" ["Nongovernmental Sector of Almaty"] in Russian, "Otsenka I monitoring SMI v predviborniy period" in Russian ["Media Assessment and Monitoring During the Election Period"]; author of research papers on history of Kazakhstan, tourism industry of Kazakhstan, mass media and public relations.

Laila Seysembekovna Akhmetova, UNESCO Department of Journalism and Communication, al-Farabi Kazakh National University, Kazakhstan. Doctor of History (since 1998), Professor of Political Science (since 2001). Author of monographs on the history of the Second World War, the history of Kazakhstan of the XX century. Researcher in the field of media, public relations, political science. Editor-in-Chief of "PR and Media in Kazakhstan. Research Papers Collection". The most noted works: "Patriotism from the Perspective of History", monograph (1996), "Journalism and Management", study guide (2000), "Conflictology", study guide (2003), "Don't Agree to Slavery", textbook of the International Organization for Migration (2005), "Russian-Kazakh dictionary of Civil Society Terms" (2006), "The Worst Forms of Child Labor in the Republic of Kazakhstan: Assessment of the Situation and Measures Taken in the Country" (2008), "Press Release. Press Conference. Work experience", study guide (2008), "Political Science", study guide (2006), "Measuring Democracy in the Elections of Almaty", study guide (2007), "First Persons of Kazakhstan in the Stalin Era", political history (2010), "The National Idea of Kazakhstan: Ways to Solve Problems", monograph (2010), "Political Advancement of Women" (2010), "Tripartite Agreement on Social Partnership" (2010), "Terrorism and Extremism: History, Theory and Practice of the Issue", tutorial (2011), "Gender Dimension and Expansion of Political Rights of Women in Kazakhstan" (2012), "Labor Collectives and Conflicts: Work experience" (2012), "Media Education in Kazakhstan" (2013), "Principles of Business Communication", study guide (2015), "1941. Brest Fortress. Kazakhstan", monograph (2016), "UNESCO International Journalism Standards in Kazakhstan" (2019), "Media and Information Literacy Among Young People in the Field of National Security" (2019), "Panfilov's Men: Our Pride, Our Glory", monograph (2020), "Military Journalism 1940-22.06.1941: on the Example of the military correspondent M. Daurembaev", educational and methodological manual (2022), etc.

B. Sharmila received her Ph.D. Degree in Electrical Engineering under Anna University, Chennai in the year of 2013. She completed her Post Graduate in Applied Electronics in 2004 and her Under Graduate in Electronics and Instrumentation Engineering in the year 2000. She has a total experience of 18 years in teaching. She holds on additional responsibility of Chief Officer (Operations) in SREC SPARK Incubation Foundation. She is the NISP Convenor, MoE of the institution. She is a active member of MHRD IIC, MoE, CoIN and Innovation Ambassador for mentoring the Stakeholders for Projects, Startups and Incubation. She is a recognized Research Supervisor under Anna University, Chennai. Currently she guides One Full Time Research Scholar and 4 Part Time Research Scholars. Her area of interest includes Control Systems, Sensor Technology, Networked Control System and Automation. he published 50 Technical Research Papers in referred International Journals and Conferences. Also presented and published papers in International Conferences held at Singapore, Dubai and Indonesia. She has published one Book and four Book Chapters (Three Springer Series).

Elizaveta Bogdanova graduated from St. Petersburg University of Architecture and Civil Engineering in the field of Land Management and Cadastre (Bachelor's degree). Graduated from St. Petersburg University of Architecture and Civil Engineering in the field of Land Management and Cadastre (Master's degree). Postgraduate student of the St. Petersburg Mining University of Empress Catherine II in the field of Land Management, Cadastre and land monitoring. Area of interest: research in the field of environmental protection measures, cadastral accounting of natural resources.

Pravinsang P. Dodia boasts over three decades of academic and research excellence, with a remarkable 40 publications in prestigious national and international journals. His expertise spans the ecology and behavior of birds and mammals, emphasizing conservation biology, toxicology, and biodiversity monitoring. As a recipient of grants from esteemed agencies, He leads pioneering studies that address critical environmental challenges. His mentorship extends to nine PhD candidates and 23 MSc dissertation students, nurturing the next generation of conservation scientists. In administrative roles like course coordinator and as a member of reputable organizations, he shapes academic landscapes. His persuasive advocacy extends to expert lectures at conferences, inspiring action on pressing conservation issues. Dr. Dodia's enduring dedication and visionary leadership inspiring positive change and shaping the future of environmental stewardship.

Alper Durmaz was born on May 18, 1990, in Istanbul, Turkey. He completed his undergraduate studies in 2013 at Ondokuz Mayıs University, where he majored in Secondary Education in Science and Mathematics with a focus on Biology Teaching. Following this, he earned an associate degree in Medical Imaging Techniques from the Department of Medical Services and Techniques at the same university in 2016. Driven by a passion for biology, Dr. Durmaz pursued a master's degree in Botany at Ondokuz Mayıs University, graduating in 2016. His master's thesis focused on the ecological characteristics of Aristolochia samsunensis Davis. He continued his academic journey at the same institution, earning his Ph.D. in Biology in 2021 with a dissertation titled "A Study on Some Edaphic Factors Affecting Zonation in Halophytic Vegetation in the Yeşilırmak Delta." His Ph.D. advisor was Prof. Dr. Hasan Korkmaz, who also supervised his master's research. Since 2023, Dr. Durmaz has been working as a postdoctoral contracted researcher at the Ali Nihat Gökyiğit Botanical Garden Application and Research Center of Artvin Çoruh University. His areas of expertise include plant ecology and plant systematics, with significant research contributions on the ecology of halophytic vegetation. Dr. Durmaz has been actively involved in various national and international projects. He has participated as a researcher in projects such as the study of edaphic factors affecting halophytic vegetation zonation in the Yeşilırmak Delta and the establishment of processing and storage facilities for aromatic plants in Artvin. He has also contributed to archaeological surface research on the Akalan Castle and its surroundings in Samsun. Additionally, he is involved in international COST projects aimed at promoting sustainable food production, agriculture, and forestry science. Dr. Durmaz has published numerous articles in national and international peer-reviewed journals and has presented his work at international conferences and symposiums. He has also authored chapters in scientific books, focusing on biochemical methods and nanotechnological applications in plant-microbe interactions. Dr. Alper Durmaz continues to make significant contributions to the field of plant ecology and systematics. Through his research and academic endeavors, he not only advances scientific knowledge but also mentors students, helping to shape the future of botanical sciences.

Vladimir Nikolaevich Goryachkin is the acting director of the Leningradsky Agrochemical Service Center. Graduated from the Far Eastern State Agrarian University in Blagoveshchensk. He is completing his postgraduate studies at the Leningrad State Agrarian University. He is one of the leading managers of the agro-industrial complex of the Leningrad region.

Evgeniy Yurievich Kartashov Candidate of Technical Sciences, Associate Professor. Specialist in the development of various models of unmanned aerial vehicles (UAVs) and installation of digital devices on them, including digital cameras. In addition, he is developing approaches to processing images obtained using UAVs. In addition to scientific work, he does a lot of methodological work, organizing the work of a children's club to develop UAV models, launch them, and also hold competitions.

Mikhail Kataev received his master's degree in 1984 from Tomsk State University. Joined the Institute of Atmospheric Optics of the Siberian Academy of Sciences. In 1993 he defended his Ph.D. thesis. In 1995 he went to work at the Tomsk University of Control Systems and Radioelectronics at the Faculty of Control Systems. In 2002 he defended his doctoral dissertation. In 2006 he became a professor. Scientific directions are associated with solving mathematical problems of data processing and analysis. He is a specialist in the development of software of various classes related to data processing and analysis, business processes. Conducts educational and scientific work. Under his leadership, 5 candidates of sciences, dozens of undergraduates and bachelors have been defended. He personally and in co-authorship has written about 300 articles, several monographs, textbooks. He is an academician of the International Academy of Informatization, an expert of the Ministry of Education and Science, an academician of the International Academy of Higher Education.

Vasily Kovyazin has Research interests in Formation of sustainable ecosystems and rational use of soil and plant resources of megacities, development of forest and city cadastres for the purposes of recreational and rational nature management.

Vishal M. Makwana is currently pursuing his PhD in Zoology at the Zoology Department, Sir P. P. Institute of Science, M. K. Bhavnagar University, specializing in the urban ecology of birds. He has made significant scholarly contributions, publishing five research articles and three book chapters in renowned national and international journals and books. His research interests span diverse and critical fields, including ornithology, entomology, and environmental toxicology. He is a sought-after speaker, having delivered 5 invited talks and presented his research at 6 prominent national and international conferences and workshops. He has reviewed for many International journals of repute.

About the Contributors

Froilan Mobo is a Doctor of Public Administration graduate from the Urdaneta City University Class of 2016 and a graduate of the 2nd Doctorate Degree (Ph.D.) in Development Education program at the Central Luzon State University, Nueva Ecija, Philippines, Class of 2022. On March 11, 2024, Dr. Mobo was accredited and reclassified by the Commission on Higher Education (CHED) to the position of Professor II in the Philippine Merchant Marine Academy (PMMA), and this allowed him to work with different international research institutions, such as the Director and Research Consultant of the IKSAD Research Institute, Turkey. At present, he is in the process of finishing his 3rd master's degree, leading to social studies education at Bicol University. Recently, Dr. Mobo passed Batch 3—Certified Research Professional—and ranked in the top 5 in the National Examination.

D. Devasena Mohan completed her Ph.D Degree in Information and Communication Engineering under Anna University, Chennai in the year 2020. She completed her post Graduate degree in Communication Systems in the year 2010 under Anna University, Coimbatore. She secured university rank during her post Graduate Programme. She completed her under graduate degree in B.E. Electronics and Instrumentation Engineering from Maharaja Engineering College, Avinashi, Coimbatore under Anna University Chennai. She has 15 years of experience in teaching. Her research interests are in Image Processing, VLSI Implementation. Sensors, Instrumentation, waste management. She has presented papers 9 papers in National and International Conferences. She published 15 technical papers in international Journals and published three book chapters. She received 3 IPR copyrights for laboratory manuals and granted 2 patents.

|Truong An Nguyen - Contributing Author|-

|Pasko Olga - Contributing Author|Pasko Olga Anatolyevna worked at the National Research Tomsk Polytechnic University, Engineering School of Natural Resources, Department of Geology, Professor. Currently, he is Vice-Rector for Research at the National Open Institute St. Petersburg. She has a work experience in the "Agrophysical Research Institute", St. Petersburg as the head of the experimental research sector. Main responsibilities: management of profiles in the areas of "Land Management" (master's degree) and "Land management, cadastre and land monitoring" (postgraduate studies); preparation and implementation of accreditation for these specialities in 2016 and 2021; planning, organization and control of educational, educational and methodological work in the disciplines provided, taking into account the specifics of the contingent of students. Olga Anatolievna created some working curricula for the provided disciplines and management of their development by other teachers. She also conducted all types of training sessions, managing course and diploma projects and research work for bachelor's and masters. There is her control of the methodological support of the provided disciplines, management of the textbooks preparation, educational and teaching aids, lecture notes and other methodological material on the disciplines provided, direct participation in their development and in preparing them for publication. Her experience is the management of research work in the scientific direction of the department; involvement in its implementation of teachers, teaching and support staff of the department; graduate students and students of the department, as well as specialists from other structural divisions of the university; management of the preparation of graduate students and applicants at the department. She is also known for her teaching experience at universities in Tomsk: National Research Tomsk Polytechnic University, National Research Tomsk State University, Tomsk University of Architecture and Civil Engineering, Tomsk Agricultural Institute – a branch of the Novosibirsk Agrarian University. More than 200 publications have been published, including in journals included in the Scopus and WoS databases – 20. Hirsch Scopus Index – 4, WoS – 3, Hirsch RSCI – 10. Awarded with thanks from the Ministry of Science and Higher Education of the Russian Federation, the Administration of the Tomsk Region, State Duma of the Tomsk Region, Administration of the Novosibirsk Region, Government of the Novosibirsk Region, Omsk State University, Tomsk State University, Tomsk Polytechnic University.

Haytham Mohamed Salem obtained his PhD from the Polytechnic University of Madrid (Spain) in 2014. His PhD thesis was about the effects of conservation tillage systems on soil properties and erosion control in arid and semi-arid regions. Currently, he is working as a postdoctoral fellow at the Department of Soil and Water Systems, University of Idaho, USA. As well as, he had worked as a researcher of Soil and Water Conservation at the Desert Research Center (DRC, Egypt) for more than fourteen years. During his work Dr. Haytham Salem has contributed significantly to reduce soil erosion and increase rainwater use efficiency and the resulting sedimentation and understanding the behavior of soil erosion, which is of particular importance to control the degradation of productive soils in the Mediterranean region. Furthermore, he has extensive experience in improving soil health and crop production using sustainable soil and water management and restoration practices under rainfed and irrigation conditions.

|Igor Sorokin - Contributing Author|The main goal of the FSBI "Agrochemical Service Station "Tomskaya" is to monitor the state of soil fertility of agricultural lands in the Tomsk region and to formulate, on its basis, scientifically based support for the preservation and improvement of soil fertility to increase the volume of production of high-quality agricultural products. Under the leadership of Doctor of Biological Sciences Igor Borisovich Sorokin, the Tomsk State Agrochemical Service monitors the condition of agricultural lands, compliance with regulations on the use of agrochemicals and organic fertilizers, and carries out comprehensive agrochemical and toxicological surveys.

|Emine Torunoğlu - Contributing Author"

Y. Dharshan has received his Master's degree in Control and Instrumentation Engineering from Regional centre of Anna University, Coimbatore in the year 2015 and Bachelor's degree in Instrumentation and Control Engineering from Tamilnadu College of Engineering, Coimbatore in the year 2011. He received his Doctoral Degree in the Faculty of Electrical Engineering from Anna University in the year 2024. He has 13 years of experience in the field of teaching. His areas of interest are Control Systems, Machine Vision. and Sensors & Automation.

Alexander Zakharchenko graduated from Tomsk State University in 1980with a degree in soil science and agrochemistry. He worked as a research assistant at the Research Institute of Biology and Biophysics at TSU from 1980 to 2005. In 2006, he defended his thesis for Doctor of Biological Sciences in the field of soil science. Since 2006, he worked at Ugra State University as a professor. In 2008, he was awarded the title of associate professor in ecology. From 2013 to 2018 he taught at the National Research Tomsk Polytechnic University as a professor. Since 2018, I have been working at the Tyumen Scientific Center of the Siberian Branch of the Russian Academy of Sciences as a chief researcher. Scientific activities are related to the problems of soil formation, monitoring of environmental conditions and reclamation of anthropogenic soils in sanitary protection zones and industrial sites. Fundamental research has been carried out on the influence of plants on the soil. It is shown that there are no significant environmental consequences of the impact on soils of the alternating electromagnetic field of overhead power lines, as well as the main patterns of changes in the morphology and chemical composition of soddy-podzolic soils in the case of their natural recovery with varying degrees of initial technogenic disturbance in the sanitary protection zone of power lines. In recent years I have been engaged in geochemical studies of mercury pollution in lakes of the Khanty-Mansiysk Autonomous Okrug at oil fields, the content of microelements in the snow cover of the Tyumen region, and assessment of atmospheric air pollution. Over the past 5 years 25 scientific articles have been published, 10 are indexed by Scopus, 1 is a monograph, 3 articles are web of science. Hirsch index – 4. Participated in the RRFFR project (This study was supported by Russian Foundation For Basic Research, grant No. 1905-50062/20).

Alexander Viktorovich Zakharchenko, born in 1955, citizenship of the Russian Federa-tion, Doctor of Biological Sciences, specialty 03.00.27 – "Soil Science", associate Professor, specialty ecology, chief researcher of the Institute of Problems of the Development of the North of the TSMC SB RAS Graduated from Tomsk State University (TSU) with a degree in soil science and agrochemistry. After graduating from TSU in 1980, he joined the Laboratory of Electromagnetic Biology at the Research Institute of Biology and Biophysics at TSU (BB Research Institute). In 2000 he defended his PhD dissertation. In 2002, he entered the doctoral program at Tomsk State University of Control Systems and Radioelectronics and graduated from it. In 2006, he defended his doctoral dissertation on the topic "Spatial organization and morphogenesis of forest and anthropogenically altered soils" at Tomsk State University. He worked as a professor at Ugra State University since 2007, received the title of asso-ciate professor in economics in 2008. Since 2014, he has been employed as a professor at the Department of General Geology and Land Management at the National Research Tomsk State University. In 2018, he was hired as a chief researcher at the Institute of Problems of the Devel-opment of the North of the Tyumen Scientific Research Center SB RAS, where I have been working until now. Research interests include soil morphology, fundamental problems of soil formation of natural and man–made natural systems, organization of environmental monitoring systems, data analysis, the problem of changing scale (MAUP) in soil science and environmental management studies.

Index

I

K

L

M

N

O

P

R

S

T

U

V

W

Z

www.ingramcontent.com/pod-product-compliance
Lightning Source LLC
LaVergne TN
LVHW060356300926
849162LV00006B/1278
* 9 7 9 8 3 6 9 3 3 3 7 4 7 *